Statistical Physics

GREGORY H. WANNIER
Late Professor of Physics, University of Oregon

Dover Publications, Inc., New York

Published in Canada by General Publishing Company, Ltd., 30 Lesmill Road, Don Mills, Toronto, Ontario.
Published in the United Kingdom by Constable and Company, Ltd.

This Dover edition, first published in 1987 is an unabridged and unaltered republication of the work first published by John Wiley and Sons, Inc., New York, in 1966.

Manufactured in the United States of America
Dover Publications, Inc., 31 East 2nd Street, Mineola, N.Y. 11501

Library of Congress Cataloging-in-Publication Data

Wannier, Gregory H.
  Statistical physics.

  Reprint. Originally published: New York : Wiley, 1966.
  Bibliography: p.
  Includes index.
  1. Statistical physics.   I. Title.
QC174.8.W35   1987   530.1′3   87-8909
ISBN 0-486-65401-X (pbk.)

Preface

The field which is comprised under the modern name "statistical physics" used to appear traditionally under three separate titles: thermodynamics, statistical mechanics, and kinetic theory. The division had its origin in divergences in outlook among the pioneers of the science in the nineteenth century. These divergences arose in connection with the then quite young atomic hypothesis, and the desire of some to separate the results which were reliable from those which were merely speculative. Today this divergence in outlook no longer exists, and the traditional division is outdated. Maintaining it in teaching seems actually an evil, because it prevents the instructor from shedding all possible light on the topic under discussion. Naturally some vestiges of the old division still exist today; in particular, the fusion of statistical mechanics and kinetic theory is the subject of a strong research effort at the present time. In a textbook, this fusion can thus only be vaguely indicated.

Beside these philosophical reasons for blending the various aspects of statistical physics there is also a practical reason which is particularly relevant in the training of graduate students in physics. It is an unfortunate trend of modern times, which probably affects all phases of scholarship, that the student is urged to absorb an ever-increasing amount of information. In physics this is particularly noticeable in the large amount of time now devoted to quantum mechanics. In connection with this phenomenon the teacher is faced with the task of presenting other material in the most efficient way possible; if a presentation has the possibility of saving the student time, it is well worth working out. It is the author's belief that a unified presentation of thermal physics is such a timesaver, and the present book is meant to implement this belief.

In conformity with the ideal just explained, this book contains within one cover the essentials of thermodynamics, statistical mechanics, and kinetic theory. It discusses even the most elementary matters briefly, but proceeds at times to quite advanced topics. It is primarily designed for instruction in a one-year course of non-specialist graduate students. In schools with a strong undergraduate physics curriculum it can perhaps be used for juniors and seniors. The material presented exceeds slightly the amount teachable in one year. It can be brought down to that format in various ways. One possibility is to leave out selected topics in Part II (applications), another to omit Part III (kinetic theory). Still another way of selecting the content is suggested by the symbol o or * affixed to certain chapters. Chapters having a o are particularly easy and can be disposed of by a well prepared student in one cursory reading; this would create a harder course. On the other hand, chapters or parts of chapters having a * are more advanced than the remainder. Leaving them out thus creates an easier course.

A liberal quantity of problems is added to the chapters. Here again, attention is paid to the limited time available to the student. A small fraction of the problems is singled out as "recommended problems." Working them out should provide adequate understanding of the text. It is hoped that this will render the book suitable for self-study. A further help in this direction is the solutions to the problems which are appended to the text.

It is a pleasure to acknowledge the assistance received in the preparation of the manuscript. Students at the University of Oregon were willing to act as audiences, and through their questions, helped remove many obscurities. Allan N. Kaufmann read the entire manuscript with a great deal of care and sympathy; his suggestions for improvements have been invaluable. George E. Uhlenbeck, Mark Kac, Robert Hardy, and James Kemp helped in the particularly difficult section on kinetics. Help in the writing of other sections was received from Berndt Crasemann, Robert Mazo, and D. ter Haar. Bette Minturn typed the entire manuscript twice, and the members of my family, Carol, Tony, and Peter, shared in the heavy labor of proofreading.

Gregory Wannier

April, 1966
Eugene, Oregon

Contents

PART I

Principles of
statistical thermodynamics

○ I

The first law of thermodynamics*

If the various parts of physics were to be placed in logical order, with each part dependent on those preceding, but not those following, then mechanics should probably be placed first. This would be proper whether the mechanics considered were to be of the classical or the quantum type. The French educational system has emphasized this position of mechanics in the past by treating it as part of mathematics. "Rational mechanics" it is then called; it is developed deductively from a limited set of axioms just as any other branch of mathematics.

There was a time when scientists and philosophers expected that all of physics, and even all of science, would finally be given axiomatic structure. This expectation has not been fulfilled. Each branch of science uses its own mixture of inductive and deductive reasoning, and the accumulation of knowledge in our own time has become much too rapid to allow the creation of a logical, all-embracing structure. We need flexibility to understand the world around us, and we gain such flexibility by having different

---

* A few of the chapters are headed by one of these symbols: ○, ★. Their meaning is the following:

○ This is material covered in a good undergraduate course. Cursory reading, coupled with some problems, should be adequate for well prepared students.

★ This is advanced material of unusual difficulty. Students with average preparation may do well to omit it.

3

viewpoints stressing different aspects. This flexibility is not easily reconciled with a rigid axiomatic framework, such as may be found in a traditional course on rational mechanics.

Statistical physics stands at the borderline of deductive science. People have tried to axiomatize parts of it and have had a fair success in doing so. However, not all conceptual difficulties have been eliminated completely. It is true in particular that some general theorems depend for their validity on complicated unphysical assumptions. In practice, the results of the theory are well verified, in spite of the weaknesses mentioned. It is therefore advisable in a first study of statistical physics to touch on "foundation problems" only lightly and to employ whatever reasoning seems most efficient. This procedure occasionally implies a shift of viewpoints which borders on the inconsistent. Thus, for instance, we shall sometimes consider our objects quantum systems so as to make the states countable. At other times we shall turn around and treat the energy as a continuous variable to allow differentiation with respect to some parameter. The student should learn early to take these shifts in his stride. By allowing them we gain an efficient and yet essentially correct approach to the subject. It goes without saying that the correspondence with physical reality has to be watched carefully in such a way of procedure.

1-1. Systems and state variables

Before we proceed with any theoretical developments it is appropriate to ask ourselves what the objects of study are in statistical physics. Statistical physics represents an effort to get away from the abstractions of mechanics and to approach more closely the realities of the world around us. Our objects are therefore to be inanimate but concrete things which we shall call *systems*. An iron bar, a block of ice, or a vessel filled with a gas mixture may serve as typical examples of systems. They are complicated and have many degrees of freedom. A complete description of such a system in the sense of rational mechanics is out of the question. We shall try to derive certain properties of these systems even though we cannot know their condition completely. This is where statistics comes into the picture. We hope to get results by averaging over parameters whose exact values are not known. The behavior in the large of our systems is thus seen as an average of some sort. The manner in which the average is taken is the central problem of statistical mechanics. We are not ready to consider it at this time. Instead, we shall try to indicate first the nature of the description attempted for a system.

Let us take as an example the iron bar mentioned above which is typical of the general case. A certain number of parameters, finite or infinite, are needed to distinguish it from another, slightly different bar. We might want to know the dimensions, chemical composition, arrangement of crystallites, etc.; many of these parameters can vary from point to point—we may have a distribution of stresses and magnetizations, foreign inclusions, etc. However, if suitable care is taken, the number of these variables can be cut down considerably. We may study a single crystal under uniform stress, bring it into a uniform magnetic field, and we can see to it that its chemical composition is uniform. For some purposes the shape of a body is of no importance; for a magnet this is not necessarily the case, but if we assume it to be long in the direction of magnetization, other details will not matter.

For such a magnet, we can say in a limited sense that its properties are determined by the pressure and the magnetic field in which it is found, and the mass numbers determining the composition: p, H, m_1, m_2, ... At least these are the electromechanical parameters which we can pick at will to suit our fancy.

One variable has not been listed so far, namely temperature. Temperature is a concept which has no direct meaning in mechanics or electromagnetism. It is usually determined by making an auxiliary mechanical measurement such as the volume of a quantity of mercury under its vapor pressure. Thus temperature lacks direct status. Yet there is little exaggeration in saying that statistical physics is written around the temperature concept. Thus, at the moment, we might think of temperature as a concept, not fully legitimate, which derives from our feeling of hot and cold and whose gradient indicates the direction of heat flow in the case of thermal contact. It will be the business of statistical physics to elucidate the obscurities of this concept.

Among the parameters describing our iron bar, those which can be varied easily by experimental manipulation are naturally preferred for examination. Thus it is true that the electrical conductivity of the bar is a function of chemical composition as well as of temperature. However, to vary the composition we shall have to make new samples each time while dependence on temperature may be checked by simpler means. Thus temperature t, pressure p, and magnetic field H have a preferred position as *state variables* for the iron bar even though we must agree that other variables have been ignored thereby. We assume that the conductivity σ of the bar stands in a functional relation to these state variables, so we can write

$$\sigma = \sigma(p, H, t) \tag{1.01}$$

Relations of the type (1.01) are rather casually written down. The implication is often made that no assumption is involved in writing them. If the state variables were giving a complete description this would be so. In the present instance this cannot possibly be the case because in a complete description the variable t is actually superfluous. Thus, from the point of view of the system the variables in (1.01) are selected even though we may not be able to control others. In physics this sort of situation can give rise to the phenomenon of *hysteresis:* because we have ignored important variables we find that the state of the system depends on past history, in addition to the chosen state variables. The example of the iron bar is very much to the point here. Iron is known for its hysteresis effects when subjected to a magnetic field. In Fig. 1.1 is drawn a magnetization versus

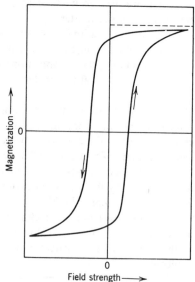

Fig. 1.1. Breakdown of the "equation of state" concept. Magnetization versus magnetic field in iron.

magnetic field curve for iron in which hysteresis is prominent. The whole idea of a functional relation between the condition of the system and selected variables crumbles here. Hysteresis effects are much more common in nature than is often supposed. Other examples of such effects are: the inhibition of the growth of the stable crystalline phase in a rigid matrix; the inhibition of a chemical reaction because of a "rate" bottleneck, the finite difference between the force necessary to move an object in the direction $+x$, as compared to moving it in the direction $-x$. In practice, such effects must be carefully circumvented to make statistical considerations applicable.

1-2. The equation of state

We may assume for the following discussion that we deal with a system for which the hysteresis difficulty can be avoided. We deal then with a substance whose behavior is truly a function of the state variables. We can then call on the mathematical concept of a function of several variables to guide our thinking. We might wish to investigate whether the function is or is not continuous or differentiable. If it is, the derivatives can also be investigated in this manner. Another aspect which must be thought about is what properties of the material should be investigated in preference to others. The conductivity chosen above is scarcely a good starting point for understanding iron because Ohmic conduction is a frictional flow phenomenon of great complexity. We might therefore try to find a "simpler" property to start a description. A more central concept than electrical conductivity is the density ρ of a material. We might therefore wish to consider the function

$$\rho = \rho(p, H, t) \tag{1.02}$$

for our iron bar. Within this relation, the dependence of ρ on H, called the *volume magnetostriction*, is a small and rather specialized effect. We are moving toward more basic things if we consider only

$$\rho = \rho(p, t) \tag{1.03}$$

Relation (1.03) is considered characteristic of a chemical substance. It is called its *equation of state*. For an actual material the equation of state represents an abstraction, as compared to its real behavior. For solid bodies the abstraction is quite drastic because shape and all that goes with it is ignored. For fluids, on the other hand, whether liquid or gaseous, the description (1.03) is reasonable. We may conceive of a fluid in which (1.03) is a rigorously correct relation, to the exclusion of all other variables. We call such a fluid a *perfect fluid*. The perfect fluid is a useful concept for theoretical reasoning.

One may wonder whether there are any instances in which the equation of state is expressible in analytic form. Such instances are of course exceptional. The one important case of an analytic equation of state is the relation which is asymptotically obeyed by very dilute gases. In the form (1.03) it reads

$$\rho = \frac{p\,\mathcal{M}}{R\,T} \tag{1.04}$$

The $\mathscr{M}$ in this formula is the chemical molecular weight; R is the universal gas constant:

$$R = 8.314 \text{ joules/degree mole} \tag{1.05}$$

and T is the so-called gas temperature. It is distinguished from the standard centigrade temperature by a shift of the origin by 273.2°.

It is worth while to reduce (1.04) to parameters which are significant on a molecular level. For this purpose we write

$$\rho = \frac{N m}{V} \tag{1.06a}$$

Here N is the number of molecules present, m their mass, and V the volume in which the gas is contained. Similarly we write

$$\mathscr{M} = \mathscr{A} m \tag{1.06b}$$

$$R = \mathscr{A} k \tag{1.06c}$$

Here $\mathscr{A}$ is Avogadro's number, and k is an atomic constant which converts the gas temperature into energy units. It is called Boltzmann's constant. Its numerical value is

$$k = 1.381 \times 10^{-16} \text{ erg/degree} \tag{1.07}$$

In all formulas of theoretical physics T appears only in combination with this constant. Substitution of expressions (1.06) into (1.04) yields

$$p V = N k T \tag{1.08}$$

This is an equation involving only pressure, temperature, and number density, that is, no individual property of the gas. It is fairly obvious that a universal equation of state such as (1.08) must have behind it a universal theory. This theory will be discussed at later occasions.

A third form of (1.04) or (1.08), which is often used, refers to one mole of gas. N then equals $\mathscr{A}$, and V becomes the molar volume, which we denote by $\mathscr{V}$. With (1.06c), (1.08) becomes then

$$p \mathscr{V} = R T \tag{1.09}$$

The molar volume of a dilute gas is thus independent of the nature of the gas. At 0° C and atmospheric pressure it equals

$$\mathscr{V}_0 = 22.4 \text{ liters/mole} \tag{1.10}$$

Equation (1.10) must not be understood to mean that one mole of an actual gas occupies the volume $\mathscr{V}_0$ at standard pressure and temperature.

The law (1.09) is a limiting law valid at high dilution, and the foregoing conditions do not satisfy this criterion. It is very closely correct to say that at $0°$ C and a pressure of 10^{-6} atm one mole of most gases occupies 22.4×10^6 liters. We idealize this situation by defining as a *perfect gas* or *ideal gas* a gas obeying relations (1.04)–(1.10) exactly. Real gases are then referred to as "imperfect."

The fact that gases come so close to obeying an analytic equation of state has naturally tempted many to find a modified equation of state for actual gases which incorporates their departure from perfection. The best known among them is the equation of van der Waals, which reads

$$\left(p + \frac{a}{V^2} \right) (V - b) = R T \tag{1.11}$$

The constant b indicates the "hard core" aspect of real gases which prevents compression beyond $V = b$ at even the highest pressures. The term a/V^2 expresses the idea of internal cohesion at moderate density; the pressure needed to hold a gas together at given temperature and density is reduced by this term. In addition the equation indicates the presence of complicated critical phenomena for fluids which will be discussed in the problems and in a later chapter.

It should be no surprise to the reader that even a very ingenious equation of state, such as (1.11) never fits an actual material exactly. A more sophisticated approach to the subject is to start out from the notion that (1.09) is valid for large V and to assume nothing beyond this except the possibility of power series expansion in $1/V$ as V becomes smaller. This leads to a parametric equation of state of the form

$$\frac{p V}{R T} = 1 + \frac{B(T)}{V} + \frac{C(T)}{V^2} + \frac{D(T)}{V^3} + \dots \tag{1.12}$$

$B(T)$, $C(T)$, $D(T)$ . . . are referred to as the second, third, fourth, etc., virial coefficient. The expansion ignores the possibility of genuine singularities in the equation of state; it is, however, excellent for the handling of data for moderately dense gases.

In most cases the equation of state is only known numerically in a limited part of its domain of definition. These empirical data are usually found to be differentiable, apart from discontinuities associated with phase changes. These discontinuities will be discussed in later chapters of this book. If we have the normal case, namely, that a relation between state variables is differentiable in a certain range, but if the analytic structure is not available, then it is still possible to use the differential form of the relation. Let us take as an example the equation of state (1.03) for a perfect fluid. It is customary to write its differential form as follows

$$\frac{1}{\rho}\, d\rho = -\alpha\, dt + \frac{dp}{B} \tag{1.13a}$$

or equivalently

$$\frac{1}{V}\, dV = \alpha\, dt - \frac{dp}{B} \tag{1.13b}$$

In this equation, the coefficients α and B are themselves functions of t and p. B is always and α usually positive. α is called the *coefficient of volume expansion* and B the *isothermal bulk modulus*. These empirical quantities are thus identified as partial derivatives by (1.13).

The choice of variables made in (1.03) and (1.13) is to some extent arbitrary. Their selection is natural for liquids and gases where the triad chosen forms a set which is relatively immune to other influences. In solids, one certainly can think of other stresses beside p. Correspondingly, one has to consider changes in shape as well as changes in density. Such modified relationships can also be brought into the differential form discussed above and will yield thereby definitions for other important empirical constants. As an example, let us write down the differential form of the equation of state for a wire pulled with a variable tensile force f. We get

$$\frac{dl}{l} = \beta\, dt + \frac{1}{q\, Y_t}\, df \tag{1.14}$$

Here l is the length of the wire and q its cross section. β is the *coefficient of linear expansion* of the material and Y_t its *isothermal Young's modulus*.

1-3. "Large" and "small" systems; statistics of Gibbs versus Boltzmann

In the preceding two sections we adopted a viewpoint which places a laboratory-sized sample into the focus of interest. We have called it a "system" and tried to approach a description of its properties directly. This is the point of view which was introduced by Gibbs into statistical mechanics. We now must come back on this definition and present a second viewpoint. A large fraction of the systems in the sense used above, when studied in detail, can be broken down with more or less accuracy into independent subsystems down to the molecular level. The earliest cases considered by Maxwell and Boltzmann were in fact of this type. It is then possible to think of the individual molecule as the system; the *sample* under consideration becomes then an assembly of systems. It so happens that many formulas apply to either kind of system with only minor changes. However, there is a profound shift in the significance of the theory which must be explained for future reference. We shall take as a typical example the hydrogen molecule H_2, which is the molecular subsystem of hydrogen gas. We shall assume the gas to contain only the hydrogen isotope of mass 1.

In the case of a single hydrogen molecule our worry about state variables and hysteresis is largely superfluous. At least according to our present understanding this system is the kind envisaged by rational mechanics, if we may look at quantum mechanics as the twentieth-century form of rational mechanics. The system consists of four particles of spin $\frac{1}{2}$. These spins are limited degrees of freedom capable of only two values. Apart from the spins the system is described by the variables appropriate for four point masses. This means twelve degrees of freedom, or twenty-four variables linked in pairs by the uncertainty principle. The state variables in this case are simply a full set of mechanical coordinates. To our knowledge, nothing more can be said about a hydrogen molecule. In particular, the temperature variable has no place in the basic description. Indeed if a hydrogen molecule is looked at individually there is very little need for statistics. Since a complete description is possible why deal with averages? In fact, averages don't tell us much in such a case, for departures from predicted average values are large for systems of molecular size, as we shall see later.

Statistics comes of course into the picture here when we are dealing with hydrogen gas rather than an individual hydrogen molecule. We shall see later that the reliability of a statistical prediction improves as the number of members for which it is made becomes large. In an ordinary container filled with hydrogen gas the number of molecules is very large, of the order 10^{20}. Thus, statistical methods are not only useful, they are necessary. In fact, the physical reason for employing statistics is the same whether the formal "system" reasoned on is laboratory-sized or molecular. It is always the large number of molecules in a sample that make statistical methods appropriate. The question whether the system reasoned upon does or does not consist of a large number of atoms is a purely formal one. In some cases we follow Maxwell and discuss a simple system, and find then that our sample consists of a large number of such systems. In other cases we follow Gibbs and reason on a system containing a large number of parts; statistics then applies actually to an individual sample. The Gibbsian device of taking a large number of large systems, called an *ensemble*, is a theoretical device without physical content and should be understood as such.

There is a small number of phenomena in which the objects of study are neither laboratory-sized nor molecular. The classical case is Brownian motion. Here the fluctuation phenomena associated with molecules become observable on larger bodies. Another example is that of macro-molecules as studied in biochemistry or polymer chemistry. This case is complementary to the first one. Macromolecules have so many degrees of freedom that some of their state variables have to be ignored in any observation. Concepts associated with large bodies, such as temperature, acquire thereby meaning for an individual molecule. We shall not elabor-ate on these exceptional though important situations in the following discussion.

1-4. The First Law; heat, work, and energy

The first law of thermodynamics is distinguished from energy conserva-tion laws in general in that the nature of heat is involved. This recognition of the nature of heat is preceded, both logically and historically, by the more elementary recognition that heat is a quantity. We shall first say a few words about this elementary aspect of the question.

The discovery that heat is a quantity belongs to the achievements of early physics; it is associated with the name of Joseph Black. If a heated copper block at $100°$ C is dipped into a beaker containing an equal weight of water at $20°$ C, the result is an adjustment of the whole not to a temperature of $60°$ C, but to a temperature much closer to $20°$. One concludes from this that a thing called heat must be effective in this adjustment, and that the heat necessary to produce a temperature change of $1°$ in a body is not the same for different objects. In other words different bodies have different *heat capacities C*. Heat capacity is easily seen to be proportional to mass so that we can define the *heat capacity per unit mass* or *specific heat c* as typical of a substance. We thus find that the heat Q necessary to raise the temperature of a substance from t_1 to t_2 equals

$$Q = m \int_{t_1}^{t_2} c(t) \, dt \qquad (1.15)$$

The above example then teaches us that the specific heat of copper is substantially smaller than that of water. A certain number of cross checks are of course necessary to verify that Q is an additive quantity which is preserved in simple heat exchanges. This is easily done by comparative experiments involving third or fourth bodies.

It was natural for early physicists to define a *heat content* for materials, that is, to assume that the quantity Q is somehow stored in a body after transfer. Such a conception allows (1.15) to be written in differential form, namely,

$$dQ = m\,c(t)\,dt \qquad (1.16)$$

Equation (1.15) is thereby transformed into

$$Q(t_2) - Q(t_1) = m \int_{t_1}^{t_2} c(t)\,dt \qquad (1.17)$$

where $Q(t)$ is now the heat content.

The concept of heat content can be extended to include the processes of melting and evaporation by the writing of a formula of the type

$$Q(\text{liquid}) - Q(\text{solid}) = m\,l \qquad (1.18)$$

where l is called the *latent heat* of fusion. The fusion process occurs of course at a fixed temperature for a pure substance. Equation (1.18) expresses the notion that at equal temperature the liquid phase of a substance has a higher heat content than the solid one. The quantitative concept of heat is again checked in phase change experiments of all sorts.

The more refined insight that heat is a quantity, but not a conserved quantity, is due to the fundamental experiments of Rumford and Joule. The former showed that heat can be produced in unlimited amounts by the performance of work, and the latter that there is an equivalence relation in such experiments. We may express this equivalence most conveniently in the form

$$1\ \text{cal} = 4.18\ \text{joules}$$

or in words

Energy is conserved. Heat is a particular form of energy.

The law of conservation of energy, generalized by the First Law, has been verified in innumerable experiments involving friction and cycles of various kinds. Cycles are experiments in which a system is finally returned to its original condition. They are particularly useful in clarifying the relationship of the First Law to the state variables defined in the preceding section. It can easily be shown that the total heat evolved in one cycle does not in general vanish. Only the sum of the heat and work does. There is therefore no such thing as a heat content depending on the state variables, and (1.17) is an incorrect equation. Insofar as (1.16) is the differential form of (1.17), it must also be considered incorrect. However, there is the phenomenon of heat passing from one body to another, and therefore (1.16) can be given a limited meaning. That kind of limitation is expressed in this book by the placing of a horizontal bar across the d of dQ as follows:

$$\bar{d}Q = m \, c \, dt \qquad (1.19)$$

The bar expresses the idea that $\bar{d}Q$ is not the differential of a quantity which is a function of the state variables. Similarly we denote by $\bar{d}W$ the work done on a system. This quantity also need not vanish when the system is taken through a cycle. However, the First Law asserts that the total amount of energy furnished does vanish. This means that we have

$$\oint (\bar{d}Q + \bar{d}W) = 0 \qquad (1.20$$

for any cyclic process. Let us call the quantity defined thereby the *energy content E*. We then have

$$dE = \bar{d}Q + \bar{d}W \qquad (1.21)$$

Here the d before E has no bar because the energy content is a function of the state variables of the system.

It is useful at this point to pause and reflect what (1.21) means in the case of "small" systems. Just as temperature has no meaning for a molecular system, so heat has no meaning either. Since the mechanical coordinates give a complete description of a "small" system, the only way to transfer energy to it is to act on these coordinates. With the term $\bar{d}Q$ absent in (1.21) the entire discussion collapses into nothing more than the statement that the increase in energy of a system equals the work done on it. In other words, the First Law asserts that all forces acting on a system are conservative if the mechanical description of that system is complete. This results suggests the notion that $\bar{d}Q$ and $\bar{d}W$ are really not as distinct in nature as traditional thermodynamics suggests. The term $\bar{d}Q$ arises in (1.21) because most coordinates necessary for a complete description of a "large" system have been ignored. However, work can be done on these coordinates in addition to work which may be performed by acting on the explicit state variables. This work on ignored coordinates is the heat $\bar{d}Q$ in equation (1.21). The fact that heat transfer is a form of work can be verified directly in certain limiting cases such as Brownian motion. The heat transfer results here at least in part from collisions with neighboring molecules and is visible as random work in the erratic zigzag motion of the particles.

The implied notion of ignored coordinates has also an effect on the left-hand side of (1.21). As a rule, we are only incidentally interested in those parts of the energy E which are easily recognized as mechanical. In fact, typical experiments relevant in statistical physics go to some trouble to avoid the appearance of such terms. We may take the case of the Joule paddle wheel experiment as an example. The arrangement is shown in Fig. 1.2. While two weights descend slowly under gravity, they drive a paddle wheel in a container with water via connecting strings passing over two pulleys. The experiment is usually interpreted by excluding the weights from the system. Gravity simply provides a conveniently precise method of measuring the work done by the paddle wheel on the water. The container is provided with baffles for the specific purpose of preventing the water

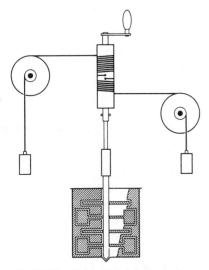

Fig. 1.2. The Joule paddle wheel experiment as a typical experiment of statistical physics.

as a whole from acquiring kinetic energy, that is, an easily recognizable form of mechanical energy. The result, a rise in the temperature of the water, could have been accomplished more easily by heating it. There must therefore be a hidden component of energy in the water which is closely linked to the temperature and which remains after all recognizable mechanical forms of energy have been subtracted out. This component is called the *internal energy U*. Thus for the Joule paddle wheel experiment the First Law takes the form

$$dU = dQ + dW \qquad (1.22)$$

and the experiment furnishes the mechanical equivalent of heat because a rise in U which is normally produced by heating is produced here by precisely measurable mechanical work. It must be remembered in the following that even though (1.22) is the equation generally employed,

(1.21) is the one generally valid. To make (1.22) valid special precautions must be taken to avoid the production of kinetic, gravitational, and other forms of mechanical energy.

The splitting off of the mechanical forms of energy to isolate an internal energy closely akin to the old concept of heat content is not always a feasible procedure. We may take as an example the change in size and shape of a body through the application of forces. In simple cases this can be handled through the theory of elasticity. A new form of potential energy, elastic energy, is defined; thereupon the deformation becomes a problem in conservative mechanics. This viewpoint has its limitations, as is well known: a deformation may be plastic rather than elastic; elasticity theory then breaks down. For us here it is more important to realize that even in the elastic range the separation of elasticity from thermal phenomena is not feasible. To take an extreme example, we may stretch a wire at a given temperature without doing any work. We simply heat it first so that thermal expansion brings it to the desired length. Thereupon we cool it in a clamped condition. The elastic energy comes here entirely from heat furnished. Even in more conventional experiments on elastic deformation the same idea applies. The elastic energy depends on the elastic constants, such as the bulk or Young's modulus, as parameters; these parameters are temperature dependent. Thus if compression at one temperature is followed by expansion at another the original elastic energy "stored" may not be returned entirely. A splitting off of this form of energy from U is therefore impossible. Once a splitting is found impossible we are forced to go to the opposite extreme and treat elastic energy as an aspect of U. U is dependent on shape and volume as well as on temperature, and this secondary dependence is tested by deformation of the body.

The internal energy U completes the triad of concepts built around the idea of hot and cold: temperature, heat, internal energy. It is also the bridge which links the field to mechanics. What has been said about heat and work applies even more to the internal energy. It is not different in principle from other forms of energy. It is simply the energy of the degrees of freedom which are ignored in our description of a "large" system. This energy is present in a random form and necessitates the use of statistical methods for its description. The idea that internal energy is reducible to other forms of energy was first enunciated by Mayer and Helmholtz as a hypothesis and is known as the *mechanical theory of heat*. According to this theory an object which has a temperature does not differ in principle from the objects of rational mechanics, and its internal energy is basically potential or kinetic energy of its mechanical degrees of freedom. The difference from mechanics arises in the way these degrees become manifest to the observer. The observer has no direct information concerning them; he is therefore forced to construct a theory which treats them statistically. The energy of these degrees of freedom is the internal energy. Temperature obviously has something to do with the abundance of this type of energy since it controls the sense of the heat exchange between two bodies in thermal contact. We cannot go beyond this point at the present time. After having implied that heat flow is an uncontrolled form of work, we turn around and say that heat flow can be controlled after all with the help of the temperature variable. We must continue for some time yet to employ temperature in this semiempirical manner.

1-5. Precise formulation of the First Law for quasistatic change

The First Law, being a form of the law of conservation of energy, has a very wide range of application. Its greatest usefulness arises, however, for a restricted class of states which we call *equilibrium states*.

Ideally equilibrium is the state reached by a system if all forces have been permitted to act until mechanical equilibrium is reached, and all opportunities for heat flow have been exhausted. The system will then no longer change in time. In practice, we must use the concept in a very much less circumspect way. After a system has been prepared in a certain way it will relax toward equilibrium in a manner which can often be analyzed in terms of a series of relaxation times $\tau_1, \tau_2, \tau_3, \ldots$. Some of these times are longer, some are shorter than the time of observation. If all times are either very much longer or very much shorter than the time of observation the equilibrium concept can be used in a relative sense: equilibrium is assumed for processes whose τ_i is short, while processes whose τ_i is long are ignored and replaced by a fixed constraint. This haphazard procedure actually works in a sufficient number of cases to be worth while as the basis of a scientific discipline.

With the help of the concept of equilibrium a second concept is introduced, that of *quasistatic change*. Quasistatic processes are an abstraction and are best defined as a locus of equilibrium states. The idea behind the concept is that if a given process, say the compression of a gas, is carried out at a certain rate the pressure must at every instant be slightly larger than would be required to hold the gas at its momentary pressure and temperature. The pressure excess is necessary to ensure a measurable rate for the process. Similarly, a pressure less than the equilibrium pressure is necessary for expansion at a finite rate. From the point of view of precise formulation this over- or underpressure is a nuisance because the amount of it depends on frictional effects which are incidental to the main processes of compression and expansion. The concept of quasistatic change implies that this pressure excess or defect can be made small if the change is imposed slowly. Finally we can proceed mentally to a limiting situation in which the process of compression has become so slow that the pressure at any given instant is indistinguishable from the equilibrium pressure at the given temperature and volume. This pressure is given by the equation of state. Similarly, in a more complicated situation when we have a body described by a set of coordinates x_1, x_2, x_3, . . . x_n and the temperature, the magnitudes of the conjugate forces X_1, X_2, X_3, . . . X_n occurring in a transformation are in general not clearly defined. But the equilibrium values of these forces are (usually!) well defined and the mental extrapolation of going to the quasistatic limit allows us to substitute these limiting values. The generalized force X_i thus becomes a function of the coordinates x_i and the temperature. Mechanical processes become thereby conservative and reversible. It should not be forgotten, however, that in many cases quasistatic change does not bring about uniqueness of forces and reversibility. The hysteresis loop of Fig. 1.1 is an example of this; in these cases production of heat cannot even in principle be separated from mechanical action.

If the concept of quasistatic change is a legitimate abstraction in a given situation, then the work term dW in (1.22) is expressible in terms of the infinitesimal displacements of the coordinates x_s. The coefficients in the expression are the conjugate forces X_s. The work term then takes the form

$$dW = X_1 \, dx_1 + X_2 \, dx_2 + \ldots + X_n \, dx_n \qquad (1.23)$$

An instance of a coordinate would be volume; its conjugate force is the negative pressure; work done on a system through variation of this variable equals

$$dW = -p \, dV \qquad (1.24)$$

In this connection, we can think of the equation of state as the relation which gives the force conjugate to volume as a function of the volume, with temperature thrown in as a parameter.

It follows from these considerations that if the First Law is applied to a system constrained to move quasistatically, the expression (1.23) can be substituted into (1.22), and we obtain

$$dQ = dU - X_1 \, dx_1 - X_2 \, dx_2 - \cdots \qquad (1.25)$$

which means that we have at least expressed dQ as a differential form. In perfect fluids, with volume as the only coordinate, the work term is given by (1.24), and (1.25) specializes into

$$dQ = dU + p \, dV \qquad (1.26)$$

It goes without saying that the correct use of (1.25) depends on a correct assignment of conjugate forces to generalized coordinates. This is the business of mechanics or electromagnetism. In a general way, an understanding of this is presupposed in statistical physics. However, the systems discussed here are so much more complicated than those usually considered in mechanics that some old formulas have to be almost rederived to be believable. We may take as an example the work of magnetization of an iron bar. It is given by the formula

$$dW = \mathbf{H} \cdot d\mathbf{M} \tag{1.27}$$

where $\mathbf{H}$ is the field externally applied to the bar and $\mathbf{M}$ its magnetic moment. The formula is trivial if the system to be magnetized is thought of as an aggregate of freely orientable dipoles of fixed strength. A real system such as our iron bar is not that simple. In fact, Fig. 1.1 implies that an irreversible phenomenon is associated with magnetization. A more sophisticated proof of (1.27) is therefore desirable which makes less specific assumptions about the nature of magnetism. We shall now carry through such a proof.

Assume that the system to be magnetized has the shape of a cylinder whose magnetization is at all times parallel to its axis. The magnetizing device shall be a permanent bar magnet which is not influenced by the system. Both objects shall be assumed lying along the x axis with their magnetization parallel to that axis (Fig. 1.3). The dimensions of the system

Fig. 1.3. Magnetization of a sample by a permanent magnet.

shall be small compared to those of the bar magnet. The magnetic field seen by the system will then be mainly in the x direction; this field will be denoted by $H(x)$. As the system is approaching the bar magnet, starting from $x = -\infty$, the attractive force exerted on it equals $M\,(dH/dx)$. If it is the only force acting, the system will be accelerated toward the magnet. We prefer the process to proceed quasistatically. For this purpose a balancing force F equal to $-M\,(dH/dx)$ must be applied to the system. This outside force does the following negative work

$$W = -\int_{-\infty}^{a} M \frac{dH}{dx}\, dx = -\int_{0}^{H(a)} M\, dH$$

where a is the value of x at which we arrest the approach of the two objects. The foregoing term is obviously quite different from (1.27), but there are two processes mixed up in the result: the magnetization of the

system and the approach of the two bodies. To obtain the work of magnetization alone we must add a further step. We must withdraw the system
to its original position at $x = -\infty$ and retain its magnetic moment. To
do this we must assume that the final magnetization can somehow be
"fixed" in the magnet by some kind of magnetic "glue" (the remanence
phenomenon in iron allows in fact something like it in practice). We then
can withdraw the magnet so "glued" to $x = -\infty$ and retain its moment.
Positive work is now done by the balancing force F in the amount

$$W' = -\int_a^{-\infty} M(a)\frac{dH}{dx}\,dx = M(a)\,H(a)$$

From this we get by addition

$$W + W' = M(a)\,H(a) - \int_0^{H(a)} M\,dH$$
$$= \int_0^{M(a)} H\,dM$$

as expressed in (1.27). Since the permanent magnet is not changed and
since the original positions are restored, the work obtained above must
be the work of magnetization of the system. Similar reasoning can
establish the correct form of dW in other cases.

We see that the First Law abolishes the idea that heat is stored in a body,
but retains the notion of heat transferred as a shorthand expression for
random work done. Let us see what happens to the concept of heat
capacity and specific heat in the course of this shift of viewpoints. The
definition generally adopted retains the relationship between heat transfer
and heat capacity. In other words one writes

$$dQ = C\,dt \tag{1.28}$$

The formula is actually not the most general one conceivable because it
assumes that a transfer of heat always produces a change in temperature.
Transfer of heat for the melting or evaporation of a body is thus left out
for the moment. Even so equation (1.28) covers a range of possibilities.
Since dQ is not a perfect differential C is not a simple derivative; something
more than the state variables are thus needed to determine it numerically.
To see what this something is we eliminate dQ from (1.28) and (1.22) and
write

$$C\,dt = dU - dW \tag{1.29}$$

Heat capacity is here seen to be reduced by work performed on the system.
This implies that the heat capacity is increased if a system performs work
at the same time as it acquires heat. The one definition that (1.29) lends

itself to is that of an *anergetic heat capacity*, that is, heat capacity without the simultaneous performance of work. It is given by

$$C_0 = \left(\frac{\partial U}{\partial t}\right)_{dw=0} \tag{1.30}$$

For some reason this possibility of a general definition is not much in use, and heat capacities are defined in connection with definite generalized coordinates. The most important case is that of a perfect fluid for which volume and temperature are the only state variables needed to determine pressure. One can then use the First Law in the form (1.26). Adopting V and t as state variables and writing

$$dU = \left(\frac{\partial U}{\partial t}\right)_V dt + \left(\frac{\partial U}{\partial V}\right)_t dV$$

one obtains

$$dQ = \left(\frac{\partial U}{\partial t}\right)_V dt + \left\{\left(\frac{\partial U}{\partial V}\right)_t + p\right\} dV \tag{1.31}$$

The heat capacity at constant volume C_v can thus be expressed as

$$C_v = \left(\frac{\partial U}{\partial t}\right)_V \tag{1.32}$$

This heat capacity is a special case of the anergetic heat capacity defined earlier, namely, when the only type of work considered is a change in volume. Unless extraordinary precautions are taken an experiment measuring heat capacity will not measure this heat capacity, but rather the heat capacity at constant pressure C_p, which, from (1.31) and (1.32), equals

$$C_p = C_v + \left\{\left(\frac{\partial U}{\partial V}\right)_t + p\right\} \left(\frac{\partial V}{\partial t}\right)_p \tag{1.33}$$

We see from this that there are at least two different types of heat capacity C_v and C_p which differ according to which side conditions are imposed during the heating process. Even more definitions exist if other state variables are introduced beside the volume.

We shall follow the precedent of Section 1-2 and specialize the results obtained for the case of perfect gases. It is a consequence of the Second Law which we cannot verify at this time that a perfect gas not only obeys its special equation of state, but also has an internal energy which is a function of temperature only. Consequently, we have the relation

$$\left(\frac{\partial U}{\partial V}\right)_T = 0 \tag{1.34}$$

If we insert this and (1.09) into (1.33) we obtain

$$\mathscr{C}_p - \mathscr{C}_v = R \tag{1.35}$$

Here $\mathscr{C}_p$ and $\mathscr{C}_v$ are the *molar heat capacities* or *molar specific heats* at constant pressure and constant volume, respectively, and R is the universal gas constant introduced in (1.05). It is seen from (1.31) that for a perfect gas the difference between $\mathscr{C}_p$ and $\mathscr{C}_v$ arises entirely from the work of expansion. For other materials there is also a contribution to this difference from the volume dependence of the internal energy.

A glance at the First Law in the form (1.25) shows that beside *anergetic change*, that is, change without work, we can also conceive the idea of *adiabatic change*, which is change without flow of heat. Adiabatic conditions are approached in practice either by careful thermal insulation or by a change of conditions in time which is sufficiently rapid to suppress heat flow. For a perfect fluid which is described by temperature and volume only, the differential equation for adiabatic quasistatic change is found from (1.26) as

$$dU + p\,dV = 0 \tag{1.36}$$

A more useful form of this relation results from (1.31), (1.32) and (1.33):

$$C_v\,dt + (C_p - C_v)\left(\frac{\partial t}{\partial V}\right)_p dV = 0 \tag{1.37}$$

which is a differential equation giving the volume as a function of temperature for adiabatic conditions. Since C_p is generally larger than C_v, the slope of the volume versus temperature curve is the reverse of what it is at constant pressure. At constant pressure volume and temperature generally rise together, while under adiabatic conditions the temperature rises as the volume is reduced.

For perfect gases equation (1.37) is integrable because we have from (1.09)

$$\left(\frac{\partial T}{\partial V}\right)_p = \frac{T}{V}$$

and therefore

$$\left(\frac{dT}{dV}\right)_{\text{adiab}} = -\frac{\mathscr{C}_p - \mathscr{C}_v}{\mathscr{C}_v}\frac{T}{V}$$

For many gases $\mathscr{C}_p$ and $\mathscr{C}_v$ are constants. The differential equation then can be integrated in closed form and yields

$$T\,V^{\gamma-1} = \text{constant} \tag{1.38}$$

where we set

$$\gamma = \frac{\mathscr{C}_p}{\mathscr{C}_v} \tag{1.39}$$

Elimination of T from (1.38) and (1.09) yields then

$$p V^\gamma = \text{constant} \tag{1.40}$$

Since γ is a number larger than unity the pressure needed for a given amount of compression is larger under adiabatic conditions than it would be according to Boyle's law. One way to see this is to compare the bulk modulus B_T which results from Boyle's law with the adiabatic bulk modulus B_{adiab}. If we take differentials of (1.08) and compare with the definition (1.13) we find

$$B_T = p \tag{1.41}$$

that is, the isothermal bulk modulus of a perfect gas equals its pressure. If we apply the same method to (1.40) we obtain instead

$$B_{\text{adiab}} = \gamma\, p \tag{1.42}$$

which is a larger value.

A. RECOMMENDED PROBLEMS

1. Suppose we assume that air is a mixture of nitrogen gas ($\mathscr{M} = 28$) and oxygen gas ($\mathscr{M} = 32$), both perfect, and that their mixture obeys the addition rule for partial pressures (Dalton's law). Calculate the mole fraction x of oxygen in air from the following data. At a total pressure of 10^6 dynes/cm^2 and 20° C the density of air is 1.189×10^{-3} gm/cm^3.

2. The bulk modulus of water at 0° C equals 2.032×10^4 atm and at 100° C 2.137×10^4 atm. The mean volume expansion coefficient at 1 atm is 4.33×10^{-4} deg^{-1}. Find the same coefficient at 1000 atm.

3. Calculate the isobaric coefficient of volume expansion of a perfect gas.

4. A cylinder contains 1 liter of a perfect gas under 3 atm pressure. How much work is done by the gas if it is expanded isothermally to 1 atm? Compare this work with the work done starting and finishing at the same points, but proceeding in a straight line in a p–V diagram.

5. Six cubic centimeters of a paramagnetic salt obeying Curie's law

$$I = \frac{C}{T} H$$

are used as a coolant through demagnetization from an applied field of 10,000 oe to 0 oe. The Curie constant of the salt equals 0.1°, and U may be assumed a function of temperature only. Find the heat absorbed by the salt if it is demagnetized isothermally (a) at 300° K and (b) at 4° K.

6. Prove from the First Law that any perfect fluid obeys the relation

$$\frac{\left(\frac{\partial V}{\partial p}\right)_T}{\left(\frac{\partial V}{\partial p}\right)_{\text{adiab}}} = \frac{C_p}{C_v}$$

Here $\left(\dfrac{\partial V}{\partial p}\right)_{\text{adiab}}$ is the derivative of V with respect to p under adiabatic conditions.

7. It is customary to carry out the integration of (1.14) at constant temperature approximately, treating all quantities as constants, except the two quantities following the differentials (Hooke's law).

(a) Show that if the work of extension of a wire is worked out on this basis it is proportional to the difference of the squares of the final and initial forces.

(b) Work out the analogous result, not treating l and q as constant. Treat the case of zero volume elasticity

$$q\, l = \Omega = \text{constant}$$

Show that the work is still the difference of an expression pertaining to the final state, and the same expression pertaining to the initial state. Work out that expression.

(c) Show that expression (b) reduces to (a) if the length in either state differs little from the length at zero force.

B. GENERAL PROBLEMS

8. Steel has a coefficient of linear expansion of 13×10^{-6} deg^{-1} and an isothermal Young's modulus of 2×10^{12} dynes/cm^2. A steel rail of cross section 100 cm^2 is laid between fixed supports at 10° C. What force does it exert on its ends at 35° C?

9. Prove that if a differentiable relationship $F(x, y, z) = 0$ holds between three variables the following identities are valid:

$$\left(\frac{\partial x}{\partial y}\right)_z \left(\frac{\partial y}{\partial x}\right)_z = 1$$

$$\left(\frac{\partial x}{\partial y}\right)_z \left(\frac{\partial y}{\partial z}\right)_x \left(\frac{\partial z}{\partial x}\right)_y = -1$$

10. What value results from the van der Waals equation for the second virial coefficient of a gas?

11. A compromise equation between the one for a perfect gas and the van der Waals equation is

$$p\,(V - b) = R\,T$$

Calculate the bulk modulus and the coefficient of thermal expansion resulting from this equation. Can you say something about the difference in behavior between this type of gas and a perfect gas?

12. Air may be taken as a perfect gas having $\gamma = 1.40$. One mole of air performs a transformation from an initial state of $291°$ K and 21 liters to a final state of $319°$ K and 22 liters. The transformation is performed along a straight line on a p-V diagram. Calculate the change in internal energy, the work performed, and the heat absorbed by the system.

13. A compressor takes air at $300°$ K at 1 atm and delivers compressed air at 2 atm, consuming 200 watts of useful power. If the compression is adiabatic and reversible, what volume will it deliver per second and at what temperature? ($\gamma = 1.40$)

14. Stressed permalloy has a rectangular hysteresis loop lying between ± 2 oe for the magnetizing field, and ± 1000 gauss for the induction. Calculate the rates of temperature rise due to hysteresis when a specimen is placed in a 60 cycle alternating field of amplitudes 1, 3, and 50 oe, respectively. The specific heat of permalloy is 3.2 j/cm³ deg.

15. According to a theory first set forth by Langevin, magnetization I, field H, and temperature T of a paramagnet are related in the following way:

$$I = n\,\mu\,L\!\left(\frac{\mu H}{kT}\right)$$

where

$$L(x) = \coth x - \frac{1}{x}$$

Here n is the number of atoms per unit volume, and μ is their magnetic moment. Show that the Langevin law reduces to the Curie law for small field, namely,

$$I = \frac{C}{T}\,H$$

and determine the Curie constant C.

16. Show that two substances obeying the van der Waals equation

satisfy the principle of corresponding states. By this we mean that pressure, volume, and temperature can be so scaled as to make the two equations look identical.

17. It is an empirical fact that steam behaves like a perfect gas mechanically, but has a specific heat depending on temperature according to the relation

$$\mathscr{C}_p = 8.62 + 0.002\, t \text{ cal/deg mole}$$

where t is in degrees centigrade. (*a*) Calculate the pressure–temperature relation which follows from this formula under adiabatic conditions. (*b*) Compare the result with the formula which would hold for a perfect polyatomic gas.

18. The van der Waals equation exhibits the phenomenon of a critical point, that is, a value of temperature and pressure at which $\left(\dfrac{\partial p}{\partial V}\right)_T$ and $\left(\dfrac{\partial p^2}{\partial V^2}\right)_T$ vanish simultaneously. Find the values of p_c, V_c, and T_c in terms of the equation parameters and form the product

$$\frac{p_c\, V_c}{R\, T_c}$$

Compare the value of this product with the one resulting from the Dietrici equation, which reads

$$p\,(V-b) = R\,T \exp\left[-\frac{a}{R\,T\,V}\right]$$

Empirical values of this product lie between 0.29 and 0.30 for common gases.

19. Show that the Dietrici equation introduced in the preceding problem has the following properties:

(*a*) It yields the same second virial coefficient as the van der Waals equation.

(*b*) For every positive temperature and every volume $V > b$ it yields a unique positive value of the pressure.

(*c*) For every temperature $T \geqq T_c$ and every pressure it yields a unique value of the volume.

(*d*) For every temperature $T < T_c$ there exists a range of pressures such that there are three possible volumes for a given pressure and V_c is always intermediate between the largest and the smallest of the three volumes.

20. A gas is found to have a coefficient of volume expansion α and an isothermal bulk modulus B which equal, respectively,

$$\alpha = \frac{1}{T}\left(1 + \frac{3a}{V T^2}\right) \qquad B = p \Big/ \left(1 + \frac{a}{V T^2}\right)$$

(a) Show that the relationships are compatible. (b) Derive the equation of state of the gas.

21. Show that a mixture of two perfect gases each having constant γ also has constant γ. Calculate this value of γ in terms of γ_1, γ_2, and the mole fractions f_1 and f_2.

22. The equation of state of many solids can be approximated by the expression

$$p V + \mathscr{G}(V) = \Gamma \mathscr{U}$$

Here $\mathscr{G}(V)$ is an unknown function of V, and Γ is a constant, the Grüneisen constant. Show that if this equation is valid we have

$$\frac{B\alpha}{C_{vv}} = \Gamma$$

Here B is the isothermal bulk modulus, α the volume expansion coefficient, and C_{vv} the heat capacity per unit volume at fixed volume.

23. Prove that the temperature rise γ per unit strain of a wire stretched adiabatically equals

$$\gamma = -\frac{1}{\beta}\left[\frac{Y_s}{Y_T} - 1\right]$$

where β is the coefficient of linear expansion at zero tension, Y_T the isothermal, and Y_s the adiabatic Young's modulus.

2

Elementary statistical methods in physics

It is generally believed that the First Law of thermodynamics is an exact law whereas the Second Law is only statistically true. The usual derivations given for the two laws do not bring out this distinction very well. This book is meant to be an improvement of this situation. Statistics is introduced here after the discussion of the First Law, and the Second Law is obtained as a by-product of statistical mechanics.

Before we introduce statistics into mechanics we must first understand statistical reasoning. Statistical reasoning makes use of the notion of probability. The notion of probability has a precise meaning for the case of repeated observations, where it expresses the relative frequency with which one obtains a certain result. In such a definition of probability the notion of repeated observations under identical conditions is assumed to be intuitively understood. However, we certainly use the idea of probability in a wider sense prior to any actual observations, and thereby make a priori statements about the results to be expected. We will, for instance, assert that the probability for a pair of ones on a pair of dice is $\frac{1}{36}$, and a number of contradictory observations will not easily shake this belief. The laws of probability are thus often accepted on other than purely empirical grounds. Such a priori acceptance can be in error in individual cases (for instance, when the dice above are loaded). Whether probabilistic reasoning is relevant in a given situation is a matter of inductive reasoning and is subject to its usual hazards.

If we take probability as equivalent to relative frequency in repeated observations, the following two theorems become obvious:

1. *Theorem of compound probabilities: If two events A and B are mutually exclusive, the probability for having either A or B is equal to the sum of the probability for A and the probability for B.*

2. *Theorem of independent probabilities: If two events A and B are independent, the probability of simultaneously having A and B is equal to the product of the probability for A and the probability for B.*

These two laws allow us to build up the probability for a complicated sequence of events from the probability for the simple events which compose the sequence. The above case of estimating the probability for two ones to be obtained with a pair of dice is of this type. On intuitive grounds the probability for a one to result on one die is thought to be $\frac{1}{6}$. Theorem 2 then predicts $\frac{1}{36}$ for the combination. The combination of a six and a one, on the other hand, is $\frac{1}{18}$ because it consists of two possible events, namely, of getting six with die α and one with die β or else one with die α and six with die β. Each event has probability $\frac{1}{36}$ by the previous argument. Furthermore the two events are mutually exclusive, and thus the probabilities add by Theorem 1. It is interesting that the argument involves the distinguishability of the two dice; we had to give them "names" α and β. This question will preoccupy us when we are applying this type of argument on an atomic or subatomic scale.

We can extend this reasoning to more complicated situations. Thus we would say that if a coin is tossed the probability of it coming up "heads" is $\frac{1}{2}$. Then by Theorem 2 the probability for any sort of pattern of ten consecutive throws is $(\frac{1}{2})^{10} = 1/1024$. However, we may not be interested in the exact pattern. We may only wish to know what the probability is for seven heads and three tails. This requires application of Theorem 1, that is, summation over all admissible patterns. In the present case we require only multiplication of 1/1024 by the number of possible patterns yielding a 7:3 ratio. One way to get this number is by permutation of the order of the 10 experiments. There are 10! such permutations. However, for every pattern there are 7! 3! permutations which do not change it. Thus the number of patterns comes out to be

$$\frac{10!}{7! \cdot 3!} = \binom{10}{3} = 120$$

A somewhat safer way to get the same answer is to concentrate on the three "tail" throws in the run of 10 and to consider the number of ways these three throws can occur in a sequence of 10. If we consider the three as objects to be placed in succession, we find 10 possible positions for the first, 9 for the second, 8 for the third. This covers all patterns because the heads have to occupy the remaining 7 positions. In fact, each pattern is covered six times by this reasoning because the three tail throws were treated as distinct objects, which they are not. Hence the number of patterns equals the above number divided by 6, or

$$\frac{10 \cdot 9 \cdot 8}{2 \cdot 3}$$

which is identical with the previous expression. Finally the probability comes out to be

$$P = \binom{10}{3}\left(\frac{1}{2}\right)^{10} = \frac{15}{128} = 0.117$$

2-1. Probability distributions; binomial and Poisson distributions

The introductory examples lead into the subject of probability distributions. A probability distribution is a sort of catalogue in which the arrangements or events are listed in one column, and the probabilities in the other. Usually the events are mutually exclusive, and the list exhaustive of all possibilities. The probabilities of the catalogue then must add up to 1. In the above example of the two dice there are 21 patterns to be listed; in 15 of them the two dice bear different numbers, and in 6 they bear equal numbers. The probability for the first type is $\frac{1}{18}$, for the second $\frac{1}{36}$. As indicated, we have

$$\sum_i P_i = 15 \cdot \tfrac{1}{18} + 6 \cdot \tfrac{1}{36} = 1$$

The probability distribution makes the probability P a *function* of certain characteristics of the events under consideration. The number of events necessary to verify a probability distribution is obviously much larger than the number of events necessary to verify an individual probability. For a complete verification, every event must occur a large number of times. In most physics problems such a detailed verification is not possible, and one is satisfied to work out certain averages. In the above case of the two dice one might wish to verify the honesty of the dice by working out the average number of eyes in a throw. This must be 7. For if we arrange the possibilities

$$1, 2, 3, 4, 5, 6$$

in a line and pick a pair at random, say 5, 3, we can match it with another pair symmetrically located about the midpoint; in this case it is 2, 4. Their average is 7, so that the average for the entire distribution is 7. Very much more elaborate tests for randomness have been devised than this one. It is possible thereby to verify on the data with a high degree of certitude whether certain probability assumptions are justified.

Physics is often concerned with the placing of events or objects in "boxes" or "slots." If the events are equivalent and the slots are also, no more physics is needed and the problem is purely one of probability calculus. We shall take up a few of these problems now.

The *normal binomial distribution* arises in the following situation. Let there be a large number of equivalent sites which may or may not be occupied by a single one of a large number of equivalent objects having no identity. Each occupation occurs independently of all the others except that multiple occupation is excluded. What is the probability that there are n objects in s chosen sites if the number n is, on the average, equal to m?

We may think of the sites as boxes, say, at the bottom of a pinball machine and the objects as balls. It is then seen that there is in the first place a combinatorial factor to be taken into account. There is only one pattern corresponding to all boxes empty. On the other hand, if one box is occupied it might be any of the s boxes, and thus our statement subsumes s different patterns. Similarly, the statement that two boxes are occupied subsumes $(1/2)s(s-1)$ different patterns. In a general way

$$\binom{s}{n} = \frac{s!}{n!(s-n)!}$$

patterns are subsumed in the statement that there are n balls in s boxes. Let us denote by $p_c(n)$ the probability of realizing a particular occupation pattern or complexion of n objects among s sites. The total probability of having n objects on s sites is then

$$p(n) = p_c(n) \binom{s}{n} \tag{2.01}$$

where $p_c(n)$ is the probability of realizing a particular complexion.

For the evaluation of $p_c(n)$ two paths can be followed. The first method is based on the theorem of independent probabilities as applied to each site separately. Let the probability of a given site being occupied be π; the probability of it being empty is then $1 - \pi$. The probability $p_c(n)$ takes then directly the form

$$p_c(n) = \pi^n (1-\pi)^{s-n}$$

and $p(n)$ becomes

$$p(n) = \binom{s}{n} \pi^n (1-\pi)^{s-n} \tag{2.02}$$

The second method avoids the assumption of independence applied to individual sites and is thereby more easily extended to statistical mechanics.

In this argument we consider simultaneously two groups of s and s' sites. We observe then that the two groups together form again a group of $s + s'$ sites. We then have, by the theorem for independent probabilities applied to the groups only,

$$p_c[n+n'; s+s', m^\star(s+s')] = p_c(n; s, m^\star s) \cdot p_c(n'; s', m^\star s')$$
(2.03)

Here $m^\star$ is the average number of objects per site; it must evidently be common to the two groups. Now let us remove an object from the first group and repeat the argument. We find

$$p_c[n+n'-1; s+s', m^\star(s+s')] = p_c(n-1; s, m^\star s) \cdot p_c(n'; s', m^\star s')$$
(2.04a)

and likewise for the second group

$$p_c[n+n'-1; s+s', m^\star(s+s')] = p_c(n; s, m^\star s) \cdot p_c(n'-1; s', m^\star s')$$
(2.04b)

The right-hand sides of the last two equations must be identical since the left-hand sides are. We obtain from this the proportion

$$\frac{p_c(n; s, m^\star s)}{p_c(n-1; s, m^\star s)} = \frac{p_c(n'; s', m^\star s')}{p_c(n'-1; s', m^\star s')}$$
(2.05)

If we call the value of this fraction ξ, we see that it can depend neither on n nor on s, but only on $m/s = m^\star$, because it is the only variable common to the two fractions. By recurrence we then obtain

$$p_c(n; s, m) = p_c(0; s, m)\, \xi^n$$
(2.06)

and finally from (2.01), since $p_c(0) = p(0)$,

$$p(n; s, m) = \binom{s}{n} \xi^n p(0; s, m)$$
(2.07)

We can eliminate the two constants $p(0)$ and ξ by using the supplementary data for the problem, namely,

$$\sum_{n=0}^{s} p(n; s, m) = 1$$
(2.08)

$$\sum_{n=0}^{s} n\, p(n; s, m) = m$$
(2.09)

The two series are expressible by the binomial formula. We find

$$\xi = \frac{m/s}{1 - m/s}$$
(2.10a)

$$p(0; s, m) = \left(1 - \frac{m}{s}\right)^s$$
(2.10b)

As predicted in connection with (2.05), ξ depends only on m/s. The substitution of the two expressions (2.10) into formula (2.07) gives finally again formula (2.02)

$$p(n; s, m) = \frac{s!}{n!(s-n)!} \frac{(s-m)^{s-n} m^n}{s^s} \tag{2.11}$$

in agreement with the first derivation. We see that

$$\pi = m^{\star} = \frac{m}{s} \tag{2.12}$$

A very important limiting case of the binomial distribution is the *Poisson distribution*. In order to obtain it we have to reformulate the previous problem slightly as follows. Let there be an infinite continuum of positions, all equivalent, which can be occupied by any of a large number of equivalent objects. Each occupation occurs independently of all the others. What is the probability $p(n; m)$ of finding n objects in a particular subregion, if the average number in this subregion is m?

This kind of problem arises, for example, if one wishes to determine the number of particles from a radioactive source arriving at a counter during a fixed time interval. The particles are presumably statistically independent, and time is a continuous variable. If, on the other hand, the question were how many mutations were induced by the radioactive source in a protein molecule of finite length, the fact that there are only s sites to act on would presumably enter and the binomial distribution would apply. Another case to which the Poisson distribution applies is that of density fluctuations in a gas. Suppose we take a subvolume of a container filled with gas, and suppose the density is such that we should expect m molecules in the subvolume. Then, space being considered continuous, the probability of finding n molecules in this volume is given by the Poisson distribution formula.

The numerical expression for the Poisson distribution arises from (2.11) by a simple passage to the limit, letting s approach infinity while n and m remain finite. Stirling's formula is then applied to the two factorials containing s and the limiting rule for forming exponentials is used. One finds thus

$$p(n; \infty, m) = \frac{e^{-m} m^n}{n!} \tag{2.13}$$

An interesting logical viewpoint associated with (2.13) is that the formula should arise from the preceding discrete problem whether or not multiple occupation of boxes is allowed. This is indeed the case. An *anomalous binomial distribution* results if multiple occupation of the boxes is allowed but all other previous conditions are maintained.

The computation for this distribution is somewhat harder. A first part, dealing with arrangements having a particular complexion, can follow exactly the equations (2.03)–(2.06) and end up with (2.06):

$$p_c(n; s, m) = p_c(0; s, m)\, \xi^n \tag{2.06}$$

But there is a second combinatorial part which requires us to find the number of ways in which n objects without identity can be placed in s boxes, with multiple occupation allowed. We solve this by denoting the boxes by letters $z_1, z_2, z_3, \ldots z_s$ and the balls by $b_1, b_2, b_3, \ldots b_n$. Then we consider ordered sequences such as

$$z_1\, b_7\, b_9\, b_3\, z_3\, z_4\, b_6\, b_{10}\, z_9 \ldots$$

They are assumed to start always with z_1, but are otherwise arbitrary. If we make the convention that the b's following a z are in that box, every permutation of the $n+s-1$ letters following z_1 represents a possible arrangement; in the sequence above, we have three balls in box 1, none in box 3, two in box 4, etc. There are $(n+s-1)!$ such arrangements. However, since the balls have no identity, any permutation of the b's yields the same arrangement;* also any permutation of the $s-1$ boxes together with the b's following represents the same arrangement. The total number of arrangements for n balls in s boxes is therefore

$$\frac{(n+s-1)!}{(s-1)!\, n!}$$

and, by (2.06), the probability equals

$$p(n; s, m) = \frac{(n+s-1)!}{(s-1)!\, n!}\, p(0; s, m)\, \xi^n \tag{2.14a}$$

The front factor is the binomial factor for the power $-s$ and is advantageously written that way:

$$p(n; s, m) = \binom{-s}{n} p(0; s, m)\, (-\xi)^n \tag{2.14b}$$

We now proceed as previously, determining the two unknowns by the side conditions (2.08) and (2.09). A short calculation yields then

$$p(n; s, m) = \binom{-s}{n} \left(\frac{s}{s+m}\right)^s \left(-\frac{m}{s+m}\right)^n \tag{2.15a}$$

or

$$p(n; s, m) = \frac{(s+n-1)!}{(s-1)!\, n!} \frac{s^s\, m^n}{(s+m)^{s+n}} \tag{2.15b}$$

* The case of the balls having identity is more realistic in simple applications. It yields expressions associated with the multinomial expansion coefficients. The case is worked out in Chapter 4 in connection with Maxwell-Boltzmann statistics.

The main difference from (2.11) is that n and m can be larger than s this time.

If we pass to the limit $s = \infty$ in (2.15), the Stirling formula and the limiting rule for forming exponentials again reproduce (2.13), as previously.

2-2. Distribution function for large numbers; Gaussian distribution

It is a very common occurrence that the numbers n and m in a distribution are quite large. The distribution then acquires a characteristic peakedness which it is important to understand. We shall restrict our calculations in the following to the Poisson distribution, but the two binomial distributions actually share this trait under the same circumstances.

In order to understand the Poisson distribution a little better we shall have a look at its moments. The Nth moment of a distribution $p(n)$ is defined as

$$\langle n^N \rangle = \sum_{n=0}^{\infty} p(n)\, n^N \tag{2.16}$$

The first three of these moments are the most significant. In our case we have

$$\langle n^0 \rangle = \sum_{n=0}^{\infty} \frac{e^{-m} m^n}{n!} = 1 \tag{2.17}$$

This is obviously right from the definition of probability. The next moment equals

$$\langle n \rangle = \sum_{n=1}^{\infty} \frac{n\, e^{-m} m^n}{n!} = m \tag{2.18}$$

This is the result of our construction. The next moment contains significant new information:

$$\langle n^2 \rangle = \sum_{n=1}^{\infty} \frac{n\, e^{-m} m^n}{(n-1)!} = m^2 + m \tag{2.19}$$

The significance of this formula shows up when the *mean square deviation* of the distribution is computed:

$$\langle (n-\langle n \rangle)^2 \rangle = \langle n^2 \rangle - \langle n \rangle^2$$

The two preceding equations yield for this

$$\langle (n-\langle n \rangle)^2 \rangle = m$$

or

$$\langle (n-\langle n \rangle)^2 \rangle = \langle n \rangle \tag{2.20}$$

The square root of this number is a fair measure of the deviation from the average to be expected in a particular "run" which is subject to the Poisson distribution. Suppose that we deal with the number of particle counts in a counter, and that the expected number of counts is 100. We must then be prepared for an actual number differing from 100 by less than 10 most of the time. As the number of counts goes up the relative deviation goes down, but the absolute deviation goes up as the square root of the number of counts. The same reasoning applies to density fluctuations of a gas in a small subvolume of a large volume.

It should be realized that for large m these distributions become very pointed. An easy way to bring this out is to approximate the function in the neighborhood of its maximum by a simpler curve. We do this for the curve (2.13) by replacing the factorial by the Stirling approximation:

$$p(n) \approx \frac{e^{-m} m^n}{n^n e^{-n} \sqrt{2\pi n}}$$

or passing to the logarithm

$$\ln p(n) \approx n - m + n \ln \left(\frac{m}{n}\right) - \ln \sqrt{2\pi n}$$

Writing for $\ln(m/n)$ its expansion

$$\ln \left(\frac{m}{n}\right) = \ln \left(1 + \frac{m-n}{n}\right) \approx \frac{m-n}{n} - \frac{1}{2}\left(\frac{m-n}{n}\right)^2 + \cdots$$

we get

$$\ln p(n) \approx \frac{1}{2}\frac{(n-m)^2}{n} - \ln \sqrt{2\pi n}$$

and

$$p(n) \approx \frac{1}{\sqrt{2\pi n}} e^{-\frac{1}{2}\frac{(n-m)^2}{n}}$$

The formula yields by far the largest values when n and m are about equal. We can therefore replace n by m in the nonessential positions and write

$$p(n; \infty, m) \approx \frac{1}{\sqrt{2\pi m}} e^{-\frac{1}{2}\frac{(n-m)^2}{m}} \tag{2.21}$$

This is the standard Gaussian probability curve. If one calculates the moments of this approximate distribution by integrating rather than summing over n, and if one integrates over the entire infinite range of the variable, then the three first moments (2.17), (2.18), and (2.19) come out as previously.

The increasing peakedness of the distribution for large m shows up if we plot $p(n; \infty, m)/p(m; \infty, m)$ against the relative displacement x, where x is defined as

$$x = \frac{n-m}{m} \tag{2.22}$$

From (2.21) we find for this ratio

$$\frac{p(n; \infty, m)}{p(m; \infty, m)} \approx e^{-\frac{1}{2}mx^2} \tag{2.23}$$

that is, a formula containing the number m in the exponent. The kind of values of m which arise in the study of gases can easily be 10^{15} to 10^{20}. Substantial deviations from normal density are extremely unlikely in such cases. This is only the first instance in which we can see how the probability predictions of statistical physics become virtual certainties for large samples.

As we leave this subject it is important to realize that the binomial formulas or the Poisson formula do only a combinatorial job on events for which we already understand intrinsically their probability of occurrence. The quantities π, $m^\star$ or m are supposed to be already known before the experiment begins, and the formula simply deduces from these numbers, supposedly known, what the probability of getting various patterns is. In the coin throwing experiment the "heads" or "tails" value was $\frac{1}{2}$. If someone asks for the probability of throwing either a six or any other number with a die, the probability for the first event would have to be assumed $\frac{1}{6}$. Similar a priori information has to be fed into the analysis in physical cases: the scatter in a Geiger counter can be calculated only if the average number is known. Naturally, in a scientific experiment we might wish to reverse this procedure. We might want to find the average from the scatter. This is of course standard inductive practice. If there is some reason unknown to us, why π or m is not a constant in the experiment, then the entire analysis will fail. Thus the Geiger counter experiment might be falsified by a normally recurring tendency of cosmic rays to appear in the morning, or by radioactive dust regularly arriving in the afternoon from a nearby factory. Many experimental data have in fact been misinterpreted by experimenters who applied the apparatus of this section to situations which did not justify it.

2-3. Statistical aspect of quantum mechanics

In the passage from classical to quantum mechanics many of the familiar classical relations acquire a statistical aspect because they are converted into relations between expectation values. Let us take as an instance the basic formula

$$\text{Force} = \text{Mass} \times \text{Acceleration}$$

Ehrenfest first pointed out the statistical aspect which this relation acquires in quantum mechanics.†

Suppose we think of a mechanical system as a set of N point masses of mass m_ν obeying Schroedinger's equation. Such a system can be represented reasonably well by a hamiltonian of the form

$$\mathscr{H} = \frac{1}{2} \sum_{\nu=1}^{N} \frac{1}{m_\nu} p_\nu^2 + \Phi(\mathbf{x}_1, \mathbf{x}_2, \ldots \mathbf{x}_N) \tag{2.24}$$

The wave function Ψ is a function of $\mathbf{x}_1, \mathbf{x}_2, \ldots \mathbf{x}_N$ and the time t; it usually cannot be written down explicitly. We shall assume for it that a boundary can be found in space such that all surface integrals vanish. This assumption is probably not necessary. It seems highly likely that all major conservation laws can be extended to portions of space, provided surface integrals are interpreted correctly (for conservation of matter this is standard). Such studies are not available in the literature and will therefore not be attempted here.

The law of acceleration can be broken down into two parts: namely, the relation between the rate of change of position of a particle and its momentum; and the relation between the rate of change of its momentum and the acting potential. Each of these relations is statistical in quantum theory.

For the first part, we supplement (2.24) by the definition of the mean position of one of the particles, as follows:

$$\langle \mathbf{x}_1 \rangle = \iint \ldots \int \Psi^\star \mathbf{x}_1 \Psi \, d\mathbf{x}_1 \, d\mathbf{x}_2 \ldots d\mathbf{x}_N \tag{2.25}$$

The mean velocity is then defined as the rate of change of this quantity. This rate of change is governed by the time-dependent Schroedinger equation

$$i \hbar \frac{\partial \Psi(\mathbf{x}_1, \mathbf{x}_2, \ldots \mathbf{x}_N, t)}{\partial t} = \mathscr{H} \Psi \tag{2.26}$$

Substitution of (2.26) into (2.25) yields

$$\frac{d}{dt} \langle \mathbf{x}_1 \rangle = \iint \ldots \int \left\{ \frac{\partial \Psi^\star}{\partial t} \mathbf{x}_1 \Psi + \Psi^\star \mathbf{x}_1 \frac{\partial \Psi}{\partial t} \right\} d\mathbf{x}_1 \ldots d\mathbf{x}_N$$

$$= -\frac{i}{\hbar} \iint \ldots \int \{ -(\mathscr{H}\Psi)^\star \mathbf{x}_1 \Psi + \Psi^\star \mathbf{x}_1 (\mathscr{H}\Psi) \} \, d\mathbf{x}_1 \ldots d\mathbf{x}_N$$

† P. Ehrenfest. *Z. Physik* **45**, 455 (1927).

We now make use of the assumption that the boundary can be placed in such a way that all surface integrals vanish. The selfadjointness of the operator $\mathscr{H}$ then enters into play, and we can write

$$(\mathscr{H}\Psi)^\star \approx \Psi^\star \mathscr{H}$$

This yields

$$\frac{d}{dt}\langle \mathbf{x}_1 \rangle = -\frac{i}{h} \int \int \cdots \int \Psi^\star (\mathbf{x}_1 \mathscr{H} - \mathscr{H}\mathbf{x}_1) \Psi \, d\mathbf{x}_1 \ldots d\mathbf{x}_N$$

The commutator is easily evaluated with the help of (2.24) and yields

$$\mathbf{v}_1 \equiv \frac{d}{dt}\langle \mathbf{x}_1 \rangle = \frac{1}{m_1}\langle \mathbf{p}_1 \rangle \qquad (2.27)$$

where $\mathbf{p}_1$ is either one of the following two operators:

$$\mathbf{p}_1 = -i\hbar \frac{\partial}{\partial \mathbf{x}_1} \quad \text{operating to the right} \qquad (2.28a)$$

$$\mathbf{p}_1 = i\hbar \frac{\partial}{\partial \mathbf{x}_1} \quad \text{operating to the left} \qquad (2.28b)$$

The operator $\mathbf{p}_1$ is also selfadjoint like $\mathbf{x}_1$. The velocity concept is used in limited portions of space in connection with the law of conservation of matter. The correct form of the operator in such cases is found to be the arithmetic mean of (2.28a) and (2.28b).

Equation (2.27) is the first part of Newton's second law. The second part connects force and rate of change of momentum. We start out with the expression

$$\frac{d}{dt}\langle \mathbf{p}_1 \rangle = \frac{d}{dt} \int \int \cdots \int \Psi^\star \mathbf{p}_1 \Psi \, d\mathbf{x}_1 \ldots d\mathbf{x}_N$$

We carry out the differentiation under the integral sign and substitute (2.26). The resultant expression is transformed as previously, neglecting all surface integrals. The result is

$$\frac{d}{dt}\langle \mathbf{p}_1 \rangle = -\frac{i}{h} \int \int \cdots \int \Psi^\star \{\mathbf{p}_1 \mathscr{H} - \mathscr{H}\mathbf{p}_1\} \Psi \, d\mathbf{x}_1 \ldots d\mathbf{x}_N$$

The commutator is easily evaluated with the help of (2.24) and (2.28); it yields finally

$$\frac{d}{dt}\langle \mathbf{p}_1 \rangle = -\left\langle \frac{\partial \Phi(x_1, x_2, \ldots x_N)}{\partial \mathbf{x}_1} \right\rangle \qquad (2.29)$$

The equations (2.27) and (2.29) together are the quantum equivalent of Newton's second law of motion; the law is reduced here to a relation between average quantities. In other words we find that the second derivative of the mean position of a particle multiplied with its mass equals the mean force it experiences. Both these mean values are taken with respect to the wave function valid at a particular time. In general the result is in semiquantitative agreement with classical theory. Exceptional situations do arise, however, in which we get results which are not expected in classical physics. Let us take as an example the barrier leakage problem shown in Fig. 2.1: a particle is confined initially to a space limited on the

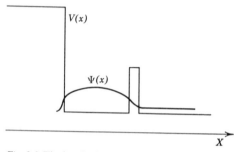

Fig. 2.1. The barrier leakage phenomenon in relation to Newton's second law.

left by an impenetrable wall and on the right by a thin wall. What will happen to the particle in time? The answer is of course that the wave function inside the space will "leak" or "tunnel" to the right. This must square somehow with (2.27) and (2.29). In particular the mean force must be positive if one takes the x direction to the right. Intuitively, we feel that the tunneling has something to do with the "thinness" of the right wall. This does not show up in any direct way in (2.29); the contributions of the right wall to the force are limited to its two step function positions, and for any reasonable leakage type wave function the contribution of the left edge of the wall will be larger. Thus the contribution of the wall to the right-hand side of (2.29) is negative; the result corresponds to the physical fact that "most of the time" the confined particle will bounce off the wall after having reached it. In order to get the effect we are looking for, we must bring in the left wall. It makes a contribution of the correct sign to the average force. As it is higher, its contribution is likely to be larger than the one of the other wall. This gives the right sign to the average force. Physically this means that if the particle approaches the right wall only a small number of times then it will be thrown back. The presence of the left wall is needed so that the particle can repeat the process a sufficient number of times to make "tunneling" statistically significant, while at the same time tunneling through the left wall remains negligible.

The moral of this example is that the quantum-mechanical relation

$$m_1 \frac{d^2}{dt^2} \langle \mathbf{x}_1 \rangle = - \left\langle \frac{\partial \Phi(\mathbf{x}_1, x_2, \ldots x_N)}{\partial \mathbf{x}_1} \right\rangle \tag{2.30}$$

is the formal extension of Newton's second law and is usually equivalent to it in a semiquantitative sense. There are, however, exceptional situations in which (2.30) is still valid, but in such a way as to depart from classical intuition. We must then look for other ideas to gain understanding. In the present instance such understanding is furnished by wave optics. Light can traverse a thin layer of totally reflecting material in a way which is essentially equivalent to the barrier penetration phenomenon of quantum mechanics. Exploiting this analogy, we usually employ plane waves to solve such problems. For such wave functions the surface integrals which cropped up in the derivation of (2.30) do not all vanish, and we must supplement the equation by considering the momentum transport across the boundary of the region under study.

Before leaving this subject we shall give a less familiar example of a classical equation which is obeyed statistically in quantum mechanics. It is the relation

Work done = Change in kinetic energy

We obtain this relation simply by asking for the time derivative of $(1/2\, m_1)p_1^2$ and proceeding as for (2.30). The commutator to be worked out is slightly more complicated. We find

$$\frac{d}{dt} \left\langle \frac{1}{2\, m_1} p_1^2 \right\rangle = - \frac{1}{2\, m_1} \left\langle \mathbf{p}_1 \cdot \frac{\partial \Phi}{\partial \mathbf{x}_1} + \frac{\partial \Phi}{\partial \mathbf{x}_1} \cdot \mathbf{p}_1 \right\rangle \tag{2.31}$$

which is the expected result. It is seen that in this case classical theory yields for the right-hand side a product of two factors which do not commute, namely, force and velocity. In such cases quantum mechanics yields something in addition to the statistical generalization of the classical answer, namely, a prescription for interpreting the classical product. The above type of product is called the symmetrized product.

The statistical relations (2.30) and (2.31) could be dismissed as exercises except that they are valid for many-body systems. For such systems direct computation of the wave function is usually impossible. Thus the equations contain information which is not easily available in any other form and which may be of value. In particular, if the potential Φ is all internal between the particles and obeys Newton's third law, we get from (2.29) by simple summation

$$\frac{d}{dt} \int\int\int \ldots \int \Psi^\star \left(\sum_\nu \mathbf{p}_\nu \right) \Psi \, d\mathbf{x}_1 \ldots d\mathbf{x}_N = 0 \qquad (2.32)$$

Simple summation does not work in the case of equation (2.31). We must add in this case that the time derivative of the potential Φ also equals the sum of the right-hand terms, but with opposite sign. We find therefore

$$\frac{d}{dt} \left\langle \sum_\nu \frac{1}{2\,m_\nu} p_\nu^2 + \Phi(\mathbf{x}_1, \mathbf{x}_2, \ldots \mathbf{x}_N) \right\rangle = 0 \qquad (2.33)$$

A simpler way to get (2.33) is to generalize the preceding considerations and to show from (2.26) that for any selfadjoint operator A which does not contain the time explicitly we have

$$i\,\hbar \frac{d}{dt} \int\int \ldots \int \Psi^\star A \Psi \, d\mathbf{x}_1 \ldots d\mathbf{x}_N =$$

$$= \int\int \ldots \int \Psi^\star \{ A \mathscr{H} - \mathscr{H} A \} \Psi \, d\mathbf{x}_1 \ldots d\mathbf{x}_N \qquad (2.34)$$

Equation (2.33) then follows from (2.34) by the simple observation that

$$\mathscr{H}\mathscr{H} - \mathscr{H}\mathscr{H} = 0$$

This operator aspect of quantum mechanics is probably familiar to the reader. What needs emphasis here is that we recover the relations of classical physics as relations between averages. These averages are averages over space, but not time. In other words, the relations are valid at every instant of time. This feature is worth some notice because there are other approaches to energy conservation in particular, which make it appear that it does not apply for short time intervals.

Even though quantum mechanics leads to a large number of relations which are statistical we are not justified in thinking of quantum mechanics as a whole as a statistical discipline. The reason is of course that there is a substrate which is deterministic. This substrate is the wave function which obeys a differential equation of the first order in the time. Thus the present condition of the wave function determines the future entirely.

2-4. Statistics dealing with averages in time; virial theorem

The easiest form of statistical mechanics arises from taking averages in time. If a mechanical relation consists of a sum of terms one of which is a time derivative, a simpler relation can often be obtained. If the quantity whose time derivative appears is bounded and if the average is taken over a sufficiently long time, the average of the time derivative vanishes. The relation then holds as a relation between averages without the time derivative term. As an example we may take equation (2.29). If a particle is in a cyclic orbit or a stationary condition, or if the momentum is bounded for some other reason, the time average of the left-hand side is zero. The equation then tells us that the time average of the force is also zero. This relation must apply to the earth in its motion around the sun. It must also apply to any electron in a stationary state in an atom or molecule. It applies even to band electrons in solids which have a nonvanishing expectation value of the velocity. This expectation value is independent of the location within the crystal, and therefore the mean force must again vanish.

Such statistical relationships are extremely numerous. We may take, for instance, (2.31) and average it over time. If the particle under consideration is in a stationary state or a state in which its kinetic energy is bounded, the time average of the left-hand side vanishes. The time average of the work done by the force must then also vanish.

There is no reason to dwell very long on such relations involving time averages. There is, however, one such relation which is generally ignored in elementary mechanics and which is quite useful in statistical mechanics. This is the *virial theorem*. We shall now derive this theorem from quantum mechanics, and then present some elementary applications.

Consider the selfadjoint operator

$$\mathcal{O} = \frac{1}{2} \sum_{\nu=1}^{N} (\mathbf{x}_\nu \cdot \mathbf{p}_\nu + \mathbf{p}_\nu \cdot \mathbf{x}_\nu) \tag{2.35}$$

and work out the commutator of this quantity with the hamiltonian (2.24). The result is

$$\mathcal{O}\mathcal{H} - \mathcal{H}\mathcal{O} = i\hbar \sum_{\nu=1}^{N} \left\{ \frac{1}{m_\nu} p_\nu^2 - \mathbf{x}_\nu \cdot \frac{\partial \Phi}{\partial \mathbf{x}_\nu} \right\}$$

Insertion of this result into (2.34) yields

$$\frac{d}{dt} \langle \mathcal{O} \rangle = \left\langle \sum_{\nu=1}^{N} \frac{1}{m_\nu} p_\nu^2 \right\rangle + \left\langle \sum_{\nu=1}^{N} \mathbf{x}_\nu \cdot \left(-\frac{\partial \Phi}{\partial \mathbf{x}_\nu} \right) \right\rangle \tag{2.36}$$

We now take the time average of this relation. The two terms on the right then become double averages over space and time. We denote the latter average by a bar over the expression. Equation (2.36) then becomes

$$\overline{\left\langle \sum_{\nu=1}^{N} \frac{1}{2 m_\nu} p_\nu^2 \right\rangle} = \frac{1}{2} \overline{\left\langle \sum_{\nu=1}^{N} \mathbf{x}_\nu \cdot \frac{\partial \Phi}{\partial \mathbf{x}_\nu} \right\rangle} \tag{2.37}$$

provided the expectation value of $\mathcal{O}$ is bounded in time. Examination of the structure of (2.35) shows that this implies not only that the momenta of the particles in the system are bounded, but also their position. This means that the system must be limited to a certain portion of space. Later on we shall apply equation (2.37) to the system of planets around the sun. In such an application we would find that the equation not only does not apply to material drifting into the solar system in hyperbolic orbits, but also that the presence of such bodies makes (2.37) slightly wrong even for the system of planets. We shall ignore these fine points and take the equation in the form (2.37). We recognize then that the left-hand side equals the mean kinetic energy of translation of the point masses making up the system. According to (2.37) this quantity equals $(-\frac{1}{2})$ times the sum of the scalar products of all position vectors with the corresponding forces, averaged over space and time. This expression is called the *virial* of the system and is often fairly easy to evaluate. It is important to notice that the theorem allows the point masses to have internal degrees of freedom provided these internal degrees are not coupled with the translational motion. The kinetic energy of these internal modes is to be ignored when the total kinetic energy is evaluated.

The virial theorem may be derived from classical mechanics by essentially the same procedure as the one given here. The relation is then simply between time-averaged quantities, without any preceding spatial average. Conversely, if we are dealing with an energy eigenstate of a quantum system, then the averages occurring in (2.37) are automatically constant in time. Averaging over time is then not necessary, and the virial theorem becomes a conservation law valid at every instant. It is then not essentially different from the quantum conservation relations of the preceding section.

The most important application of the virial theorem is to a set of particles which are held in a container at a uniform pressure. In such a case the total potential Φ in (2.24) can be divided with fair accuracy into an internal term acting between the particles and a wall term acting between the particles and the wall. Correspondingly, we can distinguish in the final result (2.37) an *internal* and an *external virial*. The only particles making a contribution to the external virial are those close to the containing wall. Rather than studying these forces in detail we rely on Newton's third law and compute the total force exerted by the gas on each section of the wall. This force is adequately represented by the pressure concept. Since the force is local, the position vector for action and reaction is substantially the same. The external virial can thus be written in the form

$$\frac{1}{2} p \int \int_A \mathbf{x} \cdot d\mathbf{A}$$

where p is the pressure exerted by the containing vessel, dA its element of area, and $\mathbf{x}$ the position vector associated with the element of area dA. The time-averaging process can be ignored here because the expression is already time independent. Or more precisely the concept of pressure is already a time average. We can transform the above surface integral to a volume integral over the volume of the container, observing that the divergence of $\mathbf{x}$ equals 3. We get then

$$\tfrac{3}{2} p \, V = \sum_{\nu=1}^{N} \frac{1}{2m_\nu} \, p_\nu^2 - (\text{Internal virial}) \tag{2.38}$$

The time-averaging process for the first term on the right is here also left off on the basis that if the number N of particles is large the fluctuation in time of their total kinetic energy is very small. We shall verify this notion in a later chapter.

Equation (2.38) forms an excellent starting point for a study of the equation of state of imperfect gases. The pressure is already isolated on the left, and the terms on the right are very likely to depend only on temperature and density. We shall use the equation for this purpose in Chapter 12. At present we shall simplify it by making the assumption that the forces between the particles are negligible. The internal virial is then zero, and equation (2.38) becomes

$$\tfrac{3}{2} p \, V = \text{Total kinetic energy of translation} \tag{2.39}$$

Equation (2.39) has obvious relevance to the equation of state for a perfect gas. If we assume that a perfect gas consists of a large number of non-interacting particles, then the similarity in the structure of (1.08) and (2.39) is striking. It leads to the conclusion that the gas temperature is a measure of the total kinetic energy of translation of the gas molecules. More precisely we find for a set of N gas molecules of equal mass m

$$\tfrac{1}{2} N \, m \langle v^2 \rangle = \tfrac{3}{2} N \, k \, T \tag{2.40}$$

where the pointed bracket indicates averaging over all molecules. Averaging in time is again not really necessary, although in principle demanded by the formula. Another well known result contained in (2.39) is Dalton's law of partial pressures for gas mixtures. If a perfect gas mixture is the same thing as a set of noninteracting but not identical particles sharing the same volume, then the total kinetic energy of translation of the particles is simply the sum of the kinetic energies for each species present. In other words, the right-hand side of (2.39) breaks up into a sum of terms, one for each species of gas molecules; moreover, these terms can be evaluated as if the other constituents were not there. We may then, if we wish, enforce the same split-up on the left of (2.39), defining the concept of partial pressure. The total pressure is simply the sum of these partial pressures, evaluated as if the other components were not there:

$$p = p_1 + p_2 + p_3 + p_4 + \cdots \qquad (2.41)$$

Application of (2.40) yields then for these partial pressures p_i

$$p_i = \frac{N_i \, k \, T}{V} \qquad (2.42)$$

Combination of (2.41) and (2.42) yields finally

$$p \, V = (N_1 + N_2 + N_3 + \cdots) \, k \, T \qquad (2.43)$$

As a second example we consider a set of particles held together by inverse square law central forces such as the electrons and nuclei of a molecule, or the planets and the sun in the solar system. Outside forces are now negligible, and the virial is to be taken over the pairwise interactions between the particles. For definiteness we shall adopt the planetary language and write the potential in the form

$$\Phi(\mathbf{x}_1, \mathbf{x}_2 \ldots, \mathbf{x}_N) = -\frac{1}{2} \sum_{\substack{\nu, \mu = 1 \\ \nu \neq \mu}}^{N} \frac{G \, m_\nu \, m_\mu}{|\mathbf{x}_\nu - \mathbf{x}_\mu|} \qquad (2.44)$$

We now form the virial, using definition (2.37),

$$\frac{1}{2} \sum_{\nu=1}^{N} \mathbf{x}_\nu \cdot \frac{\partial \Phi}{\partial \mathbf{x}_\nu} = \frac{1}{2} \sum_{\nu=1}^{N} G \, m_\nu \, \mathbf{x}_\nu \cdot \sum_{\substack{\mu=1 \\ \mu \neq \nu}}^{N} \frac{m_\mu \, (\mathbf{x}_\nu - \mathbf{x}_\mu)}{|\mathbf{x}_\nu - \mathbf{x}_\mu|^3}$$

In the double sum on the right the summation indices can be permuted. If we take the arithmetic mean of the two expressions so obtained, we get

$$\frac{1}{2} \sum_{\nu=1}^{N} \mathbf{x}_\nu \cdot \frac{\partial \Phi}{\partial \mathbf{x}_\nu} = \frac{1}{4} G \sum_{\substack{\nu, \mu = 1 \\ \nu \neq \mu}}^{N} \frac{m_\nu \, m_\mu \, (\mathbf{x}_\nu - \mathbf{x}_\mu)^2}{|\mathbf{x}_\nu - \mathbf{x}_\mu|^3}$$

which, because of (2.44), means simply that

$$\frac{1}{2} \sum_{\nu=1}^{N} \mathbf{x}_\nu \cdot \frac{\partial \Phi}{\partial \mathbf{x}_\nu} = -\frac{1}{2} \Phi(\mathbf{x}_1, \mathbf{x}_2, \ldots \mathbf{x}_N) \qquad (2.45)$$

We now insert (2.45) into (2.37) and abbreviate the total translational kinetic energy by K. We find then

$$\overline{K} = -\tfrac{1}{2} \overline{\Phi} \qquad (2.46)$$

The time-averaging procedure cannot be ignored this time because the number of particles to which (2.46) is applied is often quite small (for instance, the moons of Jupiter). In atomic applications the quantum-averaging process is also to be applied to (2.46). If the state is stationary, this is the only averaging process necessary and the time averaging can be ignored.

Equation (2.46) gains particular strength if it is combined with the law of conservation of energy:

$$K + \Phi = E \qquad (2.47)$$

It is now possible to determine $\overline{K}$ and $\overline{\Phi}$ separately in terms of E. The result is

$$\overline{K} = -E \qquad (2.48a)$$

$$\overline{\Phi} = 2E \qquad (2.48b)$$

Since K is intrinsically positive, we find that the total energy is always negative and that the total potential energy is also negative and on the average equal to twice the total energy. Kepler's planetary orbits and the negative energy wave functions of the electron in the hydrogen atom must conform to the equations (2.48). The real power of the relations (2.48) is of course that they also apply to many-body systems whose motion is not understood in detail; molecules, the system of planets, the moons of Jupiter, etc. It is to be remembered from the derivation that the relations apply only to bound states, that is, states in which the mutual distance of all interacting particles remains limited in time. Indeed, equation (2.48a) shows that the virial theorem cannot be applicable to any state for which the total energy is positive.

A. RECOMMENDED PROBLEMS

1. A declarer at bridge notices that his two opponents hold between them four trumps. Find the probabilities that they are held in the proportions 0:4, 1:3, 2:2. Solve the problem in two stages. First assume that the four cards are embedded in an infinity of irrelevant cards (popular reasoning). Then take into account that each opponent is dealt only 13 cards.

2. Find the number of ways in which N distinguishable objects can be arranged in two piles of M and $N - M$ objects, respectively ($M \neq \frac{1}{2}N$). How many ways do we get from this for arrangement in two piles of any size? Check your answer by direct reasoning.

3. (Random walk problem.) A man walks along a straight line in steps d, positive or negative, starting from the origin. His probability of taking a positive step is π, a negative step $1 - \pi$.

(a) Show that the probability of his being md steps from the origin is zero unless m is absolutely not larger than, and is of the same parity as, s. If it is not zero it is given by the binomial formula (2.02) with $n = \frac{1}{2}(s + m)$.

(b) Calculate the expectation value of the displacement from the origin and show that the mean square deviation from this value is proportional to s.

4. Derive the virial theorem from classical mechanics.

5. Verify for the ground state wave function of the hydrogen atom that it obeys the distribution law (2.48) for the mean kinetic and potential energy.

B. GENERAL PROBLEMS

6. One cubic millimeter of a suspension of bacteriophage is added to each of 100 test tubes containing a bacterium which is susceptible to the bacteriophage. Upon later inspection, six test tubes are found not to be infected. Find the density of bacteriophage in the suspension used.

7. Compute the mean number and the root mean square deviation for the number of molecules in 1 cm^3 of a perfect gas at standard pressure and temperature.

8. How small must a gas volume at standard pressure and temperature be if the root mean square deviation is to be 1% of the mean number of molecules?

9. A die is thrown four times in succession. Calculate the relative probability of getting two sixes and two ones in any order, as compared to getting three sixes and one one in any order. Calculate also the absolute probabilities.

10. A declarer at bridge notices that this opponents hold between them an ace and two kings. Compute the probabilities for the various distributions by the two methods explained in Problem 1. Add to it a third case in which a worthless card was accidentally uncovered when dealing to opponent 1.

11. A man is dealt five cards in succession at poker. Find the relative probability of getting a full house as compared to getting a flush.

12. Prove from the virial theorem that the mean potential energy of a harmonic oscillator equals its mean kinetic energy.

13. Simplify the orbit of a planet around the sun by taking it to be a circle, and verify the following:

(a) The mean force in any fixed direction is zero.

(b) The mean velocity in any fixed direction is zero.

(c) The relations (2.48) demanded by the virial theorem are valid.

14. An anharmonic oscillator has a restoring force proportional to the cube of the displacement. Show that its kinetic energy equals in the mean twice its potential energy.

15. Construct the ground state wave function for the potential

$$\Phi = \frac{A}{r^2} - \frac{B}{r}$$

and show that the mean force vanishes in the ground state whether considered as a potential in one dimension (in which case $\Phi = \infty$ for negative r) or in three dimensions. (*Hint:* The ground state wave function is of the form $r^p e^{-qr}$.)

16. For the case of the Poisson distribution (2.13), calculate the probabilities of observing 0, 1, 2, 3, 4, 5, 6, 7, 8, 9, 10 events in a time interval during which we expect 4. Repeat, giving the probabilities from 0 to 20 in an interval during which we expect 8. Plot the results in the following way. Plot for each number the probability of finding that number *or less*. Plot the two results on the same graph, reducing the abscissa scale by a factor $\frac{1}{2}$ for the second graph.

17. Prove generally from the Schroedinger equation for one particle that the mean force in any stationary state is zero if the wave function falls off sufficiently rapidly at large distances.

18. Find the relation analogous to (2.39) for a gas of photons by assuming that for them the kinetic energy term in (2.24) is to be replaced by $\sum_{\nu} c\, p_{\nu}$, with c the velocity of light.

19. Extend your verification under Problem 13 to elliptic orbits. Examine also hyperbolic orbits to see which relations remain valid. Relations (a) and (b) must be examined by components.

3

Statistical counting in mechanics

Most workers in statistical mechanics would probably agree that averaging in time is the ideal form of statistical averaging in mechanics. Any other method, if employed, should be checked against this one. We have done a limited amount of such statistics in the preceding chapter. An extension of the method is possible, particularly for gases or gas-like assemblies; it is presented in Chapter 18 of this book. We have to anticipate on this work in a negative sense by pointing out its unsatisfactory aspects which have led people to consider alternative modes of taking statistical averages.

Chapter 18 and subsequent chapters show that if one wishes to follow the behavior of a system in time, attention has to be focused on "encounters" or "collisions." These events are admittedly rare, but they lead the system from one state to another. The encounters thus play an essential coupling role in any form of statistics which tries to average over time. In this there is a great disadvantage; for it is in fact found that the exact nature of these encounters does not matter for many of the equilibrium properties of a system. And complementary to this observation is a second one that a simple mechanical representation of encounters may not be sufficiently randomizing to be of value in statistics. For instance, many billiard ball models of collision fail to flip spin, and thus will not lead us to proper averages over the spin variable.

The reader may understand the situation better if we take up first the simpler problem of evaluating certain average properties of bridge hands.

If we were to attempt to understand these averages as time averages, then we would have to pay a lot of attention to the shuffling and dealing of the cards because it is the shuffling and dealing which leads from one card pattern to the next. We would thereby be forced to study the perplexing permutations which the cards undergo in this process, and we would no doubt even find individual differences from dealer to dealer. Yet it is a fact that the nature of these processes has no influence on the probability of bridge hands if the dealer is honest. The only thing we have to know is that the shuffling takes place between two deals and that it is random. Given these assurances, we can evaluate the probability of bridge hands by a priori consideration in which the shuffling as a process plays no role whatever.

We can go on from this case to an example in physics. Hydrogen gas, H_2, exists in two modifications, orthohydrogen and parahydrogen. Ortho-hydrogen has the two nuclear spins aligned parallel, parahydrogen anti-parallel. The abundance of these two species in natural hydrogen is 3:1 at elevated temperature and 0:1 near absolute zero (because parahydrogen contains the ground state of hydrogen). The case is of interest here because the ortho-para conversion is a process of some difficulty. Normal gaseous encounters either leave the spins the same or simply exchange them. In either case, the abundance of the two modifications is not altered. Normally conversion takes place on the walls, in the neighborhood of magnetic impurities. It is possible, however, to create the physical equivalent of the dishonest card dealer, namely, a container from which all traces of paramagnetic and ferromagnetic impurities have been removed; the same precautions can be extended to paramagnetic impurity gases. As a result an anomalous abundance ratio of ortho- and parahydrogen can be stabilized for extended periods.

The example shows the subordinate role which encounters play in most of statistical physics. It is possible, by the use of certain precautions, *to prevent* the establishment of equilibrium in the spin system of hydrogen gas. On the other hand, in order to get equilibrium the experimenter has only to be reasonably careless. The detailed nature of his carelessness is of no interest. Of even less interest are the detailed processes by which equilibrium is brought about. Thus, in statistical physics, we try to avoid a detailed study of randomizing processes by substituting other averaging procedures. Again, the case of bridge hands may serve as an illustration. Instead of following the dealing and shuffling of one bridge hand we do statistics by looking at a large number of bridge hands, all dealt. Such an average, called an *ensemble average*, was first introduced into statistical mechanics by Gibbs. We must think of this chiefly as a device. Bridge hands can be contemplated indefinitely, but mechanical systems follow a natural evolution in time which is part of the nature of the state contemplated. Thus, the notion of time average can never be excluded entirely from consideration.

The counting element which is going to replace time in our formation of averages is the *element of phase space* in classical mechanics and the *quantum state* in quantum mechanics. We must not expect a watertight proof that this type of count is equivalent to a time average for any type of system. The reason for this is that stray interactions, such as the paramagnetic wall impurities discussed above, can assume an enormous importance in shaping the type of average which is established; in the above example, the stray interactions decide whether we should average over spin or treat the gas as a mixture of two gases of fixed spin. In view of such uncertain borderline cases no full equivalence theorem can be expected. What we can do is show that our counting procedure is not in contradiction with time averaging; in other words, that it is automatically stationary in time.

3-1. Statistical counting in classical mechanics; Liouville theorem and ergodic hypothesis

We start out by discussing the classical case. Suppose that we have a conservative system of N degrees of freedom obeying classical mechanics. Call its coordinates $q_1, q_2, \ldots q_N$ and the conjugate momenta $p_1, p_2, \ldots p_N$. Let $\mathcal{H}(q_1 \ldots q_N, p_1 \ldots p_N)$ be the hamiltonian of the system. Then, if we know the values of p_i and q_i at a given instant, their future values are completely determined by Hamilton's equations:

$$\frac{dq_i}{dt} = \frac{\partial \mathcal{H}}{\partial p_i} \tag{3.01}$$

$$\frac{dp_i}{dt} = -\frac{\partial \mathcal{H}}{\partial q_i} \tag{3.02}$$

These equations find a simple interpretation if the notion of phase space is introduced; this is a $2N$ dimensional space of coordinates $q_1, q_2, \ldots q_N$, $p_1 \ldots p_N$. Within this space an instantaneous state of our system is represented by a point, and the development of the state in time by a trajectory. Classical mechanics teaches us that there is a single trajectory which passes through a given point and that this trajectory is completely determined by this point.

If we consider a large number of representative points of the system on neighboring trajectories, we can treat them as a fluid if the points are sufficiently dense in phase space. For this purpose it is useful to define an element of volume. One possible choice is

$$d\Omega = dq_1 \, dq_2 \ldots dq_N \, dp_1 \ldots dp_N \qquad (3.03)$$

One might suppose that this simple-minded choice may not be retained without a factor and that this factor would depend on the system of coordinates. This is not the case. The product of differentials (3.03) is invariant with respect to all possible transformations in which the meaning of q_i and p_i as conjugate coordinates and momenta is preserved. We shall not prove this theorem for the most general transformation of this type, which is called a *contact transformation*. We shall restrict ourselves to the transformations called *point transformations*, that is, those for which the coordinates transform among themselves:

$$Q_i = Q_i(q_1, q_2, \ldots q_N) \qquad (3.04a)$$

These formulas also determine the transformations of the momenta which are linear but have coefficients depending on the q's. We obtain them easily in the following manner. We have the formula

$$\frac{\partial \mathcal{H}}{\partial p_i} = \sum_{\kappa=1}^{N} \frac{\partial \mathcal{H}}{\partial P_\kappa} \frac{\partial P_\kappa}{\partial p_i}$$

By virtue of (3.01), this may also be written as

$$\frac{dq_i}{dt} = \sum_{\kappa=1}^{N} \frac{dQ_\kappa}{dt} \frac{\partial P_\kappa}{\partial p_i}$$

We see from this equation that

$$\frac{\partial q_i}{\partial Q_k} = \frac{\partial P_k}{\partial p_i} \qquad (3.05)$$

This system of equations is immediately integrable and gives

$$P_k = \sum_{i=1}^{N} \frac{\partial q_i}{\partial Q_k} p_i \qquad (3.04b)$$

In order to prove the invariance of the volume (3.03), it is necessary to investigate the Jacobian

$$\frac{\partial(q_1, q_2, \ldots q_N, p_1 \ldots p_N)}{\partial(Q_1, Q_2, \ldots Q_N, P_1 \ldots P_N)}$$

This determinant of $2N$ rows and columns decomposes into four pieces of N rows and columns as indicated in the following:

$$\frac{\partial(q_i, p_i)}{\partial(Q_k, P_k)} = \begin{vmatrix} \dfrac{\partial q_1}{\partial Q_1} & \dfrac{\partial q_1}{\partial Q_2} & \cdots & \dfrac{\partial q_1}{\partial Q_N} & 0 & 0 & \cdots & 0 \\[2mm] \dfrac{\partial q_2}{\partial Q_1} & \cdot \cdot & \cdots & \dfrac{\partial q_2}{\partial Q_N} & 0 & 0 & \cdots & 0 \\[2mm] \cdots & \cdots & \cdots & \cdots & 0 & \cdot & \cdots & \cdot\cdot \\[2mm] \dfrac{\partial q_N}{\partial Q_1} & \cdot\cdot & \cdots & \dfrac{\partial q_N}{\partial Q_N} & 0 & \cdot & \cdots & 0 \\[2mm] \dfrac{\partial p_1}{\partial Q_1} & \cdot\cdot & \cdots & \dfrac{\partial p_1}{\partial Q_N} & \dfrac{\partial p_1}{\partial P_1} & \cdot & \cdots & \dfrac{\partial p_1}{\partial P_N} \\[2mm] \cdots & \cdot\cdot & & \cdot\cdot & \cdot\cdot & \cdot\cdot & & \cdot\cdot \\[1mm] \cdots & \cdot\cdot & & \cdots & \cdot\cdot & \cdots & & \cdot\cdot \\[2mm] \dfrac{\partial p_N}{\partial Q_1} & \cdot\cdot & \cdots & \dfrac{\partial p_N}{\partial Q_N} & \dfrac{\partial p_N}{\partial P_1} & \cdot & \cdots & \dfrac{\partial p_N}{\partial P_N} \end{vmatrix}$$

The piece in the upper right contains nothing but zeros; this arises from the special structure of the transformation (3.04). The determinant reduces therefore to the product of two determinants as follows:

$$\frac{\partial(q_i, p_i)}{\partial(Q_k, P_k)} = \frac{\partial(q_1, \ldots q_N)}{\partial(Q_1, \ldots Q_N)} \cdot \frac{\partial(p_1, \ldots p_N)}{\partial(P_1, \ldots P_N)}$$

By applying (3.05) to the second factor we obtain

$$\frac{\partial(q_1, \ldots q_N, p_1, \ldots p_N)}{\partial(Q_1, \ldots Q_N, P_1, \ldots P_N)} = \frac{\partial(q_1, \ldots q_N)}{\partial(Q_1, \ldots Q_N)} \cdot \frac{\partial(Q_1, \ldots Q_N)}{\partial(q_1, \ldots q_N)} = 1$$

$$(3.06)$$

which is what was to be proved.

We shall now prove that this volume which we have defined in an invariant way is also invariant with respect to time along a trajectory, that is, the deformations which it undergoes keep its measure constant.

For this we consider the logarithmic derivative of (3.03) with respect to time, that is,

$$\frac{1}{d\Omega}\frac{d}{dt}(d\Omega) = \lim_{\Delta t = 0} \left\{ \frac{d\Omega(t+\Delta t) - d\Omega(t)}{\Delta t \, d\Omega(t)} \right\}$$

$$= \lim_{\Delta t = 0} \frac{1}{\Delta t} \left\{ \frac{\partial(q_i(t+\Delta t), p_i(t+\Delta t))}{\partial(q_i(t), p_i(t))} - 1 \right\}$$

To evaluate this it is simplest to expand the parenthesis in powers of Δt and to stop with linear terms. We have

$$q_i(t+\Delta t) = q_i(t) + \frac{dq_i(t)}{dt}\,\Delta t + 0(\Delta t^2)$$

$$p_i(t+\Delta t) = p_i(t) + \frac{dp_i(t)}{dt}\,\Delta t + 0(\Delta t^2)$$

In these expressions the coefficients of Δt are to be replaced by (3.01) and (3.02). We are then prepared to calculate the partial derivatives which enter into the Jacobian:

$$\frac{\partial q_i(t+\Delta t)}{\partial q_k(t)} = \delta_{ik} + \frac{\partial^2 \mathcal{H}}{\partial p_i\,\partial q_k}\,\Delta t + 0(\Delta t^2)$$

$$\frac{\partial q_i(t+\Delta t)}{\partial p_k(t)} = \frac{\partial^2 \mathcal{H}}{\partial p_i\,\partial p_k}\,\Delta t + 0(\Delta t^2)$$

$$\frac{\partial p_i(t+\Delta t)}{\partial q_k(t)} = -\frac{\partial^2 \mathcal{H}}{\partial q_i\,\partial q_k}\,\Delta t + 0(\Delta t^2)$$

$$\frac{\partial p_i(t+\Delta t)}{\partial p_k(t)} = \delta_{ik} - \frac{\partial^2 \mathcal{H}}{\partial q_i\,\partial p_k}\,\Delta t + 0(\Delta t^2)$$

The Jacobian whose elements we proceed to write down then takes the following form:

$$\frac{\partial(q_i(t+\Delta t),\, p_i(t+\Delta t))}{\partial(q_i(t),\, p_i(t))} = |\mathscr{E} + \{\mathcal{M}\}\Delta t + 0(\Delta t^2)|$$

$\mathscr{E}$ is the diagonal unit matrix, and $\{\mathcal{M}\}$ is the following matrix:

$$\{\mathcal{M}\} = \begin{bmatrix}
\dfrac{\partial^2 \mathcal{H}}{\partial p_1\,\partial q_1} & \dfrac{\partial^2 \mathcal{H}}{\partial p_1\,\partial q_2} & \cdots & \dfrac{\partial^2 \mathcal{H}}{\partial p_1\,\partial q_N} & \dfrac{\partial^2 \mathcal{H}}{\partial p_1^2} & \cdots & \dfrac{\partial^2 \mathcal{H}}{\partial p_1\,\partial p_N} \\[2ex]
\dfrac{\partial^2 \mathcal{H}}{\partial p_2\,\partial q_1} & \dfrac{\partial^2 \mathcal{H}}{\partial p_2\,\partial q_2} & \cdots & & & & \\[2ex]
\cdots & \cdots & \cdots & \dfrac{\partial^2 \mathcal{H}}{\partial p_N\,\partial q_N} & \cdots & & \dfrac{\partial^2 \mathcal{H}}{\partial p_N^2} \\[2ex]
-\dfrac{\partial^2 \mathcal{H}}{\partial q_1^2} & \cdots & \cdots & -\dfrac{\partial^2 \mathcal{H}}{\partial q_1\,\partial q_N} & -\dfrac{\partial^2 \mathcal{H}}{\partial p_1\,\partial q_1} & \cdots & -\dfrac{\partial^2 \mathcal{H}}{\partial p_N\,\partial q_1} \\[2ex]
\cdots & \cdots & \cdots & \cdots & & -\dfrac{\partial^2 \mathcal{H}}{\partial q_2\,\partial p_2} & \cdots \\[2ex]
\cdots & \cdots & \cdots & \cdots & \cdots & & -\dfrac{\partial^2 \mathcal{H}}{\partial p_N\,\partial q_N}
\end{bmatrix}$$

We see that the trace of the matrix $\{\mathcal{M}\}$ is zero. But it is just this trace which forms the coefficient of the linear term in Δt if one expands the determinant written above; therefore, the value of the derivative under consideration is zero:

$$\frac{1}{d\Omega} \frac{d}{dt}(d\Omega) = 0 \qquad (3.07)$$

Since the point t is entirely arbitrary, $d\Omega$ is a constant along the trajectory.

From this demonstration, we immediately deduce *Liouville's theorem:*

Imagine that we have a sufficiently large number of mechanical systems so that their representative points in phase space may be treated like a fluid. Then the density ρ of the fluid in the neighborhood of each representative point P is constant with respect to time, if P remains in the interior of the fluid.

We derive from Liouville's theorem the following corollary:

If a density ρ of representative points in phase space is to be stationary in time, it is necessary that it be constant along every trajectory.

It is a temptation to add "and sufficient." This was done by all early investigators. The result is a density function which is not fully determined since all that is known about it is that it must be constant along a classical trajectory.

The early efforts to obtain a more sweeping result are associated with the *ergodic hypothesis.* According to this hypothesis, a representative point will traverse in time all points whose energy is equal to its own. One can prove that this hypothesis is false for a hamiltonian system. One substitutes for it a weaker hypothesis called the quasi-ergodic hypothesis. According to this hypothesis, almost all the trajectories of a system are such that the representative point approaches as closely as one wishes any accessible point in phase space provided its energy is the same as for the trajectory. One is led to the idea behind this hypothesis by observing the Lissajou figures for a two-dimensional oscillator whose frequencies are incommensurable. But this example suggests another for which even the quasi-ergodic hypothesis is broken: the oscillator with commensurable frequencies. The reason for the validity of the ergodic hypothesis cannot be found entirely in a hamiltonian framework. Over and above this framework we must introduce the notion of molecular disorder. When we pour milk into coffee there is no doubt that we shall obtain, by a slight agitation, a complete mixing of the two constituents. This mixing is not obviously contained in the laws of mechanics. Unmixing of the coffee and milk in time is conceivable mechanically, but highly improbable statistically. We wish to bring in this notion of probability in some form and thereby render statistical reasoning more powerful than it would be on mechanical grounds alone. The ergodic hypothesis is one possible form of this idea. Whatever exact form it is given, the conclusion one is trying to reach is the following: the energy occupies a special position among the constants of the motion in that the density in phase space is different for different energies but is independent of all other parameters. The implication is that stray interactions will easily shift the trajectory in phase space, but that the energy transmitted by such interactions is small.

There is no question but that the ergodic model proposed is not always entirely right. The best way to understand its limitations is to look for cases in which it must be modified. A number of applications are known where angular momentum enters in addition to energy in determining the probability density in phase space. They show fairly conclusively that the ergodic hypothesis is satisfactory in the vast majority of all applications.

3-2. Statistical counting in quantum mechanics

The classical concept of density in phase space has to be thought through again for the case of quantum systems. There is an obvious difficulty in the concept of phase space which arises from Heisenberg's uncertainty principle. According to this principle, the accuracy Δq_i with which a coordinate q_i can be observed is linked to the accuracy Δp_i with which the conjugate momentum p_i can be observed by the relation

$$\Delta q_i \, \Delta p_i \geqq h \tag{3.08}$$

where h is the quantum of action introduced by Planck. Its numerical value is

$$h = 6.626 \times 10^{-27} \text{ erg sec} \tag{3.09}$$

One should expect that the uncertainty principle introduces a graininess into phase space which makes the volume concept still usable for wave packets which extend over many grains. In fact the grain introduced by quantum theory is the individual quantum state. It is convenient to reformulate the probability concept for an ensemble of quantum systems because it comes out to be more straightforward than the classical formulation.

The quantum approach to counting is best developed in two stages. In the first stage we use a representation in which the hamiltonian is diagonal. Then we proceed to a second stage valid for an arbitrary representation. In the first stage we assume the hamiltonian diagonalized for our system. We are then given a set of wave functions Ψ which depend parametrically on a set of quantum numbers $n_1, n_2, n_3, \ldots n_N$. The natural variable which we have is the amplitude with which a particular wave function of the basic set participates in the actual wave function of the system. Let us denote the square of this amplitude by $p(n_1, n_2, n_3, \ldots n_N)$. We wish to make this quantity p the equivalent of the classical density function in phase space. To show the correctness of this we must show (a) that p goes over into the density function in the classical limit and (b) that it has ergodic properties.

To verify (a) it is advantageous to proceed by the old quantum theory, which is still valid as an approximation to the modern theory if we deal with large quantum number. According to this theory one is not to alter the classical trajectories, but one rejects among them all those which do not obey the Bohr–Sommerfeld quantum conditions:

$$\oint p_i \, dq_i = n \, h \tag{3.10}$$

with respect to all their coordinates. The integral on the left is the surface enclosed by the trajectory in the plane p_i, q_i. It follows that the surface enclosed between two successive trajectories n and $n + 1$ is equal to h. This leads to the conclusion that every permitted trajectory can be enclosed by a volume h^N in phase space. Furthermore, if one takes a volume in phase space which is large with respect to h^N, the number of states which are found within it will be approximately equal to $d\Omega/h^N$, where $d\Omega$ is defined by (3.03). ρ and p then represent identical quantities from differing points of view. They are related by

$$p = h^N \rho \tag{3.11}$$

We will use p in the future because of the advantage which it has in not possessing any dimension. We shall redefine the classical ρ correspondingly so as to suppress the factor h^N in (3.11).

The verification of (b) seems at first sight a triviality, for quantum theory tells us that each number $p(n_1, n_2, \ldots n_N)$ is a constant of the motion. However, we must repeat here the more sophisticated thinking necessary for the ergodic hypothesis. It is not reasonable to expect that an actual hamiltonian really includes all interactions of a real system. There will be stray interactions left outside the hamiltonian which will produce transitions between the stationary states of the system as defined by the approximate hamiltonian. These stray interactions can be assumed weak. We may then assume for them the result of first-order perturbation theory, according to which the transition probabilities P_{ij} that result from it obey the symmetry relation

$$P_{ij} = P_{ji} \tag{3.12}$$

and are different from zero only if the energy of the two states is equal. An ergodic assumption for the probabilities $p(n_1, n_2, \ldots n_N)$ which makes p a function of the energy only is thus not in contradiction with (3.12). We write therefore

$$p(n_1, n_2, \ldots n_N) = p[E(n_1, n_2, \ldots n_N)] \tag{3.13}$$

The usual reservation has to be made in connection with (3.13): exceptional situations may arise in which p depends on other conserved quantities beside the energy. This may even go so far that certain states are classified as "inaccessible" within the time of the experiment.

We must now follow our reasoning by a second stage in which (3.13) is cast into a form which is invariant with respect to the representation used. In the form in which it is written it will only yield the correct expectation value for quantities which are diagonal with respect to the quantum numbers $n_1, n_2, \ldots n_N$. Expectation values of operators which are not diagonal involve the phases of the quantum amplitudes as well as the intensities. Suppose we subsume the quantum numbers $n_1, n_2, \ldots n_N$ by the single index i or j. Then our wave function Ψ has the form

$$\Psi = \sum_i a_i \, \psi_i \tag{3.14}$$

If we now look for the expectation value of an operator A which is not diagonal in these wave functions, we find

$$\langle A \rangle = \sum_{ij} a_i^\star \, a_j \, \langle i | A | j \rangle \tag{3.15}$$

The role which we originally reserved for the quantities $|a_i|^2$ is thus taken over by the matrix

$$\langle j | \rho | i \rangle = a_j \, a_i^\star \tag{3.16}$$

The matrix (3.16) is called the *density matrix* of our system. With its help (3.15) takes the simple form

$$\langle A \rangle = \text{Trace} \, (\rho \, A) \tag{3.17}$$

The invariance of (3.17) can now be handled with the tools of matrix theory. The most general transformation of the matrix A is a similarity transformation which results from a shift in the base (3.14). We must now show that, although ρ is constructed in a different way, its definition (3.16) yields for it the same transformation properties. Once this is shown, the expression (3.17) is automatically invariant because it is the trace of the product of two matrices each one of which is subject to the same transformation laws.

To prove that the matrix (3.16) has the same transformation properties as the matrix of a dynamical variable we introduce a second base φ_s which is obtained from the base ψ_i in (3.14) by the transformation

$$\varphi_t = \sum_i \psi_i \, \alpha_{it} \tag{3.18a}$$

The transformation matrix α_{it} is unitary, and we have therefore

$$\psi_i = \sum_s \alpha_{is}^* \, \varphi_s \tag{3.18b}$$

Denote matrix elements in the φ system by round brackets, in the ψ system by pointed brackets. We then have

$$(s|A|t) = \int \varphi_s^* \, A \, \varphi_t = \sum_{ij} \alpha_{is}^* \, \alpha_{jt} \int \psi_i^* \, A \, \psi_j$$

$$(s|A|t) = \sum_{ij} \alpha_{is}^* \, \alpha_{jt} \langle i|A|j \rangle \tag{3.19}$$

for the transformation law of a dynamical variable. On the other hand, the new equation replacing (3.14) is

$$\Psi = \sum_s b_s \varphi_s \tag{3.20}$$

We relate the two formulas with the help of (3.18) and find

$$a_i = \sum_s \alpha_{is} \, b_s \tag{3.21a}$$

or

$$b_t = \sum_i a_i \, \alpha_{it}^* \tag{3.21b}$$

Now forming the density matrix by the definition (3.16), we find

$$(s|\rho|t) = b_s \, b_t^* = \sum_{ij} \alpha_{is}^* \, \alpha_{jt} \, a_i \, a_j^*$$

or

$$(s|\rho|t) = \sum_{ij} \alpha_{is}^* \, \alpha_{jt} \langle i|\rho|j \rangle \tag{3.22}$$

The transformation law (3.22) is identical with (3.19), and the invariance of definition (3.17) is thereby proved. The definition (3.17) automatically disposes of our second problem, namely, how to define probability ergodically when we are not given a set of wave functions which diagonalize the energy. We simply replace (3.13) by the matrix relation

$$\rho = p(\mathcal{H}) \tag{3.23}$$

Equation (3.13) can then be considered a special form of (3.23) when the matrix representing $\mathcal{H}$ is diagonal.

We conclude this section by observing that it is common practice to liberate the density matrix from the narrow definition (3.16). We may think of this by saying that one includes in the definition of the density matrix a time or ensemble average over the right-hand side of (3.16). The result of such a procedure is to slough off the off-diagonal elements in an energy representation; for we see from (3.14) that elements connecting states of different energy oscillate in time, while the diagonal ones are independent of time. Thus a time average will tend to remove the former, but not the latter. It is in fact only after this sloughing-off process has been completed that the operator relation (3.23) can be obeyed if (3.16) is obeyed initially. It is even possible to reformulate quantum mechanics, using the density matrix as the object of study instead of the wave function. Thus one may derive from (3.16) without difficulty that

$$ i \hbar \frac{d\rho}{dt} = \mathscr{H} \rho - \rho \mathscr{H} \tag{3.24} $$

The sign of this relation is the *reverse* of the sign which holds for dynamical variables. This opposition in sign points out the fact that we do not deal here with the same type of equation. Equation (3.24) was derived for the Schroedinger representation, in which the operators corresponding to dynamical variables have no time dependence. Inversely, the well known time dependence of the dynamical variables arises in the Heisenberg representation, in which the wave functions and, by implication, the density matrix are treated as constants in time. Equations (2.34) and (3.17) are valid in either case, and the reversal in the sign of (3.24) is just what is needed to make it so.

Equation (3.24) can be used as a basis for a wider study of the density matrix. If the matrix is given at one instant of time one can predict its future behavior. However, in the present context we are interested only in a density matrix which is stationary in time since it is supposed to supply our basis for counting. To make this so, ρ can only be a function of the constants of the motion; this follows from a comparison of (3.24) and (2.34). In this sense (3.24) is the quantum equivalent of Liouville's theorem. We have pointed out earlier that the theorem does not go far enough. In the ergodic hypothesis we remove all but the energy as variables on which ρ depends. This farther-reaching notion finds its quantum formulation in (3.23).

A. RECOMMENDED PROBLEMS

1. Verify explicitly the invariance of the volume of phase space in the passage from Cartesian to spherical polar coordinates.

2. Study the counting problem for a one-dimensional harmonic oscillator by adopting the known energy quanta, but retaining the classical trajectories in phase space associated with these energies. Show that each trajectory is easily enclosed in an area equal to h.

3. Investigate the problem of counting in statistical mechanics by the following study of free particles in a box.

(a) Compute the number of quantum states per unit energy range by the classical phase space method.

(b) Carry out the same calculation by enumeration of the quantum states.

(c) Show that the two enumerations are identical (not just proportional!) provided the energy can be treated as a continuous variable in case (b).

B. GENERAL PROBLEMS

4. Prove for a particle in a one-dimensional box that each quantum state occupies an area h in phase space.

5. The result of Problem 1 seems to contradict the intuitive notion of equal weight for equal solid angles because the weight factor $\sin \vartheta$ is missing. Show that if we average a quantity which depends on p_ϑ and p_φ only through the intermediary of the kinetic energy and if the integration over these momenta goes from $-\infty$ to $+\infty$ we do in fact restore the weight factor $\sin \vartheta$ through these operations.

6. Investigate the nature of the quantum condition for angular momentum

$$M^2 = J(J+1)\,\hbar^2$$

by taking the case of a rigid diatomic molecule and associating the volume in four-dimensional phase space located between $J - \frac{1}{2}$ and $J + \frac{1}{2}$ with the degenerate quantum state J. Comment on the result. What about the state $J = 0$? Can the result be fixed, and if so, how?

7. Attention was called to the fact that the equation of motion (3.24) for the density matrix has the opposite sign of the generally known equation of motion for dynamical variables in the Heisenberg representation. Substantiate the claim made in the text that this change in sign is in fact necessary to assure the validity of (2.34) in either representation.

4

The Gibbs-Boltzmann distribution law

Statistical physics was first devised by Maxwell and Boltzmann for an assembly of identical small particles moving independently from each other. Such an assembly forms a useful approximation to a gas. Subsequently, Gibbs recast the reasoning so that it applies also to large bodies with a complicated internal structure. We shall generally follow his reasoning here because it yields a better understanding of statistical physics in general, without losing touch with the early examples. A further advantage in exposition is gained if all systems are assumed quantized because of the simplicity of counting states. This advantage was of course not available to the creators of our science, but blends with it so naturally that it is hard to believe they derived their beautiful theoretical structure without it. We are thus giving a doubly unhistorical presentation but one whose clarity would perhaps appeal to the founders.

4-1. Derivation of the Gibbsian or canonical distribution

We are following Gibbs in setting ourselves the task of deriving the law of probability distribution for a system having a complicated internal structure. As indicated above we assume the system to have a quantum hamiltonian and a set of quantum states. We shall designate the system by the label 1. In order to learn something about it we embed it in a still larger system capable of giving it energy, but not acting mechanically on any of its state variables. One must think of this transmission of energy as taking place in the form of heat transfer through one of the "stray interactions" discussed earlier. If the system 1 has only a small number of degrees of freedom, for instance, if it is an atom, this transfer may have to be looked at more closely for some purposes. The large system embedding the system 1 can be viewed in various ways. We may view it thermodynamically as a "heat bath," or we may view it with Gibbs as a large number of systems essentially identical with system 1. Such a collection of systems is called an *ensemble*. We shall favor this latter viewpoint in the following discussion.

Let us give to the ensemble including the system 1 the label 0, and to the ensemble without the system 1 the label 2. The system 2 and the system 0 are then not essentially different. The system 0 has the additional property that it can be subdivided into the two subsystems 1 and 2 which are independent. Let us label the quantum states of each subsystem by a second number. The energies of the system 1 can then be written as $E_1^1, E_2^1, E_3^1, \ldots$; similarly, the energies of the system 2 as $E_1^2, E_2^2, E_3^2, \ldots$. Since the two systems are independent, the totality of all quantum states of the system 0 is obtained by combining any two substates in pairs (i, k). The energy of such a pair state is the sum of the individual energies. We have therefore the relation

$$E_{i,k} = E_i^1 + E_k^2 \tag{4.01}$$

Because the two subsystems are independent we can find also a formula for the probability that the system 0 is in such a pair state. The theorem of independent probabilities tells us that

$$p_{i,k} = p_i^1 \cdot p_k^2 \tag{4.02}$$

The formulas (4.01) and (4.02) are pretty, but essentially useless unless something like an ergodic postulate can be applied to them. It has been shown in Chapter 3 that this postulate leads us beyond mechanics and perhaps also occasionally into error. Knowing these limitations, we make it now at this point, writing

$$p_i^1 = p^1(E_i) \tag{4.03a}$$

$$p_k^2 = p^2(E_k) \tag{4.03b}$$

$$p_{i,k} = p^0(E_i + E_k) \tag{4.03c}$$

Insertion of this into (4.02) yields

$$p^0(E_i + E_k) = p^1(E_i) \cdot p^2(E_k) \tag{4.04}$$

Equation (4.04) is a functional equation for the probability functions valid for all values of the indices i and k. In its derivation the difference in structure between the systems 1 and 2 has not played any role. However, we cannot solve the equation retaining this symmetry. We must assume that the systems 2 and 0 are ensembles or reservoirs with an extremely dense and slightly unpredictable energy spectrum, for we wish to treat the variable E_k as continuous or "potentially" continuous and $p^2(E_k)$ as a differentiable function of it. Logarithmic differentiation of (4.04) with respect to E_k yields then

$$\frac{\partial \ln p^0(E_i + E_k)}{\partial E_k} = \frac{d \ln p^2(E_k)}{dE_k} \tag{4.05}$$

The derivative may be a complicated function of the energy of the states of the system 2; we shall denote it by $-\beta_2$:

$$\frac{d \ln p^2(E_k)}{dE_k} = -\beta_2 \tag{4.06}$$

The equation is therefore not directly integrable. However, (4.05) shows that β_2 does not depend on the system 1. Furthermore, we have that

$$\frac{\partial \ln p^0}{\partial E_i} = \frac{\partial \ln p^0}{\partial E_k}$$

The logarithmic derivative with respect to the energy E_i of the left of (4.04) therefore exists and equals $-\beta_2$. Consequently this derivative can also be taken on the right and yields the same value. This leaves us with the relation

$$\frac{d \ln p^1(E_i)}{dE_i} = -\beta_2 \tag{4.07}$$

notwithstanding the discrete character of the index i. The interpretation we must give to this formula is that it yields the probability of occupation of a quantum state as a function of its energy. This function is definable over the entire energy range but has a meaning only if a quantum state of the system 1 happens to lie at the energy E_i. In distinction from equation (4.06), equation (4.07) *is* integrable and yields

$$p^1(E_i) = \text{constant} \cdot \exp(-\beta_2 E_i) \tag{4.08}$$

Since the selection of the system 1 out of the ensemble 0 was entirely arbitrary, the argument can be repeated for every other system of the ensemble. Finally the β's so obtained must all be identical to satisfy (4.04). Thus β comes out to be a characteristic of the ensemble 0 only. Equation (4.08) can therefore be written in the form

$$p^1(E_i) = \exp[\beta_0 (a_1 - E_i)] \tag{4.09}$$

It is characteristic of (4.09) that by simple product formation we can show it to be also valid for the ensemble as a whole. The distinction between ensemble and subsystem disappears therefore in the final result. The integration constant a_1, on the other hand, does depend on the individual subsystem. However, it is easily eliminated; p^1 is a probability and therefore obeys the relation

$$\sum_i p^1(E_i) = 1 \tag{4.10}$$

Substitution of (4.10) into (4.09) then yields the result

$$p(E_i) = \frac{\exp(-\beta E_i)}{\sum_v \exp(-\beta E_v)} \tag{4.11}$$

This is the Gibbs distribution law. We have suppressed in it all indices referring to ensembles or subsystems because it applies to either case equally well. We must just remember that if a system obeying (4.11) is in equilibrium with another system with which it can exchange energy then the two systems will share the parameter β. The sum in the denominator of (4.11) is called the *sum of states* or *partition function* for the system. We shall see that a large number of properties of a system can be deduced from the knowledge of the partition function only.

There is a second type of argument, originally due to Boltzmann, which can be used to establish the distribution (4.09). In this argument one assumes that the total energy of the ensemble is fixed. Such an ensemble is called *microcanonical*. The argument is, strictly speaking, not compatible with the idea of discrete quantum states because the sum of the energies of two different sets of states will not usually add up to the same amount. One must therefore assume in fact that the energy of the ensemble is *approximately* constant, with an uncertainty which is very much larger than the energy spacing of an individual system. The basic idea of this second procedure is to maximize the probability of the arrangement, under the constraint that the sum of the energies of all the systems have a prescribed value.

In this approach we reason on the occupation numbers of the various quantum states. We denote the states by 1, 2, 3, . . . and the numbers of systems in these states by $N_1, N_2, N_3, \ldots$. The ensemble must be assumed large enough that even these numbers are large. The number of ways in which a given set of occupation numbers N_i can be realized is then given by the multinomial coefficient

$$C(N_1, N_2, N_3, \ldots) = \frac{N!}{N_1! \, N_2! \, N_3! \ldots} \tag{4.12}$$

where
$$N = N_1 + N_2 + N_3 + \cdots \tag{4.13}$$

Other things being equal, the probability for realizing a given set of occupation numbers is proportional to $C(N_1, N_2, N_3, \ldots)$. We maximize this quantity C, or rather its logarithm, with the microcanonical constraint on the energy and the condition (4.13) as side conditions. In other words we maximize the expression

$$\mathscr{L} = \ln C(N_1 N_2, N_3, \ldots) + \alpha \left\{ N - \sum_i N_i \right\} + \beta \left\{ E - \sum_i N_i E_i \right\}$$
(4.14)

Here α and β are Lagrangian multipliers which permit us to treat the quantities N_i as independent. For the factorials in (4.12) we use Stirling's approximation. The quantity $\mathscr{L}$ then becomes

$$\mathscr{L} = N \ln N - N - \sum_i N_i \ln N_i + \sum_i N_i$$
$$+ \alpha \left\{ N - \sum_i N_i \right\} + \beta \left\{ E - \sum_i N_i E_i \right\}$$
(4.15)

Differentiation of (4.15) with respect to N_i yields the result

$$- \ln N_i - \alpha - \beta E_i = 0$$
(4.16)

This is again the result (4.09), which is thus confirmed by an entirely different argument. The microcanonical reasoning is somewhat more natural in the Maxwell-Boltzmann type of statistics where the "systems" are atoms or molecules. The assumption that the quantities N_i are large is then not forced. However, this is exactly the situation in which quantum theory brings about a modification of the counting of states, as we shall see in Chapter 9. The second derivation of the Gibbs distribution appears thereby weakened from the modern point of view. Darwin and Fowler* have shown that this weakening is only apparent. Expectation values for all relevant quantities can be obtained for the microcanonical ensemble in the form of contour integrals. If these contour integrals can be evaluated by the method of steepest descent, equation (4.16) results for the expectation value of N_i. This type of evaluation is almost always legitimate. Troublesome exceptions to this situation do arise occasionally.†,‡ One faces then the unpleasant fact that two types of ensembles yield different answers. In these exceptional situations a deeper physical analysis is required; it usually confirms the canonical result (4.09).

* C. G. Darwin and R. H. Fowler. *Phil. Mag.* **44**, 450 and 823 (1922). See also R. H. Fowler. *Statistical Mechanics.* Cambridge: Cambridge University Press, 1936, Chapter II.

† M. Lax. *Phys. Rev.* **97**, 1419 (1955).

‡ C. C. Yan and G. H. Wannier. *J. Math. Phys.* **6**, 1833 (1965).

4-2. Elucidation of the temperature concept

In the Gibbs distribution (4.11) or (4.16) a parameter makes its appearance which controls the dependence of the probability on energy. We called it β and followed tradition in choosing the sign as indicated. If β is a positive number the probability for a given quantum state being occupied decreases as the energy increases. It was long believed that this must be the correct sign for intrinsic reasons. The belief was based on the fact that all systems known had an energy spectrum which was unlimited in the direction of high energy. To enforce the probability normalization (4.10) for such systems β has to be positive. However, in recent years, systems with a closed energy spectrum have become known, particularly spin systems. For such systems there is no reason why the state of highest energy should not be the most heavily populated one, and negative values of β must thus be considered possible. The fact that such systems are never fully isolated, and therefore not stable indefinitely, is no argument against negative β, because no real system ever fulfills these conditions of isolation anyway. We must therefore conclude that β is a parameter which is generally positive but can occasionally take on negative values.

Much more important than the question of the sign of β is that of its meaning. The meaning can be brought out by listing a series of theorems which result from the preceding equations.

(a) *Two systems which cannot exchange energy have in general different β's.* The proof of this is implicit in the preceding derivation.

(b) *If two systems are capable of exchanging energy their values of β become equal when equilibrium is reached.* This proposition is the one proved in Section 4-1; the argument is specifically worked out in the text following (4.08).

(c) *The mean energy of any system increases monotonically as β decreases.* Proof. Denote by U the average energy. We then have from (4.11)

$$U = \frac{\sum_\nu E_\nu \exp(-\beta E_\nu)}{\sum_\nu \exp(-\beta E_\nu)} \tag{4.17}$$

and therefore

$$\frac{\partial U}{\partial \beta} = \frac{\left[\sum_\nu E_\nu \exp(-\beta E_\nu)\right]^2 - \left[\sum_\nu \exp(-\beta E_\nu)\right]\left[\sum_\nu E_\nu^2 \exp(-\beta E_\nu)\right]}{\left[\sum_\nu \exp(-\beta E_\nu)\right]^2}$$

$$\tag{4.18}$$

Because of (4.11) this can be put in the form

$$\frac{\partial U}{\partial \beta} = \langle E \rangle^2 - \langle E^2 \rangle$$

or

$$\frac{\partial U}{\partial \beta} = - \langle (E - \langle E \rangle)^2 \rangle \tag{4.19}$$

Equation (4.19) yields the inequality

$$\frac{\partial U}{\partial \beta} \leqq 0 \tag{4.20}$$

with the equality sign holding only if all energies E_v are equal; such a system would be incapable of exchanging energy with other systems and thus is of no interest here. The proposition (c) is therefore proved.

(d) *In an equalization of β between two systems energy always flows from the system with the smaller β to the one with the larger β.*

This is a corollary of the preceding results.

The four propositions (a)–(d) show that β is a variable controlling heat flow between systems in thermal contact. Its reciprocal is therefore a temperature on some scale.

4-3. The perfect gas; Maxwellian distribution

It is useful to apply the Gibbsian distribution immediately to the case of perfect gases, for this application throws further light on our temperature parameter β. We have previously introduced another temperature scale, the perfect gas scale, through the definition (1.04). In order to see the relationship between β and the perfect gas temperature T we assume with Maxwell that a perfect gas is a collection of non-interacting mass points devoid of internal structure and statistically independent. These gas molecules can then be taken as independent Gibbsian systems whose only degrees of freedom are their three degrees of freedom of translation. The probability formula (4.11) then takes for them the form

$$p(E) = \frac{\exp(-\beta E)}{\frac{1}{h^3} \int\int\int\int\int\int \exp(-\beta E)\, dp_x\, dp_y\, dp_z\, dx\, dy\, dz} \tag{4.21}$$

Here the sum of states in the denominator is written as an integral, a transformation which was worked out in Problem 3 of Chapter 3. The expression (4.21) still refers to an individual quantum state, but passage to a continuum picture is obviously in order. According to the fundamental analysis of Chapter 3 the expression (4.21) gives the probability of finding a molecule in a volume in phase space whose magnitude equals h^3. We prefer to ask for the *probability density* $P(\mathbf{p},\mathbf{x})$ that the six variables of the molecule have values between p_x and p_x+dp_x, p_y and p_y+dp_y, p_z and p_z+dp_z, x and $x+dx$, y and $y+dy$, z and $z+dz$. To make this conversion we must multiply (4.21) with the ratio of the two volumes of phase space, that is,

$$\frac{1}{h^3}\, dp_x\, dp_y\, dp_z\, dx\, dy\, dz \tag{4.22}$$

This yields

$$P(p_x, p_y, p_z, x, y, z)\, dp_x\, dp_y\, dp_z\, dx\, dy\, dz =$$

$$= \frac{\exp(-\,\beta\, E)\, dp_x\, dp_y\, dp_z\, dx\, dy\, dz}{\iiint\!\!\iiint \exp(-\,\beta\, E)\, dp_x\, dp_y\, dp_z\, dx\, dy\, dz}$$

Superfluous differentials can now be dropped on either side, yielding

$$P(\mathbf{p}, \mathbf{x}) = \exp(-\,\beta\, E)/\!\iiint\!\!\iiint \exp(-\,\beta\, E)\, d\mathbf{p}\, d\mathbf{x} \tag{4.23}$$

For a start let us also neglect gravity and assume the gas molecules all of equal mass and enclosed in a box of volume V. We have then

$$E = \frac{1}{2\,m}\,(p_x^2 + p_y^2 + p_z^2) \tag{4.24}$$

and P becomes

$$P(\mathbf{p}, \mathbf{x}) = \left(\frac{\beta}{2\,\pi\,m}\right)^{3/2} \frac{1}{V}\, \exp\left[-\,\frac{\beta}{2\,m}\,(p_x^2 + p_y^2 + p_z^2)\right] \tag{4.25}$$

Equation (4.25) is the *Maxwellian velocity distribution*, showing the probability density to be Gaussian in the three components of momentum or velocity.

Formula (4.25) contains the temperature parameter β. To get an expression in terms of the perfect gas scale let us calculate the total kinetic energy U of the molecules and compare with earlier results. We find for the mean energy u of one molecule from (4.24) and (4.25)

$$u = \iiint\!\!\iiint \frac{1}{2\,m}\,(p_x^2 + p_y^2 + p_z^2)\, P(\mathbf{p}, \mathbf{x})\, d\mathbf{p}\, d\mathbf{x}$$

The integral is elementary and yields

$$u = \frac{3}{2\beta} \qquad (4.26)$$

and therefore for a gas of N molecules

$$U = \frac{3N}{2\beta} \qquad (4.27)$$

We may compare this result with previous formulas which apply to assemblies of noninteracting molecules. Such a calculation was carried out in Chapter 2, in connection with the virial theorem. A connection was established between the product pV and the mean kinetic energy. Comparing this relation with the empirical equation of state for perfect gases, we ended up with a relationship between the total kinetic energy of translation and the gas temperature. This relation is (2.40). We have now in (4.27) a relation of the same type between this energy and the temperature parameter β. This relation is not empirical, but derived. It proves incidentally the basic structural information contained in (1.08), namely, that for a perfect gas the product pV/N should be a function of temperature only. In addition, comparison of (2.40) and (4.27) yields the relationship between the gas temperature and our temperature parameter β, namely,

$$\beta = \frac{1}{kT} \qquad (4.28)$$

We shall retain the letter β in future work as a useful abbreviation which is rapidly gaining ground. One can even argue that its use would be preferable to T, now that negative temperatures are recognized as possible. Negative temperature joins onto positive temperature across $\beta = 0$, not $T = 0$. However, the point here is not to quarrel with ancient usage, but to justify the occasional use of an already common abbreviation. A more serious objection to the gas temperature scale is that it introduces a superfluous dimension, the degree Kelvin, into physics. In all theoretical work, T arises in the fixed combination kT which has the dimension of an energy. Proposals have therefore been made to use a temperature having the dimension of an energy, namely, the reciprocal of β. This proposal is not quite as convenient as the one made here, and has of course also no chance of displacing ancient usage. We shall therefore not employ it in this book.

One can take the basic probability formula (4.25) to derive simpler ones by integration over some variables. The one easiest to plot is the probability $p(v)\,dv$ that the speed of a molecule has a value between v and

$v + dv$, independently of position or direction of motion. The reader is invited to deduce from (4.25) that this probability equals

$$p(v) = \left(\frac{2}{\pi}\right)^{\frac{1}{2}} \left(\frac{m}{kT}\right)^{\frac{3}{2}} v^2 \exp\left(-\frac{mv^2}{2kT}\right) \tag{4.29}$$

A plot of this function is given in Fig. 4.1.

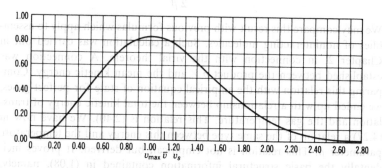

Fig. 4.1. **Maxwellian distribution of molecular speeds.** The abscissa scale is v/v_{max}, where v_{max} is the speed having highest probability. The ordinate is plotted so that the area under a segment of curve equals the probability for that range. $\bar{v}$, the mean speed, and v_s, the root mean square speed, are indicated on the abscissa.

Before leaving the subject of the Maxwellian distribution we may wish to explain the full meaning of (4.23) by bringing in gravity at the earth's surface. Then (4.24) is to be replaced by

$$E = \frac{1}{2m}(p_x^2 + p_y^2 + p_z^2) + mgz \tag{4.30}$$

The distribution factors with respect to the new variable z and the three momentum components, and the density ρ as function of height can be investigated independently of the velocity distribution. The result is

$$\rho = \rho_0 e^{-\frac{mgz}{kT}} \tag{4.31}$$

This is the so-called law of atmospheres, valid for a flat earth having an atmosphere of fixed temperature. In spite of these shortcomings, the formula indicates approximately how the pressure falls over the earth's surface A z of $10^4 m$ is needed to make the exponent about 1. This is roughly the height of Mt. Everest over sea level.

4-4. Energy distribution for small and large samples; thermodynamic limit

The Maxwellian distribution can be used for various illustrative purposes because of its simple Gaussian structure. It is possible to show in particular how the statistics of Gibbs and Boltzmann are formally similar. The discrepancy in the meanings of the two formalisms can also be brought out. It arises from the different role of the fluctuations in the two approaches.

To demonstrate these features, let us begin with the formal similarity. We might have done Gibbsian statistics on a set of gaseous samples, each of n molecules. For each of these samples the probability of finding the representative point in a volume h^{3n} in phase space would have been, by (4.11),

$$p(E) = \frac{\exp(-\beta E)}{\frac{1}{h^{3n}} \int \int \int \int \ldots \int \exp(-\beta E)\, d\mathbf{p}_1\, d\mathbf{p}_2 \ldots d\mathbf{p}_n\, d\mathbf{x}_1 \ldots d\mathbf{x}_n} \tag{4.32}$$

with

$$E = \frac{1}{2m}\{(p_1^x)^2 + (p_1^y)^2 + (p_1^z)^2 + (p_2^x)^2 + \cdots (p_n^z)^2\} \tag{4.33}$$

Obviously, the probability (4.32) is simply the product of the probabilities (4.21). This has to be so because the molecules were assumed independent to start with. Averaging (4.32) over all but one of these molecules will in turn bring us back to the starting point. Instead of proceeding in this fashion we can investigate the n molecules as a unit. To start with we can compute directly the mean energy U_n of the gaseous sample, that is, the expectation value of the quantity (4.33). We find without difficulty

$$U_n = \tfrac{3}{2} n k T \tag{4.34}$$

This formula can be obtained from (4.26) by simple summation and thus contains no new information. However, when we consider fluctuations the situation is entirely different. We may, in fact, convert the Maxwellian into a Gibbsian system step by step, adding one molecule after another. When we have reached the stage of n molecules, the probability of finding an energy value between E and $E + dE$ is obtained, according to (4.33), by multiplying the Boltzmann exponential with the volume of a spherical shell of radius $[(2/m)E]^{1/2}$ and thickness dE. The sphere is in a space of $3n$ dimensions. It then follows from dimensional considerations alone that the volume of the shell equals a constant times $E^{\frac{3}{2}n-1} dE$. Equation (4.32) yields then for the probability $b_n(E)\, dE$

$$b_n(E)\, dE \propto E^{\frac{3}{2}n-1} \exp(-\beta E)\, dE \tag{4.35}$$

As the value of n increases we find a startling change in the function $b_n(E)$. Its maximum lies at

$$E_{max} = (\tfrac{3}{2}n-1)\,k\,T \tag{4.36}$$

slightly shifted from the mean value (4.34). If we set

$$E = E_{max} \cdot x$$

and study the function

$$\frac{b_n(E)}{b_n(E_{max})} = f_n(x)$$

then we find for $f_n(x)$

$$f_n(x) = x^{\frac{3}{2}n-1} \exp[-(\tfrac{3}{2}n-1)\,(x-1)] \tag{4.37}$$

The function $f_n(x)$ always has a maximum at $x = 1$ which equals 1, and vanishes for very small and very large x. The curve is plotted in Fig. 4.2

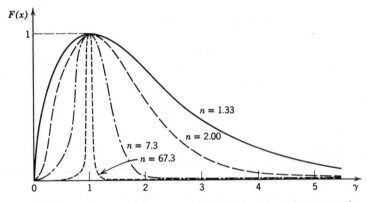

Fig. 4.2. Sharpening up of the energy distribution of a perfect gas as the number of molecules considered is increased (after R. Becker).

for $\tfrac{3}{2}n - 1 = 1, 2, 10, 100$. It is seen that the maximum sharpens up progressively as n becomes large. A simple analytical way to see this sharpening up is to evaluate the second derivative of $f_n(x)$ at $x = 1$. We find

$$f_n(1) = - (\tfrac{3}{2}n - 1)$$

This is a negative number which increases in magnitude with n.

The sharpness of the maximum of $b_n(E)$ is of fundamental importance in all of statistical physics and thermodynamics. In thermodynamics we state simply that the energy of a gas is a function of the state variables V and T. However, when we consider one single molecule at the temperature T we find for its energy the broad distribution curve

$$b_1(E) \propto E^{\frac{1}{2}} \exp(-\beta E) \qquad (4.38)$$

It is therefore quite untrue that we know its energy once we are given the temperature. A similar situation prevails for a small number of molecules. However, as we proceed to laboratory-sized samples of gas, n gets to be of the order 10^{20}. The curve $b_n(E)$ then tells us that it is practically certain that the energy of the sample has the value (4.34). Thermodynamics draws from this the ultimate consequence and simply asserts that the energy *is equal* to the expression (4.34). Thus the statement that the energy is a function of temperature acquires only a sense in the limit of large n.

It is possible to illustrate this point about the sharpness of the energy distribution for large samples without bringing in perfect gases. We can start instead with the general equation (4.19) valid for any system. It was indicated in equation (1.30) that dU/dT is one of the possible definitions of the heat capacity C. Equation (4.19) can therefore be given the form

$$(\Delta U)^2 = \langle (U - \langle U \rangle)^2 \rangle = C\,k\,T^2 \qquad (4.39)$$

Now it follows from the definition of C that if we retain only orders of magnitude the order of U must be

$$U \sim C\,T$$

and hence

$$\frac{(\Delta U)^2}{U^2} \sim \frac{k}{C} \qquad (4.40)$$

The order of magnitude of C is generally about $N\,k$, as (4.34) shows. Consequently, the right-hand side of (4.40) is of the order of the reciprocal of the number of molecules in the sample under consideration. Thus we can see independently from the case of perfect gases that the energy becomes extremely sharply defined as the number of molecules in a sample gets large.

4-5. Equipartition theorem and dormant degrees of freedom

Formula (4.26) for the energy of translation of a free particle is a special instance of the equipartition theorem of classical mechanics. The theorem states:

Every degree of freedom of a body which contributes a square term of a *coordinate or momentum to the total energy has a mean energy of $\frac{1}{2}kT$ in that degree of freedom.*

The proof is quite straightforward. Let x be such a coordinate. Its contribution to the energy is

$$\varepsilon = \tfrac{1}{2}\,\alpha\,x^2 \tag{4.41}$$

and the probability for a value between x and $x + dx$ equals

$$p(x)\,dx = \frac{\exp[-\tfrac{1}{2}\,\alpha\,\beta\,x^2]}{\displaystyle\int_{-\infty}^{+\infty} \exp[-\tfrac{1}{2}\,\alpha\,\beta\,x^2]\,dx} \tag{4.42}$$

We make use of (4.42) to work out the mean energy

$$\langle \varepsilon \rangle = \frac{\tfrac{1}{2}\,\alpha \displaystyle\int_{-\infty}^{+\infty} x^2 \exp[-\tfrac{1}{2}\,\alpha\,\beta\,x^2]\,dx}{\displaystyle\int_{-\infty}^{+\infty} \exp[-\tfrac{1}{2}\,\alpha\,\beta\,x^2]\,dx}$$

The easiest and most instructive way to do the integral in the numerator is to integrate it by parts so that it looks like the denominator. The result is

$$\langle \varepsilon \rangle = \frac{1}{2\,\beta}$$

or

$$\langle \varepsilon \rangle = \tfrac{1}{2}\,k\,T \tag{4.43}$$

For this derivation to be valid, it is not necessary that α be a constant. It can depend on other coordinates as long as x is not involved. Since it drops out of the argument in the end, such dependence will not matter. Thus we have, for instance, for the rotational degrees of freedom of a diatomic molecule

$$\varepsilon = \frac{1}{2\,I}\left\{ p_\vartheta^2 + \frac{1}{\sin^2\vartheta}\,p_\varphi^2 \right\} \tag{4.44}$$

The expression is quadratic in p_ϑ and p_φ and yields for each of the two terms a contribution $\frac{1}{2}kT$ to the energy, that is, kT altogether. The fact that p_φ^2 has a coefficient dependent on ϑ does not affect this argument in the least.

The theorem has two fields of application. The first one deals with the mean kinetic energy of a body. According to a general theorem of mechanics, the kinetic energy of any system can be reduced to a sum of squares of momenta, with coefficients depending on the coordinates. If we apply the equipartition theorem to this situation we find for the mean total kinetic energy of the system

$$K = \tfrac{1}{2} \mathcal{N} \, k \, T \qquad (4.45)$$

Here $\mathcal{N}$ is the toal number of square terms in the kinetic energy; this is equivalent to the number of degrees of freedom of the body.

The second type of application deals with cases where, in addition to the kinetic energy, the potential energy is also quadratic in some of the corresponding coordinates. This happens when a certain value of x, say x_0 is the equilibrium value of the system and the system is held at this value by harmonic restoring forces. Again, a general theorem of mechanics enhances the importance of this case. The theorem states that if a system is held in a certain equilibrium position by arbitrary forces the restoring force for very small displacements is almost always linear in the displacement. The student has become acquainted with this feature in Problem 7 of Chapter 1. Hooke's law may be said to be a crude early formulation of this general theorem. Application of the equipartition theorem to this case yields

$$U = \mathcal{N}' \, k \, T \qquad (4.46)$$

where $\mathcal{N}'$ is the total number of degrees of freedom in which an equilibrium value of the coordinate is held by elastic restoring forces.

The application of these results to nature is checkered with successes and failures. When Maxwell first devised the kinetic theory of gases he thought of gas molecules as point masses having only translational energy. This idea worked fine mechanically, as we saw in Chapter 2. Such a theory also predicts with (4.45) that

$$U = \tfrac{3}{2} N k T \qquad (4.47)$$

where N is the number of molecules. The 3 arises from the three degrees of freedom of translation. Early measurements on air did not check this prediction. The velocity of sound contains the adiabatic bulk modulus of air, through (1.42), and this, in turn, involves the specific heat of air, which was thus found to be

$$\mathscr{C}_v = \tfrac{5}{2} R \qquad (4.48)$$

Only later, the value (4.47) was checked on mercury vapor. It was then realized that (4.48) arises from the fact that air consists mostly of diatomic molecules, and that the 5 brings in the two degrees of freedom of rotation of such molecules.

Once it was realized that the rotation as well as the translation of molecules counts in their specific heat the following question arose: why do not all gases have the specific heat $3\,R$? Any extended solid object has three degrees of rotation, not two, and the idea that diatomic molecules are sticks without extension perpendicular to their length seemed slightly preposterous, particularly as gas kinetic radii could be determined by other means, such as measurement of viscosity. These measurements gave a finite extension even to atomic gas molecules, such as mercury and argon. The degrees of freedom which did not show up in the specific heat were referred to as *dormant*. The problem of the dormant degrees of freedom has no solution in classical physics. In quantum theory the energy becomes a constant independent of temperature as soon as kT becomes substantially smaller than the quantum step from the ground state to the first excited state. This explains the phenomenon of dormancy.

The same mixture of success and failure as for gases followed the application of (4.46) to solids. The formula explains one of the most venerable laws of physics, the *law of Dulong and Petit*, formulated in 1819. The law states that all simple atomic solids have the same molar specific heat. From (4.46), the value should be

$$\mathscr{C}_v = 3\,R \tag{4.49}$$

because the translational coordinates of the atoms are held at fixed values by harmonic restoring forces. This result checked well with early data but was upset when it was discovered that the specific heat of most solids drops when we go below room temperature. We have of course again a dormancy phenomenon. Its complete explanation was given by Einstein in 1908, when he computed the heat capacity of a harmonic oscillator. If a degree of freedom has a hamiltonian of the form

$$\mathscr{H} = \tfrac{1}{2}\,(p^2 + \omega^2 q^2) \tag{4.50}$$

then the heat capacity of this degree of freedom is not equal to k because p and q do not commute. The energy eigenstates of the system (4.50) are well known. They equal

$$\varepsilon(n) = (n + \tfrac{1}{2})\,\hbar\,\omega \tag{4.51}$$

We therefore find from (4.11)

$$p(n) = \frac{1}{f}\exp[-\,\beta\,(n+\tfrac{1}{2})\,\hbar\,\omega] \tag{4.52}$$

where the partition function f is defined as

$$f = \sum_{\nu=0}^{\infty}\,\exp[-\,\beta\,(\nu+\tfrac{1}{2})\,\hbar\,\omega] \tag{4.53}$$

It is easily evaluated to be

$$f = \tfrac{1}{2} \operatorname{cosech} \tfrac{1}{2} \beta \hbar \omega \qquad (4.54)$$

From (4.51) and (4.52) the mean energy of the mode is found to be

$$
\begin{aligned}
\langle \varepsilon \rangle &= \sum_{\nu=0}^{\infty} p(\nu)\, \varepsilon(\nu) \\
&= \frac{1}{f} \sum_{\nu=0}^{\infty} (\nu+\tfrac{1}{2})\, \hbar\, \omega \exp[-\,\beta\,(\nu+\tfrac{1}{2})\,\hbar\,\omega] \\
&= -\frac{1}{f} \frac{df}{d\beta}
\end{aligned}
$$

Insertion of (4.54) into this result yields for $\langle \varepsilon \rangle$

$$\langle \varepsilon \rangle = \tfrac{1}{2} \hbar\, \omega \coth \tfrac{1}{2} \beta\, \hbar\, \omega \qquad (4.55)$$

The curve is plotted in Fig. 4.3. For sufficiently large temperature it

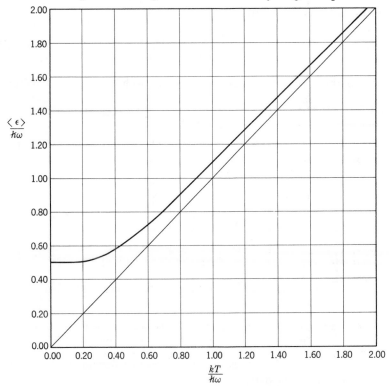

Fig. 4.3. Average energy versus temperature for the quantized harmonic oscillator.

approaches asymptotically the straight line $k\,T$. But if the temperature is such that

$$k\,T \ll \hbar\,\omega$$

the energy simply equals the ground state energy $\tfrac{1}{2}\hbar\,\omega$, and the slope of the curve becomes horizontal. Since the slope of the curve equals the heat capacity, the heat capacity approaches zero under these circumstances. The asymptotic slope is the classical slope incorporated into the law of Dulong and Petit.

Differentiation of (4.55) with respect to temperature yields explicitly the heat capacity

$$C = \frac{d\langle\varepsilon\rangle}{dT}$$

which can be given the form

$$C = k\left\{\frac{\tfrac{1}{2}\hbar\,\omega/k\,T}{\sinh\tfrac{1}{2}\hbar\,\omega/k\,T}\right\}^{2} \qquad (4.56)$$

A plot of this curve is given in Fig. 4.4 together with heat capacity data for

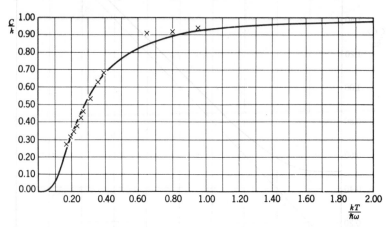

Fig. 4.4. Heat capacity versus temperature for the quantized harmonic oscillator. The experimental points shown are empirical data for diamond.

diamond as presented by Einstein.* It is seen that formula (4.56) yields quite a good fit to the data.

* A. Einstein. *Ann. Physik* **22**, 186 (1907).

A. RECOMMENDED PROBLEMS

1. Atomic oxygen has a ^3P ground state with levels $J = 2$ at 0 cm^{-1}, $J = 1$ at 158.5 cm^{-1}, and $J = 0$ at 226.5 cm^{-1}. Calculate the fraction of the total population in the three sublevels at room temperature.

2. Deduce the Maxwellian distribution (4.29) for the speed of a gas molecule from the Gibbs–Boltzmann distribution law.

3. Calculate the mean speed in a Maxwellian distribution and compare it with the root mean square velocity.

4. Derive the law of atmospheres (4.31) from hydrostatics.

B. GENERAL PROBLEMS

5. The acceleration of gravity at the moon's surface is 167 cm/sec^2 and its diameter 3476 km. What molecular weight must a gas have so that a molecule can escape from the moon with the r.m.s. velocity of a Maxwellian distribution? Take the mean temperature of the moon as 300° K.

6. Calculate the most probable speed in the Maxwellian distribution and compare with the r.m.s. velocity.

7. A monatomic gas is contained in a vessel from which it leaks through a fine hole. Show that the mean kinetic energy of the molecules leaving the vessel is $2kT$.

8. Show that the mean relative speed of two molecules in a Maxwellian distribution equals $\sqrt{2}$ times the mean speed of a single molecule.

9. Prove from general principles that for any velocity distribution the inequality $\langle \frac{1}{v} \rangle \langle v \rangle > 1$ must hold.

10. The Debye temperature Θ of a monatomic solid may be taken as a measure of the mean frequency f of vibration according to the relation

$$hf = \tfrac{3}{4} k \Theta$$

If we accept this formula, at what temperature are 50% of the oscillators in a solid excited out of the ground state?

11. Although internal free rotation in molecules (such as the methyl

group in ethyl alcohol $H—\overset{\displaystyle H}{\underset{\displaystyle H}{C}}—\overset{\displaystyle H}{\underset{\displaystyle H}{C}}—OH$) is not as common as once sup-

posed, it is a convenient object for the study of the Boltzmann distribution law because the mathematical identity

$$\sum_{-\infty}^{+\infty} e^{-\pi \alpha n^2} = \frac{1}{\sqrt{\alpha}} \sum_{-\infty}^{+\infty} e^{-\frac{\pi}{\alpha} n^2}$$

can be applied to it. Verify with the help of this identity how the quantum expression for the mean energy of free internal rotation approaches the equipartition value at high temperature. Does the heat capacity reach its limiting value from above or below as T increases?

12. Compute and plot the curve for heat capacity versus temperature for free internal rotation.

13. Show that if the energy ε depends on a generalized coordinate q and its conjugate momentum p in such a way that $\varepsilon \to \infty$ as p or $q \to \pm \infty$ the following generalization of the equipartition theorem is valid:

$$\left\langle q\, \frac{\partial \varepsilon}{\partial q} \right\rangle = \left\langle p\, \frac{\partial \varepsilon}{\partial p} \right\rangle = k\,T$$

14. Derive a law of atmospheres for a spherical earth with a gravitational acceleration varying inversely as the square of the distance. Show that the calculation yields a finite particle density at infinity. Give the numerical value of that density.

15. A realistic approach to the true condition of the atmosphere is to assume it in hydrostatic equilibrium but to assume temperature equilibration only up to the point that it can be carried out by adiabatic expansion of rising air masses. Calculate the temperature gradient at the earth's surface on the basis of these assumptions.

16. Calculate the total number of molecules striking unit area of wall in unit time, given a Maxwellian distribution. Show that it can be given the form $\frac{1}{4}\, n\, \bar{v}$, where $\bar{v}$ is the mean speed.

5

Statistical justification of the Second Law

Having worked out the application of the Gibbs distribution to the study of individual degrees of freedom, we return to our over-all emphasis on macroscopic systems. We have seen in Section 4-4 that for such systems the probability distribution (4.11) is of secondary importance. The average quantities which result from the distribution form the important result, and their fluctuations are sufficiently small to be negligible for many purposes. Correspondingly, a shift in viewpoint will be adopted and maintained until Chapter 22. Attention will be focused on the properties of these averages and the relationships which they have to the state variables.

5-1. Definition of entropy; entropy and probability

It goes without saying that the most important average to be studied is the average energy (4.17), since the whole discussion of the preceding chapter revolved around the energy concept. In accordance with our shifted viewpoint we shall call this quantity U simply the energy or, in thermodynamic parlance, the internal energy of the system. In order to calculate this energy for an actual system, it is not necessary to know the distribution of quantum states exactly. It is sufficient to know the *sum of states* or *partition function F*, defined by

$$F = \sum_{\nu} \exp(-\beta E_{\nu}) \qquad (5\,01)$$

Formula (4.17) for U is then expressible in terms of F through the closed operation

$$U = -\frac{d \ln F}{d\beta} \tag{5.02}$$

The partition function is essentially a controlling function which determines the average energy and, as we shall see later, the average of other quantities as well. However, we have just learned that for a "large" system the energy fluctuation is very small. It follows then that the part of the energy spectrum which is effective within the sum (5.01) under given external conditions is also small. Thus, notwithstanding the formal structure of (5.01), all effective values of E_v in this sum are actually very close to U.

Given this situation, what extra information does (5.01) convey which is not contained in U? To get a feeling for this let us first look at a simplified model in which all energies in the sum of states are equal; such a system is called "microcanonical". Then F takes the form

$$F = M \exp(-\beta E)$$

where M is the degeneracy or spectral multiplicity of the energy level. Equation (5.02) supplements this by the statement

$$U = E$$

which leads to the relation

$$F = M \exp(-\beta U) \tag{5.03}$$

Thus F simply consists in this special case of a spectral multiplicity factor times the Boltzmann exponential of the internal energy.

A real sum of states is admittedly not of the simple structure (5.03), but the complication is in many cases not as great as one would guess from appearances. As the range of contributing energies is usually small, the Boltzmann exponential in (5.01) cannot depart much from $\exp(-\beta U)$. Expression (5.01) thus differs significantly from this exponential primarily because there is a number of states in the contributing energy range which is a function of the system and, for each system, a function of the energy and the state variables. By an unfortunate choice of language most books refer to this number as a probability even though it is a number larger than 1. To avoid this confusion and to emphasize the analogy to (5.03) we shall call the number the *statistical multiplicity*. The quantity plays a role analogous to the spectral multiplicity in the simple model treated earlier, and we shall denote it by the same letter, M. With the help of this definition, (5.01) takes the form (5.03) also in the more general case, and we have always

$$F = M \exp(-\beta U) \tag{5.03}$$

It must be understood of course that a certain amount of temperature dependence is now inherent in M, although this dependence is usually weaker than the dependence on β directly visible in the formula.

Before we proceed any further with the logical development it may be worthwhile to obtain a feeling for the real behavior of the system partition function F of a "large" system, for it is F which controls the contribution of various configurations to the behavior of the system. By (5.01), F is the "sum of states," and equation (4.11) shows that the probability for any subset among these states is computed simply by taking the sum of states over the subset and dividing by F. Now equation (5.03) suggests that F can be understood as the product of a Boltzmann exponential and a statistical multiplicity factor. The dependence of the exponential on the energy is obvious: it is falling with increasing U. It is the factor M which needs a closer look. We called it statistical multiplicity, emphasizing the discrete character of the energy states perhaps unduly. If one thinks of F as being an integral over the energy, then M becomes associated with the "density of states," that is, the number of states per unit energy range. Offhand, one might think of this as a very complicated function. We can get a feeling for it, however, by going back to the case of perfect gases, whose energy distribution was completely analyzed both from the point of view of an individual molecule and that of a larger aggregate. The result was the probability distribution (4.35). The structural similarity of (4.35) and (5.03) is obvious. Both have a Boltzmann exponential and a front factor which indicates the statistical weight of various energies. We infer from this comparison that we have the order-of-magnitude relation

$$M \sim \alpha\, U^f \tag{5.04}$$

where f is the number of degrees of freedom of the system: $\frac{3}{2} N$ for a simple perfect gas. We have seen in Section 4-4 how the interplay of this factor with the Boltzmann exponential gives rise to the narrow spread of the probability distribution. Equation (5.03) suggests a similar interplay in the general case. Naturally, equation (5.04) must be taken as merely suggestive, with the possibility of exceptions. But it does suggest that M is apt to depend exponentially on the size of the system and that it might be useful to use its logarithm as a convenient variable. We shall now verify this point in a more rigorous way, by consideration of the union of two weakly coupled systems.

Suppose we denote the two systems by 1 and 2, and their union by 0, as we did on a similar occasion in Chapter 4. It was explained at that time that the totality of all states resulting from the union of the two subsystems is the combination of all pairs of states (i, k) of the systems 1 and 2.

According to (4.01) the energy of such a state is the sum of the energies, and therefore

$$F_0 = \sum_{i,k} \exp[-\beta(E_i + E_k)]$$
$$= \sum_i \exp(-\beta E_i) \cdot \sum_k \exp(-\beta E_k)$$

which means simply that

$$F_0 = F_1 \cdot F_2 \tag{5.05}$$

The partition function of the union of two uncoupled systems equals the product of their partition functions. This result confirms the qualitative argument (5.04). If we observe further that it is the logarithm of F which appears in formula (5.02) for the energy, the logarithm of F clearly appears as the quantity of primary interest. Equation (5.03) then takes the form

$$\ln F = \ln M - \beta U \tag{5.06}$$

and (5.05)

$$\ln F_0 = \ln M_1 + \ln M_2 - \beta(U_1 + U_2) \tag{5.07}$$

The logarithm of the statistical multiplicity is seen to have additive properties, as might be inferred from our introductory simple example. It is therefore a good extensive parameter to work with. It is called the *entropy* S. For historical reasons entropy is given the same dimension as Boltzmann's constant so that the actual formula reads

$$S = k \ln M \tag{5.08}$$

Equation (5.06) then takes the form

$$k \ln F = S - \frac{U}{T} \tag{5.09}$$

and the relevant aspect of (5·07)

$$S_0 = S_1 + S_2 \tag{5·10}$$

We have thus gained a new function of the state variables of a system, the entropy. It is additive in situations in which other extensive functions such as volume and energy are additive. The meaning of entropy is that it measures the number of ways in which a system can realize a given condition characterized by fixed state variables. It is therefore a measure of the disorder in that condition. Opinions are divided as to whether entropy can be, strictly speaking, defined on an individual system. It is certainly easier to visualize its meaning when the system is part of an ensemble, and the ensemble contains members which together realize all possible conditions leading to the same macroscopic state. There can be no doubt, on the other hand, that entropy is a concept which is of use for an individual system. How this can be is not entirely understood at present. The author inclines toward an explanation according to which the time average of an individual system is equivalent to an ensemble average. This explanation has its difficulties, as we shall see. Another more radical explanation is to deny that a system is in a "pure state" which can be mechanically described. This type of explanation is preferred by those who are satisfied to have an operational definition of physical concepts.

A disadvantage of the definition of entropy given in (5.09) is that it is linked to the existence of an equilibrium state and in particular the existence of a temperature β in the sample. If we wish to extend the entropy concept to more general situations, we must be able to get along without the Gibbs formula (4.11). To obtain such an extension we think of the entropy as the property of an ensemble, and take the ensemble to be microcanonical for simplicity. The number of ways to realize a given set of occupation numbers of states is then given by (4.12). From (5.08) the entropy S of the ensemble equals k times the logarithm of that number, or

$$S = k \ln \frac{N!}{N_1! N_2! N_3! \ldots} \tag{5.11}$$

As previously, we approximate the factorials by Sterling's approximation. We get then

$$\frac{S}{k} \approx N \ln N - \sum_i N_i \ln N_i$$

$$= - \sum_i N_i \ln \frac{N_i}{N} = - N \sum_i \frac{N_i}{N} \ln \frac{N_i}{N}$$

Our introductory Chapter 2 on probability has stressed that probability is essentially relative frequency of occurrence. We may therefore set

$$p_i = \frac{N_i}{N} \tag{5.12}$$

and write

$$\frac{S}{k} = - N \sum_i p_i \ln p_i$$

If entropy is additive between uncoupled systems (as we shall prove below), the share of entropy of an individual member of the ensemble equals then

$$S = - k \sum_i p_i \ln p_i \tag{5.13}$$

The argument (5.11)–(5.13) is meant to be suggestive rather than deductive, and we shall now verify independently that the result (5.13) is reasonable. First of all we may verify that the definition reduces to (5.09) at equilibrium. Indeed, substitution of (4.11) into (5.13) yields

$$S = -\frac{k}{F}\sum_i \exp[-\beta E_i]\{-\beta E_i - \ln F\}$$

with F given by (5.01). The definition (4.17) brings this into the form

$$S = \frac{U}{T} + k \ln F$$

in agreement with (5.09). As our next check we want to verify the additivity of S for two uncoupled systems, even if they are not in equilibrium. By (4.04) the entropy S^0 of the union of two such systems equals

$$\begin{aligned}
S^0 &= -k\sum_{i,k} p_i^1 p_k^2 \ln (p_i^1 p_k^2) \\
&= -k\sum_{i,k} p_i^1 p_k^2 \{\ln p_i^1 + \ln p_k^2\} \\
&= -k\sum_i p_i^1 \ln p_i^1 - k\sum_k p_k^2 \ln p_k^2 \\
&= S^1 + S^2
\end{aligned}$$

as desired. Further checks of the definition (5.13) are that it equals zero if a system is certain to be in a single quantum state and that it is positive for any more disordered condition. Furthermore, it is defined for any state of the system.

As the final step we can liberate ourselves from the restriction that S must be expressed in terms of quantum states for which the density matrix ρ is diagonal. In passing from a special to a general reference frame we apply the transformation (3.22) to the diagonal density matrix ρ. For equation (5.13) the result will be

$$S = -k \operatorname{Tr} \{\rho \ln \rho\} \tag{5.14}$$

where the symbol Tr mean taking the trace of the matrix product in curly brackets.

5-2. Proof of the Second Law for "clamped" systems

Having obtained a satisfactory definition of entropy under arbitrary conditions, we can now proceed to prove the Second Law for it by enunciating the theorem:

The entropy of a system, as defined by (5.14), is larger if the density matrix ρ has the Gibbsian structure, that is, is diagonal in the energy with diagonal elements as given by (4.11), than for any other possible form of the density matrix obeying the same side conditions, namely,

$$\text{Tr}(\rho) = 1 \qquad (5.15)$$

$$\text{Tr}(\rho \mathscr{H}) = U \qquad (5.16)$$

Proof. Proving a quantity positive often involves expressing it as a sum of squares. This is not possible here. There is, however, another function which performs the same service here

$$f(x) = x \ln x - x + 1 \qquad (5.17)$$

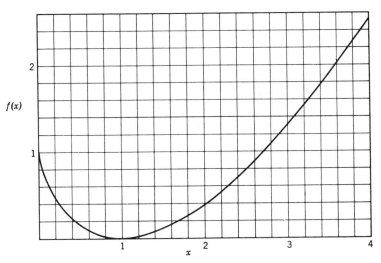

Fig. 5.1. Plot of the function $f(x)$ employed in making entropy estimates.

whose plot is shown in Fig. 5.1. The function vanishes, together with its derivative, at $x = 1$ and has a positive second derivative for all positive x. It is therefore positive if x is positive or zero, with the exception of $x = 1$, at which point it is zero.

The proof of the theorem proceeds preferably in two stages which we call stage A and stage B. In stage A we prove the theorem for forms of the density matrix which are also diagonal in the energy, but for which the diagonal elements do not have the magnitude (4.11). Let the competing probabilities have the value q_i. Then we are to prove that

$$\Delta S = k \sum_i \{q_i \ln q_i - p_i \ln p_i\} \qquad (5.18)$$

is greater than zero if, in accordance with (5.15) and (5.16),

$$\sum_i q_i = \sum_i p_i = 1 \tag{5.19}$$

$$\sum_i q_i U_i = \sum_i p_i U_i = U \tag{5.20}$$

unless all quantities q_i are equal to p_i. The estimate involves consideration of the mixed expression

$$\sum_i q_i \ln p_i$$

Using (4.11), (5.19), and (5.20), we transform this as follows:

$$\sum_i q_i \ln p_i = \sum_i q_i \{- \beta U_i - \ln F\}$$

$$= \sum_i p_i \{- \beta U_i - \ln F\}$$

$$= \sum_i p_i \ln p_i$$

Inserting this into (5.18), we obtain

$$\Delta S = k \sum_i q_i \ln \left(\frac{q_i}{p_i}\right) \tag{5.21}$$

We add to (5.21) the following expression, which vanishes identically because of (5.19):

$$0 = \sum_i p_i \left\{1 - \frac{q_i}{p_i}\right\}$$

and get

$$\Delta S = k \sum_i p_i \left\{\frac{q_i}{p_i} \ln \frac{q_i}{p_i} - \frac{q_i}{p_i} + 1\right\} \tag{5.22}$$

The curly bracket is the function (5.17) and is positive unless $q_i = p_i$ for all i. This finishes stage A of the proof.

In stage B we drop the hypothesis that the density matrix is diagonal in the energy. There exists, however, an orthonormal set of wave functions in which ρ is diagonal, and thus the definition (5.14) reduces to a sum over diagonal terms.* The basic entropy difference discussed here then takes the form

$$\Delta S = k \left\{\sum_n q_n \ln q_n - \sum_i p_i \ln p_i\right\} \tag{5.23}$$

* J. von Neumann. *Mathematische Grundlagen der Quantenmechanik.* Berlin: Julius Springer-Verlag, 1932, p. 202.

Here the sum over n goes over a set of orthonormal states which differs from the set i. To avoid repetition, we assume positively that set n is non-diagonal in the energy. This means that there is at least one state n, say $n = 1$, whose wave function is composed of at least two wave functions of different energy, say $i = 1$ and $i = 2$. The side conditions (5.15) and (5.16) now take the form

$$\sum_n q_n = 1$$

and

$$\sum_{i,n} \alpha_{in} q_n (\alpha^{-1})_{ni} U_i = U$$

where $\|\alpha_{in}\|$ is the unitary transformation matrix from the base n to the base i. For later use, we prefer the side conditions in the form

$$\sum_{i,n} |\alpha_{in}|^2 q_n = \sum_{i,n} |\alpha_{in}|^2 p_i \tag{5.24}$$

and

$$\sum_{i,n} |\alpha_{in}|^2 q_n U_i = \sum_{i,n} |\alpha_{in}|^2 p_i U_i \tag{5.25}$$

Here the unitarity of the transformation matrix $\|\alpha_{in}\|$ both with respect to its lines and its columns has been employed. In the same way expression (5.23) can be converted into a double sum

$$\Delta S = k \sum_{i,n} |\alpha_{in}|^2 \{q_n \ln q_n - p_i \ln p_i\} \tag{5.26}$$

As in stage A of the proof we now consider the mixed expression

$$\sum_{i,n} |\alpha_{in}|^2 q_n \ln p_i$$

which becomes by (4.11), (5.24), and (5.25)

$$\sum_{i,n} |\alpha_{in}|^2 q_n \ln p_i = \sum_{i,n} |\alpha_{in}|^2 q_n \{- \beta U_i - \ln F\}$$

$$= \sum_{i,n} |\alpha_{in}|^2 p_i \{- \beta U_i - \ln F\}$$

$$= \sum_{i,n} |\alpha_{in}|^2 p_i \ln p_i$$

We use this identity to bring (5.26) into the form

$$\Delta S = k \sum_{i,n} |\alpha_{in}|^2 q_n \ln \frac{q_n}{p_i} \tag{5.27}$$

We add to (5.27) the following expression, which vanishes identically because of (5.24):

$$0 = k \sum_{i,n} |\alpha_{in}|^2 p_i \left\{ 1 - \frac{q_n}{p_i} \right\}$$

and get

$$\Delta S = k \sum_{i,n} |\alpha_{in}|^2 p_i \left\{ \frac{q_n}{p_i} \ln \frac{q_n}{p_i} - \frac{q_n}{p_i} + 1 \right\} \tag{5.28}$$

The curly bracket is again the function (5.17) which is positive or zero, and vanishes only if $q_n = p_i$. In this stage B the condition $q_n = p_i$ is impossible to satisfy for all relevant pairs of variables. By assumption, p_1 and p_2 are different because they refer to states of different energy; also by assumption α_{11} and α_{21} both do not vanish. Since q_1 cannot simultaneously be equal to p_1 and to p_2, at least one of the two terms in the double sum is positive. We have therefore under all circumstances

$$\Delta S > 0 \tag{5.29}$$

which was to be proved.

The theorem just proved establishes the second law of thermodynamics for a "clamped" and adiabatically isolated system, that is, one on which no work is done and for which the internal energy is fixed. Within these limits the proof is an extremely wide one because it compares the Gibbsian probability distribution with any other possible probability distribution and proves that the entropy in the former is larger. In particular, the proof does not involve the assumption that the comparison state obey conditions of partial or local equilibrium, or depart from the Gibbsian state only infinitesimally. The comparison state might, for instance, be air in turbulent motion in a closed container, as compared to air at equilibrium in the same container and having the same energy. The question how fast such a state evolves toward the equilibrium state (if it does evolve at all) is of course not answered by the theorem. Such questions belong to the field of chemical and physical kinetics and have their peculiar theoretical difficulties. We shall discuss these difficulties in connection with some selected equilibration processes to be studied in Chapter 18 and following. These studies can of course not match the generality of the theoretical proof given here. The same is true of the examples and applications which follow, where some attention must be paid to computability.

5-3. The Ehrenfest or adiabatic principle

The proof of the Second Law given in Section 5–2 cannot be considered final because of the "clamping" condition used in the derivation. To prove the Second Law under more general conditions we must first take time out to consider the behavior of a Gibbsian equilibrium system under applied forces. The central concept in this wider view is the concept of work. Work as a general feature is very difficult to comprehend, and some restricted forms of work have been considered in Chapter 1. Instead of considering all possible forms of work we considered only work performed in changing the state variables $x_1, x_2, \ldots x_n$; this involves the conjugate forces $X_1 \ldots X_n$, and work takes then the special form (1.23). It was then considered useful to restrict the possible forms of work still further by the concept of quasistatic change, that is, changes in which the state variables are altered so slowly that the forces acting are at all times practically indistinguishable from their equilibrium values. In many (but not all) practical instances, the forces $X_1, X_2, \ldots X_n$ can then be treated as functions of the state variables $x_1, x_2, \ldots x_n$ and the temperature. The equation of state (1.03), which expresses the pressure of a fluid as a function of its volume and temperature, may be considered the simple prototype of such relations.

It must not be imagined that we can restrict the dynamics of our systems under consideration so severely and still come up with a sweeping general formulation of the Second Law. We shall find a Second Law which is similarly restricted but which is nevertheless extremely useful.

In Chapter 4, the question concerning the energy of a system was reduced into questions about individual quantum states. What these states are cannot always be known, but their existence is asserted and results are obtained by applying probabilistic reasoning to them. When we now talk about the state variables x_s in this connection, we must return to these quantum states. What happens to them when such a variable is changed? The answer is, of course, that the system is changed, and with it its quantum states, and the energy of these quantum states. However, in the overwhelming majority of cases the quantum state persists as such and its label can be retained.† We refer to this extension of the quantum state as its *adiabatic continuation*. For this quantum state and its continuation the *Ehrenfest principle* asserts:

The quasistatic forces acting on a system in an individual quantum state are conservative.

The principle is also called the *adiabatic principle*. The origin of the name is cleared up by the following argument. Suppose we specialized the definition (1.23) of work so as to make it apply to a system in an individual quantum state. Let us denote by $(X_s)_i$ the forces acting on the system in the

† Apparent exceptions to this rule are particularly common in the theory of magnetism. They arise from a sloppy handling of the boundary conditions of the Schroedinger equation involved.

state i. The total work done is then the same as for (1.23) except for the index i. The Ehrenfest principle then asserts that

$$dE_i - \sum (X_s)_i \, dx_s = 0 \qquad (5.30)$$

The equation obviously looks like (1.25) with the heat term suppressed; in other words (5.30) looks like the First Law for adiabatic changes. We can thus gain a more colorful formulation of the Ehrenfest principle by giving it the form:

Any system confined to a single quantum state reacts adiabatically to external forces.

It is tempting to put this principle on a footing similar to that following (1.21), namely, that on an atomic scale all mechanics is conservative, and that consideration of a single quantum state brings out this atomic aspect even in a large system. To some extent, this is of course true; the process called "heat flow" is obviously meaningless when a single quantum state is considered, just for the same reason as this term disappears in the description of an individual atom or molecule. However, the physics behind the Ehrenfest principle is much more subtle than this simple analogy might indicate, for it is always true that work and energy must balance in an adiabatically isolated system. But the corresponding equation (5.30) holds only if the forces $(X_s)_i$ are the *quasistatic* forces. The reason for this greater flexibility is that a system in a given quantum state has two options how to react to a change in external constraints. It can either remain in this quantum state, which is altered parametrically as external conditions change, or it can make a transition to another state. Conservation of energy, being statistical in nature, as explained in Chapter 2, applies only to the totality of all these possibilities. Within this wide spectrum of probable paths the Ehrenfest principle now asserts in particular that quasistatic forces and the quantum state on which they act are singled out for a special kind of law of energy conservation. The reason for this is that the probability for *all* transitions to other states becomes negligibly small as the rate of change approaches the quasistatic limit. Thus, when the limit is reached, conservation of energy must apply to the individual quantum state and the work done by the quasistatic forces, taken by themselves alone.

The justification of this idea may be found in the perturbation calculus of quantum mechanics. Suppose we impose on a system a change of the variable x_s from the value ξ to the value η. We can do this in a variety of ways, but we may take most simply the change as linear in t over the necessary time interval:

$$x_s = \xi + \alpha t$$

Here $\alpha\, t$ is to be considered a perturbation, and α the "slowness parameter" of the perturbation. It controls the rate at which the change in x_s is brought about. Now perturbation theory tells us that the perturbation $\alpha\, t$ has two effects: it induces a change per unit time in the energy level which is proportional to α, and it causes transitions to other levels; the probability for these processes is found to be proportional to α^2. If one now allows $\alpha\, t$ to increase to its final value $\eta - \xi$, after which it is arrested, one will find a change in the level equal to

$$c_1 \int_0^{(\eta - \xi)/\alpha} \alpha\, dt = c_1\, (\eta - \xi)$$

while the probability for a transition becomes

$$c_2 \int_0^{(\eta - \xi)/\alpha} \alpha^2\, dt = c_2\, \alpha\, (\eta - \xi)$$

Note that the energy level displacement does not contain the slowness parameter α, while the transition probability does contain it. Hence, the number of transitions for a fixed change in the coordinate x_s becomes zero in the limit $\alpha \to 0$. This is the quasistatic limit. By the law of the conservation of energy it is then necessary that the displacement $c_1\, (\eta - \xi)$ be equal to the work supplied. Therefore, the change in the energy of the level is equal to work supplied quasistatically, which is the relationship asserted by Ehrenfest's principle.

5-4. Extension of the Second Law to general systems

The Ehrenfest principle provides the necessary information which allows the extension of the Second Law to systems whose state variables are subject to change. To apply the principle we exploit (5.30) for the differential equations it provides. We find

$$(X_s)_i = \left(\frac{\partial E_i}{\partial x_s} \right)_{x_1, x_2, \ldots x_{s-1}, x_{s+1}, \ldots x_n} \tag{5.31}$$

In other words the generalized force may be expressed as the partial derivative of the energy with respect to the corresponding state variable of the system. The formula is a microformula, that is, one valid for an individual quantum state. To make from it a formula for a statistically variable system we combine (5.31) with the probability expression (4.11), obtaining

$$X_s = \frac{1}{F} \sum_v \frac{\partial E_v}{\partial x_s} \exp(- \beta\, E_v) \tag{5.32}$$

In agreement with the usage employed in (4.17) we leave no averaging sign on the left of (5.32) because only the average force is significant in most cases. We must remember, however, that it is an average and that deviations from it may become significant in special circumstances. A second feature to be emphasized in (5.32) is that in general

$$X_s \neq \frac{\partial U}{\partial x_s}$$

because the processes of differentiation and average formation do not commute.

Formula (5.32) can be given a form which allows us to compute the generalized force from the knowledge of the partition function only. This point was first emphasized in connection with (5.02), but it now takes a wider significance. Since the energy levels E_i depend parametrically on the state variables, the partition function F also becomes dependent on these same variables. Equation (5.32) can then be written as a simple derivative in the form

$$X_s = -\frac{1}{\beta} \left(\frac{\partial \ln F}{\partial x_s} \right)_{\beta, x_1, x_2, \ldots x_{s-1}, x_{s+1}, \ldots x_n} \tag{5.33}$$

The form which this equation takes for perfect fluids is particularly useful. It is, from (1.26),

$$p = k T \left(\frac{\partial \ln F}{\partial V} \right)_T \tag{5.34}$$

By the same token, U can now no longer be treated as a total derivative as in (5.02) because F has become a function of many variables. The formula must be changed into

$$U = - \left(\frac{\partial \ln F}{\partial \beta} \right)_{x_1, x_2, \ldots x_n} \tag{5.35}$$

Equations (5.33) and (5.35), taken together, form a total differential equation for $\ln F$ in terms of the state variables, including temperature. The equation reads

$$d \ln F = - U \, d\beta - \beta \sum_s X_s \, dx_s \tag{5.36}$$

Equation (5.36) is one of the ways to express the second law of thermodynamics when it is valid with equality sign. Technically speaking, it is in fact its most convenient form, as it permits rapid solution of many practical problems. It indicates in particular that, if work is done on a system at constant temperature, it does not necessarily go into increasing the internal energy by a corresponding amount. To bring the equation into its generally recognized form we eliminate $\ln F$ between (5.36) and (5.09). We thus get

$$\frac{1}{k} dS = \beta \, dU - \beta \sum_s X_s \, dx_s \tag{5.37}$$

This may be given the form

$$dS = \frac{1}{T} \left\{ dU - \sum_s X_s \, dx_s \right\} \tag{5.38}$$

or with (1.23)

$$dS = \frac{1}{T} \{ dU - dW \} \tag{5.39}$$

Application of the First Law in the form (1.22) reduces this finally to

$$dS = \frac{dQ}{T} \tag{5.40}$$

What is remarkable about (5.40) is that the differential on its right-hand side has a cross bar, but the one on its left does not. We have therefore the following theorem:

For any system subject to quasistatic change the gas temperature forms an integrating denominator of the heat it absorbs. The state variable whose change is given by the integral is the entropy.

The theorem implies that we have the generally valid equation

$$S(P_2) - S(P_1) = \int_{P_1}^{P_2} \frac{C(T, x_1, x_2, \ldots)}{T} \, dT + \sum_v \frac{L_v}{T_v} \tag{5.41}$$

Here the quantities L_v are the latent heats of the phase transformations occurring, and T_v the transformation temperatures. The quantity C can be any sort of heat capacity effective in the process under consideration. For an impure substance a phase transformation may extend over a certain temperature interval and will appear partly or entirely as an enhanced heat capacity. Irrespective of these complications, the theorem states that the right-hand side of (5.41) depends only on the initial and final states as shown. In fact the theorem allows us to say that, while the concept of heat as a conserved quantity has failed, the entropy acts as a very respectable substitute. The weakness of (5.40) is not in the equation itself, but in the restrictions which must be placed on the processes to which it applies. We have already seen in the preceding chapter that (5.40) cannot be universally true. For in that special case of clamped systems of fixed energy the right-hand side of (5.40) necessarily vanishes by the First Law. Yet the entropy of the equilibrium state was shown to be larger than that of all other possible states. Clearly, we have obtained an equality sign in (5.40) simply by the device of suppressing nonequilibrium states from consideration.

If a process is adiabatic as well as quasistatic, then the integration of equation (5.40) simply yields zero. We may therefore express the Second Law in this form:

If an adiabatically isolated system is subject to quasistatic change its entropy remains a constant.

Therefore S is a proper parameter for labeling the adiabats of a system. We shall demonstrate this property in the applications.

We are now in a situation where we have proved the Second Law with inequality sign for "clamped" systems, and with equality sign for quasistatic change generally. The proofs were based on quantum mechanics. The question now arises how far we can go to give the Second Law its full range of validity. This does not go entirely without the use of some axiomatic principle, but the principle invoked can be kept very weak. What we want to do is to compare some unknown process with a quasistatic process. For the purpose of this comparison the following postulate is needed:

Between two states of fixed energy and coordinates the quasistatic path always requires a smaller amount of work than any other path.

The postulate is essentially contained in Section 1-5, where the idea of quasistatic change was introduced. The idea explained there was that to drive a process at a finite rate a certain force is needed. Almost always, a reduction of this force is possible by reducing the rate, and even when it is not the case (as in solid–solid friction), a postulate of this sort does not violate physical intuition. The quasistatic form of a process was therefore immediately conceived as the form requiring the least work. The postulate above simply formalizes this notion. With the help of it we can prove the theorem:

If an adiabatically isolated system undergoes a change which brings it into equilibrium its entropy cannot decrease during that change.

Proof. As seen from the theorem, we are constrained to restrict consideration to final states which are equilibrium states, for the preceding discussion has not given us information about any other final states. The final state P_2 must therefore be assumed Gibbsian. The initial state P_1 and the path followed, on the other hand, may be of great complexity.

The path is indicated in Fig. 5.2 in a space having horizontal axes U and x (the latter standing for several coordinates x_i) and S along the vertical axis plotted upward. We now associate with each point P of the path the corresponding equilibrium point P', constructed according to the rule of Section 5-2. The point lies vertically above P and may coincide with it as a limiting case. This coincidence always occurs at P_2. The path $P_1'P_2$

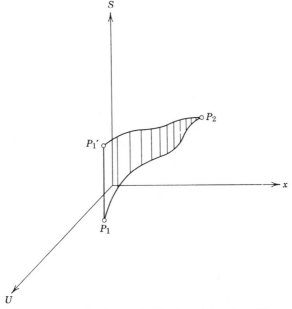

Fig. 5.2. Diagram for the proof of the Second Law. An arbitrary adiabatic path is "projected" on a quasistatic path of equal energy and coordinates.

so obtained is a quasistatic but usually not adiabatic path. We shall now compute the entropy difference between P_1 and P_2 by following the path $P_1P_1'P_2$.

First by the theorem of Section 5-2 the point P_1' lies higher than P_1. To show that P_2 lies higher still we investigate the path $P_1'P_2$, making use of the postulate formalized above. It states that the work done on the system is smaller along each piece of the quasistatic path than along the corresponding piece of the actual path of which it is a projection. Since the energy change is the same it follows that more heat has to be furnished along the quasistatic path than along the actual path. However, the actual path being adiabatic, the heat furnished along any piece of it is zero,

Consequently, the heat furnished along the quasistatic path is always positive or zero, and thus by (5.40) the entropy increases. It follows that the path from P_1' to P_2 is monotonically not-falling. Consequently P_2 lies at least as high as P_1', and higher than P_1. In other words we have that

$$S(P_2) - S(P_1) \geq 0 \qquad (5.42)$$

which was to be established.

Extension of the Second Law to nonadiabatic processes would presumably result in a formula of the form

$$dS \geq \frac{dQ}{T} \qquad (5.43)$$

for any sort of change. Other versions can be derived from this by substituting for dQ the differential form (1.25) or (1.26). The formula is right in some sense, but the question arises whether it can be given a meaning. The difficulty resides in the appearance of T, which has, strictly speaking, only a meaning for equilibrium systems. A formal proof of (5.43) therefore cannot be expected. We shall use (5.43) only in the limited sense that the stationary condition (5.39) is to be taken so that S is a maximum, U a minimum, etc. Other maximum and minimum conditions of this sort will be introduced in Chapter 7.

5-5. Simple examples of entropy expressions

It was shown in (5.40) that for systems in equilibrium there is a simple connection between the change of their entropy and the heat furnished. The formula which results from this connection is (5.41). In the following examples we shall not be concerned with phase transistions, and (5.41) takes then the simpler form

$$S(P_2) - S(P_1) = \int_{P_1}^{P_2} \frac{C}{T} \, dT \qquad (5.44)$$

The interesting feature about this formula is that it applies to any sort of heat capacity occurring in an actual process and to any sort of system. Under all circumstances the temperature T defined originally for perfect gases is supposed to supply an integrating denominator and to furnish a perfect differential. The formula or the more general equation (5.41) also provides a prescription for computing entropy from experimental data, independently of its fundamental definition.

Clearly, for perfect gases with their analytical equation of state it should be easy to carry out the integration for S analytically. This is indeed the case. We take in this case (5.40) with dQ in the form (1.26), valid for perfect fluids. This yields

$$dS = \frac{1}{T}\{dU + p\,dV\} \tag{5.45}$$

Substitution of p from the equation of state (1.08) brings this into the form

$$dS = \frac{1}{T}\left(\frac{\partial U}{\partial T}\right)_V dT + \left\{\frac{1}{T}\left(\frac{\partial U}{\partial V}\right)_T + \frac{Nk}{V}\right\} dV \tag{5.46}$$

The right-hand side must be a perfect differential. Therefore the derivative of the first coefficient with respect to V must equal the derivative of the second coefficient with respect to T. When one works this out one finds easily that

$$\left(\frac{\partial U}{\partial V}\right)_T = 0 \tag{5.47}$$

which implies also that

$$C_v = C_v(T) \tag{5.48}$$

We have introduced in previous work the relation (5.47) as a separate property of perfect gases, over and above the equation of state. That relation is now seen to be a consequence of it. If C_v is approximately a constant, as is often the case, the integration of (5.46) becomes elementary. The entropy of 1 mole of gas comes out to be

$$\mathscr{S} = \mathscr{C}_v \ln T + R \ln V + \text{constant} \tag{5.49}$$

or with (1.09) and (1.35)

$$\mathscr{S} = \mathscr{C}_p \ln T - R \ln p + \text{constant} \tag{5.50}$$

In these formulas for the entropy there are three aspects which are in need of further discussion. The first is that we may employ the entropy as a "label" for the adiabats of a system. In fact the relation $S = \text{constant}$, applied to (5.45), is identical with equation (1.36) characterizing the adiabats of a perfect fluid. The only advance that we have made in going from (1.36) to (5.49) is that we know exactly how much entropy (and therefore how much heat) we have to furnish to the gas to get from one adiabat to the other. Working this out, we find that the constant on the right-hand side of equation (1.38) equals

$$\exp\frac{\mathscr{S}}{\mathscr{C}_v}$$

The "parameter" of the adiabats is thus a function of the entropy, as indicated. This observation is of course valid in general, not just for perfect gases.

The second aspect characteristic of all the equations (5.44)–(5.50) is that they define entropy only up to an additive constant. All procedures making use of the Second Law to measure or compute entropy are subject to this shortcoming. Since entropy is conceived as an absolute quantity by either one of the alternative definitions (5.08) and (5.13), there can be in principle no such arbitrary constant in it. The determination of the constant in the practical formulas written down here thus is a legitimate problem of some importance. A great deal of light is shed on this question by the Nernst heat theorem, which will be discussed in Section 5-7. An actual determination of the entropy constant requires quantum theory, even in the classical range of behavior of a system. Such a determination will be carried out for selected gases in Chapter 11.

The third observation, which affects particularly (5.49) and (5.50), is the following. The formal structure of these equations is such that it appears that $\mathscr{S}$ approaches $-\infty$ as T approaches zero. The obvious conclusion to draw is that they are not valid in that region. This conclusion is in fact wrong. Gases near absolute zero are actually very close to perfect so that the formulas do not become invalid. However, they exist only at very low pressure because if the pressure exceeds the vapor pressure of one of the condensed phases of the substance, the material condenses out. It is therefore the rapid decrease in p which overbalances that of T and thus keeps the total entropy positive, in accordance with the absolute definitions.

The energy–entropy situation of paramagnetic materials has some points in common with that of gases. If we insert the work term (1.27) into (5.36) we get, neglecting angular effects,

$$d \ln F = - U \, d\beta - \beta \, H \, dM \tag{5.51}$$

A *perfect paramagnet* is characterized by a magnetic equation of state of the form

$$M = f\left(\frac{H}{T}\right) \tag{5.52}$$

where f is some unspecified function of its argument. This means that the coefficient of dM in (5.51) is a function of M only. The cross derivative of $\ln F$ with respect to β and M thus vanishes, as does the corresponding cross derivative with respect to β and V for perfect gases. We therefore get an equation analogous to (5.47), namely,

$$\left(\frac{\partial U}{\partial M}\right)_T = 0 \tag{5.53}$$

and U becomes a function of the temperature exclusively. It follows from this in either case that isothermal work does not affect the energy of the system but influences only the entropy. The sign of the effect is evident from (5.39): work done on a system decreases its entropy. For gases the decrease in entropy is associated with the decrease in volume, as seen from (5.49), for a decrease in volume diminishes the "available states." The corresponding case of paramagnets is even more instructive. Paramagnets consist of many small magnetic dipoles of fixed strength but variable orientation. In the absence of an applied magnetic field the resultant magnetic moment M is zero. The zero resultant arises from the random orientation of these dipoles. This is a situation of great disorder, that is, high entropy. As a magnetic field is applied to the sample, these dipoles are forced into a preferred orientation, and the entropy of the sample is thereby diminished. According to (5.53), the work of isothermal magnetization on a perfect paramagnet is exclusively performed to diminish this entropy, leaving the energy alone. The energy balance of the situation is obviously taken care of here by a withdrawal of heat from the sample into a temperature bath with which it is in contact. Inversely, demagnetization of a paramagnetic substance tends to withdraw heat from a body with which it is in contact. This effect is widely employed at very low temperatures; it replaces pumping on liquid helium, which ceases to be effective because of the low vapor pressure of the gas.

Lest the point of view of the reader be distorted by the exclusive consideration of "perfect gases" and "perfect paramagnets" a word should be added here that these materials are quite exceptional, and that there is a common-sense relationship of mechanics and heat which is quite different and whose structure should be recalled here. This common-sense point of view is always adopted in mechanics and electricity courses: it implies that temperature can be ignored when dealing with these disciplines. In order to shed some light on these questions we cross-differentiate U and X_s in (5.36), the first and most convenient form of the Second Law. We find

$$\left(\frac{\partial U}{\partial x_s}\right)_{x_t, T} = X_s - T\left(\frac{\partial X_s}{\partial T}\right)_{x_t}$$

Combining this with the definition (1.30) of the anergetic heat capacity C_0, we find

$$dU = C_0\, dT + \sum_s \left(X_s - T\frac{\partial X_s}{\partial T}\right) dx_s \tag{5.54}$$

We may insert this expression for dU into the differential equation (5.38) for the entropy. We find then

$$dS = \frac{C_0}{T} dT - \sum_s \frac{\partial X_s}{\partial T} dx_s \tag{5.55}$$

Equations (5.54) and (5.55) accomplish a split-up of the isothermal work into a part which augments energy and another which diminishes entropy. They form a proper starting point for the discussion of the point raised.

The common-sense experience, particularly for solids, may be summed up in the statement "A hot rock is still a hard rock." Mathematically, this means that we have

$$\left| \frac{\partial X_s}{\partial T} \right| \ll \left| \frac{X_s}{T} \right| \tag{5.56}$$

Most of the work term then occurs in (5.54) and only a small fraction in (5.55). That work term is temperature independent. Then cross differentiation enforces also the relation

$$C_0 = C_0(T) \tag{5.57}$$

The energy breaks up into a thermal and a mechanical component which are not cross-coupled:

$$U = U_{\text{thermal}} + U_{\text{mechanical}} \tag{5.58a}$$

The coupling of the entropy with the mechanical properties becomes negligible, and we have simply

$$S = S_{\text{thermal}} = \int \frac{C_0}{T} dT \tag{5.58b}$$

Whenever this situation prevails, mechanics and electricity can be discussed without reference to temperature. At the same time $\int C_0 \, dT$ has a meaning: heat content becomes the thermal part of the internal energy. It goes without saying that this common-sense type of situation is very widespread for mechanical properties. It even occurs in magnetism. Diamagnetic materials have generally a susceptibility which is nearly temperature independent, and there exists even a temperature-independent form of paramagnetism. In these cases magnetic properties can be discussed without reference to temperature, following the model (5.58).

We may pass from this situation to the one of our "perfect gases" and "perfect paramagnets." We now have relations of the type

$$\frac{p}{T} = \varphi(V)$$

or

$$\frac{H}{T} = \varphi(M)$$

which can be generalized to

$$\frac{\partial X_s}{\partial T} \approx \frac{X_s}{T} \qquad (5.59)$$

in contrast to (5.56). Consultation of (5.54) and (5.55) shows that now the split-up of the isothermal work between energy and entropy is reversed. The energy equation simply reduces to

$$U \approx U_{\text{thermal}} = \int C_0 \, dT \qquad (5.60a)$$

The entropy now splits up. Equation (5.59) means that

$$\frac{X_s}{T} = \phi_s \approx \text{independent of } T$$

and consequently

$$S \approx S_{\text{thermal}} + S_{\text{mechanical}} \qquad (5.60b)$$

or

$$S = \int \frac{C_0}{T} \, dT + \int \sum_s \phi_s \, dx_s \qquad (5.60c)$$

The distinction between the two limiting situations is made perfectly in the case of the dielectric constant. In magnetism the two effects also have the opposite sign, which is confusing. Paraelectric and dielectric susceptibility, on the other hand, have the same sign, but are distinguished by their thermal coupling. Dielectricity is a temperature-independent electric induction phenomenon, while paraelectricity arises from the ordering of random dipoles. Correspondingly, the susceptibility is a constant in the first case, and varies inversely as T in the second.

It is natural that a course in thermal properties tends to neglect the common-sense situation in favor of cases in which mechanical or electromagnetic behavior is cross coupled with entropy. This remark is meant to rectify this situation somewhat. It should also be mentioned that the cross coupling of thermal and mechanical properties can be of a more complicated nature which fits neither of the two limiting situations worked out above.

5-6. Examples of entropy-increasing processes

As illustrations of the entropy-increasing property of irreversible adiabatic processes two such processes will be analyzed now; a third one will be found among the problems.

Our first example is temperature equilibration. Two bodies 1 and 2 are originally at temperatures T_1 and T_2. Their heat capacities C_1 and C_2 may be taken as constants independent of temperature. The problem is to find the entropy change which results from their thermal contact and to show that it is positive.

It is easily found that the final equilibrium temperature of the two bodies equals

$$T = \frac{C_1 T_1 + C_2 T_2}{C_1 + C_2} \tag{5.61}$$

The total change in entropy results from application of (5.44) to each body separately. The entropy difference comes out to be

$$\Delta S = C_1 \ln \frac{T}{T_1} + C_2 \ln \frac{T}{T_2} \tag{5.62}$$

It is a priori obvious that (5.62) must be positive because quantities of heat dQ are transferred without loss from one body to the other; the entropy dQ/T removed thereby from the hot body is smaller than that gained by the cold one. However, in (5.62) this is not directly obvious because the two terms are of opposite sign. To analyze (5.62) directly one must proceed in a more subtle way. If one introduces the abbreviation

$$\frac{C_1}{C_1 + C_2} = p$$

then one has

$$\frac{C_2}{C_1 + C_2} = 1 - p$$
$$0 < p < 1$$

Expression (5.62) can then be given the form

$$\Delta S = (C_1 + C_2) \ln \frac{pT_1 + (1-p)\,T_2}{T_1^p\, T_2^{(1-p)}} \tag{5.63a}$$

or

$$\Delta S = (C_1 + C_2) \ln \frac{p\dfrac{T_1}{T_2} + (1-p)}{\left(\dfrac{T_1}{T_2}\right)^p} \tag{5.63b}$$

The proof thus depends on the proposition that any weighted arithmetic mean is larger than the corresponding weighted geometric mean, which is easily demonstrated. Consider the function

$$f(x) = p\,x + (1-p) - x^p$$

Here $f(x)$ is the difference of the numerator and the denominator in (5.63b). It obeys the relations

$$f(1) = 0$$
$$f'(1) = 0$$
$$f''(x) = \frac{p\,(1-p)}{x^{2-p}} > 0$$

Hence $f(x)$ is greater than zero for all positive x except 1, and (5.63) is positive unless $T_1 = T_2$.

If T_1 is very nearly equal to T_2, say smaller, an estimate of the entropy increase can be obtained by setting

$$\frac{T_1}{T_2} = 1 - r$$

substituting this into (5.63b), and expanding in powers of r. The first two terms are equal in the numerator and the denominator, and the quadratic term gives the signficant answer, which comes out to be

$$\Delta S \approx \frac{1}{2} \frac{C_1 C_2}{C_1 + C_2} \frac{(T_2 - T_1)^2}{T^2} \tag{5.64}$$

As our second example we take the mixing of two substances. This process requires a more intimate knowledge of details than the first one. We shall therefore take the two substances to be two different perfect gases without chemical affinity which are at equal pressure and temperature. The irreversible form of the process is illustrated in Fig. 5.3. The mixing is initiated simply by breaking down a partition in a box. The process is adiabatic and isoenergetic, but the intermingling of the molecules is an irreversible process which proceeds with an increase in entropy. To estimate this increase we must find a reversible way of attaining the same final state, and compute the entropy change by using the Second Law

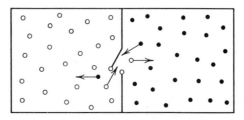

Fig. 5.3. Mixing of two gases as an irreversible process.

with equality sign. The reversible way will involve the furnishing of a certain amount of heat, and the entropy increase can then be obtained from (5.40). Finding a reversible path to replace the irreversible phenomenon of Fig. 5.3 requires a certain amount of ingenuity and general knowledge.

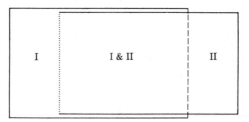

Fig. 5.4. Illustration of the "mixing box," a device to estimate theoretically the entropy of mixing of two gases.

The basic notion needed is the fact that instances are known of *semi-permeable materials* which let one gas pass, but not the other. Hot palladium, for instance, is permeable to hydrogen, and glass to helium and sodium vapor. On this basis we design the following "thought experiment."

Let there be two boxes of equal volume which can be pushed into each other or withdrawn, as shown in Fig. 5.4. The left wall of the right container shall be permeable to the molecules I but not II, and the right wall of the left container shall be permeable to the molecules II but not I. The other walls shall be impermeable to both types of molecules. We see that such a device permits in principle the mixing of equal volumes of gas in a reversible way: pushing the boxes into each other will mix the gases, pulling them apart will unmix them again. If we assume the device to be placed in a vacuum, then the total force on each box is zero because the pressure difference acting across the two side walls of the same box is the same, and therefore the two forces are equal and opposite. Thus the work done in displacing the two boxes with respect to each other is only of the incidental frictional type, which we neglect in the quasistatic approximation. Since there is also no change in the energy of the gas molecules, the heat furnished is zero and the entropy does not change during the operation of the mixing box. There is, however, an entropy change previous to its use, for before the machine can be used each gas must be expanded isothermally from its initial volume to the final volume which the two gases will occupy together.

We arrive then at the following computational procedure. Let there be N_1 molecules of the first gas and N_2 molecules of the second gas, and let p and T be given. The initial volumes are then from (1.08)

$$V_1 = \frac{N_1 k T}{p}$$

$$V_2 = \frac{N_2 k T}{p}$$

The entropy increase resulting from the mixture is then equal to the increase resulting from expanding each gas isothermally to the final volume $V_1 + V_2$. From equation (5.49) this comes out to be

$$\Delta S = N_1 k \ln \frac{V_1 + V_2}{V_1} + N_2 k \ln \frac{V_1 + V_2}{V_2}$$

$$= N_1 k \ln \frac{N_1 + N_2}{N_1} + N_2 k \ln \frac{N_1 + N_2}{N_2}$$

Let us introduce the mole concentrations c_1 and c_2 into this expression:

$$c_1 = \frac{N_1}{N_1 + N_2} \tag{5.65a}$$

$$c_2 = \frac{N_2}{N_1 + N_2} = 1 - c_1 \tag{5.65b}$$

The preceding expression then becomes

$$\Delta S = (N_1 + N_2) k \{-c_1 \ln c_1 - c_2 \ln c_2\} \tag{5.66}$$

The entropy of mixture of 1 mole of the final gas mixture is therefore equal to

$$\Delta \mathscr{S} = R \left[-c_1 \ln - c_2 \ln c_2 \right] \qquad (5.67)$$

Since c_1 and c_2 are proper fractions, the square bracket is intrinsically positive as expected.

Expression (5.67) has been derived for perfect noninteracting gases. It gives an indication of the entropy of mixture which might be involved in the mixing of other fluids, but it is by no means an exact expression in those cases. If we are, for instance, to determine the entropy of mixture of water and alcohol at room temperature, the correct procedure is to heat both the water and the alcohol separately from room temperature to such a high temperature that both behave as nearly perfect gases, and then mix them and cool the vapor mixture back to room temperature. The entropy changes in the heating and cooling can then be determined from (5.41), and the entropy of mixing in the gas phase from (5.66). Naturally, the entropy change in the heating stage will cancel approximately with that of the cooling stage, but a significant discrepancy from (5.66) is often left.

It is of course much simpler to calculate the entropy of mixing of two gases from the partition function than from any imaginary experiment. Such a calculation shows immediately that we calculate these quantities for each gas component, assigning it the common temperature and the total volume, while before the mixing only the partial volume occupied could be counted. Therefore (5.49) is the fundamental formula, and (5.50) is derived. In this way (5.66) is reproduced for a mixture of two gases. For a mixture of several gases we get in the same way

$$S = \left(\sum_i n_i \mathscr{C}_{v_i} \right) \ln T + R \sum_i n_i \ln \mathscr{V}_i + \text{constant} \qquad (5.68)$$

where i are the constituents of the mixture, and n_i the number of moles of the gas i present. If we want to bring the pressure in we can do so as follows. We may introduce Dalton's idea of partial pressure and define, as previously in (2.42),

$$p_i \mathscr{V}_i = R T$$

This yields

$$S = \left(\sum_i n_i \mathscr{C}_{p_i} \right) \ln T - R \sum_i n_i \ln p_i + \text{constant} \qquad (5.69)$$

If we now want the total pressure in the picture, we make use of Dalton's law (2.41). Defining

$$c_i = \frac{N_i}{\sum\limits_{\nu} N_\nu} \tag{5.70}$$

we obtain

$$p_i = c_i p \tag{5.71}$$

Substitution of (5.71) into (5.69) yields

$$\mathscr{S} = \left(\sum_i c_i \mathscr{C}_{p_i} \right) \ln T - R \ln p - R \sum_i c_i \ln c_i + \text{constant} \tag{5.72}$$

The entropy of mixing is now apparent in the third term on the right; it is positive as expected. If (5.72) is compared with the earlier formula (5.50), which should also apply to a gas mixture, it is seen that (5.50) is not actually wrong because it describes the entropy correctly as a function of the state variables T and p. It is the additive constant which hides the mixing term. This term is not simply obtained by adding the entropy constants of the individual gases because the term is nonlinear in the mole concentrations. It remains, of course, homogeneous of the first degree in the mole numbers; however, this does not imply linearity, as one might at first suppose.

5-7. Third law of thermodynamics

The definitions (5.08) and (5.13) of entropy lead to results having a significant simplicity in the neighborhood of the absolute zero of temperature. If the temperature of a body is lowered sufficiently so that the energy separation between the ground state and the first excited state is large compared to kT, then all exponential terms in (5.01) become small in comparison to that referring to the state of lowest energy. Similarly, all probability expressions have a small exponential except the probability for the ground state. If the degeneracy factor is not too unfavorable the system will then be almost certain to be in the ground state. If the system is in that state, either one of the definitions (5.08) or (5.13) yields for its *zero point entropy*

$$S(0) = k \ln M(E_0) \tag{5.73}$$

Here $M(E_0)$ is the dengeneracy or multiplicity of the ground state. It is suspected by mathematical physicists that the ground state of every quantum-mechanical system is nondegenerate. If this suspicion is correct the entropy of every system is zero at the absolute zero of temperature. This type of argument forms the foundation of a theorem which was enunciated by Nernst long before the advent of quantum mechanics. It is referred to sometimes as the *Nernst heat theorem*, and sometimes as the *third law of thermodynamics*. It states:

The entropy of every chemically simple perfectly crystalline body equals zero at the absolute zero of temperature.

At first sight the statement seems to require such a strong condition, namely, nondegeneracy of the ground state, that one would scarcely expect the law ever to be valid for an actual body. A paramagnetic impurity atom somewhere in the sample, free to rotate, is sufficient to give the ground state some degeneracy. What saves the law from being a perfectionist's dream is that contributions to S of the order k cannot be measured. S is measured in units of R, which is a unit 10^{23} times larger. Degeneracies of the order of a hundred or even a million therefore make no significant contributions to S when inserted into (5.73). Degeneracies which have to be watched are those to which every atom or molecule of the sample makes a contribution. As an example, we may take a crystal of paraffin chemically pure, say ikosane, $C_{20}H_{42}$, all of the same branching structure. All its electronic spins are locked into covalent bonds. Therefore it forms a reasonable candidate for Nernst's theorem, even though it is somewhat sluggish to crystallize. To make sure that we have no impurity trouble we make the paraffin out of the isotopes C^{12} and H^1. The paraffin so obtained has, however, some zero point entropy left, namely, that due to the spin of the proton which is $\frac{1}{2}$ and capable of two orientations. If these spins remain uncoupled and free to move, 1 mole of ikosane has the statistical multiplicity of

$$M = (2)^{42 \mathscr{A}} \tag{5.74a}$$

where $\mathscr{A}$ is Avogardo's number. The contribution of this multiplicity to the molar entropy is

$$\Delta \mathscr{S} = 42\, R \ln 2 \tag{5.74b}$$

which is very substantial on any scale. We may try to take care of such an argument by saying that if the proton spin remains unbound throughout an experiment, the fact that every mole of protons makes a contribution of 0.693 R to the entropy can perhaps be ignored. This point of view has been adopted in a weaker form of the Third Law which reads:

The entropy of every chemically simple perfectly crystalline body may be considered zero at the absolute zero of temperature.

This second form of the law can only be accepted as an expedient to get around spin degeneracy in certain practical problems. Spins, even nuclear spins, are being oriented today partly through cooling, partly through magnetization, so that the second formulation cannot be accepted as satisfactory. The first one does in fact go more directly to the heart of the problem.

One advantage of spin degeneracy is that, although it may spoil the Third Law, it spoils it by an amount which is readily computed. There are other cases which create more difficulty. Careful measurements of specific heats from the absolute zero to the gas phase allow a determination of the zero point entropy which is independent of (5.73). This possibility arises because the entropy constant of gases can be computed theoretically; we shall do this in Chapter 11. If we combine this with measurements handled through (5.41), we get an indirect but very reliable measurement of the zero point entropy. The result is that quite a few substances have zero point entropies over and above the spin type, whose origin is sometimes hard to trace. Ice may serve as an example. It has a zero point entropy whose origin was finally traced by Pauling. The ice crystal allows each of its hydrogen atoms two alternate positions within the oxygen framework. The protons cannot fill these positions independently, but they fill them cooperatively in such a way that each oxygen atom has two protons attached to itself. The result is a ground state degeneracy of approximately‡

$$M = (\tfrac{3}{2})^N \tag{5.75a}$$

where N is the number of molecules of water present. Application of (5.73) to this yields a molar zero point entropy equal to

$$\mathscr{S} = R \ln \tfrac{3}{2} = R \cdot 0.4055 \tag{5.75b}$$

Ice thus is shown to be an exception to the Third Law. We have just seen that all mixtures are automatically exceptions if the components are mixed at random because there is an entropy of mixture associated with this type of disorder. It is of course conceivable that such mixtures, as well as the low-temperature ice discussed above, would transform into ordered arrangements if given a chance so that the zero point entropies observed do not refer to the true ground state. The sluggishness of most transformations at low temperature enters here as a hindering circumstance, preventing us from deciding whether Nernst's law is or is not a law of nature for all materials.

A third and even weaker form of the Third Law is universally true and follows from either the definition (5.08) or (5.13). It states:

The entropy of every system is positive or zero.

‡ L. Pauling. *J. Am. Chem. Soc.* **57**, 2680 (1935).

In thermodynamics where entropy is often known only up to an unknown additive constant, the law may have to be given this form:

The thermodynamically defined entropy of any system has a finite lower limit.

The law is far from trivial and connects in fact directly with the problems that Nernst faced when he enunciated the Third Law. He observed what seemed to be a universal rule, namely, that the heat capacity of all substances declines as the temperature is lowered. He correctly deduced that one of the keys to this phenomenon lies in equation (5.44), which makes S tends to $-\infty$ at absolute zero unless C tends to zero at absolute zero. Happily, this feature of the heat capacity is nailed down even by the weakest formulation of the Third Law, namely, the one just given. We therefore derive the corollary:

The heat capacity of every material tends to zero at the absolute zero of temperature.

Even this law, whose universal validity is here asserted, must be applied with care. The decline of the heat capacity, observed in a certain range, may not be the final decline. An unsuspected degree of freedom may be frozen out at some lower temperature. The associated entropy change produces a rise in the specific heat, followed soon by a (perhaps) final decline. Thus even this weakest of the formulations of the third law of thermodynamics conceals some surprises for the experimenter.

A. RECOMMENDED PROBLEMS

1. In simple cases, the Ehrenfest principle applies directly to adiabatic macroscopic changes. This means that not only is the reacting force equal to the mean rate of change of the energy levels with the coordinate, but also no rearrangement is necessary in the population of the levels after the change has taken place. Show that this is the case for a monatomic perfect gas in a cubic box of variable volume.

2. Follow the model calculation of Section 5-5 to write down expressions for the molar energy and entropy of a van der Waals gas. Also write down an equation for its adiabats, making the supplementary assumption that $\mathscr{C}_v$ is independent of T.

3. Two volumes of the same perfect gas are at equal temperature. They are adiabatically isolated from the outside, but connected with each other by a tube closed with a stopcock.

(*a*) Show that the entropy of the whole system is increased when the stopcock is opened unless the initial pressures were equal.

(*b*) Compare the energy and entropy balance as it results from the opening of the stopcock, as compared to a reversible path which leads to the same final state.

4. The high-temperature behavior of iron may be described as follows. Below 900° C and above 1400° C α-iron is the stable modification, and between these temperatures γ-iron is stable. The specific heat of each phase is approximately constant, being 0.775 j/gm for α-iron and 0.690 j/gm for γ-iron. Find the latent heat at each transition.

B. GENERAL PROBLEMS

5. Show that the general definition (5.13) of entropy does indeed contain the low-temperature expression (5.73) as a special case.

6. Suppose we modify Problem 1 by placing the gas in a rectangular box and varying only one dimension. What goes on on a microscopic scale when the gas is compressed adiabatically?

7. Derive the adiabatic equation of state for diatomic molecules by first applying the Ehrenfest principle to find the infinitesimal change in energy for an infinitesimal change in volume, and then working out the subsequent rearrangement in the energy distribution.

8. What happens to the relation derived in Problem 6, Chapter 1, if we deal with a body for which mechanical and thermal properties are decoupled, as described on page 104?

9. A ferromagnet consists of N atoms of spin $\frac{1}{2}$. At sufficiently high temperature the spins are randomly oriented, but at low temperature their coupling causes them to exhibit ferromagnetism; as a result all spins are aligned parallel at the absolute zero of temperature. Suppose the associated heat capacity can be roughly approximated as

$$C(T) = C_0 \frac{T}{T_c} \quad \text{if } 0 < T < T_c$$

and

$$C(T) = 0 \quad \text{if } T > T_c$$

The temperature T_c separates ferromagnetic from nonferromagnetic behavior (Curie point). Use the basic principles about entropy developed in this chapter to give an estimate for the constant C_0.

10. It is possible to imagine a union of two systems, 1 and 2, such that

the energy levels $\varepsilon_r^{(1)}$ and $\varepsilon_s^{(2)}$ are not disturbed, but the occupation probabilities are, owing to correlation effects:

$$P_{rs} \neq \left(\sum_\sigma P_{r\sigma} \right) \left(\sum_\rho P_{\rho s} \right)$$

Show that the entropy of such a correlated system is smaller than that of an uncorrelated system for which the brackets on the right-hand side of the above inequality have the same value.

°6

Older ways to the Second Law

We have established in the previous chapter that the Second Law is a consequence of statistics. Like all proofs of the law, the one given also consists of two parts. In Section 5-4, we proved the Second Law with equality sign for systems subject to quasistatic change. In Section 5-2 we proved it with the inequality sign for any system for which the external constraints and the energy are fixed. A combination of the two proofs establishes the Second Law with inequality sign in general, provided one can return from the final to the initial values of the constraints and the energy by quasistatic means and define temperature at least piecewise for the initial state. This will take care of a great number of practical situations.

However, one must not read more into these proofs than is actually there. After having made plausible by other arguments that there is an equilibrium state toward which matter tends when not externally disturbed, we proved that this state is also a state of maximum entropy. We have no indication of the rate at which matter tends to approach this state; in fact, we do not know that matter approaches it at all. There are numerous cases, particularly in chemistry, in which no measurable rate can be found at which a system not in equilibrium approaches the equilibrium state. This empirical situation is matched by others which yield a zero rate of change theoretically, when in fact a measurable shift toward equilibrium is observed. There is a very awkward "conservation law" for the entropy of a Hamiltonian system which can be proved from either quantum or classical mechanics. For the quantum proof we proceed from (5.14) and (3.24). The latter equation implies that

$$i \hbar \frac{d}{dt} (\rho \ln \rho) = \mathcal{H} \rho \ln \rho - \rho \ln \rho \mathcal{H} \qquad (6.01a)$$

The right-hand side has no trace. Thus, if the trace of this equation is taken, and it is observed that trace taking and differentiation with respect to time commute, we obtain instantly

$$\frac{dS}{dt} = 0 \qquad (6.01b)$$

The difficulty arises also classically where the entropy is essentially statistical multiplicity, that is, the volume in phase space occupied by members of the ensemble. By Liouville's theorem, this volume is a constant. There are various ways to argue oneself out of the difficulty. One is based on the observation that the volume in phase space assumes very weird shapes in time, with many indentations, so that in a "coarse grained" observation the volume occupied becomes in fact larger. One might call this the "milk poured in the coffee" argument. Another type of argument lays stress on the environment, sometimes called "the observer." This environment is thought to be so fluctuating that no hamiltonian can ever describe it completely. One might call this the "sand in the machine" argument. Neither argument is entirely satisfying, yet there is no question about the point under consideration. The conservation law (6.01) is *not* a valid physical law, for reasons which are not entirely clear. In view of these difficulties the older style proofs of the Second Law retain a certain attractiveness even today. A short discussion of these procedures, together with their modern refinements, will therefore be presented in this chapter.

6-1. Proof by the method of Carnot cycles

The older discussions of the Second Law approach their subject axiomatically. One starts out with a principle which is generally accepted and derives from it results of wider consequence than those which could be derived by any statistical procedure. The reader is no doubt familiar with the idea of perpetual motion machines and the fact that the First Law can be said to assert that the most common type of these machines ("work-for-nothing") is impossible. The law leaves open the possibility of a more refined type of such machines ("work-for-heat"); we owe it to the genius of Carnot that he realized that, although the conversion of work into heat is in fact unlimited and the conversion of heat into work is in fact possible, the latter is severely limited in principle. Kelvin summed up the situation in a basic postulate which states:

A transformation whose only result is the conversion into work of heat extracted from a body at uniform temperature is impossible.

First let us verify that, indeed, a contradiction to Kelvin's postulate would allow something closely approaching perpetual motion. An ocean-going vessel, for instance, could extract its power for propulsion simply

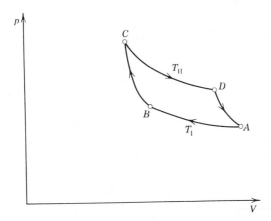

Fig. 6.1. Cycle of a Carnot engine operating with a perfect gas.

from the heat of sea water and cool the water in the process. Even this cooling of the water would not really be a practical limitation to this type of perpetual motion, for almost the entire power output of the motors of the ship returns to the surroundings as frictional heat, the only limitation being the small-scale destruction of the ship's structure incidental to its motion. This type of limitation affects of course the efficiency of all machine processes and can safely be omitted from the discussion. We thus can say that Kelvin's postulate denies in principle the existence of a type of perpetual motion machine, called perpetuum mobile of the second kind, whose convenience, if existing, would in no way differ from that of the more usual perpetuum mobile of the first kind, which violates the first law of thermodynamics.

The Carnot cycle permits a general exploitation of the Kelvin postulate because it is concerned with the conversion of heat into work. In the course of this study we learn what is meant by reversible transfer of heat; the concept of entropy is thereby formulated. To begin, let a Carnot engine be a device operating quasistatically between two heat reservoirs at temperatures T_I and T_{II}, with $T_{II} > T_I$. At first we shall particularize the engine still further as a vessel filled with a perfect gas and capable of variable volume because of a movable piston. The vessel and piston shall have no heat capacity, but they shall permit the vessel to be held either in adiabatic isolation or in perfect thermal contact with either reservoir. As shown in Fig. 6.1 the engine operates in four stages.

1. At the beginning of stage 1 it is at temperature T_I and in contact with the reservoir at the same temperature (point A). It is thereupon compressed isothermally to point B.

2. At stage 2, it is adiabatically isolated and compressed further until it reaches the higher temperature T_{II} (point C).

3. At stage 3, it is expanded in contact with the reservoir at T_{II} up to a point D, which is determined indirectly by the requirements of stage 4.

4. At stage 4 it is expanded adiabatically until the temperature has reached T_I again. The requirement for the break-off point D is that the cycle close at point A.

The Carnot engine performs work because the heat delivered to reservoir I at stage 1 is smaller than the heat given up by reservoir II at stage 3. Indeed we have from (1.08), (1.24), and (1.34)

$$Q_I = \int_A^B p \, dV = N k T_I \ln \frac{V_B}{V_A} \tag{6.02}$$

and

$$Q_{II} = \int_C^D p \, dV = N k T_{II} \ln \frac{V_D}{V_C} \tag{6.03}$$

Here we followed the convention of this book, counting Q as positive when furnished to the engine. We have now to exploit the adiabatic legs of the cycle. According to (1.38), we have

$$T_I V_B^{\gamma-1} = T_{II} V_C^{\gamma-1}$$

and

$$T_I V_A^{\gamma-1} = T_{II} V_D^{\gamma-1}$$

Forming ratios, we find

$$\frac{V_B}{V_A} = \frac{V_C}{V_D} \tag{6.04}$$

Substitution of (6.04) into (6.02) and (6.03) gives the chief result

$$\frac{Q_I}{T_I} + \frac{Q_{II}}{T_{II}} = 0 \tag{6.05}$$

The efficiency η of the Carnot engine is defined as the ratio of the available work to the heat delivered by reservoir II. We find for it from (6.05)

$$\eta = \frac{Q_{\mathrm{II}} + Q_{\mathrm{I}}}{Q_{\mathrm{II}}} = 1 - \frac{T_{\mathrm{I}}}{T_{\mathrm{II}}} \tag{6.06}$$

The Carnot reasoning now proceeds to argue that all legs of the Carnot cycle are reversible, and it thus might have been driven as a refrigerator with all Q's having the opposite sign. Relation (6.05) would still hold. It then follows that any other reversible engine operating between the two reservoirs I and II must have the same efficiency. Otherwise the more efficient of the two engines could be driven in tandem with the other as a

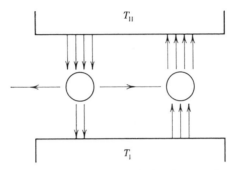

Fig. 6.2. Proof of Carnot's relation, by operation in tandem of two reversible heat engines.

refrigerator, as shown schematically in Fig. 6.2. This figure shows two engines, schematized as circles, operating between the same heat reservoirs, represented by rectangular blocks. The heat inputs and outputs are represented by vertical arrows and the work by horizontal arrows. For simplicity the left engine is given an efficiency of 50% and the right, which operates as a refrigerator, of 25%. The in-tandem operation is read off the figure. The arrangement is to leave reservoir II undisturbed, by having one engine extract exactly as much heat as the other returns. The energy balance is then between the work and the exchange of heat with reservoir I. The efficient engine delivers two units of heat to the reservoir and gains two units of work. The inefficient engine, on the other hand, extracts three units of heat and needs only one unit of work. As a result, one unit of work has been gained, and reservoir I has lost a corresponding amount of heat. This is a violation of Kelvin's axiom. The assumption that the two reversible engines have different efficiency is thus shown to be impossible, which means that any such engine, however constructed, must have the efficiency (6.06). A further refinement of this same argument shows that any real irreversible engine operating between the two reservoirs I and II must obey the inequality

$$\eta \leqq 1 - \frac{T_I}{T_{II}} \tag{6.07}$$

which can also be written as

$$\frac{Q_{II}}{T_{II}} + \frac{Q_I}{T_I} \leqq 0 \tag{6.08}$$

One can repeat the argument for inefficient refrigerators and show that while the sense of the inequality (6.07) is then reversed the inequality

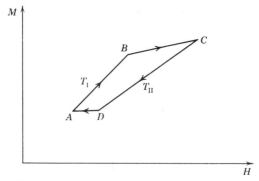

Fig. 6.3. Carnot cycle for a paramagnetic substance.

(6.08) remains the same. We may therefore consider it the fundamental result for operations of a cyclic engine operating between two heat reservoirs.

There is little point in following the classical argument further, deriving the relationship which must hold for an arbitrary reversible and irreversible engine. The procedure is simply to place such engines in tandem with a large number of Carnot engines. The result is that the reversible extraction of heat is always carried out in such a way that

$$\oint \frac{dQ}{T} = 0 \tag{6.09}$$

over a closed cycle. The concept of entropy is therefore essentially defined.

Lest anybody believe that the Carnot cycle depends somehow on perfect gases, we shall repeat the argument for perfect paramagnets. In Fig. 6.3

the cycle is shown as a diagram in the H-M plane. We have again two reservoirs at temperatures T_I and T_{II} and a working substance which we may visualize as a paramagnetic salt. We magnetize the salt at the lower temperature T_I, then break thermal contact and magnetize further. The result is a heating of the salt which is such that the loss of entropy of its magnetic system is compensated by an increase in the entropy of its other degrees of freedom. Adiabatic magnetization proceeds to a temperature T_{II}. Here the process is arrested and isothermal demagnetization follows. Finally the cycle is closed by adiabatic demagnetization in such a way that the starting point of the cycle is reached again. If we call the corners of the diagram in H-M space A, B, C, D as previously, we have, using (1.27) and (5.53),

$$Q_I = -\int_A^B H \, dM \tag{6.10a}$$

and

$$Q_{II} = -\int_C^D H \, dM \tag{6.10b}$$

Equation (6.10a) is a heat loss for the paramagnetic salt acting as "engine," while (6.10b) is positive. The signs are therefore the same as in the gas engine. If we now compute the entropy absorbed by the salt, that is, the quantity (6.05), we obtain

$$\frac{Q_I}{T_I} + \frac{Q_{II}}{T_{II}} = -\int_A^B \frac{H}{T} \, dM - \int_C^D \frac{H}{T} \, dM \tag{6.11}$$

By (5.52) these integrals depend on the end points only, and we can transform the right-hand side to

$$-\int_A^D \frac{H}{T} \, dM - \int_D^B \frac{H}{T} \, dM - \int_C^B \frac{H}{T} \, dM - \int_B^D \frac{H}{T} \, dM$$

The second and the fourth integrals cancel. The first and the third refer to the adiabatic legs of the cycle, for which we have

$$dU - H \, dM = 0$$

Hence the above expression becomes

$$-\int_A^D \frac{dU}{T} - \int_C^B \frac{dU}{T}$$

where U is only a function of T as stated in (5.53). Hence we find for the above expression

$$-\int_{T_I}^{T_{II}} \frac{1}{T}\frac{dU}{dT}\, dT - \int_{T_{II}}^{T_I} \frac{1}{T}\frac{dU}{dT}\, dT = 0$$

This establishes that (6.11) is zero, yielding

$$\frac{Q_I}{T_I} + \frac{Q_{II}}{T_{II}} = 0 \tag{6.12}$$

which is a result completely equivalent to (6.05).

This magnetic argument frees us to some extent from the perfect gas as a working material, but it needs the substitution of another substance which is also perfect in some sense. Actually the argument that these substances for which we compute the efficiency of Carnot cycles do not exist is not very important. We prove that the cycles provided by these ideal substances place an upper limit on the efficiency of any cycle, and for this purpose the idealized substance ought to be good enough. In passing from (6.05) to (6.09) we essentially carry out a passage to the limit. The consideration of irreversible engines adds to this the inequality sign evident in (6.08). Thus we get finally

$$dS \geqq \frac{dQ}{T} \tag{6.12}$$

This can be derived from (6.08) by an easy but long-winded argument.

There have been some presentations of thermodynamics which make far-reaching use of the method of cycles. The procedure is the inverse of the one used in the two preceding examples: a Carnot cycle is set up operating with an unknown material, the efficiency condition (6.05) is invoked, and a relationship governing the equation of state is the result. The method is less efficient than the differential methods we shall develop later, but it retains a certain experimental "feel" which appeals to some readers. We shall not make use of the method in general, but we shall give an example here. We shall derive the equation which governs the equilibrium of a vapor with its fluid and which is known as the *equation of Clausius and Clapeyron*. We shall do it here by constructing a Carnot cycle operating with the saturated vapor in equilibrium with the liquid phase. In other words we shall discuss a schematized form of a steam engine to obtain the result.

The cycle is shown in Fig. 6.4 on a $p-V$ diagram. The isotherms are horizontal lines, while the adiabats have a more complicated form. Within the entire diagram the properties of the system are affected very simply by a horizontal displacement: they change because the relative abundance of the two phases changes, but the intensive parameters of each phase remain constant. We shall therefore adopt a slightly different

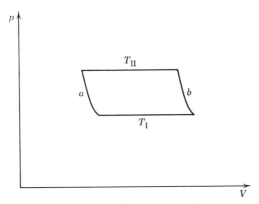

Fig. 6.4. Carnot cycle for a saturated vapor in equilibrium with its fluid phase.

labeling of the cycle from that used for the other cases. The right-hand adiabat shall be called b and the left-hand one a. We have then along every horizontal line (isotherm) the First Law

$$dQ = dU + p \, dV$$

This is integrable as p is a constant. With the help of the abbreviation

$$H = U + pV \tag{6.13}$$

where H is called the *enthalpy* of the system, the First Law integrates along the isotherms to give

$$Q_{ab} = H_b - H_a \tag{6.14}$$

This relation is differentiable vertically (for varying pressure and temperature), and the dQ resulting needs no bar. We now must shift our attention to the two adiabats. By integration by parts the First Law is transformed into a form appropriate for this case

$$dH_a = V_a \, dp_a \tag{6.15a}$$

or

$$dH_b = V_b \, dp_b \tag{6.15b}$$

The index on p is actually superfluous as p depends on vertical position only. We now subtract the two equations for the same vertical positions on the two adiabats and find

$$d(H_b - H_a) = (V_b - V_a)\, dp \tag{6.16}$$

and hence with (6.14)

$$dQ_{ab} = (V_b - V_a)\, dp \tag{6.17}$$

Now we use the efficiency condition (6.05) for Carnot cycles in a slightly more sophisticated form. Starting from

$$\frac{Q_{ab}(T_2)}{T_2} = \frac{Q_{ab}(T_1)}{T_1}$$

we let T_2 tend toward T_1 and find

$$\frac{dQ_{ab}}{dT} = \frac{Q_{ab}}{T} \tag{6.18}$$

which reduces (6.17) to

$$\frac{dT}{dp} = \frac{T(V_b - V_a)}{Q_{ab}} \tag{6.19}$$

We shall have occasion to come back to this equation.

6-2. Proof of Caratheodory

Many physicists have had a feeling that the Kelvin postulate is too specific as a basic axiom for a theory, and that the Carnot cycle is at the same time too "experimental" and too idealized to provide the kind of derivation which one would like to see for the Second Law. We shall now proceed to give such a proof which was developed by Caratheodory.* It will be seen that a more abstract proof like this one resembles the statistical proof more than the picturesque arguments we have developed up to this point.

The starting postulate of Caratheodory is right away more abstract than the postulate of Kelvin but contains the latter as a special case. It reads as follows:

There exist arbitrarily close to any given state of a system other states which cannot be reached from it by adiabatic processes alone.

* C. Caratheodory, *Math. Ann.* **67**, 355 (1909).

As usual the deduction has two parts. First the Second Law is derived with the equality sign for quasistatic processes; then the inequality is obtained for irreversible processes.

For the first part we take the First Law in the form (1.25), with the inequality (4.20) thrown in as a supplementary assumption. The supplementary assumption allows a mutual substitution of U and T or β. The differential equation for adiabatic quasistatic processes then takes the form

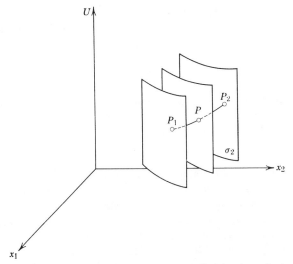

Fig. 6.5. Development of adiabats by Caratheodory's method.

$$dU - \sum_i X_i(x_1, x_2, \ldots x_n, U)\, dx_i = 0 \qquad (6.20)$$

In the neighborhood of a particular point P_1 such a differential equation defines an element of area in space. (The geometrical language will be as if there were only two coordinates x_1 and x_2; for one coordinate there is nothing to prove, for in that case an integrating denominator of (6.20) always exists, as is proved in the theory of ordinary differential equations.) Now according to our postulate there is at least one line P_1P_2 along which there are points which cannot be reached while having the above form equal to zero. For every point P along that line equation (6.20) defines an element of surface as shown in Fig. 6.5. These elements can be labeled according to the distance P_1P measured along the line P_1P_2. Let the parameter be denoted by σ. Then if we extend these elements of area throughout space we feel the weight of Caratheodory's axiom. Because of this axiom, these areas *cannot cross*, that is, a point belonging to the manifold σ_1 cannot also belong to the manifold σ_2. These manifolds are as numerous as the points on a line and have differentially the structure of surfaces at every point. Hence they form a one-parameter family of surfaces filling space. Because of our postulate this family has the particular property that no two members of it have any point in common.

Now let us move from one of these surfaces to the other along the original line P_1P_2. The form $dU - \sum_i X_i \, dx_i$ does not vanish along that line by assumption, and neither does the differential $d\sigma$. By a general property of differential geometry the two expressions are then proportional and we can write

$$dU - \sum_i X_i \, dx_i = \tau \, d\sigma \qquad (6.21)$$

which, because of (1.25), can also be written as

$$\frac{dQ}{\tau} = d\sigma \qquad (6.22)$$

This is the first part of Caratheodory's theorem:

For every system in thermal equilibrium there exists an integrating denominator τ which will make dQ a complete differential.

Equation (6.22) still contains no structural information concerning τ and σ, and it is in fact very likely that we have not chosen them wisely by this blind procedure. But it does define at any rate a state variable $\sigma(x_1, x_2, \ldots, U)$ or $\sigma(x_1, x_2, \ldots, t)$ for the system which we shall call the *empirical entropy*. Entropy appears in this approach first as a convenient variable to label the adiabats of a system.

In order to gain structural information concerning σ and τ we must repeat *mutatis mutandis* the argument of Chapter 4: we must consider the union of two systems and exploit the fact that it forms again a system. Let the first system be characterized by the variables $U_1, x_1, x_2, \ldots$ and the second by $U_2, y_1, y_2, \ldots$, and let their common temperature be t. Then we can write relation (6.21) three times, once for the first system, once for the second system, and once for the two combined:

$$dU_1 - \sum_i X_i \, dx_i = \tau_1 \, d\sigma_1 \qquad (6.23a)$$

$$dU_2 - \sum_k Y_k \, dy_k = \tau_2 \, d\sigma_2 \qquad (6.23b)$$

$$d(U_1 + U_2) - \sum_i X_i \, dx_i - \sum_k Y_k \, dy_k = \tau \, d\sigma \qquad (6.23c)$$

where of course

$$\tau_1 = \tau_1(x_1, x_2, \ldots, x_n, t)$$
$$\tau_2 = \tau_2(y_1, y_2, \ldots, t)$$
$$\tau = \tau(x_i, y_k, t)$$
$$\sigma_1 = \sigma_1(x_i, t)$$
$$\sigma_2 = \sigma_2(y_k, t)$$
$$\sigma = \sigma(x_i, y_k, t)$$

This general dependence can be simplified greatly if we observe that (6.23c) is the sum of (6.23a) and (6.23b). We have therefore

$$\tau\, d\sigma = \tau_1\, d\sigma_1 + \tau_2\, d\sigma_2 \tag{6.24}$$

which means that

$$\sigma = \sigma(\sigma_1, \sigma_2) \tag{6.25}$$

The entropy of the union thus depends only on the entropies of the two parts. Furthermore

$$\left(\frac{\partial\sigma}{\partial\sigma_1}\right)_{\sigma_2} = \varphi(\sigma_1, \sigma_2) = \frac{\tau_1}{\tau}$$

This means that τ depends on the variables of system 2 only through t and σ_2, since τ_1 contains only t and the variables of the first system and φ contains only the two entropies. Repetition of this argument for $(\partial\sigma/\partial\sigma_2)_{\sigma_1}$ yields that

$$\tau = \tau(\sigma_1, \sigma_2, t)$$

This imposes an even greater restraint on τ_1 and τ_2 which cannot depend on variables of the other system. We must therefore have

$$\tau_1 = \tau_1(\sigma_1, t)$$
$$\tau_2 = \tau_2(\sigma_2, t)$$

and hence by implication

$$\tau = \tau(\sigma, t) \tag{6.26}$$

Relation (6.24) also yields information concerning dependence on temperature. We notice that the differential dt is absent and therefore

$$\left(\frac{\partial^2\sigma}{\partial t\, \partial\sigma_1}\right)_{\sigma_2} = 0$$

or

$$\left(\frac{\partial(\tau_1/\tau)}{\partial t}\right)_{\sigma_1, \sigma_2} = 0$$

This relation can be given the symmetric form

$$\left(\frac{\partial \ln \tau_1}{\partial t}\right)_{\sigma_1} = \left(\frac{\partial \ln \tau_2}{\partial t}\right)_{\sigma_2} = \left(\frac{\partial \ln \tau}{\partial t}\right)_{\sigma}$$

This logarithmic derivative, which is the same for all systems in equilibrium, depends only on t and σ_1 for system 1, only on t and σ_2 for system 2. Hence it depends only on t and is a universal function of t:

$$\left(\frac{\partial \ln \tau_1}{\partial t}\right)_{\sigma_1} = g(t)$$

Integrating, we find

$$\ln \tau_1 = \int^t g(t)\, dt + \ln \Sigma_1(\sigma_1) \tag{6.27}$$

The integrating denominator τ of dQ is the product of a universal function of the temperature of the system and a specific function of its empirical entropy.

The remainder of the work consists in providing suitable definitions for the quantities involved. We define the combination

$$T(t) = \exp \int^t g(t)\, dt \tag{6.28}$$

which is also a universal function of temperature (apart from a factor), as the *absolute thermodynamic temperature T*. With its help (6.22) takes the form

$$d\sigma = \frac{dQ}{T(t)\,\Sigma(\sigma)} \tag{6.29}$$

As the last step we introduce *absolute entropy* by the definition

$$dS_1(\sigma_1) = \Sigma_1(\sigma_1)\, d\sigma_1 \tag{6.30}$$

which brings the Second Law into the usual form

$$dS(\sigma) = \frac{dQ}{T(t)} \tag{6.31}$$

It incidentally also makes entropy differentially additive because (6.24) now reduces to

$$dS = dS_1 + dS_2 \tag{6.32}$$

The question whether this relation can be integrated to yield additivity of entropy in general is very delicate. The phenomenon of entropy of mixture shows that the question cannot always be answered in the affirmative.

It is to be noted that the thermodynamic definitions (6.28) and (6.30) for absolute temperature and entropy leave an undetermined factor in the former and an undetermined additive constant in the latter. This last constant does not arise when entropy is defined statistically. It is essentially the subject of the third law of thermodynamics, discussed at the end of the preceding chapter.

Caratheodory's axiomatic approach has the agreeable feature of yielding the increasing property of the entropy quite easily. While this is the difficult part for statistics, there may be good reasons for this difficulty. It may very well be that Caratheodory's axiom itself is only statistically valid and thus helps us right at the start over the profound problematics of the Second Law.

To apply Caratheodory's method to two states of a system linked by an irreversible path we must assume them linked in addition by a reversible path. Otherwise the entropy difference between the states cannot be defined, and no further statement can be made about the situation. If such a reversible path exists, however, then the entropy change on the irreversible path can be estimated by computing it along the reversible path by standard methods. We have in fact already used this procedure twice in previous chapters. Now it might of course also occur that these new paths lead us around on one and the same adiabat. No new results would then be obtained. We know, however, that real irreversible processes have a wider range. In the Joule paddle wheel experiment, for instance, irreversible work carries out the function usually accomplished by a supply of heat. There exist therefore irreversible adiabatic processes which permit us to leave a given adiabat. It is now easy to make more and more precise statements concerning these irreversible processes:

(*a*) It is not possible that both states of higher and lower entropy can be reached irreversibly from a given point, for this would violate Caratheodory's axiom.

(*b*) It is not possible to reach higher entropies from one section of an adiabat and lower ones from another, for by an obvious detour Caratheodory's axiom could again be violated.

(*c*) It is not possible to reach higher entropies irreversibly from the adiabats of one body, and lower ones from the adiabats of another, for if we could increase irreversibly the entropy of *A* and decrease the entropy of *B* we could first bring the bodies into thermal contact and carry out a change which decreases the entropy of *A* and increases that of *B*. Thereafter, we could by assumption restore body *A* to its original condition by irreversible means, increasing its entropy. The net result would be an adiabatic irreversible process which would increase the entropy of B and leaves A unchanged. This is contrary to the starting assumption.

All this gives us the theorem:

In all adiabatic processes the entropy is either never decreasing or else never increasing.

A single adiabatic irreversible process will decide the direction left open in the theorem. The heating of a body by friction and the temperature equilibration of two bodies in thermal contact are both processes which increase entropy. Entropy in all adiabatic processes must therefore be never decreasing, yielding

$$dS \geq 0 \tag{6.33}$$

Finally for general irreversible processes we must have that

$$dS \geq \frac{dQ}{T} \tag{6.34}$$

for if we had for any irreversible process $dS < (dQ/T)$ we could first add dQ according to this relation and then remove it reversibly so that the equality sign holds as in (6.31). The result would be an irreversible adiabatic change violating (6.33).

This recapitulation of older methods to establish the Second Law gives results more easily but has the great disadvantage that the concept of entropy as the measure of statistical multiplicity is not brought out. We have therefore preferred to introduce entropy statistically and give a short review of the older viewpoints concerning the subject in a supplementary chapter.

A. RECOMMENDED PROBLEMS

1. A perfect gas for which $\gamma = 1.5$ is used as a working substance in a Carnot cycle operating between 500° and 250° K. In contact with the high-temperature reservoir the gas is expanded from a pressure of 10 atm and a volume of 2 liters to a pressure of 4 atm and a volume of 5 liters. Between what limits of pressure and volume does it operate in contact with the low-temperature reservoir?

2. A *heat pump* is essentially a Carnot refrigerator, for which the heating of the high-temperature reservoir is of primary practical interest. Suppose a building is heated to 25° C by a heat pump operating between the building and an underground spring at 5° C. By how much is the energy input multiplied in this device, as compared to a simple electric heating system which converts the energy input directly into heat?

B. GENERAL PROBLEMS

3. Prove that if we deal with a system whose work term involves only one coordinate (such as $-p\,dV$ for the perfect fluid), then an integrating denominator of $dU - dW$ always exists trivially, without assumption.

4. Apply the theoretical understanding acquired in Problem 3 to find all possible integrating denominators for the following differential forms:

$$x\,dy - y\,dx$$
$$(y-x)\,dx - dy$$
$$(x^2+y^2)\,dy - x\,y\,dx$$

5. Repeat the argument of page 121 for inefficient refrigerators and show that the sign of the inequality (6.08) is in fact not changed as compared to the case of inefficient engines.

6. Three identical bodies of constant thermal capacity are at temperatures $450°$ K, $450°$ K, and $150°$ K. If no work or heat is supplied from the outside, what is the highest temperature to which any of these bodies can be raised by the operation of heat engines?

7. A paramagnetic salt has a Curie constant of $0.01°$ and a specific heat at constant field of 10^6 ergs/cm³ deg. To what temperature can we bring it by adiabatic demagnetization from an applied field of 10^4 oe to zero, starting at $4.2°$ K?

8. The Otto cycle approximates the operation of a gasoline engine. It consists of two adiabats and two isochores and an exhaust cycle, as

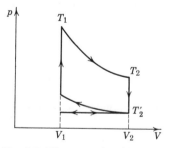

Fig. 6.6. Diagrammatic representation of the Otto cycle for a gas.

shown in the figure. Prove that the efficiency of the air standard Otto cycle has the following two properties:

(*a*) It depends only on the compression ratio V_2/V_1; derive this dependence.

(*b*) It is smaller than the efficiency of a Carnot engine operating between the extreme temperature points of the cycle.

°7

Thermodynamic exploitation of the Second Law; mass transfer problems

Although the inequality sign in the Second Law provides a great deal of the fascination of statistical physics, the situation is entirely different when it comes to applications. In bread-and-butter work an equality is worth very much more than an inequality. We shall review briefly in this chapter how this equality is employed to analyze the behavior of physical and chemical systems.

7-1. Legendre transformations and thermodynamic potentials

The second law of thermodynamics with equality sign is a total differential equation which can be put into a variety of forms. Already in our past work we have encountered some of these forms, namely (5.36), (5.38), (5.54), and (5.55). Many other forms have been written down, together with a catalogue of names and definitions. These modified forms are based on a type of transformation invented by Legendre in mechanics, which interchanges the role of a force and an associated coordinate; they are known as Legendre transformations.

The principle of Legendre transformations in thermodynamics is quite straightforward. Suppose we deal with the Second Law in the form

$$dF = x_1 \, dy_1 + x_2 \, dy_2 \qquad (7.01)$$

then an equivalent and just as simple form reads

$$d(F - x_1 y_1) = - y_1 \, dx_1 + x_2 \, dy_2 \tag{7.02}$$

This new form is particularly convenient if the law is applied to an experiment in which x_1 is to be kept constant. Since we want to make use of transformations of this type, we are placing the variables x_i and y_i into a particularly close relationship. We might call them conjugate. There is, however, a certain arbitrariness in the association because it depends on the differential which is isolated on the left-hand side of (7.02). Contemporary custom is to isolate energy-like expressions. Then x_i and y_i are like a coordinate and its conjugate force, and we shall use the term *conjugate variables* for this pair when the context eliminates possible confusion. Older usage is to isolate an entropy-like quantity on the left. This usage should not be entirely abandoned because it leads often to simpler results than the more modern convention.

Let us start out with the Second Law in its classical form (5.38)

$$dU = T \, dS + \sum_s X_s \, dx_s \tag{7.03}$$

Its most obvious shortcoming is that it is suitable for the study of adiabatic rather than isothermal changes. To find a form adapted to the latter purpose we need only follow the derivation of (5.38) backward and return to the form (5.36). This return is accomplished by the *Massieu potential J*, which is defined as the right-hand side of (5.09)

$$J = S - \frac{U}{T} \tag{7.04}$$

It is, by (5.09), directly connected with the canonical system partition function F. In terms of J, the Second Law (7.03) takes the form

$$dJ = \frac{U}{T^2} dT - \frac{1}{T} \sum_s X_s \, dx_s \tag{7.05}$$

This reads for a perfect fluid

$$dJ = \frac{U}{T^2} dT + \frac{p}{T} dV \tag{7.06}$$

Useful relationships are obtained from a total differential equation such as (7.06) by the device of cross differentiation: the derivative with respect to V of the coefficient of dT must equal the derivative with respect to T of the coefficient of dV. We shall make no attempt here to make a systematic list of such relations. Some of them will be found in the problem section. However, a formula of exceptional importance is obtained in this way from (7.06), namely,

$$\left(\frac{\partial U}{\partial V}\right)_T = -p + T\left(\frac{\partial p}{\partial T}\right)_V \tag{7.07}$$

The usefulness of this relation can be gauged by the student if he remembers that for perfect gases both sides of (7.07) vanish. The relation is thus particularly useful in exploring the properties of imperfect gases. In such cases it is an advantage that the right-hand side involves the equation of state only. Information about the energy of a system can thus be obtained from the equation of state. Some problems at the end of the chapter will follow up this aspect of (7.07).

There is a second way, in widespread use, which transforms (7.03) into a form suitable for isothermal processes. If the *Helmholtz free energy A* (often called simply "free energy") is introduced through the definition

$$A = U - TS \tag{7.08}$$

then the Second Law (7.03) takes the form

$$dA = -S\,dT + \sum_s X_s\,dx_s \tag{7.09}$$

In (5.54) and (5.55) we went through an elaborate analysis to separate out the part of the isothermal work which increases the internal energy from the part which diminishes the entropy. Such an analysis is often neither feasible nor desirable. Equation (7.09) simply states that isothermal work increases the free energy; this is sufficient for many purposes. In addition to this advantage A has the merit of being closely associated with the canonical partition function F. We find from (5.09)

$$F = \exp[-\beta A] \tag{7.10}$$

Now formulas such as (5.34) reduce simply to

$$p = -\left(\frac{\partial A}{\partial V}\right)_T$$

This replaces the more naive conception which would replace A in this formula by U. The "energy" which appears in textbooks on electromagnetic theory is in fact usually the Helmholtz free energy A.

Some thermodynamic potentials are adapted to the special case of perfect fluids when the work term takes the form (1.24). The first of these is the *enthalpy H*, which is defined as

$$H = U + pV \tag{7.11}$$

Equation (7.03) then takes the form

$$dH = T\,dS + V\,dp \tag{7.12}$$

The formula takes into account the fact that most experiments are carried out at constant pressure, while rigid maintenance of size is usually very difficult. From this point of view (7.11) is simply a more adequate definition of energy for many experiments. This idea can be combined with the one previously introduced concerning temperature to lead to the definition of the *Gibbs free energy G*:

$$G = U + pV - TS \tag{7.13}$$

for which the Second Law takes the form

$$dG = -S\,dT + V\,dp \tag{7.14}$$

An obvious implication of (7.14) is that if a fluid is not "clamped" and adiabatically isolated but rather is held at constant pressure and temperature it is not the entropy which has a stationary value, but the Gibbs free energy. A further look into the Second Law in the inequality form (5.43) shows that the stationary value must be a minimum.

The advantage of enthalpy-like functions over the internal energy is not restricted to simple fluids and the case of pressure and volume. It is generally so that, experimentally, we can more easily hold the forces at certain prescribed values than the corresponding coordinates. The theory, on the other hand, prefers to deal with prescribed values of the coordinates. As an example we may take up the case of magnetizable materials dealt with extensively in Chapter 1. The work term is of the form (1.27); the Second Law thus takes the form

$$dU = T\,dS + \mathbf{H} \cdot d\mathbf{M} \tag{7.15}$$

Clearly the use of an energy function responding to a change in the applied field is greatly preferable in such cases. We may call the resultant expression *magnetic enthalpy E*. It is defined by

$$E = U - \mathbf{H} \cdot \mathbf{M} \tag{7.16}$$

and (7.15) reads for it

$$dE = T\,dS - \mathbf{M} \cdot d\mathbf{H} \tag{7.17}$$

We have now the magnetic field as an independent variable. Clearly, we can proceed still further and construct a Gibbs-like function which is stationary at a fixed temperature and fixed magnetic field. The necessary transformations are obvious and need not be elaborated here any further.

A remark should be added here concerning a point which can give a great deal of trouble. Everybody knows that a magnetic dipole of moment M capable of free rotation has a potential energy in a field $\mathbf{H}$ which equals

$$- \mathbf{H} \cdot \mathbf{M}$$

It is therefore natural to add such a term directly to the left-hand side of (7.15). Only a glance at the equations which follow (7.15) is needed to realize that such an addition produces utter nonsense out of (7.16) and (7.17). When you place a phenomenon into the work term you must let the equations themselves tell you what the potential energy is; in other words you will find the work terms partially integrable and you can subtract the integrated part from U. In the present instance (7.16) does this automatically. If you want to check that this is the correct potential energy for the case mentioned, integrate the second term in (7.15) for fixed M and H and variable angle, and assume also that S is independent of that angle. The result will be a potential energy of the form $- \mathbf{M} \cdot \mathbf{H}$, a result which is made explicit in (7.16). Of course equation (7.15) or (7.17) is much more flexible than our result just quoted for the magnetic potential energy. In some cases the potential energy has a different form: a factor $\frac{1}{2}$ is, for instance, likely to appear in the potential energy if the magnetic moment is linearly induced by the field. In other cases, no potential energy can be defined because of the hysteresis phenomenon. All this is taken care of by the work term (1.27). It is therefore preferable, if there is any doubt about a phenomenon, not to count its potential energy in U, but rather to write down the correct work term in dW. The equations will then indicate what, if any, potential energy terms can be added to U in a particular instance.

7-2. Thermodynamics of bulk properties; extensive and intensive variables

A very large part of thermodynamics deals with bulk or volume properties of materials and neglects surface effects. In this type of work it is useful to make the distinction between extensive and intensive variables and to reduce the former to density functions. The physical basis for this viewpoint lies in the fact that size is often a very superficial property of a material, and structure can be discussed apart from size. We can then classify the variables of statistical physics into two classes: variables such as pressure, temperature, and electric and magnetic fields, whose magnitude has intrinsic importance for the sample, and variables like volume, energy, entropy, and magnetic moment, which are proportional to the size of the sample, and for which only the constant of proportionality has structural significance. If we examine the basic equations written thus far, we find that we have neglected this feature entirely, that is, we have always assumed the amount of matter present as fixed. We must amend our past equations before proceeding any further. If we deal with a simple chemical substance this amendment is relatively simple. We may add to equation (7.03) a single term and write

$$dU = T\,dS + \sum_s X_s\,dx_s + \mu\,dN \tag{7.18}$$

Here μ is called the *chemical potential* of the material. It can be written either as

$$\mu = \left(\frac{\partial U}{\partial N}\right)_{S,x_s} \tag{7.19a}$$

or

$$-\frac{\mu}{T} = \left(\frac{\partial S}{\partial N}\right)_{U,x_s} \tag{7.19b}$$

A more complicated situation arises when the system under study contains several substances. We have then to prescribe several numbers $N_1, N_2, \ldots$, giving the number of molecules of the various types. Equation (7.18) becomes then

$$dU = T\,dS + \sum_s X_s\,dx_s + \sum_t \mu_t\,dN_t \tag{7.20}$$

Equation (7.20) has the added interest over (7.18) that it brings back the question of composition, which is of course of prime importance in all chemical studies. The quantities μ_t are called the *chemical potentials* of the respective species. They are expressible as partial derivatives, as in the preceding case of a chemically pure substance:

$$-\frac{\mu_k}{T} = \left(\frac{\partial S}{\partial N_k}\right)_{U,x_s,N_t,t\neq k} \tag{7.21}$$

If we deal with simple fluids there is a specially close relationship between the chemical potentials and the Gibbs free energy. This relationship is based on Euler's identity for homogeneous functions. Euler's identity deals with the following situation. Let

$$f(x, y, z, u, v, w)$$

be a homogeneous function of degree n in the variables x, y, z, that is,

$$f(\lambda x, \lambda y, \lambda z, u, v, w) = \lambda^n f(x, y, z, u, v, w) \tag{7.22}$$

then we have the identity

$$n f(x, y, z, u, v, w) = x\frac{\partial f}{\partial x} + y\frac{\partial f}{\partial y} + z\frac{\partial f}{\partial z} \tag{7.23}$$

If we take equation (7.20) for the case of perfect fluids only and write it as

$$dU = T \, dS - p \, dV + \sum_t \mu_t \, dN_t \qquad (7.24)$$

then we have a total differential equation for U in terms of S, V, N_1, N_2, ... All these quantities are extensive if we deal with bulk properties, that is, U is homogeneous of degree 1 in these variables. Equation (7.23) is therefore applicable and we find

$$U = T \, S - p \, V + \sum_t \mu_t \, N_t \qquad (7.25)$$

or in terms of the definition (7.13)

$$G = \sum_t \mu_t \, N_t \qquad (7.26)$$

This is the so-called *Gibbs-Duhem relation* which appears here as an identity valid for a potential defined in another way, and useful only in the discussion of bulk properties of fluids. However, since this conventional type of procedure is barred to us in many interesting applications, it is often preferable to adopt (7.26) as the defining equation of G. In the case of a chemically simply substance this definition amounts simply to saying that

$$\mu = g = \frac{G}{N} \qquad (7.27)$$

that is, the chemical potential is the Gibbs free energy per molecule.

The preceding analysis suggests that the chemical potentials μ_t stand in a closer relation to the Gibbs free energy than to the entropy. Indeed equation (7.24) can be combined with (7.13) to yield

$$dG = - S \, dT + V \, dp + \sum_t \mu_t \, dN_t \qquad (7.28)$$

Thus, the quantities μ_t control the change of the Gibbs free energy with composition at fixed pressure and temperature:

$$\mu_i = \left(\frac{\partial G}{\partial N_i} \right)_{T, p, N_t, t \neq i} \qquad (7.29)$$

Obviously, the Gibbs-Duhem identity can also be derived from (7.28) with the help of the Euler identity (7.23).

The importance of the quantities μ_t can be sensed from their alternative appearance in either (7.20) or (7.28). Both these equations are forms of the Second Law, amended to take care of a possible variation in the quantity of matter present. The front part of the equation is not the same. However, the factor μ_t in front of dN_t has remained. Thus the indication is that, whenever a thermodynamic potential is considered as a function of the amount of the various chemical substances present, the multipliers entering are always the same, namely, the chemical potentials μ_t or, in the case of entropy-like potentials, the combination $\beta \mu_t$.

The use of the Euler relation is not restricted to simple fluids although this is its primary field of usefulness. To discuss a more involved example we might take a magnetizable material consisting of a single substance. We may then specialize (7.18) with the help of (1.24) and (1.27) to read

$$dU = T\, dS - p\, dV + \mathbf{H} \cdot d\mathbf{M} + \mu\, dN \tag{7.30}$$

It so happens that the magnetic moment $\mathbf{M}$ is also an extensive quantity, and that $\mathbf{H}$ is intensive. The Euler relation (7.23) is therefore applicable. It yields

$$\mu = \frac{U + p V - \mathbf{H} \cdot \mathbf{M} - T S}{N} \tag{7.31}$$

It is in the spirit of this book to define G in this case by the Gibbs-Duhem relation (7.26), and not by (7.13). We would therefore call the numerator on the right of (7.31) the Gibbs free energy of this system.

One can always convert extensive into intensive quantities by constructing density-like functions. These are formed by dividing each of them by one chosen one among them. One gets thus quantities per gram, mole, atom, unit volume, etc. Habits are often conflicting here. Energies are usually given per gram or mole, while magnetic moment is given per unit volume; it is then called magnetization. As far as possible we shall avoid such conflicting definitions. In equation (7.31) the natural unit of reduction would appear to be the molecule. We could thus define reduced quantities as follows:

$$\text{Energy per molecule} = \frac{U}{N} = u \tag{7.32a}$$

$$\text{Volume per molecule} = \frac{V}{N} = v \tag{7.32b}$$

$$\text{Magnetic moment per molecule} = \frac{M}{N} = m \tag{7.32c}$$

$$\text{Entropy per molecule} = \frac{S}{N} = s \tag{7.32d}$$

Equation (7.31) then becomes a relation between intensive quantities:

$$\mu = u + p v - \mathbf{H} \cdot \mathbf{m} - T s \tag{7.33}$$

7-3. Equilibrium of two phases; equation of Clausius and Clapeyron

In order to gain a good understanding of the thermodynamics of mass transfer we shall begin this section by applying bulk thermodynamics to the simplest of all mass transfer problems, the equilibrium between two phases of a pure substance. Thereafter, we shall generalize the method to multiphase multicomponent systems. This will culminate in Gibbs' phase rule. Finally, in Section 7-5 we shall review again the original problem of equilibrium between two phases, going beyond the bulk approximation. The result so obtained should shed light on the question of phase equilibrium in general.

For the first simple formulation of two-phase equilibrium we imagine two phases of a chemically pure substance adiabatically isolated and enclosed in a container of fixed size. We shall talk about the two phases as if they were water and steam, but we shall make no specific assumption which would limit us to this particular system. Because of the assumptions made, the state of this system is covered by the powerful theorem of Section 5-2, according to which a "clamped" adiabatically isolated system will have a unique equilibrium state in which the entropy is a maximum. An obvious consequence of this will be that each phase by itself must be in a state of maximum entropy. In addition, however, the entropy must also be stationary with respect to the extra degree of freedom present, namely, the possibility of the transfer of a water molecule from the interior of the water phase to the interior of the steam phase. Denote by 1 the water and by 2 the steam phase. We then apply (7.18) or (7.24) to each phase separately, writing

$$dS_1 = \frac{1}{T_1} dU_1 + \frac{p_1}{T_1} dV_1 - \frac{\mu_1}{T_1} dN_1 \qquad (7.34)$$

$$dS_2 = \frac{1}{T_2} dU_2 + \frac{p_2}{T_2} dV_2 - \frac{\mu_2}{T_2} dN_2 \qquad (7.35)$$

Our maximum condition on the entropy states that

$$d(S_1 + S_2) = 0 \qquad (7.36)$$

under the side conditions

$$U_1 + U_2 = \text{constant} \qquad (7.37a)$$

$$V_1 + V_2 = \text{constant} \qquad (7.37b)$$

$$N_1 + N_2 = \text{constant} \qquad (7.37c)$$

The condition (7.37c) alone is new, indicating that the entropy must also be a maximum with respect to exchange of matter between the two phases. The result is

$$T_1 = T_2 \tag{7.38a}$$

$$p_1 = p_2 \tag{7.38b}$$

$$\mu_1 = \mu_2 \tag{7.38c}$$

Again, the first two equations (7.38) are well known equilibrium conditions while (7.38c) is new. It states that the chemical potential of the water in the two phases must be equal.

The handling of this simple problem may appear somewhat artificial. The balancing of two phases is very difficult in a rigid container, with enforced absence of heat flow through the walls of the container. It would be much more natural to think of the experiment as taking place in a heat bath which fixes the temperature, and with suitable devices attached regulating the pressure. Under these conditions the total Gibbs free energy should be a minimum, and (7.28) should be applied. Applying it to each phase, we get

$$dG_1 = - S_1 \, dT - V_1 \, dp + \mu_1 \, dN_1 \tag{7.39a}$$

$$dG_2 = - S_2 \, dT - V_2 \, dp + \mu_2 \, dN_2 \tag{7.39b}$$

We now demand that

$$d(G_1 + G_2) = 0 \tag{7.40}$$

under the side conditions

$$dT = 0 \tag{7.41a}$$

$$dp = 0 \tag{7.41b}$$

$$N_1 + N_2 = \text{constant} \tag{7.41c}$$

The result is the same, namely,

$$\mu_1 = \mu_2 \tag{7.42}$$

The fact that μ remains the multiplier of dN thus leads to this stable result: regardless of the side conditions imposed, water and steam in equilibrium across a plane surface adjust to equality in pressure and temperature, and to equality of the chemical potential. This last equality arises specifically because water molecules are free to evaporate and condense from one phase to the other. If the reaction is stopped, for instance, by insertion of an impenetrable film at the interface, then equation (7.42) need not be obeyed.

Let us list some of the alternative forms which can be given to (7.42). According to (7.27), the relation reads

$$g_1 = g_2 \tag{7.43a}$$

where g is the Gibbs free energy per molecule.

To start with, (7.43a) can be used directly, in conjunction with the definitions (7.11) and (7.13). It reads then

$$T = \frac{H_2 - H_1}{S_2 - S_1} \tag{7.44}$$

The numerator on the right is the latent heat of the phase transformation. Formula (7.44) then conveys the notion that if one understands the heat of transformation and the entropy change independently, perhaps from direct theoretical considerations, one can predict the transformation temperature. In this sense, (7.44) is not limited to phase transformations, but is also useful for a gradual structural change which takes place over a finite temperature range. The usefulness of (7.44) is usually qualitative, but the indication which it furnishes can be of great value for structural understanding.

A more precise result is obtained if (7.43a) is differentiated, yielding

$$dg_1 = dg_2 \tag{7.43b}$$

If (7.14) is now employed on each phase one gets

$$- (s_1 - s_2)\, dT + (v_1 - v_2)\, dp = 0 \tag{7.45}$$

Substitution of (7.44) into this yields finally

$$\frac{dT}{dp} = \frac{T(V_2 - V_1)}{H_2 - H_1} \tag{7.46}$$

This is the *equation of Clausius and Clapeyron*. It is an ordinary differential equation for the transformation temperature as a function of the vapor pressure. The denominator $H_2 - H_1$ is the heat of transformation. Equations (7.44) and (7.46) are both constructed in such a way that H, V, or S can be per gram, mole, or molecule, for the specificity drops out of the equation. Equation (7.46) has been derived earlier in a different way, as equation (6.19).

The Clausius-Clapeyron equation is the first instance of a reduction in the "degrees of freedom" of a system by the presence of more than one phase. The equation predicts the pressure if the temperature is given or vice versa. The original possibility of picking pressure and temperature independently has therefore been lost by our assumption that two phases of the same substance are simultaneously present. In the language of Gibbs, the presence of two phases has reduced the number of degrees of freedom of the system from 2 to 1.

7-4. Equilibrium of multiphase, multicomponent systems; Gibbs' phase rule

In order to develop the notion of a multicomponent system we shall start with the case of a homogeneous fluid (single "phase"), but one which consists of a number of compounds. Let us assume that, within the fluid, chemical reactions can take place which increase the amount of some of the compounds and diminish that of others. The equilibrium of the system is given by setting expression (7.28) equal to zero for all possible variations of the parameters. For us, the essential part of the equation is the sum at the end, namely,

$$\mu_1 \, dN_1 + \mu_2 \, dN_2 + \ldots + \mu_T \, dN_T = 0 \tag{7.47}$$

This is the answer to our question, but the answer will acquire meaning only if we are aware of the chemical reactions which can take place freely within the mixture. Let us assume as an example that we have a mixture of gases and that constituent 1 is I, and constituent 2 is I_2. Thus we have the chemical equation

$$2I - I_2 = 0$$

or

$$2 \, dN_1 = - \, dN_2$$

as a possible chemical reaction with all other dN's zero. Hence we get

$$2 \, \mu_I - \mu_{I_2} = 0 \tag{7.48}$$

as the equilibrium condition between monomolecular and bimolecular iodine. Similarly, if the temperature is sufficiently high to let the water reaction proceed,

$$2H_2 + O_2 - 2H_2O = 0$$

we have the possibility

$$dN_3 = dN_{O_2} = dx$$
$$dN_4 = dN_{H_2} = 2 \, dx$$
$$dN_5 = dN_{H_2O} = - \, 2 \, dx$$

and hence

$$2 \, \mu_{H_2} + \mu_{O_2} - 2 \, \mu_{H_2O} = 0 \tag{7.49}$$

One should not forget that the relation will not hold if the reaction time is slow compared with the time of measurement, or if the reaction is completely inhibited.

The general law of formation of these equilibrium equations follows obviously from the above examples. One writes down the chemical equations for the reactions taking place freely in the mixture. If one replaces in those equations each chemical symbol by the chemical potential for the substance, one gets the equilibrium condition in question. The equations thus obtained involve the relative concentrations of the various substances together with the pressure and temperature, but the total mass is of course not involved; this is to be expected for physical reasons.

The equilibrium equations can be used to express the number of molecules of certain compounds in terms of the number of molecules of certain others which we may call the independent constituents. In the case just discussed I_2, O_2, and H_2 would presumably be called independent while I and H_2O would be dependent on them with the help of two equilibrium equations. In general, the number of independent constituents is never less than the number of different elements present. It may be more, however, if compounds are present which do not react under given conditions. The most important consequence of equations such as (7.48) and (7.49) is the laws of mass action. These laws will be discussed in Chapter 11 and Chapter 17.

We may now put several phases together, each one containing several substances. Such a contact will generally produce changes involving mutual diffusion of products and quantitative adjustments. These processes will lower the total Gibbs free energy because they are irreversible. We talk in such cases sometimes about evaporation, sometimes about solution, sometimes about admixture, but there is no essential difference between the three cases. This will proceed as long as the Gibbs free energy can decrease. Suppose, for instance, we have air over water with oxygen present in both phases. The change of the Gibbs free energy then contains a term

$$dG = \{\mu_{O_2}^{(1)} - \mu_{O_2}^{(2)}\}\, dN_{O_2} \qquad (7.50)$$

between two phases 1 and 2. This expression must vanish in equilibrium. Therefore equilibrium will be reached when the chemical potential of oxygen is the same in the gaseous and the liquid phase. By the same reasoning we find that the chemical potential μ must be the same for each substance in all phases. In addition, expression (7.47) has to vanish in each phase; this leaves only the independent constituents for explicit consideration.

Suppose we apply this rule to a system composed of water, iodine crystals, and a gas mixture in which the two reactions (7.48) and (7.49) proceed freely. We would then get the following equations for the independent constituents:

$$\mu_{O_2}^{(1)} = \mu_{O_2}^{(2)} = \mu_{O_2}^{(3)}$$
$$\mu_{H_2}^{(1)} = \mu_{H_2}^{(2)} = \mu_{H_2}^{(3)}$$
$$\mu_{I_2}^{(1)} = \mu_{I_2}^{(2)} = \mu_{I_2}^{(3)}$$

This gives us then six conditions. In general, the number of conditions for coexistence of f phases is $n\,(f-1)$, where n is the number of independent constituents. And since these equations are to determine $f(n-1) + 2$ variables ($n-1$ relative concentrations in f phases and in addition pressure and temperature), we can pick the difference of these two numbers, namely,

$$v = n + 2 - f \tag{7.51}$$

variables independently.

Equation (7.51) is known as *Gibbs' phase rule*. It predicts the number of degrees of freedom or variability of the system. In addition one can of course vary the total mass of each phase without disturbing the equilibrium, so that

$$v' = n + 2 \tag{7.52}$$

has perhaps more right to be called the degree of variability (one picks all independent constituents and temperature and volume). But equation (7.51) refers to the internal constitution of the phases and the number of ways in which it can be varied, and this type of variability drops as the number of phases increases. In fact there is a limit to the number of phases one can have in equilibrium since v must be a positive number or zero. Equation (7.51) thus implies that

$$f \leq n + 2 \tag{7.53}$$

The applications of (7.51) are numerous. In the vapor pressure equation (7.46) the pressure is a function of temperature because only

$$1 + 2 - 2 = 1$$

variable can be picked independently. Three phases of the same substance in equilibrium demand a given pressure and temperature (triple point):

$$1 + 2 - 3 = 0$$

But as soon as a foreign gas is present this is no longer the case: the temperature for which the three phases of water are in equilibrium in the presence of air is a function of pressure (one may here treat air as a single substance, because all but one of the additional degrees of variability which seem to result from the high number of components in air are taken up by the internal constitution of the air itself).

Similarly there is only one temperature for a given pressure where water, ice, and crystalline salt are in equilibrium (eutectic point) because

$$2 + 2 - 3 = 1$$

7-5. Refined study of the two-phase equilibrium; vapor pressure of small drops

It is a very important question whether Gibbs' phase rule and the equations which lead to it are absolute laws of nature or convenient rules which are generally but not universally valid. The answer will be found to

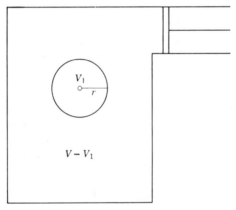

Fig. 7.1. Phase equilibrium of water and steam when the former is a sphere suspended in the latter.

be the latter. Gibbs' phase rule is based on the distinction of extensive and intensive parameters worked out in Section 7-2 and thus takes only bulk properties into account. Phenomena involving interfaces obliterate this distinction, and thus must be able to break down Gibbs' phase rule at least partially. Rather than sticking to this general abstract statement we shall take up such a case in detail. We return to our starting problem of water in equilibrium with steam, assuming the water phase to be a sphere held together by its surface tension. This is a valid assumption for small droplets of fog in the atmosphere. It is also correct for larger masses of water in a space ship. It is known that in the absence of gravity water floats as a sphere through the surrounding atmosphere. This modified problem offers an interesting challenge in that some of the simplifications used previously are no longer valid. The problem must therefore be thought through again from first principles.

Let us therefore assume that we are in a space ship, observing a ball of water floating in steam. The container holding the two-phase system shall have volume V. The volume of the sphere of water shall be denoted by V_1 or by $\frac{4}{3} \pi r^3$, according to circumstances. It is assumed not to touch the walls of the container, as shown in Fig. 7.1.

The phenomenon we wish to study is the phenomenon of surface tension, and we should perhaps emphasize *only* surface tension. We shall therefore continue to assume that both phases 1 and 2 remain perfect fluids away from this surface. In accordance with the warning in Section 7-1 the new effect will be brought in as an added term in the work only. In other words we write

$$dU_1 = T_1 \, dS_1 - p_1 \, dV_1 \tag{7.54a}$$

$$dU_2 = T_2 \, dS_2 - p_2 \, d(V - V_1) \tag{7.54b}$$

and

$$dU = dU_1 + dU_2 + \sigma \, dA$$

Here σ is the surface tension of the sphere, and A its surface area. We may as well suppress the superfluous distinction of two temperatures and write

$$dU = T \, dS + (p_2 - p_1) \, dV_1 - p_2 \, dV + \sigma \, dA \tag{7.55}$$

Both

$$V_1 = \tfrac{4}{3} \pi r^3 \tag{7.56}$$

and

$$A = 4 \pi r^2 \tag{7.57}$$

refer to the same internal coordinate r not subject to outside manipulation. The entropy must therefore be maximized with respect to it, which means that the coefficient of dr in (7.55) must vanish. This yields the well known relation

$$p_1 - p_2 = \frac{2 \sigma}{r} \tag{7.58}$$

and leaves (7.55) to read simply

$$dU = T \, dS - p_2 \, dV$$

It is seen that we have regained equation (5.45) with p_2 as the effective pressure. This is an obviously sensible result because it is phase 2 which is in contact with the walls of the container, as shown in Fig. 7.1.

So we have made no mistake. Nevertheless the calculation is a disappointment at this stage because we have obtained only the mechanical condition (7.58), but not the looked-for modification of the Clausius-Clapeyron equation. The reason is the one which was taken up in Section 7-2. Equation (5.45) does not take into account the possibility that the amount of matter in either phase may vary. In other words equations (7.54) must be amended as shown in (7.18), yielding

$$dU_1 = T\,dS_1 - p_1\,dV_1 + \mu_1\,dN_1 \qquad (7.59a)$$

$$dU_2 = T\,dS_2 - p_2\,d(V-V_1) + \mu_2\,dN_2 \qquad (7.59b)$$

Together with the surface tension term this yields

$$dU = T\,dS + (p_2-p_1)\,dV_1 - p_2\,dV + \mu_1(T,p_1)\,dN_1 \\ + \mu_2(T,p_2)\,dN_2 + \sigma\,dA \qquad (7.60)$$

Equation (7.19) shows specifically that μ_1 and μ_2 are the derivatives of U with respect to N when the two respective volumes, and hence r, are kept fixed. (In an offhand argument it might be expected that N_1 and V_1 are essentially the same variable.) Hence the mechanical condition (7.58) is maintained, and the remaining condition for evaporation

$$dN_1 = -\,dN_2$$

becomes, when applied to (7.60),

$$\mu_1(T, p_1) = \mu_2(T, p_2)$$

which means also from (7.27)

$$g_1(T, p_1) = g_2(T, p_2) \qquad (7.61)$$

The relation is identical with (7.43) except for the difference in pressure between the two phases. We proceed as previously, differentiating and using (7.14). We get

$$v_1\,dp_1 - v_2\,dp_2 = (s_1- s_2)\,dT$$

or, upon eliminating p_1 by (7.58),

$$(v_1- v_2)\,dp_2 + v_1\,d\left(\frac{2\sigma}{r}\right) = (s_1- s_2)\,dT \qquad (7.62)$$

This is the expected modification of the Clausius-Clapeyron relation (7.45), into which it goes over in the limit $r = \infty$. For finite r, the vapor pressure p_2 is a function of the two variables T and r. By equation (7.62) the original degree of variability of the liquid–vapor equilibrium is restored back to two, namely, pressure and temperature. It is only a matter of adjusting the third variable r, that is, the size of the droplets, to the other two. However, a common-sense examination of (7.62) shows that the degree of variability in p at fixed T is so small that Gibbs' phase rule remains practically valid. Let us examine (7.62) as an equation yielding a relationship between vapor pressure and radius at fixed temperature. To simplify analysis let us also assume that

$$v_1 \ll v_2$$

and that the variability of σ can be neglected. Let us assume furthermore that the steam phase is a perfect gas, obeying

$$v_2 = \frac{k T}{p_2}$$

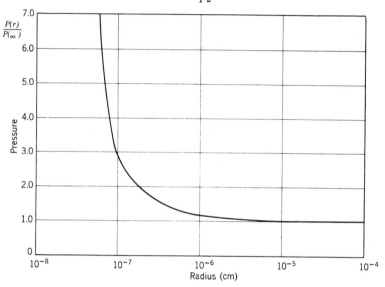

Fig. 7.2. Vapor pressure of water for small drops. Percentage increase in vapor pressure versus drop radius at room temperature.

We find then from (7.62)

$$k T \frac{dp_2}{p_2} + \frac{2 v_1 \sigma}{r^2} dr = 0$$

which integrates out to yield

$$\ln \frac{p_2(r)}{p_2(\infty)} = \frac{2 v_1 \sigma}{k T r} \tag{7.63a}$$

or in more usual units

$$\ln \frac{p_2(r)}{p_2(\infty)} = \frac{2 \mathcal{M} \sigma}{\rho_1 R T} \frac{1}{r} \tag{7.63b}$$

Here ρ_1 is the liquid density and M the molecular weight of the gaseous phase. Since R is of the order of a joule per mole per degree and σ only of the order of ergs per square centimeter, a drop has to be very small to yield a measurable increase in the vapor pressure. Figure 7.2 shows the situation for water at room temperature.

The detailed discussions of phase and chemical equilibrium just carried through should convince the reader that these equilibria with respect to transfer of matter do not introduce new physical principles, but are simply applications of the Second Law as formulated in Chapters 5 or 6. However, the circumstances under which it is applied are rather different from the ones envisaged on that occasion, for we have been feeding in information about matter existing in different phases and different chemical compositions. We were led thereby to a formal restatement of the Second Law which is suitable for mass transfer problems.

A. RECOMMENDED PROBLEMS

1. Given the density of water and ice at $0°$ C as 1.000 and 0.917 gm/cm^3, respectively, and the heat of fusion of ice as 80 cal/gm, calculate the melting point of ice at 100 atm.

2. Show that for any magnetizable material the heat capacities at constant field H and at constant moment M are linked by the relation

$$C_H - C_M = - T \left(\frac{\partial M}{\partial T}\right)_H \left(\frac{\partial H}{\partial T}\right)_M$$

3. What are the possible forms of the equation of state of a perfect fluid

$$F(p, V, T) = 0$$

if one is given that $V = V(T)$ only?

4. Calculate the initial rate of change of the boiling point of water with altitude at the earth's surface. The heat of vaporization of water is 538.7 cal/gm, and steam may be assumed a perfect gas. The average surface temperature of the earth may be taken as $300°$ K.

5. Prove Euler's identity (7.23) for homogeneous functions.

B. GENERAL PROBLEMS

6. Prove the following relation for a perfect fluid:

$$\left(\frac{\partial H}{\partial p}\right)_V = C_v \left(\frac{\partial T}{\partial p}\right)_V + V$$

7. Prove the following relation for a perfect fluid:

$$\left(\frac{\partial U}{\partial p}\right)_T = -T\left(\frac{\partial V}{\partial T}\right)_p - p\left(\frac{\partial V}{\partial p}\right)_T$$

8. Show that the Second Law for a perfect fluid can be put into the form

$$T\,dS = C_v\left(\frac{\partial T}{\partial p}\right)_V dp + C_p\left(\frac{\partial T}{\partial V}\right)_p dV$$

Compare this with the relation

$$dT = \left(\frac{\partial T}{\partial p}\right)_V dp + \left(\frac{\partial T}{\partial V}\right)_p dV$$

for the particular case of a body whose mechanical and thermal properties are not coupled.

9. Show that for a paramagnetic material obeying Curie's law we have

$$C_H = C_M(T) + \frac{V\,C\,H^2}{T^2}$$

The last C is the Curie constant.

10. The melting point of lead at atmospheric pressure is $327.0°$ C; its density decreases from 11.01 to 10.65 gm/cm³ and the latent heat of fusion is 24.5 j/gm. Find its melting point at a pressure of 400 atm.

11. The bulk modulus of a material is most easily determined from the velocity of sound within it, but the value so obtained is the adiabatic value. Establish the widely used correction formula

$$\frac{1}{B_T} = \frac{1}{B_S} + \frac{\alpha^2\,T}{\rho\,c_p}$$

Here c_p is the specific heat at constant pressure, α the volume expansion coefficient, and ρ the density.

12. At $300°$ K and atmospheric pressure water and mercury have the following properties:

		Water	Mercury
$\mathcal{M}$	Molecular weight	18.01	200.6
c_p	Specific heat at constant pressure, j/deg gm	4.179	0.1389
α	Volume expansion coefficient, deg⁻¹	2.81×10^{-4}	1.82×10^{-4}
u	Velocity of sound, m/sec	1503	1448

Calculate the specific heat at constant volume for the two materials, and give their molar heat capacity in units of R. Comment on the results.

13. The shape of the vapor pressure curve of a material comes out approximately right if the Clausius-Clapeyron equation is integrated under the following three assumptions:

(a) The heat of vaporization is a constant.

(b) The volume of the condensed phase is negligible compared to that of the vapor.

(c) The vapor is a perfect gas.

Write the equation for the vapor pressure as a function of temperature which results from this. Apply it to estimate the vapor pressure of water at $90°$ C, assuming the value at $100°$ C.

14. Radiation in thermal equilibrium can be viewed as a gas of photons which obeys the following two relations:

$$U = V u(T)$$
$$p = \tfrac{1}{3} u$$

Derive from these two relations the Stefan–Boltzmann law

$$u(T) = \text{constant} \cdot T^4$$

15. Defining the surface tension σ of a liquid through the work term as

$$dW = \sigma \, dA$$

where A is the magnitude of the surface, show that if σ is a function of temperature, but not of pressure, the internal energy U of the liquid may be given the form

$$U = U_0 + \left(\sigma - T \frac{d\sigma}{dT} \right) A$$

Here U_0 is the bulk value of the internal energy as defined for a perfect fluid.

16. The surface tension of water is approximately independent of pressure, and linear in the temperature, with a law of the form

$$\sigma = 0.204 \, (641 - T) \, \text{dyne/cm}$$

($641°$ K is the critical temperature of water). Deduce from this that the part of the internal energy which depends on the surface is proportional to the surface and independent of temperature. Find the constant of proportionality.

8

The grand ensemble; classical statistics of independent particles

We saw at the end of Chapter 7 that the possibility of transferring atoms and molecules between compounds and phases leads to a reformulation of thermodynamics. This reformulation contains no ideas which are physically new, but presents certain formal aspects which facilitate consideration of such mass transfers. A similar situation prevails in statistics. There exists a special form of statistics, the statistics of the grand ensemble. It also contains nothing physically new, but facilitates the explicit consideration of mass numbers as part of a statistical problem. Concurrently with this, we shall have a closer look in the next two chapters at the statistics of independent particles. We shall find that there are different forms of particle statistics. Even though these forms are all contained in the general theory of Chapter 4 they have their own peculiarities which need separate investigation.

8-1. Statistics of the grand ensemble

The reformulation of statistical mechanics for problems involving transfer of matter is due to Gibbs. The device he introduced is the grand canonical ensemble. In the canonical ensemble we dealt with a number of systems of identical size and composition, capable of exchanging energy. In the grand canonical ensemble the restriction to identity in size is dropped. Instead we permit the systems to exchange freely their constituent atoms or molecules, as well as their energy. We arrive thus at an ensemble of systems which differ in size and composition as well as in energy. The hope is of course that the systems thus liberated will cluster with respect to these variables about a mean, and that the fluctuation about the mean will be small. The first half of this assumption will be verified in this chapter. As for the second half the situation is more delicate than before. Systems of different size and composition also have different quantum states; this renders comparison difficult. It should be remembered in this connection that the evaporation problem studied in the preceding chapter implies density fluctuations from one spot to the other which are not small. We should therefore not ask too much from the new formalism.

It is seen from this that in the statistics of the grand ensemble energy and particle numbers are simultaneously treated as statistical elements. The mathematics needed for this purpose is not new. We only have to combine energy statistics, as developed in (4.01)–(4.11), with particle statistics, as developed in (2.03)–(2.11). In either case the basic idea was that the union of two uncoupled systems is again a system. This is again the basic idea here. The result will be a set of formulas which are slightly more cumbersome, but in no way more difficult, than those previously obtained.

In accordance with these introductory remarks we consider an ensemble of systems which vary slightly in energy and composition, and which are capable of exchanging energy and constituent atoms or molecules. Let T be the number of independent constituents. Select two particular systems out of the ensemble for closer description. In the first system let $N_1, N_2, \ldots N_T$ be the number of constituent atoms or molecules of type $1, 2, \ldots T$, and let U_i be the energy of its quantum state i. Of course U_i is in general a function of the numbers N_t. Call these same quantities for the second system $N_1', N_2', \ldots N_T', U_i'$. Then we have a function

$$p(N_1, N_2, \ldots N_T, U_i)$$

which is the probability that the particular values $N_1, N_2, \ldots N_T$ are realized for the first system, and that the system is in the quantum state i. According to the quasi-ergodic hypothesis the latter probability depends only on the energy U_i of this state. Similarly we have

$$p' = p'(N_1', N_2', N_3', \ldots N_T', U_i')$$

If we call p'' the probability that the combination of the two systems is in the particular conditions stated, and if we assert as previously that the probabilities are independent, we can write the equation

$$p''(N_1 + N_1', N_2 + N_2', \ldots N_T + N_T', U_i + U_i')$$
$$= p(N_1, N_2, \ldots, N_T, U_i) \cdot p'(N_1', N_2', \ldots N_T', U_i') \quad (8.01)$$

which is again a functional equation similar to, but more complicated than (2.03) and (4.04). In the argument which follows, it is necessary that we adopt the point of view that the p's are continuous functions of U_i, as we assumed before in order to derive (4.09). There is no need to repeat the argument of Chapter 4 here, as it is not new. If one wishes to be very precise, one says that p is the probability for a quantum state with energy U_i, if there is one. Then we proceed for N_t as in Chapter 2 and for U_i as in Chapter 4. In other words, we write

$$p''(N_1 + N_1', \ldots N_t + N_t' - 1, \ldots N_T + N_T', U_i + U_i')$$
$$= p(N_1, N_2, \ldots N_t - 1, \ldots N_T, U_i) \cdot p'(N_1', \ldots N_t', \ldots N_T', U_i')$$
$$= p(N_1, \ldots N_T, U_i) \cdot p'(N_1', N_2', \ldots N_t' - 1, \ldots N_T', U_i')$$

and

$$\frac{\partial p''(N_1 + N_1', \ldots N_T + N_T', U_i + U_i')}{\partial U}$$

$$= \frac{\partial p(N_1, \ldots N_T, U_i)}{\partial U_i} p'(N_1', \ldots N_T', U_i')$$

$$= p(N_1, N_2, \ldots N_T, U_i) \frac{\partial p'(N_1', \ldots N_T', U_i')}{\partial U_i'}$$

Identifying each time the right-hand sides, we obtain

$$\frac{p(N_1, \ldots N_t, \ldots N_T, U_i)}{p(N_1, \ldots N_t - 1, \ldots N_T, U_i)} = \frac{p'(N_1' \ldots N_t' \ldots N_T', U_i')}{p'(N_1' \ldots N_t' - 1, \ldots N_T', U_i')} = \exp \beta \mu_t \quad (8.02)$$

and

$$\frac{\partial \ln p(N_1, \ldots N_T, U_i)}{\partial U_i} = \frac{\partial \ln p'(N_1', \ldots N_T', U_i')}{\partial U_i'} = -\beta \quad (8.03)$$

All the constants defined by (8.02) and (8.03) are common to the whole ensemble. Since the μ_t's result from an equilibrium assumed with respect to a transfer of particles they must have a close relation to the chemical potentials defined in (7.20).

From (8.02) and (8.03) we derive

$$p(N_1, \ldots N_T, U_i) = \frac{1}{\mathscr{F}} \exp \beta[\mu_1 N_1 + \mu_2 N_2 + \ldots + \mu_T N_T - U_i] \quad (8.04)$$

with

$$\mathscr{F} = \sum_{N_1=0}^{\infty} \sum_{N_2=0}^{\infty} \sum_{N_3=0}^{\infty} \cdots \sum_{N_T=0}^{\infty} \sum_i \exp\left[\beta \left(\sum_t \mu_t N_t - U_i\right)\right] \quad (8.05)$$

Expression (8.04) is the Gibbs distribution law for the grand ensemble. The normalizing denominator $\mathscr{F}$, given by (8.05), is called the *grand partition function*. $\mathscr{F}$ is a function of β and of the μ_t's, and also of the "external coordinates" discussed in Chapter 1; it is distinguished from F in that it does not contain the numbers N_t. In order to reveal the nature of $\mathscr{F}$, we shall proceed in analogy with Chapter 5. First we obtain the averages by differentiation, leaving off average marks on the left in accordance with previous practice:

$$U = \left(-\frac{\partial \ln \mathscr{F}}{\partial \beta}\right)_{\beta\mu_t=\text{constant}} \quad (8.06)$$

$$X_s = -\frac{1}{\beta} \frac{\partial \ln \mathscr{F}}{\partial x_s} \quad (8.07)$$

$$N_t = \frac{\partial \ln \mathscr{F}}{\partial(\beta\mu_t)} \quad (8.08)$$

Equation (8.06) is the same as (5.35), and (8.07) is the same as (5.33). Only formula (8.08) for the average number N_t of atoms of type t is new. Following the precedent of Chapter 5, we deduce from this a total differential equation for $\ln \mathscr{F}$; this equation is a generalization of (5.36) and reads

$$d\ln \mathscr{F} = -U \, d\beta - \beta \sum_s X_s \, dx_s + \sum_t N_t \, d(\beta\mu_t) \quad (8.09)$$

From this we obtain

$$d\left(\ln \mathscr{F} + \beta U - \beta \sum_t \mu_t N_t\right) = \beta \left\{dU - \sum_s X_s \, dx_s - \sum_t \mu_t \, dN_t\right\} \quad (8.10)$$

It follows then directly from comparison with (7.20) that the left-hand side equals $(1/k)S$, and that μ_t has the same meaning as in the preceding chapter.

$$\ln \mathscr{F} = -\beta U + \beta \sum_t \mu_t N_t + \frac{1}{k} S \quad (8.11)$$

Further reduction with (7.08) and (7.26) brings it into the final form

$$\ln \mathscr{F} = \beta (G - A) \quad (8.12)$$

It is worth while to remark here that this result might have been directly inferred from (8.05) once the identity of the present μ_t with the μ_t of the preceding chapter was ascertained through (8.08); for we might have carried out first only the summation over i in (8.05) in order to make contact with the definition (5.01) of the canonical partition function. This would have yielded

$$\mathscr{F} = \sum_{N_1=0}^{\infty} \sum_{N_2=0}^{\infty} \cdots \sum_{NT=0}^{\infty} \exp\left[\beta \sum_t \mu_t N_t\right] \cdot F(N_1, N_2, \ldots N_T) \quad (8.13)$$

The expression for F is reduced by (7.10), and the remaining exponential is the Gibbs-Duhem expression (7.26). Thus the grand partition function might have been given the form

$$\mathscr{F} = \sum_{N_1=0}^{\infty} \sum_{N_2=0}^{\infty} \cdots \sum_{N_T=0}^{\infty} \exp[\beta \{G(N_t) - A(N_t)\}] \quad (8.14)$$

If the final fluctuation in the N_t's is small, terms in the neighborhood of the most probable values are much larger than all others. Thus it may be said that $\mathscr{F}$ is essentially equal to the largest term in the sum. If all other terms are thrown away, the results (8.12) and (8.14) are identical. It is to be noticed that at the corresponding point in the study of the canonical partition function a weight factor entered whose logarithm came out to be the entropy. The weight factor arose essentially from the irregular distribution of the quantum states on an energy scale. No such corresponding factor appears here in the summation over the N_t's. The implication is that the partition function F for N_t does not carry information which differs essentially from that conveyed by the function F for N_t-1.

We may specialize the general formula (8.12) to the case of a perfect fluid. The Helmholtz and the Gibbs free energies are then given by (7.08) and (7.13). The two expressions produce an almost complete cancellation, leaving only

$$\ln \mathscr{F} = \frac{p\,V}{k\,T} \quad (8.15)$$

This result is generally quoted in the textbooks.

It is seen that the grand ensemble places μ_t into the same relationship with respect to exchange of matter as β with respect to exchange of energy. The possibility of free exchange renders the quantity equal in different systems. It may be noticed in this connection that $\beta\mu_t$, not μ_t, is the basic variable controlling the distribution. This is particularly noticeable in (8.06), (8.08), and (8.09).

8-2. Other modified statistics; Legendre-transformed partition functions

The grand partition function represents a modified form of counting of great importance, because it is designed for the study of mass transfer. There are other modified statistics possible for other special purposes. In particular it is possible to design modifications of statistical counting which correspond to the Legendre transformations in thermodynamics, as discussed in Section 7-1. We shall first carry out such a modified statistical analysis for a perfect gas, holding the pressure constant. Thereafter, we shall generalize the procedure and indicate the circumstances in which such modified partition functions are in actual use.

A simple design which allows the study of a gas at constant pressure is shown in Fig. 8.1. A gas is contained in a vertical cylinder which is sealed

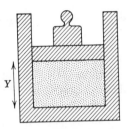

Fig. 8.1. Arrangement for
a statistical study of a gas
at constant pressure.

by a piston at the top end. The piston is weighted down by a weight placed on it; the combined mass of the piston and the weight equals M. Y shall be the height of the piston above the bottom of the cylinder, and P its momentum in the vertical direction. Outside the cylinder there shall be vacuum. The pressure of the enclosed gas is then given by

$$p = \frac{M g}{q} \tag{8.16}$$

where q is the cross section of the piston and the cylinder. Similarly the volume of the gas equals

$$V = Y q \tag{8.17}$$

We now consider the total system: the enclosed gas consisting of N molecules plus the piston and weight. It is a system of $3N+1$ degrees of freedom. Its energy equals

$$E = \sum_{i=1}^{N} \frac{1}{2 m} \mathbf{p}_i^2 + \frac{1}{2 M} P^2 + M g Y \tag{8.18}$$

The probability of finding the representative point in a volume h^{3N+1} in phase space is now given by the formula of Gibbs. The partition function $F^\star$ associated with (8.18) reads

$$F^\star = \sum_\nu \exp[-\beta E_\nu]$$

or more precisely

$$F^\star = \frac{1}{N!} \frac{1}{h^{3N+1}} \int \cdots \int \exp[-\beta E] \, d\mathbf{p}_1 \cdots d\mathbf{p}_N \, d\mathbf{x}_1 \cdots d\mathbf{x}_N \, dP \, dY \qquad (8.19)$$

Here E is the expression (8.18). In formula (8.19) the factor $1/N!$ is not in harmony with previously discussed theory; its presence will be explained in Section 8-4. It can be omitted for the purposes of the present argument; the only reason it was included is to avoid the presence of incorrect formulas in the book.

We shall carry out all integrations in (8.19) except the one over Y. Formula (8.19) becomes then

$$F^\star = \frac{1}{N!} \left(\frac{2\pi m}{h^2 \beta} \right)^{(3N+1)/2} \sqrt{\frac{M}{m}} \, q^N \int_0^\infty Y^N e^{-\beta M g Y} \, dY \qquad (8.20)$$

Here the connection (8.17) between volume and Y has been used in an essential way; it leads to the remaining integral over Y which is novel in structure. The final expression for $F^\star$ is

$$F^\star = \left(\frac{2\pi m}{h^2 \beta} \right)^{(3N+1)/2} \sqrt{\frac{M}{m}} \left(\frac{q}{\beta M g} \right)^{N+1} \qquad (8.21)$$

Formula (8.21) lends itself to the computation of two averages, the average energy and the average volume. Examination of (8.17), (8.18), and (8.19), shows that both averages can be expressed as logarithmic derivatives. For the mean energy $\langle E \rangle$ we get

$$\langle E \rangle = - \frac{\partial \ln F^\star}{\partial \beta} = \frac{5N+3}{2} kT \qquad (8.22)$$

and for the volume, after simplification with (8.16),

$$V = - \frac{q}{\beta g} \frac{\partial \ln F^\star}{\partial M} = (N + \tfrac{1}{2}) \frac{kT}{p} \qquad (8.23)$$

Apart from a correction of the order $1/N$ (8.23) is the equation of state of a perfect gas. However, $\langle E \rangle$ cannot be the internal energy U because it has a factor $\frac{5}{2}$ instead of $\frac{3}{2}$. In fact, one would suspect it to be the enthalpy

H. This is confirmed by writing down the total differential equation for $\ln F^\star$. Equations (8.22) and (8.23) yield for it

$$d \ln F^\star = - \langle E \rangle \, d\beta - \beta \, V \, dp \tag{8.24}$$

or

$$d \{ \ln F^\star + \beta \, p \, V \} = - \{ \langle E \rangle - p \, V \} \, d\beta + \beta \, p \, dV \tag{8.25}$$

This is equation (5.36), which means that

$$\langle E \rangle - p \, V = U$$

or

$$\langle E \rangle = H \tag{8.26}$$

and

$$\ln F = \ln F^\star + \beta \, p \, V \tag{8.27}$$

which, from (7.10) and (7.13), means that

$$\ln F^\star = - \beta \, G \tag{8.28}$$

Indeed the so-called *Planck potential* $-\beta \, G$ satisfies the equation

$$d(-\beta G) = - H \, d\beta - \beta \, V \, dp \tag{8.29}$$

as is verified from (7.14). This is a reproduction of (8.24).

The final result (8.27) for the partition function $F^\star$ could have been inferred directly by a cursory examination of (8.18) and (8.19). The "energy" E in (8.18) differs from the one previously used by the thermal agitation of the piston (which comes out to be unimportant) and the term $M \, g \, Y$, which, by (8.16) and (8.17), means $p \, V$. Expression (8.19) for $F^\star$ thus differs from F by a multiplier $\exp[- \beta \, p \, V]$. This is exactly what is stated in the end result (8.27).

The example given represents an amusing exercise which shows that one can do statistics of fluids at constant pressure. All that is needed is to devise an arrangement which holds the pressure constant. The integral $\int p \, dV$ can then be carried out, and a term $p \, V$ is added to the energy. In other words the enthalpy H becomes the energy which goes into the Boltzmann exponent. In the present instance, this is an artificial procedure. It becomes essential, however, when we deal with fixed field forces. Then part of the problem is to determine the energy states in the presence of such forces.

Generalizing the procedure of the preceding example, we may assume that X_1 is such a fixed field force or, more precisely, a parameter which can be freely chosen by the experimenter. The work term $\int X_1 \, dx_1$ is then integrable and can be added to the energy, yielding

$$E = U - X_1 x_1 \tag{8.30}$$

We now construct a new partition function $F^{\star}$

$$F^{\star} = \sum_{j} \exp[-\beta E_j] \tag{8.31}$$

or

$$F^{\star} = \sum_{j} \exp[-\beta(U_j - X_1 \cdot (x_1)_j)] \tag{8.32}$$

Consideration of all possible states at fixed X_1 means consideration of all possible values of x_1, for we can no longer hold x_1 fixed once we have decided to hold the conjugate force fixed. If we are able to compute this kind of partition function $F^{\star}$, we get by differentiation

$$U - X_1 x_1 = -\frac{\partial \ln F^{\star}}{\partial \beta} \tag{8.33}$$

and

$$x_1 = \frac{1}{\beta}\frac{\partial \ln F^{\star}}{\partial X_1} \tag{8.34}$$

The remaining derivatives will be as previously in (5.33). The total differential equation for $F^{\star}$ becomes then

$$d \ln F^{\star} = -(U - X_1 x_1)\, d\beta + \beta x_1\, dX_1 - \beta \sum_{s \neq 1} X_s\, dx_s \tag{8.35}$$

Comparison with (5.35) shows that

$$\ln F^{\star} = \ln F + \beta X_1 x_1 \tag{8.36}$$

which is a return to the starting equation (8.32).

Application of this kind of partition function will be found chiefly in Chapters 15 and 16, where we shall consider a system in a fixed acting magnetic field $\mathbf{H}$. We proved in Chapter 1 that the moment $\mathbf{M}$ is the coordinate conjugate to $\mathbf{H}$. The quantity E which appears as energy will then be the magnetic enthalpy

$$E = U - \mathbf{H} \cdot \mathbf{M} \tag{8.37}$$

and, from (7.10) and (8.36), the partition function $F^{\star}$ will be given by

$$\ln F^{\star} = -\beta A + \beta \mathbf{H} \cdot \mathbf{M} \tag{8.38}$$

The total differential equation obeyed by $F^{\star}$ will be

$$d \ln F^{\star} = -E\, d\beta + \beta \mathbf{M} \cdot d\mathbf{H} \tag{8.39}$$

The magnetic moment can thus be obtained by differentiation with respect to the parameter $\mathbf{H}$.

8-3. Maxwell–Boltzmann particle statistics

It is a truism that, if we desire information about a system going beyond general principles, the statistics of the system has to be brought into computable form. Most of the methods to accomplish this will be discussed in connection with their particular application. However, one form of statistical breakdown is so important and so widespread that it is proper to treat it as part of the fundamental theory. This is the statistics of independent particles. This form of statistics is of course also the oldest type. For many decades after its invention it was the only type conceived of.

The essence of the independent particle model is the following. The macroscopic body, whose properties we wish to study in detail, is assumed to consist of a set of identical particles which are independent. The thing which is specific for the macroscopic body is the set of one-particle quantum states which it provides. The quantum state of the body as a whole is specified if each of the individual particles is assigned to one of these states. The energy is simply the sum of the energies of the individual particles. Many others of its properties can be obtained by a similar summation process.

Offhand, the case of separate one-particle states does not seem very difficult. It would appear, if a set of molecules is truly independent, that one should be able to identify these very molecules with the "systems" of Gibbsian statistics. This is in fact the point of view that was taken in Chapter 4 in the discussion of the Maxwellian velocity distribution. The reasoning employed there for free particles may be transferred with only changes of notation to a general set of levels. Adoption of this point of view produces the so-called Maxwell–Boltzmann statistics.

Denote by $i, j, k, \ldots z$ the one-particle levels available to the individual particles. Label by $1, 2, 3, \ldots N$ the N particles filling these levels. (This is an important point, for the two quantum statistics deny such a possibility of individual labeling.) A general state can then be denoted by $i(1)$, $j(2)$, $k(3), \ldots z(N)$, meaning that molecule 1 is in state i, molecule 2 in state k, etc. The Gibbs law (4.11) is directly applicable to such a situation. The probability of such a state being occupied by the system is

$$P(i(1), j(2), k(3), \ldots, z(N))$$
$$= \frac{\exp[-\beta(\varepsilon_i + \varepsilon_j + \varepsilon_k + \cdots \varepsilon_z)]}{\sum_{i'} \sum_{j'} \sum_{k'} \ldots \sum_{z'} \exp[-\beta(\varepsilon_{i'} + \varepsilon_{j'} + \varepsilon_{k'} \ldots \varepsilon_{z'})]} \quad (8.40)$$

Since the whole argument for (4.11) was originally based on the idea of statistical independence and the resultant possibility of factoring probabilities, it should cause no surprise that expression (8.40) factors into N

factors, namely, the independent probabilities for the N subsystems or particles. In other words, we have

$$P(i(1), j(2), k(3), \ldots, z(N)) = p(i) \cdot p(j) \ldots p(z) \qquad (8.41)$$

with

$$p(i) = \frac{\exp[-\beta\,\varepsilon_i]}{\sum_{\nu} \exp[-\beta\,\varepsilon_{\nu}]} \qquad (8.42)$$

as in (4.11).

Examination of (8.40) shows that the identity of the particles 1, 2, 3, . . . plays a very subordinate role in the final result; the formula which is obtained from the exchange of any two of them is exactly the same as the original one, and the physical state also. It is therefore natural to lump such states together and call them a degenerate state when viewed from the system as a whole. From such a viewpoint the important aspect of the state $i(1), j(2), k(3), \ldots$ is how many particles occupy each quantum state. It is convenient to remake the notation and to make these *occupation numbers* the basic statistical element. We thus discard the labels for the particles, but retain the label $s = 1, 2, 3, 4$ for the one-particle states. The state of the system as a whole is then described by the occupation numbers $n_1, n_2, n_3, \ldots$; these numbers state how many particles are in states $1, 2, 3, \ldots$. Many of these numbers may of course be zero. Each of these composite states characterized by a given set of occupation numbers has now an intrinsic multiplicity M which arises from the possible number of ways N particles can be permuted among themselves to produce the same set of occupation numbers. This multiplicity has been written down in (4.12) and employed in an alternative derivation of the Gibbs–Boltzmann distribution law. It equals

$$M = \frac{N!}{n_1!\,n_2!\,n_3! \ldots} \qquad (8.43)$$

where of course the restriction

$$\sum_i n_i = N \qquad (8.44)$$

must be assumed obeyed. The probability formula for occupation numbers results from combining (8.41), (8.42), and (8.43). It reads

$$P(n_1, n_2, n_3 \ldots)$$
$$= \frac{N!}{n_1!\,n_2!\,n_3! \ldots} \frac{\exp[-\beta\,(n_1\,\varepsilon_1 + n_2\,\varepsilon_2 + n_3\,\varepsilon_3 + \ldots)]}{\left[\sum_{\nu} e^{-\beta\varepsilon_{\nu}}\right]^N} \qquad (8.45)$$

Equations (8.42) and (8.45) bring in another partition function, the *one-particle partition function*, which we shall denote by f. It is given by

$$f = \sum_s \exp[- \beta \, \varepsilon_s] \tag{8.46}$$

It is a sum of exponentials running over all one-particle states s of the problem. It acts as the normalizing denominator not only for the one-particle probability (8.42) but also for the expression (8.45) involving occupation numbers; in this latter expression it assures normalization through the multinomial expansion theorem. Whenever we deal with a problem for which a single-particle partition function exists effectively, we have the opportunity to do statistics on the entire system while considering just a single particle. This creates a vastly simplified problem; naturally the method has only a limited range of use because it assumes the existence of one-particle states. We employed it informally in Chapter 4, where f appears as the denominator in (4.21) or the reciprocal of the front factor in (4.25). Full use will be made of its possibilities in Chapter 11, which deals with the perfect molecular gas.

From the modern point of view the cases in nature for which Boltzmann statistics is truly the correct form of particle statistics are rare. More commonly one of the two quantum statistics to be discussed in the next chapter applies. To these two statistics Boltzmann statistics forms an approximation which is valid when the probability of a quantum state being occupied by a particle is small. As Boltzmann statistics is mathematically easier, it is always used when approximately valid. For Boltzmann statistics to be truly valid special circumstances are required. Boltzmann statistics is distinguished from the two quantum statistics in that it assumes the distinguishability of the particles. Fundamentally, this is never correct, but it can become correct in practice when some external circumstance makes particles distinguishable. To give an instance, the probability of excitation of gaseous molecules is subject to Boltzmann statistics; for although the excitation is due to electrons which are not distinguishable the electrons are located on different molecules and thereby acquire a "label." A similar situation prevails for electrons which are excited at physical defects or impurities in solids. Even though these defects may have identical structure they are distinguishable because they are localized. An exchange, whereby an electron is lowered from the first excited state to the ground state in impurity atom a, while at the same time another electron is raised from the ground state to the first excited state of impurity atom b, is thus a physically recognizable change, even though degenerate in energy. The multiplicity factor (8.43) is therefore correct, and localized electrons of this type do obey Boltzmann statistics. Other examples in which Boltzmann statistics applies can be constructed, but they are all rather specialized. For the simple independent particle problems of modern statistical mechanics the weight factor (8.43) is not correct.

8-4. Particle versus system partition function; Gibbs paradox

A glance at equation (8.41) shows that there is a trivially simple relationship between the probabilities for the individual particles to be in a given state and the probability for the system to be in a given state. The latter probability follows from the former by simple product formation, as is proper for independent events. Correspondingly the denominator in (8.40) is the Nth power of the denominator in (8.42). From this one comes almost inevitably to the conclusion that the system partition function (5.01) and the particle partition function (8.46) are related through

$$F = (f)^N \tag{8.47}$$

This formula is *wrong*. The correct formula is instead

$$F = \frac{1}{N!} (f)^N \tag{8.48}$$

Since (8.48) is in conflict with the principle of simple product formation, Gibbs arrived at it only after realizing that (8.47) gives physically incorrect answers. We shall first follow his reasoning in this respect, and then present the modern arguments in support of the second formula.

In many applications the difference between (8.47) and (8.48) is not noticeable because most equations such as (5.33) and (5.35) yield observable quantities through the derivative of the logarithm of F. For such cases no difference between the two expressions results. In order to observe a difference one must take an expression containing f itself. One such expression is the entropy, and we must therefore look in that direction if we wish to make the correct assignment.

The example to study is the monatomic gas. The energy of a molecule of such a gas equals

$$\varepsilon = \frac{p^2}{2m} \tag{8.49}$$

For the sum of states the summation is advantageously transformed into an integral as indicated in (4.21). We then obtain for the one-particle partition function (8.46)

$$f = \frac{1}{h^3} \int\int\int\int\int\int dp_x\, dp_y\, dp_z\, dx\, dy\, dz \exp\left(-\frac{\beta p^2}{2m} \right)$$

which integrates out to

$$f = \left(\frac{2\pi m k T}{h^2}\right)^{3/2} V \qquad (8.50)$$

Equation (5.35) tells us how to get the energy U of the system from the partition function, and either (8.47) or (8.48) will give the right answer, namely,

$$U = \tfrac{3}{2} N k T \qquad (8.51)$$

For the entropy we note that, according to (5.09), F is related to the energy and entropy through

$$\frac{S}{k} = \beta U + \ln F \qquad (8.52)$$

This formula will give a different answer according to whether (8.47) or (8.48) is valid. In the case of (8.47) we find with (8.50) and (8.51)

$$S = \tfrac{3}{2} N k \left(1 + \ln \frac{2\pi m k}{h^2}\right) + \tfrac{2}{3} N k \ln T + N k \ln V \qquad (8.53)$$

All terms are proportional to N, making S extensive, except the last one, which contains $N \ln N$. The reason for the appearance of this term is fairly clear. We have assigned to each atom the entire space V as its degree of freedom in space. This assignment appears as $k \ln V$ in the entropy. With N particles present the term becomes $N k \ln V$. Because of the presence of this term, S, as given by (8.53), is not an extensive variable. Some people might object that this is only an esthetic question, but this is not so. The assignment (8.47) would actually violate the Second Law. This is illustrated in the following simple thought experiment. Assume two adjoining chambers filled with the same ideal gas at the same pressure and temperature, and assume them to be separated by a removable wall. Removal of the wall can be thought of as a reversible process requiring no heat or work. Hence the entropy must remain constant. However, according to (8.53), the entropy increases upon removal of the wall by

$$(N_1 + N_2) k \ln (V_1 + V_2) - N_1 k \ln V_1 - N_2 k \ln V_2$$

or

$$- (N_1 + N_2) k \{c_1 \ln c_1 + c_2 \ln c_2\}$$

We recognize in this last term the entropy of mixture derived in equation (5.66); it is indeed present if the two gases are different. However, here the gases are the same, and the term should be absent. If we adopt formula (8.48) this will be the case. Instead of (8.53) we get

$$S = \tfrac{3}{2} N k \left(1 + \ln \frac{2\pi m k}{h^2}\right) + \tfrac{3}{2} N k \ln T + N k \ln V - k \ln (N!) \qquad (8.54)$$

According to Stirling's formula, we have

$$\ln N! \approx N \ln N - N - \ln \sqrt{2\pi N}$$

The leading term is the first one, the only one growing faster than proportionally to N. It combines with the volume term into

$$N k \ln \frac{V}{N}$$

The term under the logarithm equals the specific volume per molecule and is intensive and equal in the two chambers. Removal of the wall will thus not change it but will simply produce the term

$$(N_1 + N_2) k \ln \frac{V}{N}$$

that is, the sum of the terms arising from each chamber. The entropy of mixture comes out thereby equal to zero as required.

To make quite sure that (8.48) is correct let us verify that the entropy of mixture term has not been lost if we deal with two different gases. Two different gases occupying the same volume are entirely independent; hence F will generalize from (8.48) to

$$F = \frac{(f_1)^{N_1}}{N_1!} \frac{(f_2)^{N_2}}{N_2!} \tag{8.55}$$

The entropy follows from this as

$$S = \tfrac{3}{2} N_1 k \left(1 + \ln \frac{2\pi m_1 k}{h^2} \right) + \tfrac{3}{2} N_2 k \left(1 + \ln \frac{2\pi m_2 k}{h^2} \right)$$
$$+ \tfrac{3}{2} (N_1 + N_2) k \ln T + (N_1 + N_2) k \ln V - k \ln (N_1)! - k \ln (N_2)! \tag{8.56}$$

The crucial part is the last three terms, which are not linear in N_1 and N_2. The quantity V is best eliminated by Dalton's law:

$$V = \frac{(N_1 + N_2) k T}{p}$$

The part $k T/p$ gives rise to linear terms only. The three nonlinear terms are therefore

$$\Delta S = (N_1 + N_2) k \ln (N_1 + N_2) - N_1 k \ln N_1 - N_2 k \ln N_2$$
$$= (N_1 + N_2) k \{ - c_1 \ln c_1 - c_2 \ln c_2 \}$$

in agreement with equation (5.66).

The assignment (8.48) is thus in agreement with the facts, at least in the sample case of perfect gases. The reader may, however, still feel uneasy about this point because the expression was assigned, not derived, and was verified only in a special case. It is also not in harmony with the basic ideas of Gibbs–Boltzmann statistics, which is based on the principle of combining the probabilities for independent systems through product formation. The difficulty is a very deep one. In Boltzmann statistics all particles are in principle distinguishable. The molecules in the thought experiment following equation (8.53) are therefore distinguishable even if they are of the same kind. We could think of painting the molecules on one side of the partition red, and on the other blue. The entropy of mixture (5.66) would then be justifiably associated with the removal of the partition.

The fact that there are particles which are indistinguishable in principle, and for which mixture means no entropy increase, has therefore to be added to the basic structure of Maxwell–Boltzmann statistics as an after-thought. Fortunately, this is easily done with the help of the factor $1/N!$ in (8.48). The factor asserts explicitly that all $N!$ permutations of the N particles must be counted as one and the same physical state. We shall see that the two quantum statistics deny distinguishability from the beginning. Consequently they also do not give rise to logical difficulties of the type met in the Gibbs paradox.

8-5. Grand ensemble formulas for Boltzmann particles

As a preparation for the other two statistics we shall go through the formulas for Boltzmann statistics if the grand partition function is used. It is understood that the novel features are purely formal, since the same physics is being discussed. It does, however, demonstrate the place of the chemical potential μ in classical statistics, in preparation for its role in the other two cases.

The basic formula in this case is (8.04):

$$P(N, i) = \frac{1}{\mathscr{F}} \exp \beta(\mu\, N - U_i) \tag{8.57}$$

where U_i is the sum of all one-particle energies. We expedite the calculation by not recomputing everything from the probabilities (8.45), but using the intermediate formula (8.13) containing the canonical partition function F:

$$\mathscr{F} = \sum_{N=0}^{\infty} \exp(\beta\, \mu\, N) \cdot F(N) \tag{8.58}$$

We reduce this further to the particle partition function by using (8.48), which yields

$$\mathscr{F} = \sum_N \frac{1}{N!} e^{\beta\mu N} (f)^N$$

or

$$\mathscr{F} = \exp\{e^{\beta\mu} f\} \tag{8.59}$$

$\mathscr{F}$ can be expressed as a product of factors, each pertaining to one one-particle quantum state. We get this form by inserting (8.46) into (8.59). We find

$$\mathscr{F} = \exp\left\{ e^{\beta\mu} \sum_s e^{-\beta\varepsilon_s} \right\}$$

or

$$\mathscr{F} = \prod_s \exp\{\exp \beta (\mu - \varepsilon_s)\} \tag{8.60}$$

The form (8.60) is most readily compared with the expressions for $\mathscr{F}$ in the other two statistics, which can also be expressed as a product of contributions from one-particle quantum states.

Factorization of the grand partition function must have as a consequence the possibility of factorizing the probability (8.57), this time not from the standpoint of the individual particles, but from the standpoint of the one-particle quantum states and the degree to which they are occupied. The probability is basically given by (8.57), where U_i is given by

$$U_i = \sum_s n_s \varepsilon_s \tag{8.61}$$

Since all permutations of the particles are to be counted as the same state of occupation, the probability has to be multiplied with the weight factor (8.43). This factor must be modified, however, by the suppression of the $N!$ in the numerator. For we have resolved the Gibbs paradox by multiplying the weight of all states having N particles with the factor $1/N!$; this procedure was incorporated into $\mathscr{F}$ by (8.48) and (8.58) and must not now be forgotten. The probability $P(n_1, n_2, n_3, \ldots)$ for having the occupation numbers $n_1, n_2, n_3, \ldots$ in states 1, 2, 3 is therefore

$$P(n_1, n_2, n_3, \ldots) = \frac{1}{\mathscr{F}} \sum_s \frac{1}{n_s!} \exp \beta n_s (\mu - \varepsilon_s)$$

Because of (8.60) this takes the form

$$P(n_1, n_2, n_3, \ldots) = p_1(n_1) \, p_2(n_2) \, p_3(n_3) \ldots \tag{8.62}$$

where we have

$$p_s(n_s) = \frac{1}{n_s!} \frac{\exp \beta n_s (\mu - \varepsilon_s)}{\exp\{\exp \beta (\mu - \varepsilon_s)\}} \tag{8.63}$$

This formula is also in a condition which permits direct comparison with the modern quantum statistics. We verify readily that

$$\sum_{n_s=0}^{\infty} p_s(n_s) = 1 \tag{8.64}$$

as is required for a probability. More interesting is of course the expectation value of n_s, which results from the equation

$$\langle n_s \rangle = \sum_{n_s=1}^{\infty} \frac{1}{(n_s-1)!} \frac{\exp \beta n_s(\mu-\varepsilon_s)}{\exp\{\exp \beta(\mu-\varepsilon_s)\}}$$

With a relabeling of the running index n_s this reduces easily to

$$\langle n_s \rangle = \exp \beta(\mu-\varepsilon_s) \tag{8.65}$$

Equation (8.65) is a reproduction of the very first distribution formula we had, namely (4.09). Now, however, the constant has a meaning. In Boltzmann statistics the chemical potential is the value of the energy for which the probable number of particles in a quantum state equals unity; this energy has to be the same across any phase boundaries or in different compounds. It can happen, of course, that the probable number in any state is less than 1, in which case μ is a number smaller than the smallest energy state occurring in the problem.

The other formulas derivable from the grand partition function are more routine and lead back to previous results. Summation of (8.65) over all states yields

$$N = \sum_s \exp \beta(\mu-\varepsilon_s)$$

This can be simplified with the help of (8.46) to read

$$N = \exp \beta\mu \cdot f \tag{8.66}$$

The same result can be derived from (8.08) and (8.59):

$$N = \frac{1}{\beta} \frac{\partial \ln \mathscr{F}}{\partial \mu} = \exp \beta\mu \cdot f$$

Similarly, we get from (8.06) and (8.59)

$$U = - \exp \beta\mu \cdot \frac{\partial f}{\partial \beta} \tag{8.67}$$

Elimination of the chemical potential from the results of Boltzmann statistics is a simple matter. One simply solves (8.66) for $\exp \beta \mu$

and substitutes it into the other results, such as (8.65) and (8.67). This yields

$$\langle n_s \rangle = N \frac{1}{f} \exp[- \beta \, \varepsilon_s] \qquad (8.68)$$

and

$$U = - N \frac{\partial \ln f}{\partial \beta} \qquad (8.69)$$

These formulas are equivalent to (4.11) and (5.35), the only new feature being the explicit appearance of the multiplier N. The ease with which the elimination of μ proceeds is a special property of this type of statistics. Physically, it means that the Boltzmann distribution does not depend on particle density in an essential way, but simply contains the total number of particles as a factor. This is of course implicit in the statement on statistical independence which lies at the base of the Maxwell–Boltzmann distribution law.

A. RECOMMENDED PROBLEMS

1. Gases consisting of dipolar molecules are polarized in an electric field mainly through the orientation of these dipoles. This orientation proceeds freely but is opposed by thermal agitation. Calculate on this basis the dielectric constant of steam at $100°$ C and atmospheric pressure. The water molecule has an electric moment $\mu = 1.853 \times 10^{-18}$ esu.

2. If dipolar gases are placed in an electric field they develop a slight dependence of the density on the magnitude of that field. Show that the density is proportional to

$$\frac{\sinh \beta \, \mu \, E}{\beta \, \mu \, E}.$$

B. GENERAL PROBLEMS

3. A new kind of partition function can be constructed for a perfect gas if we suppose it to be contained in a cylinder closed by a weightless piston, which is attached to the bottom of the cylinder by a spring. The spring shall be under no tension when the volume is zero. Indicate which results come out as usual and which come out differently.

4. Orienting nuclear spins directly by an externally applied field is referred to as the "brute force" method of nuclear alignment. How low

must the temperature be to align two thirds of the proton spins in a paraffin sample parallel to a field 20000 oe? The magnetic moment of the proton is 2.793 nuclear magnetons.

5. What general prediction can be made for the equation of state of a gas of classical hard spheres of radius a if there is no other interaction between the spheres?

6. A Boltzmann gas of electrons of number density n_0, which is compensated in the mean by positive charges, tends to screen out perturbing potentials by an adjustment in its density. At finite temperature this screening is imperfect and leads to an exponential dying away of the potential at large distance from the disturbance. Derive this exponential decay law.

9

Quantum statistics of
independent particles

We have seen in Section 8-4 that the classical statistics of independent particles has an inherent logical flaw, which is known as the Gibbs paradox. The flaw arises if the particles are indistinguishable; for such particles, classical statistics yields wrong answers unless rectified. The rectification is easy: a factor $1/N!$ must be added to the Nth power of the particle partition function, where N is the number of such undistinguishable particles. This factor is not part of the classical reasoning. We shall now see that the difficulty never arises if statistics is properly applied to all possible system wave functions which can be constructed from single-particle wave functions.

9-1. Pauli exclusion principle

The easiest application of the independent particle model is the perfect gas, but the concept arises also in other contexts. In particular, the idea applies to electrons in atoms in a semiquantitative way, and now is being applied even to protons and neutrons in nuclei. In the case of electrons in atoms the notion is essential to explain the shell structure of the atom and the periodic table of the elements. In this explanation the idea of one-electron quantum states must be coupled with the prescription that only two electrons be placed in any one such state. With the discovery of electron spin this idea could be simplified into the statement:

Two different electrons cannot occupy the same quantum state.

This is the Pauli exclusion principle. The question arises whether this principle is or is not easily reconciled with quantum mechanics, and the answer is that it is. The reason for this is the following. The hamiltonian for a number of electrons, say 5, in an atom or any other system, $\mathscr{H}(1, 2, 3, 4, 5)$, is obviously symmetric in these five particles. Therefore, if $\psi(1, 2, 3, 4, 5)$ is a solution of the resultant Schroedinger equation so will be $\psi(3, 1, 4, 5, 2)$ or any other ψ with the arguments permuted; a linear combination of such wave functions is of course also acceptable. Among these combinations there are two outstanding types: the totally *symmetric* combination

$$\Psi_s = \sum_P P \, \psi(1, 2, 3, 4, 5) \tag{9.01}$$

and the totally *antisymmetric* combination

$$\Psi_a = \sum_P (-)^P P \, \psi(1, 2, 3, 4, 5) \tag{9.02}$$

Here P denotes permutation of the five arguments, and $\sum_P$ denotes summation over all 120 of these permutations. We shall assume here that the reader knows that permutations can be classified as even or odd, according to the parity of the number of two-particle interchanges necessary to reach them, starting from a standard arrangement. The factor $(-)^P$ appearing in (9.02) is meant to signify that all ψ functions in the sum containing the five arguments in an odd permutation shall be taken with negative sign. We call the resultant wave function (9.02) antisymmetric in the particles. The important thing to realize is that once a wave function is either symmetric or antisymmetric it will remain so in time. This is verified from the Schroedinger equation

$$i\hbar \frac{\partial \Psi}{\partial t} = \mathscr{H} \Psi$$

In this equation the symmetry type of $\mathscr{H}\Psi$ will be the same as that of Ψ because $\mathscr{H}$ is totally symmetric; hence $\partial\Psi/\partial t$ will be of the same symmetry type as Ψ, and finally the same will be true for Ψ at all times.

One may well ask what this has to do with the Pauli principle. The answer becomes clear when these observations about symmetry are combined with the independent particle model. Let us take as an example two particles in two states a and b. We can form four wave functions for the two particles, namely,

$$\psi_a(1)\,\psi_a(2)$$
$$\psi_b(1)\,\psi_b(2)$$
$$\psi_a(1)\,\psi_b(2)$$
$$\psi_b(1)\,\psi_a(2)$$

This number also equals the number of partitions for two particles in two boxes, namely, 2^2. However, if we look at these wave functions from the point of view of symmetry, only the first two are satisfactory as they are symmetric in the two particles. The other two go over into each other upon permutation of 1 and 2. A better choice from the point of view of symmetry would have been

$$\Psi_A = \psi_a(1)\,\psi_a(2) \tag{9.03a}$$

$$\Psi_B = \psi_b(1)\,\psi_b(2) \tag{9.03b}$$

$$\Psi_C = \psi_a(1)\,\psi_b(2) + \psi_b(1)\,\psi_a(2) \tag{9.03c}$$

$$\Psi_D = \psi_a(1)\,\psi_b(2) - \psi_b(1)\,\psi_a(2) \tag{9.04}$$

The three wave functions (9.03) are symmetric, and (9.04) is antisymmetric. The two types are rigidly separated because no perturbation of any kind can convert a wave function (9.03) into (9.04) or vice versa. However, the three wave functions (9.03) can of course be converted into each other.

Our first concern is with the wave function (9.04). It is of such a structure that if a and b are identical states it vanishes identically. Thus the requirement of antisymmetry is an automatic way of making a set of particles occupying the same set of one-particle states obey the Pauli exclusion principle. This is even more evident if the set of particles and wave functions is larger. For instance, if we take a set of five particles in five one-particle states and subject their product wave function to the procedure (9.02), we find the determinantal wave function

$$\Psi = \begin{vmatrix} \psi_a(1) & \psi_a(2) & \psi_a(3) & \psi_a(4) & \psi_a(5) \\ \psi_b(1) & \psi_b(2) & \psi_b(3) & \psi_b(4) & \psi_b(5) \\ \psi_c(1) & \psi_c(2) & \psi_c(3) & \psi_c(4) & \psi_c(5) \\ \psi_d(1) & \psi_d(2) & \psi_d(3) & \psi_d(4) & \psi_d(5) \\ \psi_e(1) & \psi_e(2) & \psi_e(3) & \psi_e(4) & \psi_e(5) \end{vmatrix} \tag{9.05}$$

Quite obviously, by the rule for determinants, if any two states are the same the wave function is zero. Therefore the principle of antisymmetry for the wave function is the proper invariant way of taking care of the Pauli exclusion principle. The derivation also shows that the principle cannot be limited to certain situations such as atoms but that once it is admitted for electrons within atoms it becomes automatically valid for electrons everywhere. In other words the statistics is a fundamental characteristic of the particle, in this case the electron.

It follows from this that other fundamental particles also have their statistics. Protons, neutrons, and all particles of half odd integral spin obey the Pauli principle, while particles of itegral spin have totally symmetric wave functions of the type (9.01). The argument for this situation is rather recondite and will not be given here.

9-2. Fermi–Dirac statistics

It is now our business to apply the Gibbs formula to a system consisting of a set of identical particles placed into one-particle states and obeying the Pauli principle. It is advantageous for this purpose to use the grand ensemble. The chemical potential μ of fermions is usually called the *Fermi energy* of the particles. The probability of a system having N particles in it and being in the quantum state i thus equals, according to (8.57),

$$P(N, i) = \frac{1}{\mathscr{F}} \exp[\beta (\mu N - U_i)] \qquad (9.06)$$

Here U_i is the sum of the energies of the occupied one-particle states; the sum consists of exactly N terms. We express this by writing

$$U_i = \sum_{\nu}^{N} \varepsilon_\nu$$

so that the grand partition function takes the form

$$\mathscr{F} = \sum_{N=0}^{\infty} \exp \beta \mu N \cdot \sum_{S_N(\nu)}^{N} \exp\left[- \beta \sum_{\nu}^{N} \varepsilon_\nu \right] \qquad (9.07)$$

Here the summations have the following significance. The sum farthest to the left is a straight sum over all integers; the central sum is a sum over all possible selections $S_N(\nu)$ of N different one-particle quantum states $\nu_1, \nu_2, \ldots \nu_N$; the sum on the right is the sum of the energies of these N selected states. In the probability (9.06) only this last sum appears so that we have

$$P(N, S_N(\nu)) = \frac{1}{\mathscr{F}} \exp\left[\beta \sum_{\nu}^{N} (\mu - \varepsilon_\nu) \right] \qquad (9.08)$$

which is an expression of the standard Boltzmann form.

The information usually desired in Fermi statistics is not the probability written out in (9.08) but the probability that a given state which we call 1 be occupied. With respect to this state (9.07) has a peculiar structure. There is a one-to-one match of the terms $N = M$ having ε_1 empty and the terms $N = M+1$ having ε_1 occupied: namely, we may select some term making a particular selection $S_M(\nu)$ not containing state 1 and the term making exactly the same selection $S_{M+1}(\nu)$, except that state 1 is added to the previous list. At the same time an extra factor $\exp \beta \mu$ appears in the second instance because the total number of particles is larger by 1. In this way, all terms in the state sum are matched one-to-one with respect to state 1, and the partition function $\mathscr{F}$ can be given the form

$$\mathscr{F} = \{1 + \exp \beta(\mu - \varepsilon_1)\} \mathscr{F}' \qquad (9.09)$$

Here $\mathscr{F}'$ is a sum like $\mathscr{F}$ except that all reference to state 1 has disappeared. We can extend this argument by recurrence to any other state, and end finally with the completely factored form

$$\mathscr{F} = \prod_s \{1 + \exp \beta(\mu - \varepsilon_s)\} \qquad (9.10)$$

Formula (9.09) also answers the question what the probability of occupation of state 1 is. We must sum in the numerator over all states having 1 filled, and place in the denominator the sum over all states, whether state 1 is empty or filled. In this process $\mathscr{F}'$ drops out, and we are left with

$$p(1) = \frac{\exp \beta(\mu - \varepsilon_1)}{1 + \exp \beta(\mu - \varepsilon_1)}$$

or generally

$$p(s) = \frac{1}{\exp \beta(\varepsilon_s - \mu) + 1} \qquad (9.11)$$

This is the *Fermi distribution law*. It makes explicit reference only to the energy ε_s of the state in question. In fact, the factorization of $\mathscr{F}$ into contributions from individual one-particle energy states means that occupation probabilities are also factorized and hence statistically independent, and a more complicated problem of multiple occupation can be handled by the usual product formation. The statistical independence of the occupation probabilities must not be interpreted, however, as implying some sort of statistical independence of the particles making up the system. The particles exert a very strong influence on each other because a particle occupying a state excludes the others from it. This is equivalent to a strong repulsive force comparable to the strongest forces occurring in the problem.

In formula (9.11) the influence of the particles on each other gives rise to the appearance of the Fermi energy. This parameter cannot be eliminated as was the corresponding constant in the passage from (8.65) to (4.11). It represents therefore a nontrivial structural feature in the distribution law.

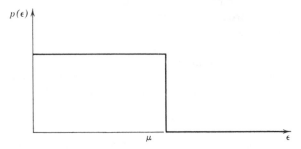

Fig. 9.1. Probability of occupation of a quantum state versus energy in Fermi statistics; situation at the absolute zero of temperature.

The Fermi distribution law depends on two parameters β and μ instead of just one, as the Boltzmann distribution does. The interplay of these two parameters is illustrated in Fig. 9.1 and Fig. 9.2. If one plots the quantity

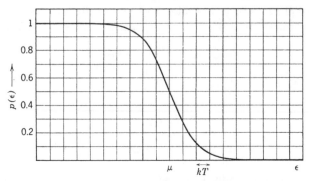

Fig. 9.2. Probability of occupation of a quantum state versus energy in Fermi statistics; situation at finite temperature; the magnitude of kT is marked on the abscissa.

(9.11) as a function of ε_s for a very low temperature, that is, a large value of β, then one finds the curve shown in Fig. 9.1, which has the shape of a rectangle. For $\varepsilon < \mu$ the probability that a level is occupied is 1, for $\varepsilon > \mu$ it is zero. If the temperature rises, that is, β falls, the transition becomes less abrupt as shown in Fig. 9.2. The probability of occupation at $\varepsilon = \mu$ always remains $\frac{1}{2}$.

A numerical accident, which gives rise to some confusion, is the fact that, in Fermi statistics, the probable number of particles in a quantum state is numerically identical with the probability of occupation of the state. This follows from the formula

$$\langle n(s) \rangle = p(s) \cdot 1 + (1 - p(s)) \cdot 0 \qquad (9.12)$$

which means that

$$\langle n(s) \rangle = \frac{1}{\exp \beta(\varepsilon_s - \mu) + 1} \qquad (9.13)$$

which is numerically the same expression as (9.11). However, formula (9.13) can be extended to groups of states and thereby can lead to numbers larger than 1. One common case of this is the appearance of a factor 2 for spin if the problem is otherwise independent of spin. Another common case arises if the energy spectrum is continuous or quasi-continuous. One introduces then a *density of states function* $g(\varepsilon)$ which gives the number of one-particle states per unit energy range. With the help of this function and formula (9.13) the number of particles in the energy range $d\varepsilon$ becomes

$$n(\varepsilon) \, d\varepsilon = \frac{g(\varepsilon) \, d\varepsilon}{\exp \beta(\varepsilon - \mu) + 1} \qquad (9.14)$$

Here the probability aspect of the number emphasized on the left side of (9.13) can as a rule safely be discarded, because the numbers handled in (9.14) are usually large and the corresponding fluctuations small. A corresponding expression can be written for the Boltzmann formula, starting from (8.65). It differs from (9.14) only by the missing 1 in the denominator. Thus the two statistics are equivalent if the chemical potential is substantially smaller than any energy in the one-particle energy spectrum. The interest in Fermi statistics is of course primarily in cases in which the equivalence does not hold. For any state having an energy less than μ, formula (8.65) predicts a probable number of occupants in excess of 1, while (9.13) yields a number larger than $\frac{1}{2}$ but smaller than 1. Another way to look at this difference is to say that relative population is not sensitive to total density in Boltzmann statistics while in Fermi statistics it is: a high density tends to push the Fermi level up so as to avoid multiple occupation.

The relationship between the total number of particles and the Fermi level is obtained by integration of (9.14). It yields

$$N = \int_{-\infty}^{+\infty} \frac{g(\varepsilon) \, d\varepsilon}{\exp \beta(\varepsilon - \mu) + 1} \qquad (9.15)$$

In the classical limit this reduces to the more trivial relation

$$N = \exp \beta \mu \cdot f$$

which was already noted in (8.66). This latter relation permits elimination of μ from any problem. In the more difficult relationship (9.15) this is not the case, and the Fermi energy μ must usually be carried as a parameter.

9-3. Theory of the perfect Fermi gas

We shall now illustrate these general principles by considering a perfect gas subject to the Pauli principle. We have given in (4.22) the number of

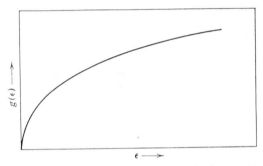

Fig. 9.3. Density of states versus energy for free particles.

states in a volume of phase space. The only form of energy present is the translational kinetic energy (8.49). A straightforward fivefold integration then permits computation of $g(\varepsilon)$ from (4.22). An extra factor 2 must be added for spin. We find

$$g(\varepsilon) = \frac{2^{1/2} \pi m^{3/2} V}{h^3} \varepsilon^{1/2} \tag{9.16}$$

where V is the volume of the gaseous sample. The behavior of this function of energy is shown in Fig. 9.3. With (9.16) the side condition (9.15) takes the form

$$N = \frac{2^{1/2} \pi m^{3/2} V}{h^3} \int_0^\infty \frac{\varepsilon^{1/2} \, d\varepsilon}{\exp \beta(\varepsilon - \mu) + 1} \tag{9.17}$$

and the energy becomes

$$U = \frac{2^{1/2} \pi m^{3/2} V}{h^3} \int_0^\infty \frac{\varepsilon^{3/2} \, d\varepsilon}{\exp \beta(\varepsilon - \mu) + 1} \tag{9.18}$$

According to equation (2.39), expression (9.18) also yields the pressure of the gas by the virial theorem.

Equations (9.17) and (9.18) are not expressible analytically in the general case. We shall therefore be satisfied to discuss limiting cases. If μ is negative and in magnitude large compared to kT, the 1 in the denominators can be left off and the integrals can be done. We find a repetition of (8.66) and (8.67), namely,

$$N = 2 \left(\frac{2\pi m k T}{h^2} \right)^{3/2} V \exp \beta \mu$$

$$U = \left(\frac{2\pi m k T}{h^2} \right)^{3/2} V \exp \beta\mu \cdot \frac{3}{\beta}$$

or more briefly

$$U = \tfrac{3}{2} N k T$$

Equation (2.39) yields then

$$p = \frac{2}{3} \frac{U}{V} = \frac{N k T}{V}$$

These are the usual formulas for the Maxwell gas. More interesting is the other limiting case, in which the occupation number varies as shown in Fig. 9.1. The denominator of the Fermi function can then be considered 1 up to the Fermi energy μ and thereafter equal to zero. In other words we assume

$$\mu \gg k T \tag{9.19}$$

and obtain then from (9.17) and (9.18)

$$N = \frac{2^{9/2} \pi m^{3/2} V \mu^{3/2}}{3 h^3} \tag{9.20}$$

$$U = \tfrac{3}{5} N \mu \tag{9.21}$$

and from (2.39)

$$p = \frac{2}{5} \frac{N \mu}{V} \tag{9.22}$$

The proper way to read these equations is to invert (9.20) so as to get μ in terms of the particle density. This yields

$$\mu = \frac{3^{2/3} h^2}{8 \pi^{2/3} m} \left(\frac{N}{V} \right)^{2/3} \tag{9.23}$$

To illustrate the working of these formulas in practical situations we may take as examples electrons in metals and He³ atoms at solid or liquid densities. In the first case

$$\mu \sim 10^{-12} \text{ erg}$$

and in the second

$$\mu \sim 10^{-16} \text{ erg}$$

Thus we find ourselves in the limiting case (9.19) for electrons in metals at all reasonable temperatures, while for He³ we have to go to 1° K or lower. Correspondingly, the energy (9.21) of an electron gas at metallic densities is about two orders of magnitude larger than thermal energy. Even more extreme is the result (9.22) for the pressure of such a gas. It is about 10^6 atm, that is, of the order of the elastic moduli of a solid. It follows from (9.22) and (9.23) that for a gas in the condition described by (9.19) the pressure varies as the 5/3 power of the density. A gas obeying condition (9.19) is said to be *degenerate*. The results derived here for a degenerate Fermi gas must be applied with some caution to electrons in metals. Such electrons do not constitute a perfect Fermi gas. In fact, a perfect degenerate Fermi gas has never been observed in nature. The reason is that the conditions for degeneracy are also those which permit the coming into play of the interparticle forces.

There is an independent and very simple way to derive the degeneracy criterion which arises from elimination of μ between (9.19) and (9.23). Classical particle physics forms a physically valid approximation to quantum physics whenever the mean interparticle distance is large compared to the De Broglie wave length which one would assign to the particles on the basis of their temperature; for the wave nature of the particles will then not interfere with the behavior of a suitably constructed wave packet. Inversely, specific quantum phenomena such as Fermi degeneracy can only occur if the condition is reversed. The criterion for degeneracy we get out of this is

$$\left(\frac{V}{N}\right)^{1/3} \ll \frac{h}{m\,v}$$

or with (4.26)

$$k\,T \ll \frac{h^2}{3\,m}\left(\frac{N}{V}\right)^{2/3} \tag{9.24}$$

This is essentially equivalent to (9.19) and (9.23) except for being a little less stringent in the numerical factor. The degeneracy condition has thus a very direct explanation in terms of the wave nature of particles in quantum mechanics.

9-4. Bose–Einstein statistics

The quantum-mechanical basis of Bose statistics has been laid in equations (9.01) and (9.03). That this type of statistics must occur is implicit in the work on Fermi statistics. Whenever we deal with composite particles which consist of an even number of particles obeying Fermi statistics, these composite units must obey Bose statistics. As mentioned above, this partial statement seems to generalize to the rule that all particles, composite or not, which have half odd integral spin obey Fermi statistics and all particles having integral spin obey Bose statistics.

The example (9.03) shows us that the requirement of total symmetry of the wave function leads to a count of states which differs from the count in Boltzmann and also from the count in Fermi statistics. In fact, equation (9.01) shows that, once the occupation numbers are specified, only one wave function can be constructed for the state, because all permutations of the starting arrangement are included in the total wave function. This time, however, multiple occupation is allowed. In fact, it is favored, for the new statistics discards the permutational multiplicity factor (8.43). This factor favors a wide distribution of particles among states to keep the denominator small. Leaving off the factor thus favors multiple occupation. Such a trend is equivalent to some sort of attractive force since Boltzmann statistics represents randomness and absence of any force. In the example of two particles distributed among two states, for instance, we find from equation (9.03) that the probability of finding the two particles in the same state is $\frac{2}{3}$ in Bose statistics while the random value is $\frac{1}{2}$. This must be kept in mind in the discussion of the Einstein condensation phenomenon discussed below.

In discussing the statistics of a set of identical particles obeying Bose statistics we start out again with the Gibbs formula in the form (9.06). This time any occupation number is allowed, but the state is completely specified when these numbers are given. Thus (9.06) becomes

$$P(N; n_1, n_2, n_3, \ldots) = \frac{1}{\mathscr{F}} \exp\left[\beta \left(\mu N - \sum_s n_s \varepsilon_s\right)\right] \qquad (9.25)$$

with the side condition

$$\sum_s n_s = N \qquad (9.26)$$

The side condition can be used to eliminate N from (9.25) so that we have simply

$$P(n_1, n_2, n_3, \ldots) = \frac{1}{\mathscr{F}} \exp\left[\beta \sum_s n_s \left(\mu - \varepsilon_s\right)\right] \qquad (9.27)$$

Here every combination of numbers n_s is present and all combinations have weight 1. It follows that the sums over n_s in the grand partition function $\mathscr{F}$ are independent, and we can simply write

$$\mathscr{F} = \sum_{n_1=0}^{\infty} \sum_{n_2=0}^{\infty} \cdots \exp\left[\beta \sum_s n_s (\mu - \varepsilon_s)\right] \tag{9.28}$$

This expression factors into one separate factor for each quantum state. It then reads

$$\mathscr{F} = \prod_s \left\{ \sum_{n_s=0}^{\infty} \exp \beta n_s(\mu - \varepsilon_s) \right\}$$

The summation on n_s converges only if μ is smaller than ε_s. It yields then

$$\mathscr{F} = \prod_s \frac{1}{1 - \exp \beta(\mu - \varepsilon_s)} \tag{9.29}$$

We see that $\mathscr{F}$ factors again in terms of one-particle states as it does in the other statistics.

With $\mathscr{F}$ factored and the exponential in (9.27) factored, the entire occupation probability (9.27) is factored into independent probabilities for each state as in the other statistics. From (9.27), the probability $p_s(n_s)$ that the occupation number for the state s has the value n_s is equal to

$$p_s(n_s) = \exp[\beta n_s (\mu - \varepsilon_s)] \left\{ 1 - \exp \beta(\mu - \varepsilon_s) \right\} \tag{9.30}$$

It is to be observed that this occupation probability has Boltzmannian form, just as in the preceding case of fermions where only $n_s = 0$ and $n_s = 1$ were allowed. One can therefore take the point of view that the two quantum statistics are a form of Boltzmann statistics in which the occupation numbers instead of the particles are the basic element of independent counting.

When Bose statistics is applied, we find as previously that the probability of occupation (9.30) is not, as a rule, the quantity wanted. One usually wants the probable number of particles occupying the state s. In Fermi statistics a numerical coincidence arose between this number and the probability of occupation, namely, the numerical identity of (9.11) and (9.13). This is not the case here. We have

$$\langle n_s \rangle = \sum_{n_s=0}^{\infty} n_s \, p_s(n_s)$$

which, from (9.30), comes out to be

$$\langle n_s \rangle = \frac{1}{\exp \beta(\varepsilon_s - \mu) - 1} \tag{9.31}$$

This is the *Bose distribution law*. Just like the Fermi law, it depends on two parameters, the temperature β and the chemical potential μ. In principle the latter quantity could be eliminated by the side condition on the number of particles

$$N = \sum_s \frac{1}{\exp \beta(\varepsilon_s - \mu) - 1} \tag{9.32}$$

In practice this equation can be solved for μ only in exceptional cases, and μ has to be carried as a parameter.

Because of the statistical independence of different quantum states, equation (9.31) can be transformed like (9.13) to yield the number $n(\varepsilon)$ of particles per unit energy range. We find

$$n(\varepsilon) \, d\varepsilon = \frac{g(\varepsilon) \, d\varepsilon}{\exp \beta(\varepsilon - \mu) - 1} \tag{9.33}$$

The probability aspect of the left-hand side is again suppressed because (9.33) is usually applied to large numbers of states and particles.

Comparison of the three forms of particle statistics shows that only Boltzmann statistics has statistical independence of the particles, as expressed in (8.41). From the standpoint of the one-particle quantum states, however, all three statistics exhibit statistical independence. In other words the probability $P(n_1, n_2, n_3, \ldots)$ that the one-particle quantum states $1, 2, 3, \ldots$ be occupied by $n_1, n_2, n_3, \ldots$ particles obeys the decomposition rule

$$P(n_1, n_2, n_3, \ldots) = p_1(n_1) \, p_2(n_2) \, p_3(n_3) \ldots \tag{9.34}$$

Expressions (8.63), (9.11), and (9.30) for $p_s(n_s)$ are different in the three cases. They are actually not well known and rarely used directly. The more widely known expressions are for $\langle n_s \rangle$, which is derived from $p_s(n_s)$. The three formulas (8.65), (9.13), and (9.31) are plotted in Fig. 9.4 as functions of ε for the same values of β and μ. It is seen that when ε is larger than μ by an amount which is large compared to kT the difference between the three statistics is unimportant and most states are empty. We have seen that this situation prevails for the entire energy spectrum in many cases of interest. If this is not the case the three statistics predict different results. In particular, the situation for bosons bears no resemblance to either one of the other two statistics. In those cases, μ could sit in the middle of the energy spectrum, indicating the point at which the mean number of particles in a quantum state is either 1 or $\frac{1}{2}$. For bosons, however, if μ coincided with any quantum state the number of particles in it would be infinite. A μ greater than any one-particle energy is definitely excluded altogether. We had to assume this formally to carry out the summation in (9.29); in addition, the end results (9.30) and (9.31) also demand it because violation of the condition would yield meaningless negative numbers.

It follows from the above that a system of bosons cannot exhibit the Fermi type of degeneracy discussed earlier in the chapter. There remains, however, a more subtle type of degeneracy to be investigated. In certain

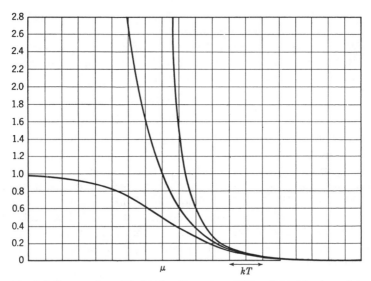

Fig. 9.4. Comparative plot showing the mean number of particles in a state against the energy for the three statistics; the values of μ and kT are chosen to be the same in three cases.

cases, the chemical potential μ can come extremely close to the state of lowest energy ε_0 of the particles, making the quantity $\langle n_0 \rangle$ in (9.31) abnormally large. This type of anomaly is discussed most readily for the case of free particles.

9-5. The perfect Bose gas; Einstein condensation

For free particles obeying Bose statistics, the density of states is given by (9.16), modified by a factor $\frac{1}{2}$ for zero spin. Equation (9.32) then yields for the total number of particles

$$N = \frac{2^{5/2}\pi\, m^{3/2}\, V}{h^3} \int_0^\infty \frac{\varepsilon^{1/2}\, d\varepsilon}{\exp\beta(\varepsilon-\mu)\, -1} \tag{9.35}$$

Reduction in the limiting case of low densities proceeds as for fermions and yields the perfect gas laws. Einstein was the first to notice, however, that the case of high density and low temperature has a degeneracy phenomenon all its own which in no way resembles the degeneracy of a Fermi gas. The integral on the right-hand side has an upper limit which is reached when the negative quantity μ reaches the value zero. This implies an upper limit for the density or a lower limit for the temperature of the gas. These critical values N_c, V_c, T_c are related by the formula

$$N_c = \frac{2\pi m k T_c^{3/2}}{h^2} V_c \, \zeta(\tfrac{3}{2}) \qquad (9.36a)$$

where the zeta function of $\tfrac{3}{2}$ coming in equals

$$\zeta(\tfrac{3}{2}) = 1 + \frac{1}{2^{3/2}} + \frac{1}{3^{3/2}} + \frac{1}{4^{5/2}} + \ldots = 2.61 \qquad (9.36b)$$

However, a real gas cannot be subject to such limitations, and therefore a computational error must have been committed in getting these results. Einstein traced this error to the substitution of (9.35) for (9.32). This substitution is generally all right except when μ comes extremely close to the state of lowest energy $\varepsilon_0 = 0$. In these circumstances the first term in the series takes the form

$$N_0 = \frac{1}{\exp \beta(\varepsilon_0 - \mu) - 1}$$

or approximately

$$N_0 = \frac{kT}{\varepsilon_0 - \mu} \qquad (9.37)$$

This term can become arbitrarily large and must become so for particle numbers larger than N_c. Then μ must come so close to ε_0 that N_0 becomes a finite fraction of all the particles. Numbers of particles in excess of N_c are thus accommodated, and (9.36) is replaced by

$$N = N_0 + N' \qquad (9.38)$$

where N_0 is computed from (9.37) and N' from (9.36). This reads explicitly

$$N = \frac{kT}{-\mu} + \left(\frac{2\pi m k T}{h^2}\right)^{3/2} V \, \zeta(\tfrac{3}{2}) \qquad (9.39)$$

The way this formula is used is that one first computes the second term in (9.39) to find the number N' of normal particles. Thereupon the chemical potential μ is adjusted to make up the difference between the computed and the actual number. This yields

$$- \mu = \frac{k\,T}{N - N'} \qquad (9.40)$$

The number so obtained for $- \mu$ can be extremely small. If this is the case, the ground state accommodates the particles which have no room in the "normal" fluid. These particles make no contribution to the energy and pressure and form a special sort of "condensed" phase. The onset of this phenomenon occurs when the external parameters have reached the condition described by (9.36). We shall apply (9.36) to a gas having the density of liquid helium, $\rho = 0.178$ gm/cm³. We then find a critical temperature T_c equal to

$$T_c = 3.2° \text{ K}$$

The theory predicts that below this temperature a perfect gas having the density of liquid helium consists of two fluids, one of which has no energy except zero-point motion. The fraction of the gas in this condensed condition increases as the temperature is lowered.

Now it is indeed a fact that liquid helium undergoes a transition of this sort at 2.2° K, namely, the transition from liquid He I to liquid He II. The low-temperature form, liquid He II, acts as a mixture of a superfluid and a normal component. The superfluid component has no heat capacity or viscosity, and the fraction of normal fluid tends to zero at absolute zero. Both these results are in agreement with the theory presented above, and the value of T_c comes out to be approximately right. It is therefore believed that the interpretation of this phenomenon as Einstein condensation is correct. There are of course complications in the case of liquid He II which arise from the fact that it is not a gas but a liquid. The interatomic forces can therefore not be neglected. Nevertheless, the calculation yields a very good, if qualitative, picture of the phenomenon.

A. RECOMMENDED PROBLEMS

1. A set of independent particles is all in the same type of state having the same energy. By an external agency (magnetic field) this state is split into two nondegenerate states having $\varepsilon = + 10^{-16}$ erg and $\varepsilon = - 10^{-16}$ erg. The probabilities of occupation of the two substates are found to be $\frac{1}{3}$ and $\frac{2}{3}$, respectively. Find the temperature and chemical potential of the particles according to the three statistics.

2. Show, by following in detail the argument of Section 9–5, that there is *no* Einstein condensation for a boson gas of free particles in one or two dimensions.

B. GENERAL PROBLEMS

3. Show for a perfect gas that its adiabats obey the relation

$$p \, V^{5/3} = \text{constant}$$

independently of the statistics of the particles composing the gas. The only restriction is that the particles have no internal degrees of freedom making a contribution to their thermal properties.

4. Suppose we deal with a set of fermions distributed among a set of one-particle states s. Show from (5.13) that, if the probability of the state s being filled equals p_s, the entropy of the fermion system equals

$$S = - k \sum_s \{p_s \ln p_s + (1 - p_s) \ln (1 - p_s)\}$$

5. Show for all three statistics that the entropy of a system of uncoupled particles distributed over one-particle states s equals

$$S = - k \sum_s \sum_{n_s = 0}^{\infty} p_s(n_s) \ln p_s(n_s)$$

Here n_s is the number of particles in the state s, and $p_s(n_s)$ the probability that the state s is so occupied.

6. Show for a set of bosons distributed over one-particle states s that in *thermal equilibrium* their entropy equals

$$S = k \sum_s \{\langle n_s + 1 \rangle \ln \langle n_s + 1 \rangle - \langle n_s \rangle \ln \langle n_s \rangle\}$$

7. The formula of Problem 6 does not yield zero if there is only a single occupied state, and thus is not always correct. Trace the error in this formula and related expressions, and indicate under which conditions the results are valid.

PART II

Equilibrium statistics
of special systems

10

Thermal properties of electromagnetic radiation

It is fitting that a survey of applications of statistical mechanics start with the properties of electromagnetic radiation in thermal equilibrium. It is true that this is not the simplest application of the general theory. But it is the domain in which simple assumptions are more nearly valid than elsewhere. We are thus led to very precise theoretical predictions, many of which were confirmed by experiment in the second half of the nineteenth century. On the other hand, discrepancies between theory and experiment arose in the same period whose solution required the quantum assumption for light. This assumption spread and broadened thereafter into modern quantum theory.

Equilibrium theory of electromagnetic radiation is essentially equilibrium theory of "empty space." This comes from the great ease with which photons are created and destroyed and from the small energy of quanta of long wave lengths. Thus a region of space cannot really be said to be in thermal equilibrium if its radiation field is not in thermal equilibrium. We shall refer to such equilibrium radiation as *heat radiation.*

10-1. Realization of equilibrium radiation; black body radiation

It is not easy in practice to create conditions which bring the electromagnetic field into true thermal equilibrium. This is because the radiation field is influenced quite readily by material bodies in the neighborhood, which emit light selectively or reflect it selectively as it arrives from other sources. These sources and the reflecting bodies all have their own temperature; their high heat capacity easily swamps the heat capacity of the "empty space" surrounding them and thereby produces a pattern of radiation very far from equilibrium. We need therefore special experimental conditions to observe true heat radiation. The best arrangement to maintain radiation at thermal equilibrium is the hollow interior of a body kept at constant temperature, a so-called *isothermal enclosure*. In order to observe the nature of the radiation in such an enclosure a small hole is made in its side and the light escaping through it observed. Presumably, if the hole is very small in comparison to the enclosure, the equilibrium within will not be seriously disturbed.

Some properties of heat radiation follow directly from the concept of thermal equilibrium. The radiation must be isotropic and unpolarized, and have the same intensity and frequency composition at every point. For if this were not the case a Carnot engine could be constructed to exploit these irregularities. Such an engine would extract work from a source at constant temperature and thus would be in violation of Kelvin's postulate. It follows that there must exist a universal function of the frequency ν and the temperature T, which gives the energy density of heat radiation per unit frequency range at a given temperature. We shall denote this function by $\Psi(\nu, T)$.

The first step in elucidating the nature of $\Psi(\nu, T)$ was taken by Kirchhoff, who observed that the radiation in an isothermal enclosure should not be disturbed by the insertion of an arbitrary body, provided the body is at the same temperature as the light "bath" in which it rests. Kirchhoff restricted his reasoning to opaque bodies, that is, bodies which transmit no light. Such a body can still have very complicated properties as regards reflection and emission of light. Kirchhoff then reasoned that the emission of light by such a body must be linked to its absorption because the emission must be exactly such as to make up the deficit left by its absorption in the heat radiation spectrum. The fraction of light absorbed at a certain point of the surface of a body is called the *absorptivity* $A(\nu)$ at that point. For if the body is opaque then any light not absorbed is reflected, and an element of surface is characterized as far as radiation is concerned by this quantity $A(\nu)$ and the *emitting power* $E(\nu)$, which is measured in ergs/cm² sec. Both $A(\nu)$ and $E(\nu)$ should be broken down into quantities dependent on angle, but we shall ignore this fine point in the present discussion. We can then say globally that if an opaque body is in thermal equilibrium with the radiation field, its emitting power and its absorptivity must be proportional at every point of its surface

$$\frac{E(\nu)}{A(\nu)} = \text{constant} \tag{10.01}$$

This constant is independent of the position and orientation of the surface, or the nature of the inserted body. It can still be a function of temperature and frequency with respect to which it acquires a universal character. Kirchhoff then went on to think of a "perfectly black" body, that is, a body which absorbs completely in all wave lengths and thus reflects nothing. For such a body the absorptivity $A_b(\nu)$ is identically equal to unity by assumption, and thus its emitting power $E_b(\nu)$ will be equal to the constant in equation (10.01), which was just shown to be a universal function of frequency and temperature. This function is linked in a simple way to the other universal function, the energy density of heat radiation. We must only remember (a) that all light travels at the velocity c, (b) that our surface radiates only into one hemisphere, and (c) that the aspect of a surface becomes smaller when viewed at a glancing angle.

The three items taken together yield

$$\frac{E(\nu)}{A(\nu)} = E_b(\nu) = \tfrac{1}{4} c \, \Psi(\nu, T) \tag{10.02}$$

The problem of heat radiation may thereby be formulated as the problem of the emitting power $E_b(\nu)$ of a perfectly black body, or the problem of *black body radiation*. The experimental device mentioned earlier, namely, the small hole leading to the interior of an isothermal enclosure, is a close approximation to an element of surface of a perfectly black body; it absorbs almost all the radiation falling upon it from the outside, and transmits and reflects little. The approximation can be improved by coating the interior of the enclosure with soot, for it is important experimentally to have no light coming from the outside return to the hole after repeated reflection inside the enclosure. Observation of such a hole is thus the correct way to elucidate the nature of heat radiation.

10-2. Thermodynamics of black body radiation; laws of Stefan-Boltzmann and Wien

It is possible to derive all relevant properties of heat radiation from modern quantum theory and to dispose thereby of the entire subject in a couple of pages. Such an approach does not do justice to the subject. The physicists at the close of the nineteenth century were intensely interested in it but found that they could not get sensible results by applying statistical mechanics. However, many of these men, such as Kirchhoff, Boltzmann, Rayleigh, Wien, and Nernst, were very ingenious. Lacking a theory, they squeezed the two laws of thermodynamics harder than they have ever been squeezed before or since, and established thereby an astonishing number of properties of heat radiation. We shall follow their way of thinking here, and derive in this section those properties of heat radiation which can be derived from thermodynamics. The proper statistical approach will then follow in Section 10-3.

For a logically clear study of heat radiation it is preferable to think of the isothermal enclosure not as being coated with soot or some other absorber, but as having on the contrary perfectly reflecting walls (no hole in the side, of course, either). Many metals come close to this condition for wave lengths equal to or longer than that of visible light. The intensity of light in the interior of such a cavity can then be analyzed into *normal modes* by well known methods. Study of these normal modes should be equivalent to study of heat radiation as long as the wave length is small compared to the size of the cavity. In such a situation the modes are almost uncoupled. We deal then with a cavity containing a set of uncoupled "particles" exerting pressure on the wall. The virial theorem should therefore apply to this situation. To apply it we have first to rederive it, because of the modified relationship between energy and momentum.

We may accept here without proof the well established result of electromagnetic theory according to which the relationship between the energy and the momentum of a light wave reads

$$E = c\,p \qquad (10.03)$$

This is to replace relation (4.24) for material particles. Actually either relation is a limiting form of the relativisitic formula

$$E^2 = m^2\,c^4 + p^2\,c^2$$

according to whether the mass term or the momentum term predominates in the energy. Light, having no mass, always obeys (10.03) for any value of p.

We now can take our case of light enclosed in a cavity, and apply to it the reasoning of the virial theorem in Chapter 2. Let the volume of the cavity be V, and let there be "particles" in the interior of the cavity which interact with the wall, but not with each other. Furthermore, let their kinetic energy and their momentum be related by equation (10.03) instead of (4.24). With this modification we can still proceed with the argument following (2.35). The result is a modified equation (2.37) which reads

$$U = \overline{\sum_{\nu} c_{\nu}\,p_{\nu}} = \overline{\sum_{\nu} \mathbf{x}_{\nu} \cdot \frac{\partial V}{\partial \mathbf{x}_{\nu}}} \qquad (10.04)$$

Reduction of the right-hand side to a wall effect exclusively goes through as before and yields finally in place of (2.39)

$$p \, V = \tfrac{1}{3} \, U \qquad (10.05)$$

This equation differs from the earlier result by a factor $\tfrac{1}{2}$ on the right.

Boltzmann first derived equation (10.05) and from it the law giving the dependence of the energy of heat radiation on temperature. This energy is related to the universal function $\Psi(\nu, T)$ introduced earlier by the formula

$$U(T, V) = V \int_0^\infty \Psi(\nu, T) \, d\nu \qquad (10.06)$$

It therefore has the form

$$U(T, V) = V \, u(T) \qquad (10.07)$$

where the energy density $u(T)$ is the integral of $\Psi(\nu, T)$ over frequency. It is therefore, like Ψ, a universal function of temperature and independent of volume. To derive a relation involving temperature we must bring the Second Law into the picture. We use it in the form (7.07)

$$\left(\frac{\partial U}{\partial V} \right)_T = - \, p + T \left(\frac{\partial p}{\partial T} \right)_V$$

Combining this equation with (10.05) and (10.07) we get

$$\frac{du}{dT} = 4 \, \frac{u}{T}$$

and hence

$$u(T) = u_0 \, T^4 \qquad (10.08)$$

This is the *Stefan–Boltzmann law* giving the variation of energy with temperature for electromagnetic radiation in thermal equilibrium. The law is usually stated via equation (10.02) in terms of the emitting power E_b of the surface of a black body

$$E_b = \sigma \, T^4 \qquad (10.09)$$

The constant σ is then called the *Stefan constant*. It is one of the fundamental constants of nature. Its numerical value has been measured as

$$\sigma = 5.67 \times 10^{-5} \text{ erg/cm}^2 \text{ sec deg}^{-4} \qquad (10.10)$$

From (10.02) the constant in formula (10.08) for the energy density comes out to be

$$u_0 = 7.56 \times 10^{-15} \text{ erg/cm}^3 \text{ deg}^{-4} \qquad (10.11)$$

One verifies easily that even at 1000° K the heat capacity of 1 cm³ of empty space is very small, of the order 10^{-11} that of 1 cm³ of water. However, it rises faster with temperature than that of material objects.

This seems to be about all that one can reasonably expect from thermo-dynamics concerning heat radiation. But heat radiation was a field which preoccupied the physicists of the later nineteenth century deeply because they felt they had a complete theoretical understanding of electromagnetism and statistics, and yet they were unable to make complete predictions, particularly as regards spectral distribution. Into this situation stepped Wien, the "last of the classical physicists," who looked over the field and squeezed still another result out of the classical theory.

We have seen in Problem 1 of Chapter 5 that for simple monatomic gases adiabatic expansion stands in a specially simple relationship to the energy states of gas molecules. Expansion of simple gases, such as neon, argon, and mercury, shifts the entire energy spectrum to lower energies, as determined by the law $T^{3/2} V =$ constant. No internal rearrangement of velocities is needed to re-establish equilibrium: adiabatic expansion leads directly from one equilibrium state to the other. This fact goes beyond the Ehrenfest principle, which asserts only that the *initial* force is controlled by the potential adiabatic response; the principle does not deny that, usually, thermal equilibrium might be disturbed by a finite change of the external variables, and that rearrangement of energy among the degrees of freedom might follow. This is in fact the case for the diatomic gases O_2, N_2, H_2, in which adiabatic expansion results at first in loss of energy of the translational degrees of freedom; thereafter this process is followed almost instantaneously by a sharing of this loss with the rotational spectrum through inelastic collisions. However, simple gases have this simple response. So Wien hypothesized that very probably heat radiation might also be of this simple type. He was supported in this guess by the fact that no structural peculiarity of radiation is known which could produce the kind of complica-tion occurring in diatomic gases.

Pursuing this idea further, Wien observed that Boltzmann's derivation, although naturally meant for heat radiation integrated over all frequencies as shown in equation (10.06), actually never made explicit use of this assumption. All parts of his derivation apply in fact to any portion of the total spectrum, provided it is large enough to be reasonably isotropic and is roughly uniform in intensity over the volume of the cavity. The only possible exception is the use of (7.07), which brings in the concept of temperature. However, this need not disturb us. If heat radiation is simple in the sense in which the noble gases are simple, then the internal coupling of the modes is entirely superfluous and even equation (7.07) will apply to portions of the spectrum. This is implied in the notion that each portion of the spectrum adjusts by itself alone into its new condition upon adiabatic expansion.

The first step we have to make to carry out Wien's program is to find the equation obeyed by adiabats if a container encloses only radiation. The relevant equations are (1.36) and (10.05), which solve without difficulty to yield

$$p^{3/4} V = \text{constant} \tag{10.12}$$

The relation between temperature and volume results from this by (10.05), (10.07), and (10.08). It reads

$$T^3 V = \text{constant} \tag{10.13}$$

The next step is to find the frequency shift upon expansion, or more precisely what frequency is imaged on what other frequency by the expansion process. Wien did this by a Doppler shift argument. We shall accomplish the same end by an argument based on geometry. If we expand uniformly all dimensions of the isothermal enclosure, the wave length of every mode must increase as the linear dimension. This means that

$$\lambda \propto V^{1/3}$$

or

$$\nu^3 V = \text{constant} \tag{10.14}$$

Combining (10.13) and (10.14), we get upon adiabatic expansion

$$\frac{\nu}{T} = \text{constant} \tag{10.15}$$

Equation (10.15) is the essence of the *Wien displacement law*. In adiabatic change the frequencies are shifted in such a way that frequency is proportional to temperature. There remains the job of drawing the conclusions concerning the spectral density function $\Psi(\nu, T)$. If an adiabatic expansion of an isothermal enclosure starts out at a temperature T and ends with T', then we have clearly from (10.06) and (10.08)

$$\int_0^\infty \Psi(\nu', T') \, d\nu' = \frac{T'^4}{T^4} \int_0^\infty \Psi(\nu, T) \, d\nu$$

By Wien's hypothesis we remove the integral signs from this equation and write

$$\Psi(\nu', T') \, d\nu' = \frac{T'^4}{T^4} \Psi(\nu, T) \, d\nu \tag{10.16}$$

Here ν and ν' are understood to be corresponding frequency intervals in the sense of the preceding analysis. This means equation (10.15) is relevant, yielding

$$\nu' = \frac{T'}{T}\,\nu, \qquad d\nu' = \frac{T'}{T}\,d\nu$$

Insertion of this into (10.16) yields

$$\Psi\left(\frac{T'}{T}\,\nu,\, T'\right)\frac{T'}{T}\,d\nu = \frac{T'^4}{T^4}\,\Psi(\nu,\,T)\,d\nu \tag{10.17}$$

This expression can be simplified in obvious ways. Thereafter, we get the end result by setting $T' = 1$ and defining

$$\Psi\left(\frac{\nu}{T},\, 1\right) = \psi\left(\frac{\nu}{T}\right) \tag{10.18}$$

We then find

$$\Psi(\nu,\,T) = T^3\,\psi\left(\frac{\nu}{T}\right) \tag{10.19}$$

This is the Wien displacement law in slightly modernized form. It states that there is only one distribution law valid at all temperatures. If frequency is plotted along the abscissa and $\Psi(\nu,\,T)$ along the ordinate, the curve expands to the right linearly with the temperature, and at the same time expands as the cube of the temperature in the vertical direction. Both these changes can be nullified by a simple change in scale which brings all the curves to coincidence. The most conspicuous feature of (10.19) is the shift with temperature of the frequency having maximum intensity in the black body spectrum. This frequency $\nu_{\max}$ obeys the relation

$$\frac{\nu_{\max}}{T} = \text{constant} \tag{10.20}$$

The frequency is close to the visible part of the spectrum for the surface of the sun which is approximately at $5500°$ K, and approximates a perfect black body fairly closely.

The Wien displacement law is the capstone of classical physics. Immediately after its enunciation it was checked and confirmed by experiment with great precision. This law closing an era was published, not inappropriately, in 1899. Only two years later, in 1901, Planck enunciated the quantum hypothesis.

10-3. Statistics of black body radiation; Planck's formula

By dint of ingenuity the derivation of the Wien displacement law side-steps the problem of counting states. Because of this an unknown function is left in the final answer. The contemporaries of Wien knew that to determine this function they would have to do statistics, which means counting states. This problem of counting is fundamental to most of statistical physics, as was explained in Chapter 3. We shall have to follow the contemporaries of Wien in their endeavor to learn to do this counting properly.

An electromagnetic oscillation in a cavity with conducting walls acts like a harmonic oscillator. Just as the harmonic oscillator passes energy back and forth between kinetic and potential, so an electromagnetic oscillation passes energy back and forth between electric and magnetic. If we write the hamiltonian of such an oscillation in the form

$$\mathscr{H} = \tfrac{1}{2}\,\omega^2 p^2 + \tfrac{1}{2}\,q^2 \tag{10.21}$$

then the coordinate q equals essentially the product of the electric field amplitude and the square root of the volume, apart from a factor of the order unity. The boundary condition for the modes is that the electric field at the surface of the cavity be either zero, that is, have a node, or else normal and have an antinode. The count resulting from this is the same as if the standing waves had a node on all surfaces and a factor 2 were added for polarization, which we shall do in the following. Counting modes presented no difficulties for the physicists of 1900 because it had been done successfully for the scalar sound field in pipes and other acoustical resonators. The general functional form for q in a rectangular box of sides a, b, c equals

$$q = q_0 \sin \frac{\pi l x}{a} \sin \frac{\pi m y}{b} \sin \frac{\pi n z}{c} \sin (\omega t - \varphi) \tag{10.22}$$

where l, m, n are positive integers. $\pi\, l/a$, $\pi\, m/b$, $\pi\, n/c$ are the x, y, z components of the wave vector $\mathbf{k}$, respectively. ω is related to $\mathbf{k}$ by the wave equation

$$\nabla^2 q - \frac{1}{c^2}\frac{\partial^2 q}{\partial t^2} = 0$$

which yields

$$\omega^2 = c^2\,(k_x^2 + k_y^2 + k_z^2) \tag{10.23}$$

The number wanted is the number $g(\nu)$ of modes per unit frequency range. The starting point is preferably a Cartesian space of coordinates l, m, n in which all modes are represented by points of integer coordinates in the first octant. The number of modes in any region is twice the volume of that region. By the definition of $\mathbf{k}$ this yields then

$$N = 2 \iiint dl\, dm\, dn = \frac{2}{\pi^3} V \iiint dk_x\, dk_y\, dk_z \qquad (10.24)$$

Equation (10.23) gives then without trouble the totality of all modes with a frequency below ν

$$N(\nu) = \frac{2}{\pi^3} V \tfrac{1}{8} \tfrac{4}{3} \pi\, k^3$$

which works out to be

$$N(\nu) = \frac{8}{3} \pi\, V \frac{\nu^3}{c^3} \qquad (10.25)$$

$g(\nu)$ is the derivative of this function with respect to ν

$$g(\nu) = 8\, \pi\, V \frac{\nu^2}{c^3} \qquad (10.26)$$

This formula was well known to Rayleigh and transferred by him from sound to light. All that remained to be done is to find the mean energy of a harmonic oscillator of frequency ν. The result of this in classical physics, namely, the equipartition theorem, is found in Chapter 4. Since the hamiltonian for each mode has the form (10.21), that is, consists of the sum of two squares, its mean energy is $k\, T$. With (10.26) we get therefore

$$\Psi(\nu, T) = \frac{1}{V} g(\nu)\, k\, T$$

or

$$\Psi(\nu, T) = \frac{8\, \pi\, \nu^2\, k\, T}{c^3} \qquad (10.27)$$

This is the *Rayleigh–Jeans law*, the inevitable result of statistical counting in classical physics. The result is structurally in agreement with the Wien displacement law (10.19) and is reasonably accurate for frequencies below the frequency maximum indicated in (10.20). But the formula shows no such maximum. Even worse, the total energy U at any temperature, defined by (10.06) comes out to be infinite, because the contribution per unit frequency range increases indefinitely as the frequency increases. This result was referred to at the time as the "ultraviolet catastrophe." It represented something worse than a disagreement between theory and experiment, namely, a contradiction within the theory itself. Perfectly logical application of established physical theory led in this instance to results which were not just wrong, but entirely unthinkable.

There is no point in our following Planck through the timid steps by which he introduced the quantum hypothesis into physics. We can assert immediately that a harmonic oscillator having a hamiltonian of the form (10.21) has a series of nondegenerate quantum states whose energy equals

$$\varepsilon(n) = (n+\tfrac{1}{2})\, h\, \nu \qquad (10.28)$$

where n is any nonnegative integer. This formula still leads to trouble in the case of light because of the zero-point energy term. One can argue that the expression

$$\varepsilon(n) = n\, h\, \nu \qquad (10.29)$$

is also consistent with the hamiltonian (10.21), the entire difference being one of the order of some factors. Passage from (10.28) to the mean energy as function of temperature was accomplished in Chapter 4. Modifications introduced by (10.29) are superficial. We find

$$\langle \varepsilon \rangle = \frac{d}{d\beta} \ln \{ 1 - \exp(-\beta\, h\, \nu) \}$$

which becomes

$$\langle \varepsilon \rangle = \frac{h\, \nu}{\exp \beta h \nu - 1} \qquad (10.30)$$

Combination of (10.30) with the Rayleigh–Jeans density of states (10.26) yields

$$\Psi(\nu, T) = \frac{8\, \pi\, h\, \nu^3 / c^3}{\exp \beta h \nu - 1} \qquad (10.31)$$

This is the *Planck radiation law* which gave to the problem of black body radiation its final solution. It obeys the Wien displacement law (10.19). It also contains the Rayleigh–Jeans law (10.27) as a limiting case for low frequency. This means that it also satisfies the Stefan–Boltzmann law (10.08) and furnishes a numerical value for the Stefan constant. We find from (10.31)

$$
\begin{aligned}
u(T) &= \int_0^\infty \Psi(\nu, T)\, d\nu \\
&= \frac{8\, \pi\, h}{c^3} \int_0^\infty \nu^3 \{ e^{-\beta h \nu} + e^{-2\beta h \nu} + e^{-3\beta h \nu} + \ldots \}\, d\nu \\
&= \frac{48\, \pi\, h}{c^3\, (\beta\, h)^4} \left\{ 1 + \frac{1}{2^4} + \frac{1}{3^4} + \frac{1}{4^4} + \ldots \right\}
\end{aligned}
$$

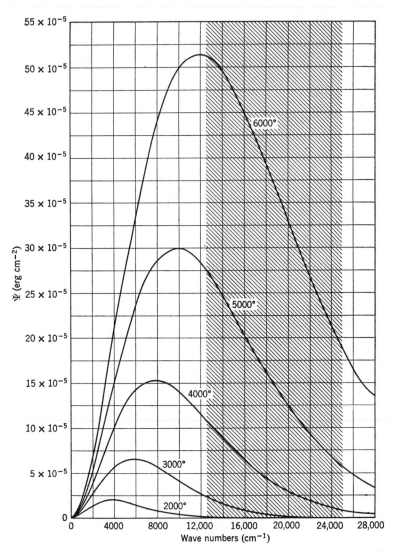

Fig. 10.1. Plot of Planck's law for black body radiation. Energy density per unit wave number versus wave number for 2000°, 3000°, 4000°, 5000°, and 6000° K. The visible range of the spectrum is cross-hatched. The area under any portion of the curve gives the energy density in erg cm^{-3} contained in that portion.

It is seen that the expression is proportional to T^4 as anticipated. The sum is known to equal $\pi^4/90$ so that we finally get for the constant u_0 in (10.08)

$$u_0 = \frac{8 \pi^5 k^4}{15 c^3 h^3} \tag{10.32}$$

This is in close agreement with the experimentally measured value (10.11). The Planck radiation law as a whole also fits the experiment very well. It pretty much had to, in view of the revolutionary assumption contained in its derivation.

Planck was a natural conservative and was reluctant to draw extreme conclusions from the quantum assumption which he had been forced to make. On the contrary, he tried for several years to reconcile his hypothesis with the established laws of physics. In this he did not succeed. He left it to his junior contemporary Einstein to draw from his work the radical conclusion that the corpuscular theory of light ought to be revived. Einstein came to this notion when studying the photoelectric effect. This effect is outside our interest here. We only want to show at this point that a statistical theory of photons as particles also leads to the Planck formula.

Einstein made the assumption that photons are particles whose energy equals $h\nu$:

$$\varepsilon = h\nu \tag{10.33}$$

It follows then from (10.03) that their momentum equals

$$p = \frac{h\nu}{c} \tag{10.34}$$

To get a count on the possible states of the photon we now proceed according to corpuscular theory. According to that theory, the number of possible states of a photon in any range equals the volume occupied by these states in phase space, measured in units of h^3. A factor 2 has to be added for the two polarizations. We thus have for the number of states

$$\mathcal{N} = \frac{2}{h^3} \int\int\int\int\int\int dp_x\, dp_y\, dp_z\, dx\, dy\, dz$$

We want to integrate this over the isothermal enclosure and the angles, leaving only frequency. This yields with (10.34)

$$\mathcal{N} = \frac{8 \pi V}{c^3} \int \nu^2\, d\nu$$

Refraining from integrating over frequency, we find indeed the Rayleigh–Jeans density of states (10.26)

$$g(\nu) = \frac{8 \pi V \nu^2}{c^3}$$

To proceed any further Einstein had to invent a new form of statistics, namely, Einstein–Bose statistics. In this statistics, interchange of two particles does not lead to a new arrangement. In other words, photons have no identity. We have to assume further that the chemical potential of photons is identically zero. There are two ways to understand this. One way is to say that if we take the Second Law in the form (8.10) we have to make the left-hand side stationary for the case of particles which can be created in arbitrary numbers. This forces us to set $\mu = 0$ to get rid of that differential. The other way to look at it is to say that in Einstein–Bose statistics μ is the energy at which the number of particles per state is effectively infinite; this is shown in Fig. 9.4. In this picture a photon which was absorbed has not been destroyed but has become a photon of zero frequency, energy, and momentum. Thus the number of such photons is effectively infinite and μ is again zero.

The upshot of this is that we combine formula (9.31) giving the number of particles per state with (10.26), the density of states function for free massless particles. In equation (9.31) we substitute ε from (10.33) and set $\mu = 0$ as explained above. The result is

$$\Psi(\nu, T) = \frac{8 \pi h \nu^3/c^3}{\exp \beta h \nu - 1}$$

which is again (10.31).

A. RECOMMENDED PROBLEMS

1. Justify the factor $\frac{1}{4} c$ in (10.02), which converts the energy density of black body radiation into the emitting power of a black body.

2. Apply the reasoning of Wien to the molecules of a perfect gas, and show that for thermodynamic reasons alone the distribution function for their speeds [previously written down in (4.29)] must have the form

$$p(v) = \frac{1}{\sqrt{T}} \, f\!\left(\frac{v}{\sqrt{T}}\right)$$

B. GENERAL PROBLEMS

3. From the point of view of radiation the sun is a circular disk acting as a black body having a temperature of $5500°$ K. Its diameter subtends an angle of $\gamma = 1°$ at the earth's surface. Compute from these data the power per unit area received in the mean at the earth's equator.

4. Assuming that the earth absorbs and re-emits as a black body the entire radiation incident upon it from the sun, calculate the resultant temperature of the earth's surface. Use the data of Problem 3.

5. Find the relation between pressure and volume which is valid for adiabatic compression of a cavity containing only electromagnetic radiation.

6. Einstein derived the Planck radiation law by placing a two-level "atom," rather than a black body, into an isothermal enclosure and postulating equilibrium for the atom and the radiation. He assumed that the transitions from the lower level 1 to the upper level 2 and vice versa are induced by radiation obeying the condition

$$h \nu = \varepsilon_2 - \varepsilon_1$$

In detail

$$\text{Rate of absorption} = B \, N_1 \, \Psi(\nu, T)$$
$$\text{Induced rate of emission} = B \, N_2 \, \Psi(\nu, T)$$

The equality of the two coefficients B follows from quantum mechanics. In addition there is a spontaneous emission process:

$$\text{Spontaneous rate of emission} = A \, N_2$$

A and B are independent of T. Show that if these assumptions are combined with the Wien displacement law, the Planck formula follows, apart from a constant.

11

Statistics of the perfect molecular gas

Many results germane to the following two chapters are scattered through Part I of this book, where they serve to illustrate some point about the general theory. It is not intended here to repeat this work but rather to go beyond it. However, it is worth while to begin the discussion by a survey of this material.

In Chapter 1 empirical equations of state were introduced for perfect and imperfect gases, and a superficial examination of the critical pheno-menon was suggested in the problem section. Introduction of the work term allowed a study of different kinds of specific heats for perfect gases, as well as of their adiabats. A straightforward application of the elementary statistical methods of Chapter 2 was made to density fluctuations and to the equation of state of perfect gases. Chapter 4 brought the Maxwellian distribution of velocities as the most important special case of the Gibbs law of energy distribution; it also contains the derivation of the classical values of gaseous specific heats. In Chapter 5 expressions were derived for the entropy of perfect gases and gas mixtures. Chapter 6 was concerned with a study of the Carnot cycle and other gaseous cycles. Chapter 7 brought in the thermodynamic potentials, all of which have their special form for gases. Chapter 8 defined the particle partition function, in addition to the system partition function and the grand partition function used up to that point; the most important application of this type of partition function is to gases; some simple cases were treated in the problems.

It may be observed that the preceding enumeration contains no reference to the degeneracy phenomena of quantum gases, which are discussed in Chapter 9. There is a good reason for this omission. The pressures and temperatures available in a terrestrial laboratory are such that gaseous form and degeneracy are mutually exclusive. The degeneracy phenomena discussed in Chapter 9 are nevertheless important to us, but in an indirect sense, through the indications they give about analogous phenomena in liquids and solids. The author is not familiar enough with cosmological theory to be able to assert positively that quantum degeneracy of gases does or does not arise in that field.

The calculations on gases can be divided into two classes, those which do, and those which do not, fit into a perfect gas framework. In this chapter, we shall take up the problems which fit into the perfect gas theory.

"Real, but perfect" gases constitute a sensible field of study because the perfect gas model is only specific about the interaction of the molecules with each other: the interaction must have a negligible effect on the averages computed. This assumption leaves open the possibility of a complicated internal structure. Most actual gases have such a structure. This structure can thus be investigated without the need to give up or modify the convenient perfect gas formalism. We shall refer to a gas of this type as a "perfect molecular gas."

11-1.   Decomposition of the degrees of freedom of a perfect molecular gas

In the new context it is convenient to think of a molecule as a collection of point particles, endowed with spin; the particles are the nuclei and the electrons. They are held together by electrical forces and act as a unit in long-range translation. Under these conditions we can avail ourselves of a rigorous theorem in mechanics (either classical or quantum) which states that the center of mass motion is separable from the internal degrees of freedom. This means also that the total kinetic energy splits into two disconnected terms. This, in turn, leads to a split-up of the partition function into a product of independent factors, as explained in equation (5.05). Once the separation is assured the translational part is simply given by (8.50). The internal part is not so immediately available. It has, however, the important property that of the two constraints on the gas, temperature and volume (or pressure), it depends only on temperature. We can therefore write the particle partition function f in the form

$$f(T, V) = f_{tr}(T, V) \cdot f_{int}(T) \tag{11.01}$$

Reduction of the partition function for internal motion into further parts is not an exact process. It is based on a general principle of mechanics, which permits approximate separation of two modes whose frequencies are of different orders of magnitude, even though they are tightly coupled. The reasoning will be given here in qualitative form, with an exactly soluble case among the problems. When two oscillators are coupled by a coupling term which is comparable to their own restoring force, they still will not mix appreciably if their frequencies are of different orders of magnitude. For, if we start with the high-frequency mode we see that it is not affected much because during its fast oscillations the slow system cannot follow and remains at rest. On the other hand, for every small piece of the oscillation of the slow system the fast system, making many oscillations, has the time to adjust to a new equilibrium. Thus the frequency of the slow system is affected, but only statically, through continuous adiabatic adjustment of the fast system to a new equilibrium position. In the theory of the internal motion of molecules this idea is applied twice, first in passing from the rotation to the vibration of the nuclei, then in passing from the vibration to the electronic motion. For these three types of motion we have

$$\omega_{\text{rot}} \ll \omega_{\text{vib}} \ll \omega_{\text{el}} \tag{11.02}$$

The factor is each time about one hundred, with ω_{el} in the visible, and the two others in the near and the far infrared, respectively. The argument applies most directly to the electronic and nuclear motions. The electric forces controlling them also cross-couple them since the same electric charges are involved. The different orders of magnitude of the masses make the basic frequencies different. We can therefore solve the problem of their motion by first analyzing the motion of the electrons, treating the nuclei as fixed. Then, having found the wave function of the electrons for every nuclear position and every quantum state, we find the force on the nuclei by straight electrostatics. This procedure for analyzing molecular motion is known as the *Born–Oppenheimer approximation*. The principle is obeyed even better in the passage from vibration to rotation. A symmetry principle is superimposed here on the Born–Oppenheimer principle: the rotational eigenfunctions, being determined by symmetry, can mix with other wave functions only under certain conditions. In the case of diatomic molecules, in particular, no coupling to the radial vibration, however strong, can alter the angular wave functions, which are the well known spherical harmonics.

The inequality (11.02) is only a section of the complete hierarchy of frequencies governing a molecule. A more complete list would be as follows:

Nuclear spin.
Electron spin and orbit.
Rotation.
Vibration.
Electron motion.
Nucleon motion.

In going down this list the frequency generally increases, there being a factor 10^{15} between the end members. The energy intervals of the first member of the sequence are small compared to kT at room temperature. The nuclear spin levels are thus equally populated and contribute to the partition function only through their total multiplicity. The same is true for the degeneracy factor of the extranuclear angular momentum J. This means that these two provide simple constant factors M_n and M_j, which we can write in the form

$$M_n = \prod_I (2I + 1) \qquad (11.03)$$

where the product goes over the nuclei of the molecule having a spin, and

$$M_j = 2J + 1 \qquad (11.04)$$

where J is the total extranuclear angular momentum. The multiplet structure due to spin-orbit splitting is on the borderline; the Boltzmann factor due to the upper levels of a multiplet sometimes does make a contribution to M_j. Finally, rotation and vibration must always be analyzed if we deal with molecules. The result will be an f_{int} of the form

$$f_{int}(T) = M_n \cdot M_j \cdot f_{rot}(T) \cdot f_{vib}(T) \qquad (11.05)$$

Breakdowns of this factorization principle do arise for a few molecules, but rarely because the coupling constants between the motions are too large to permit factorization. The usual cause of breakdown of (11.05) comes from symmetry principles in quantum mechanics; these principles sometimes force the association of certain states with each other, to the exclusion of others. We shall treat these modifications at the appropriate place.

11-2. Center-of-mass motion of gaseous molecules

The split-off of the translational motion from the other degrees of freedom is a rigorous property of molecules. By (11.01) its partition function is the only factor depending on volume. Its form (8.50) derived for atoms is unaltered, so that we have

$$f_{tr} = \left(\frac{2 \pi m k T}{h^2}\right)^{3/2} V \qquad (11.06)$$

where m is the total mass of the molecule. The equation yields, through (5.34), the equation of state for perfect gases $p V = N k T$; it also reproduces the value $\frac{3}{2} N k T$ for the kinetic energy of translation. There is a particularly simple relation between f and the Gibbs free energy; it results from (7.27) and (8.66) and reads

$$- \beta \mu = - \frac{\mathcal{G}}{R T} = \ln \frac{f}{N} \qquad (11.07)$$

It is convenient, when splitting up f, to associate the factor $1/N$ with the translational component. Thus we get for $\mathcal{G}_{tr}$ from (11.06) and (11.07)

$$- \frac{\mathcal{G}_{tr}}{R T} = \tfrac{3}{2} \ln \mathcal{M} + \tfrac{3}{2} \ln T + \ln \frac{V}{N} + \tfrac{3}{2} \ln \frac{2 \pi k}{\mathcal{A} h^2} \qquad (11.08)$$

where $\mathcal{M}$ is the molecular weight and $\mathcal{A}$ Avogadro's number. For practical purposes it is somewhat more convenient to eliminate V/N through the equation of state and write

$$- \frac{\mathcal{G}_{tr}}{R T} = \tfrac{3}{2} \ln \mathcal{M} + \tfrac{5}{2} \ln T - \ln \frac{p}{p_0} + \Psi_0 \qquad (11.09a)$$

where

$$\Psi_0 = \ln \left(\frac{2 \pi k}{\mathcal{A} h^2}\right)^{3/2} \frac{k}{p_0} \qquad (11.09b)$$

Here p_0 may be any standard pressure. Very commonly it is taken as 1 atm. The constant (11.09b) has then a universal value for all gases which equals

$$\Psi_0 = - 3.665 \qquad (11.09c)$$

The corresponding formula for the entropy follows from the definition (7.13)

$$\frac{\mathcal{S}_{tr}}{R} = \tfrac{3}{2} \ln \mathcal{M} + \tfrac{5}{2} \ln T - \ln \frac{p}{p_0} + \Psi_0 + \tfrac{5}{2} \qquad (11.10)$$

The formula is essentially a repetition of (5.50). Together with (5.72) it shows that the entropy constant has the form

$$\frac{\mathcal{S}_{tr}^0}{R} = \tfrac{3}{2} \sum_t c_t \ln \mathcal{M}_t - \sum_t c_t \ln c_t - 1.165 \qquad (11.11)$$

which is valid if the pressure is expressed in atmospheres. Other constants result from other conventions. A complete list of these constants is found in Appendix IX of *Statistical Mechanics* by Mayer and Mayer. It is seen that the simple "entropy constant" envisaged in (5.50) comes out to be rather more elaborate than one might have expected. Expression (11.11) is of course only the translational contribution, with further contributions to be expected from the internal degrees of freedom.

11-3. Rotation of gaseous molecules

For reasons of convenience we shall limit our discussion to diatomic molecules.

The motion of the two nuclei of a diatomic molecule about their center of mass represents three degrees of freedom, which are best taken as the length and the two polar angles of the radius vector pointing from one nucleus to the other. It follows then from symmetry alone that the angular wave functions are the spherical harmonics $Y_{JM}(\vartheta, \varphi)$. The angular momentum going with these wave functions is $\hbar^2 J(J+1)$ and the degeneracy $2J+1$. If the molecule were rigid the energy going with these states would be

$$\varepsilon_{\text{rot}} = \frac{\hbar^2}{2I} J(J+1) \tag{11.12}$$

where I is the moment of inertia of the molecule about the center of mass. We shall accept this expression at the moment. A discussion of its limits of validity is best deferred until we discuss the internuclear potential in Section 11-4. The classical molar specific heat arising from the two degrees of freedom is R. However, formula (11.12) shows a discrete spectrum. Consequently, we must have a dormancy phenomenon at low temperature. The dormancy may differ in detail from the one shown in Figs. 4.3 and 4.4.

These qualitative conclusions can be verified in detail by study of the rotational partition function which, from (11.12), equals

$$f_{\text{rot}} = \sum_{J=0}^{\infty} (2J+1) \exp[-J(J+1)y] \tag{11.13}$$

where

$$y = \frac{\hbar^2}{2I} \beta \tag{11.14}$$

Formula (8.69) yields from this for the energy

$$U_{\text{rot}} = \frac{N\,\hbar^2}{2\,I} \frac{\sum\limits_{J=1}^{\infty} J(J+1)\,(2J+1)\,\exp[-\,J\,(J+1)\,y]}{\sum\limits_{J=0}^{\infty} (2J+1)\,\exp[-\,J\,(J+1)\,y]} \tag{11.15}$$

and the specific heat

$$\mathcal{C}_{\text{rot}} = R\,y^2 \frac{1}{\left\{\sum\limits_{J=0}^{\infty} (2J+1)\,e^{-J(J+1)y}\right\}^2} \left[\sum\limits_{J=0}^{\infty} (2J+1)\,e^{-J(J+1)y} \right.$$

$$\left. \sum\limits_{J=1}^{\infty} (2J+1)\,J^2\,(J+1)^2\,e^{-J(J+1)y} - \left\{\sum\limits_{J=1}^{\infty} (2J+1)\,J\,(J+1)\,e^{-J(J+1)y}\right\}^2 \right] \tag{11.16}$$

Formulas (11.13), (11.15), and (11.16) are convergent for all positive values of y, and the convergence is rapid if y is of the order 1 or larger. We may then write, approximately,

$$f_{\text{rot}} \approx 1 + 3 \exp(-\,2\,y) \tag{11.17}$$

$$U_{\text{rot}} \approx \frac{3\,N\,\hbar^2}{I} \exp(-\,2\,y) \tag{11.18}$$

$$U_{\text{rot}} \approx 12\,R\,y^2 \exp(-\,2\,y) \tag{11.19}$$

Equations (11.18) and (11.19) show that both the energy and the heat capacity of the rotational degrees of freedom rise very slowly from zero at low temperature. However, unless hydrogen is involved in making up the moment of inertia, the quantity y, as defined in (11.14), is small compared to 1 in the easily accessible range of temperatures. The formulas just given are thus of little use, and the series (11.13), (11.15), and (11.16), while valid, need a large number of terms to produce a result. In particular, the limiting result of equipartition is not obvious for small y. To make this feature more apparent it is convenient to transform (11.13) to the corresponding integral by the Euler summation formula

$$\sum\limits_{\nu=0}^{\infty} \varphi(\nu) = \int_0^{\infty} \varphi(x)\,dx + \frac{1}{2}\varphi(0) - \frac{1}{12}\,\varphi'(0) + \frac{1}{720}\,\varphi'''(0) + \ldots \tag{11.20}$$

Applied to the sum (11.13), this yields

$$f_{\text{rot}} \approx \frac{1}{y} + \frac{1}{3} + \frac{1}{15}\,y + 0(y^3) \tag{11.21}$$

with y given by (11.14), or

$$\ln f_{\text{rot}} \approx -\,\ln y + \tfrac{1}{3}\,y + \tfrac{1}{90}\,y^2 + 0(y^4)$$

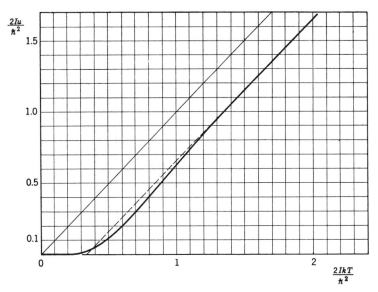

Fig. 11.1. Energy versus temperature for the rigid rotator. Both variables are in units $\hbar^2/2I$. The equipartition value appears as a 45° straight line through the origin. It is to be noted that the actual curve is asymptotically parallel to this line, but never approaches it.

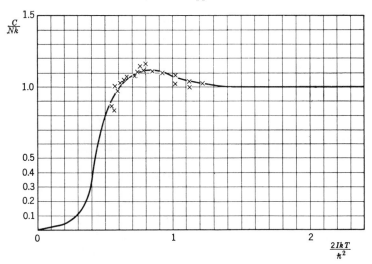

Fig. 11.2. Heat capacity versus temperature for the rigid rotator. The experimental points shown are those of Clusius and Bartholomé for HD. [*Z. Elektrochemie*, **40**, 524 (1934)]. For this molecule, the maximum of the curve lies close to 50° K.

From this, (8.69) yields for the energy

$$u_{\text{rot}} \approx k T - \frac{1}{3} \frac{\hbar^2}{2 I} - \frac{1}{45} \left(\frac{\hbar^2}{2 I} \right)^2 \frac{1}{k T} + 0 \left(\frac{1}{T^3} \right) \qquad (11.22)$$

The second term of the expansion, which shows a permanent discrepancy from equipartition, is rather surprising. The molar heat capacity resulting from this equals

$$\mathscr{C}_{\text{rot}} \approx R + \frac{1}{45} R \left(\frac{\hbar^2}{2 I k T} \right)^2 + 0 \left(\frac{1}{T^4} \right) \qquad (11.23)$$

It is seen that $\mathscr{C}_{\text{rot}}$ approaches its limiting high-temperature value R from above. Plots of these results are shown in Fig. 11.1 and Fig. 11.2.

For the high-temperature expansion of the entropy we get from (11.21) and (11.22)

$$\frac{\mathscr{S}}{R} = - \ln y + 1 - \frac{1}{90} y^2 + \cdots \qquad (11.24)$$

11-4. The rotational heat capacity of hydrogen

The formulas of 11-3 apply to diatomic molecules in general; they must be modified, however, for molecules such as H_2, N_2, O_2 provided the two atoms are also isotopically identical. There is then true symmetry with respect to exchange of the two nuclei. This symmetry is reflected in the nature of the wave function and thereby also the thermal quantities. This modification can amount to very little in favorable cases. The molecule N_2^{14}, for instance, is in a $^1\Sigma^+$ state, that is, an electronic state of maximum symmetry. To preserve this character for the rotating molecule the rotational eigenfunctions must also have this symmetry with respect to interchange of the two N^{14} nuclei. This means that only rotational wave functions of even J are allowed. If this modification is introduced into (11.20) we get

$$\sum_{\nu=0}^{\infty} \varphi(2\nu) = \frac{1}{2} \int_0^{\infty} \varphi(x) \, dx + \frac{1}{2} \varphi(0) - \frac{1}{6} \varphi'(0) + \cdots$$

In other words, the integral, which represents the classical limit, has half its previous value. Consequently the partition function in the high-temperature limit is divided by 2. This does not affect the energy or the specific heat, but subtracts a term $R \ln 2$ from the molar entropy.

A considerably more interesting case is H_2^1, a molecule consisting of two protons and two electrons. The two electronic spins are coupled in the covalent bond to a resultant spin zero, but the two proton spins are not.

The singlet and the triplet proton configuration thus appear with substantially the same energy; however, the latter is made up of three quantum states, the former of one. The requirement that the total wave function be antisymmetric in the two protons now becomes mixed up in this. It provides a direct association between nuclear spin and rotational quantum number. The nuclear singlet is antisymmetric in the two protons and must be associated with rotational functions which are symmetric; this means

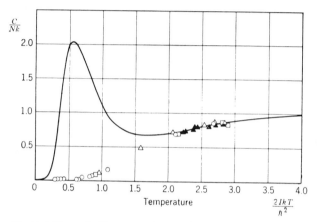

Fig. 11.3. Theoretical heat capacity versus temperature of hydrogen gas in equilibrium, and comparison with experimental points. It is seen that there is no fit between theory and experiment.

$J = 0, 2, 4, 6, \ldots$. The nuclear triplet, on the other hand, is symmetric with respect to the two protons and needs association with an antisymmetric rotational function; this means $J = 1, 3, 5, \ldots$. The rotational partition function thus comes out to be

$$f_{\text{rot}} = \sum_{J \text{ even}} (2J+1) \exp[- J(J+1)\,y] + 3 \sum_{J \text{ odd}} (2J+1) \exp[- J(J+1)\,y]$$

(11.25)

Equation (11.25) has obvious consequences for the other thermal properties. A theoretical curve for the heat capacity of hydrogen gas is shown in Fig. 11.3, together with experimental points. It is seen that the theory provides no fit whatever to the data.

The solution to the riddle of the heat capacity of hydrogen was found by Dennison. Chemists had observed that the adjustment to equilibrium of the two species of hydrogen is sluggish, except in the presence of certain catalysts. Now formula (11.25) assumes their free conversion at all times. Instead of this we must assume that we are dealing with two species:

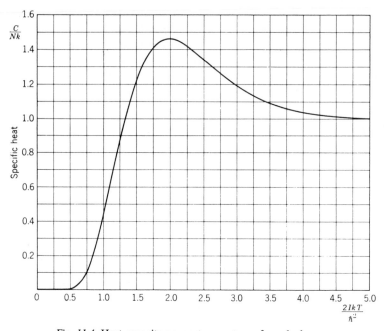

Fig. 11.4. Heat capacity versus temperature of parahydrogen.

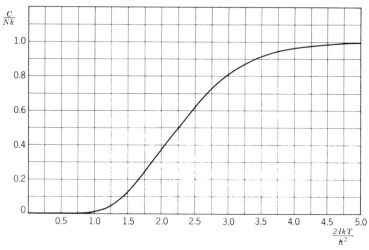

Fig. 11.5. Heat capacity versus temperature of orthohydrogen.

parahydrogen with nuclear spin zero, and *orthohydrogen* with nuclear spin 1. The rotational partition functions f of the two species are, respectively,

$$f_{\text{rot}}^{\text{para}} = \sum_{J \text{ even}} (2J+1) \exp[-J(J+1)y] \qquad (11.26)$$

and

$$f_{\text{rot}}^{\text{ortho}} = \sum_{J \text{ odd}} (2J+1) \exp[-J(J+1)y] \qquad (11.27)$$

We now construct the system partition function F by assuming that the high-temperature abundance ratio $1:3$ of the two species is frozen in.

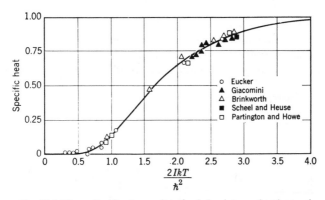

Fig. 11.6. Theoretical heat capacity of a $3:1$ mixture of ortho- and parahydrogen, and comparison with experimental points. It is seen that there is perfect agreement between them.

The gas then acts like a gas mixture, and the partition function has to be computed from (8.55). This means that we write

$$F_{\text{rot}} = \frac{(f_{\text{rot}}^{\text{para}})^{\frac{1}{4}N}}{(\frac{1}{4}N)!} \frac{(f_{\text{rot}}^{\text{ortho}})^{\frac{3}{4}N}}{(\frac{3}{4}N)!} \qquad (11.28)$$

This yields for the heat capacity

$$\frac{C_{\text{rot}}}{k} = y^2 \frac{d^2}{dy^2} \{ \tfrac{1}{4} N \ln f_{\text{rot}}^{\text{para}} + \tfrac{3}{4} N \ln f_{\text{rot}}^{\text{ortho}} \} \qquad (11.29)$$

which means simply that

$$\mathscr{C}_{\text{rot}} = \tfrac{1}{4} \mathscr{C}_{\text{rot}}^{\text{para}} + \tfrac{3}{4} \mathscr{C}_{\text{rot}}^{\text{ortho}} \qquad (11.30)$$

as one would expect for a gas mixture. The resulting specific heat curve is plotted in Fig. 11.6. It agrees perfectly with the experimental data.

11-5. Vibrational motion of diatomic molecules

We have seen in the preceding section that the internal degrees of freedom of a diatomic molecule separate by symmetry into an angular and a radial part, and that the angular wave functions are, again, by symmetry, the spherical harmonics. This leaves only one place for the coupling of rotation and vibration to make its appearance, namely the radial part of the Schroedinger equation. This equation reads

$$-\frac{\hbar^2}{2\mu}\frac{1}{r}\frac{d^2}{dr^2}(r\,R) + \left\{V(r) + \frac{\hbar^2}{2\mu}\frac{J\,(J+1)}{r^2}\right\} R = \varepsilon\,R \qquad (11.31)$$

Here μ is the reduced mass of the molecule, and $V(r)$ the internuclear potential when the molecule is at rest. This potential has a minimum at a distance r_0, the equilibrium distance of the two nuclei. At a larger distance it flattens out, leaving the two atoms independent. The resultant curve is

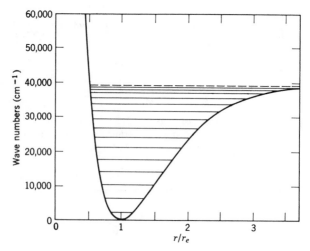

Fig. 11.7. Internuclear potential and vibrational energy levels for the hydrogen molecule. The equilibrium distance is $r_e = 0.742$ Å.

qualitatively similar to Fig. 11.7, which applies to the case of hydrogen. If the molecule is reasonably stable, statistics need not be concerned with the high-lying bound states. For the low-lying states the actual curve can be approximated by a parabola osculating to the true curve at $r = r_0$, and the simple mathematics of the harmonic oscillator can be employed. Within (11.31) the term in $J\,(J+1)$ represents the coupling of rotation and vibration; it is simply the centrifugal stretching force which is to be added to the force arising from $V(r)$. Now, this force is weak in the first place in comparison with interatomic forces. It is rendered weaker yet by the fact that it can be expanded in power series about r_0 and that the first two terms in the expansion produce no significant modification in the preceding analysis. The expansion reads, up to square terms,

$$\frac{\hbar^2}{2\mu}\frac{J(J+1)}{r^2} = \frac{\hbar^2}{2\mu}J(J+1)\left\{\frac{1}{r_0^2} - 2\frac{r-r_0}{r_0^3} + 3\frac{(r-r_0)^2}{r_0^4}\right\} \quad (11.32)$$

The first term in this expansion is just the energy (11.12) employed in the previous two sections to analyze rotational motion. It is therefore taken care of. The second term produces primarily a shift in the equilibrium distance r_0 and has no direct consequences for the energy spectrum. The third term produces a shift in the frequency of vibration, which we can estimate as

$$\frac{2\,\Delta\omega}{\omega} = \frac{\dfrac{6\,\hbar^2}{\mu}\dfrac{J(J+1)}{r_0^4}}{\kappa}$$

Here κ is the "spring constant" arising from $V(r)$. In estimating κ, it is important to realize that it is stiff even on an atomic scale: it does not permit nuclei to wander about the molecule as electrons do. We symbolize this by an arbitrary factor 10 so that

$$\kappa \sim 10\,\frac{e^2}{r_0^3}$$

and thus

$$\frac{\Delta\omega}{\omega} \sim \frac{3\,\hbar^2}{10\mu e^2 r_0}J(J+1) = \frac{3\times 10^{-54}\,J(J+1)}{10\times 10^{-22}\times 25\times 10^{-20}\times 10^{-8}} \sim 10^{-6}\,J(J+1)$$

This is a small effect indeed. Only the cubic term in (11.32) would truly modify results gained from the harmonic oscillator approximation. However, the cubic term has to compete with the natural anharmonicity of the potential pictured in Fig. 11.7 and generally loses out. Thus we can say, finally, that the centrifugal force contributes primarily the energy term (11.12) supplied by the rigid rotator model, and that all other effects of the rotational motion are negligible, except in refined spectroscopic work.

The detailed working out of the vibrational modes of polyatomic molecules presents few surprises. By a general theorem of mechanics a reduction to independent normal modes is possible if one expands the potential about the equilibrium positions of the nuclei and proceeds only to second-degree terms in the displacements. For molecules having elements of symmetry, group theory is helpful in determining these modes. In the case of diatomic molecules, to which we shall restrict the discussion here, there is only one such mode, namely, the one described in (11.31). These modes are rather higher in frequency than the rotational quantum steps discussed earlier. We have from (11.12)

$$\frac{\omega_{\text{rot}}}{c} \sim \frac{\hbar}{2 \mu d^2 c} \sim \frac{10^{-27}}{2 \times 10^{-22} \cdot 10^{-16} \cdot 3 \times 10^{10}} \sim 1 \text{ cm}^{-1}$$

while, with the previously used estimate for the spring constant κ, we get for the vibrational frequency

$$\frac{\omega_{\text{vib}}}{c} = \left(\frac{\kappa}{\mu c^2} \right)^{\frac{1}{2}} \sim \left(\frac{10 e^2}{r_0^3 \mu c^2} \right)^{\frac{1}{2}} \sim \left(\frac{10 \times 25 \times 10^{-20}}{10^{-24} \times 10^{-22} \times 10^{21}} \right)^{\frac{1}{2}} \sim 5000 \text{ cm}^{-1}$$

This result means that vibrational modes cannot be approximated classically at room temperature because the Kelvin and the wave number scales of energy more or less run together. Values of vibrational frequencies lower than the one worked out in the above estimate arise for so-called

Fig. 11.8. Bending mode of the water molecule.

bending modes, such as the one shown for water in Fig. 11.8. Here the oscillation is not directly against the valence bond between atoms, and frequencies one order of magnitude lower may arise.

The mathematics for handling the statistics of a vibrational mode is straightforward and has been carried out in Section 4-5 for the purpose of illustrating the dormancy phenomenon. If we set

$$z = \beta \hbar \omega \tag{11.33}$$

the partition function for the mode is given by (4.54) as

$$f_{\text{vib}} = \tfrac{1}{2} \operatorname{cosech} \tfrac{1}{2} z \tag{11.34}$$

The mean vibrational energy is given by (4.55):

$$u_{\text{vib}} = \tfrac{1}{2} \hbar \omega \coth \tfrac{1}{2} z \tag{11.35}$$

and the contribution to the moloar heat capacity by (4.56)

$$\frac{\mathscr{C}_{\text{vib}}}{R} = \tfrac{1}{4} z^2 \operatorname{cosech}^2 \tfrac{1}{2} z \tag{11.36}$$

For the entropy we employ (5.09). The partition function (11.34) and the energy (11.35) then yield

$$\frac{\mathscr{S}_{\text{vib}}}{R} = -\ln\left(2\sinh \tfrac{1}{2}z\right) + \tfrac{1}{2}z \coth \tfrac{1}{2}z \qquad (11.37)$$

Limiting formulas can be made up here simply from the analytical expressions. For high temperature, that is, low z, the limiting expression of u_{vib} is kT, of $\mathscr{C}_{\text{vib}}$ R, and of $\mathscr{S}_{\text{vib}}$ $R(-\ln z + 1)$. However, it is to be remembered that the high-temperature limit rarely applies to the vibration of diatomic molecules.

11-6. The law of mass action in perfect molecular gases

The general laws of chemical equilibrium take a particularly simple form if we deal with a possible chemical reaction taking place in a mixture of gaseous constituents. It was shown in Chapter 7 that a possible chemical reaction engenders an equilibrium relation among the chemical potentials. We shall now discuss the explicit and extremely convenient form which this type of equation takes for reactants in a gas phase which can be considered perfect.

Our best starting point is the proper generalization of (8.55) for a mixture of perfect gases. The relationship between the canonical partition function F and the particle partition functions f_i of the individual components equals

$$F(T, V) = \prod_i \frac{[f_i(T, V)]^{N_i}}{N_i!} \qquad (11.38)$$

Here N_i is the number of molecules of the species i present in the mixture. We have of course from (7.10) that this expression equals $\exp[-\beta A]$. If we now compute G from the formula

$$G = A + pV \qquad (11.39)$$

and express pV through Dalton's law (2.43) we get

$$\exp[-\beta G] = \prod_i \frac{[f_i(T, V)]^{N_i} e^{-N_i}}{N_i!}$$

or with Stirling's approximation for the factorials

$$\exp[-\beta G] = \prod_i \left(\frac{f_i(T, V)}{N_i}\right)^N \qquad (11.40)$$

Expression (11.40) is the logical generalization of (11.07). We know from Chapter 7 that T and p are the natural variables for G. In the present case the substitution of p for V is quite easy; for by (11.01) and (11.06) V enters as a simple factor into f_i. If we denote by $f_i^0(T)$ the particle partition function without the volume factor we can use Dalton's law (2.43) and the definition (5.70) of the mole concentration c_i to give f_i/N_i the form

$$\frac{f_i(T, V)}{N_i} = \frac{f_i^0(T)\, k\, T}{c_i\, p}$$ (11.41)

Equation (11.40) takes then the form

$$\exp[-\beta\, G] = \prod_i \left(\frac{f_i^0(T)\, k\, T}{c_i\, p} \right)^N$$ (11.42)

Taking logarithms on both sides, we get

$$-\beta\, G = \sum_i N_i \left\{ \ln \left[\frac{k\, T}{p}\, f_i^0(T) \right] - \ln \frac{N_i}{N_1 + N_2 + \ldots} \right\}$$ (11.43)

We are now in a position to compute the chemical potentials by using (7.29). Upon careful differentiation, we find the result

$$-\beta\, \mu_i = \ln \left[\frac{k\, T}{p}\, f_i^0(T) \right] - \ln c_i$$ (11.44)

The decomposition of the chemical potential into a part depending only on pressure and temperature and a part depending only on concentration is an advantage in formulating equilibrium conditions for chemical reactions in a gaseous mixture. Suppose we have a chemical reaction of the form

$$\sum_i \nu_i\, ((i)) = 0$$ (11.45a)

where ν_i are the coefficients of the chemical equation, and $((i))$ is the chemical symbol of the species i. As an example we may take

$$2\, H_2 + O_2 - 2\, H_2O = 0$$ (11.45b)

then, by (7.49), an equation of the same form holds between the chemical potentials:

$$\sum_i \nu_i\, \mu_i = 0$$ (11.46a)

or in our example

$$2\, \mu_{H_2} + \mu_{O_2} - 2\, \mu_{H_2O} = 0$$ (11.46b)

When the chemical potentials have the special form (11.44) this takes the form

$$\prod_i (c_i)^{\nu_i} = \prod_i \left[\frac{k\,T}{p} f_i^0(T)\right]^{\nu_i} \tag{11.47a}$$

which reads in our example

$$\frac{(c_{H_2})^2 c_{O_2}}{(c_{H_2O})^2} = \frac{k\,T}{p} \frac{(f_{H_2}^0)^2 f_{O_2}^0}{(f_{H_2O}^0)^2} \tag{11.47b}$$

Equation (11.47) is the *law of mass action*: the concentration product on the left is a function of temperature and pressure only as expressed on the right. This product is readily computable if only the particle partition functions for the components are computable; the exponents occurring on the two sides are the same.

There is a second form of the mass action law which takes advantage of the simple dependence on pressure of the result (11.47). If one introduces the concept of partial pressure by (5.71) one gets the result

$$\prod_i (p_i)^{\nu_i} = K(T) \tag{11.48}$$

where $K(T)$, the *equilibrium constant*, is a function of temperature only. It has the value

$$K(T) = \prod_i [k\,T f_i^0(T)]^{\nu_i} \tag{11.49}$$

This result is sometimes rendered more flexible by the introduction of a standard pressure p_0, usually 1 atm. Then (11.48) can be written in the form

$$\prod_i \left(\frac{p_i}{p_0}\right)^{\nu_i} = K^\star(T) \tag{11.50a}$$

with

$$K^\star(T) = \prod_i \left[\frac{k\,T}{p_0} f_i^0(T)\right]^{\nu_i} \tag{11.50b}$$

This is the form which will be used in the example below.

When the law of mass action is to be implemented in a particular case, all the information about partition functions accumulated in the preceding sections of this chapter must be employed. The decomposition rules (11.01) and (11.05) are helpful in breaking the functions into factors which can be discussed separately. For the most important pieces we have (11.06), (11.13), and (11.34). Eventually, more complicated expressions must be worked out in special cases. The most important factor is missing, however, from this list. The definition (8.46) of the particle partition function contains the energy in such a form that a shift in the energy reference zero contributes a factor to the partition function. The factor is such that it drops out of expressions giving the population of levels, such as (8.68). To energy expressions such as (8.69), the energy shift is simply passed on. In neither case does the choice of the reference zero have any physical significance. This is no longer the case for the equilibrium constant (11.49). If we denote by χ_i the energy of the ground state of the molecule i, then the constant $K(T)$ has a factor K_0 of the form

$$K_0 = \exp\left[- \beta \sum_i \nu_i \chi_i\right] \tag{11.51a}$$

which in the case of the water reaction (11.45b) reads

$$K_0 = \exp[- \beta (2\chi_{H_2}+\chi_{O_2}-2\chi_{H_2O})] \tag{11.51b}$$

The expression in parentheses is the difference between the final and the initial energy of the reactants, that is, the heat of reaction at absolute zero. It is thus obvious that the use of a common energy reference level is essential in the evaluation of an equilibrium constant. In the case of the water reaction, for instance, the energy difference in (11.51b) equals 58 kcal/mol or 2.5 eV. This produces a factor K_0, which equals

$$K_0 = e^{-(29,000/T)} \tag{11.51c}$$

The factor K_0 is always decisive for the temperature dependence of the equilibrium. It is also so large or so small in most cases that it determines which side is favored by the equilibrium constant. This is the situation, for instance, for the water reaction above. The other parts of the equilibrium constant deal mostly with entropy differences, favoring the side with more components, higher multiplicity, or more degrees of freedom. These factors are not large enough to reverse the effect of a K_0 having the magnitude (11.51c).

We shall conclude the discussion of the mass action law by applying the formalism to an example which permits a detailed comparison of theory and experiment, namely, the dissociation reaction of iodine vapor. The reaction is represented by the equation

$$2 I - I_2 = 0$$

In the form (11.50) the equilibrium relation reads

$$\frac{p_I^2}{p_{I_2}} = \frac{k T}{p_0} \frac{[f_I^0(T)]^2}{f_{I_2}^0(T)} \tag{11.52}$$

We shall compute the partition functions, using spectroscopic data.[*][†]

* C. E. Moore. *Atomic Energy Levels*, Vol. III. Washington, D.C.: National Bureau of Standards, 1958.

† G. Herzberg. *Molecular Spectra and Molecular Structure*, Vol. I: *Spectra of Diatomic Molecules*. Princeton, N.J.: Van Nostrand, 1950, Table 39.

Starting with atomic iodine, we have to know the atomic weight of its leading isotope I^{127}, which equals 126.96, and the nature of its ground state. It is a ^2P state with the doublet inverted. The ground state thus has $J = \frac{3}{2}$ and a multiplicity 4. The upper member of the doublet lies too high to have any effect at the temperatures considered. Thus, apart from this factor 4 the partition function of I has only a translational part of the form (11.06).

The partition function of I_2 is more complicated. The most important factor is contributed by the energy exponential (11.51). The dissociation energy from the ground state is 1.5417 eV. Our formulas are such that we must take it from the bottom of the vibrational parabola. We must therefore add $\frac{1}{2} \hbar \omega$, which changes the dissociation energy to 1.5550 eV. The rotational factor can be taken in the classical approximation; it then equals the first term of (11.21), divided by 2 because of molecular symmetry. The interatomic distance enters into the moment of inertia; it equals 2.667 Å. The vibrational partition function has the form (11.34) with a vibrational frequency equal to 214.57 cm^{-1}.

Putting all this information together, we find for $K^\star$, as defined in (11.50), with $p_0 = 1$ atm:

$$K^\star = 44.54 \, T^{3/2} \sinh \frac{154.4}{T} \exp\left[- \frac{18{,}046}{T} \right] \qquad (11.53)$$

This function is plotted on a semilog scale against temperature in Fig. 11.9. Comparison is made with experimental data by Perlman and Rollefson.* The authors infer the degree of dissociation of the molecule from the departure of the pressure from the value it would have for a perfect gas of diatomic molecules. The comparison should be made in two stages. First of all they have, for each temperature, values of the constant taken at several pressures. These pressures cover one decade; the equilibrium constant, on the other hand, varies by less than 5 % except at the lowest temperature taken. In a second stage of comparison, we can place these equilibrium constants on Fig. 11.9. It is seen that the agreement is very nearly perfect; the small discrepancy existing is reduced even further if data at the lowest pressure are compared with the theory. In other words, the nonideal character of the gas I_2 (which was corrected for, but perhaps not in the right way) places a limit on the accuracy of the experimental measurements. To get an idea of the quality of this agreement it may be recalled that the degeneracy of the ground state of atomic iodine contributes a factor 16 to the equilibrium constant. This factor is quite uniquely determined by the experimental measurements.

* M. L. Pearlman and G. K. Rollefson. *J. Chem. Phys.*, **9**, 362 (1941).

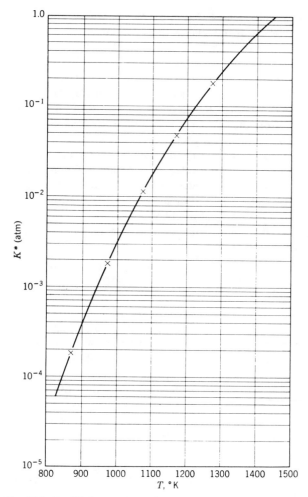

Fig. 11.9. Equilibrium constant for the dissociation reaction of iodine. Theoretical curve and experimental points of Perlman and Rollefson.

A. RECOMMENDED PROBLEM

1. According to Section 11-1 the Born–Oppenheimer approximation can be traced to a simple property of two coupled harmonic oscillators: if the oscillator frequencies are of different order of magnitude to start with, the effect of a coupling can be estimated simply even if it is strong. Verify this by computing the two frequencies arising from a hamiltonian of the form

$$\mathcal{H} = \tfrac{1}{2}(\omega_1^2 p_1^2 + \omega_2^2 p_2^2) + \tfrac{1}{2}(q_1^2 + q_2^2 + 2\lambda q_1 q_2)$$

with

$$\omega_1 \ll \omega_2$$

and

$$\lambda \sim 1$$

(a) by following the recipe given in the text and (b) by solving the problem exactly.

B. GENERAL PROBLEMS

2. Prove that the kinetic energy of a set of point masses can be rigorously decomposed into a term referring only to the motion of the center of mass, plus a term referring only to relative motion.

3. Calculate the change in the equilibrium interatomic distance of the hydrogen molecule with angular momentum quantum number. Use the data of Fig. 11.7, or Problem 7 below.

4. The deuteron has a nuclear spin 1. Write down the rotational part of the partition function of deuterium gas, assuming quantum conditions and rapid interconversion among the spin states.

5. By the law of mass action a small fraction of the water molecules in steam are dissociated into molecular hydrogen and molecular oxygen. How does this fraction f vary with pressure at fixed temperature?

6. Suppose we investigate the hydrogen chlorine reaction

$$H_2 + Cl_2 = 2\,HCl$$

and we distinguish the two isotopic species Cl^{35} and Cl^{37} but assume all hydrogen nuclei to be protons. Show that if the direct effect of mass on the partition functions is neglected, but symmetry factors are counted, a mass action law results for which the relative abundance of the two isotopes in Cl_2 is the same as in HCl. Show also that this would not be the case if the symmetry factors were omitted.

7. Compute the equilibrium constant for the reaction

$$H_2 + I_2 - 2\,HI = 0$$

from the following table of constants:

	H_2	I_2	HI
D, cm^{-1}	38,707	12,546	25,815
ω, cm^{-1}	4,395	214.6	2,309
r, Å	0.7417	2.667	1.604

The first two lines are in reciprocal wave length units. Compute the constant as a function of temperature, and give its value at 500° K.

*12

The problem of the imperfect gas

We have seen in the preceding chapter that there are a number of structural properties of gaseous substances, notably thermal and chemical, which fit into the perfect gas framework. The advantages of being able to fit them in this way are extremely great. One can advance on a broad front, dealing with many substances and many phenomena. Because of the firm theoretical framework, experimental results obtained on substances when in gaseous form give structural information about the molecules composing them, which is almost as reliable as results obtained on the molecules individually. This picture changes entirely when we deal with departures from the perfect gas laws. We now find ourselves without an obvious theoretical method. Research becomes reoriented by this lack. Instead of dealing with a wealth of experimental evidence we must investigate the theoretical procedures themselves. All extraneous complications must be avoided, and we must try to construct a theory for the simplest possible case of N identical particles linked into a unit by their mutual interactions. In constructing such a theory we cannot give much weight to experimental evidence until the theory has progressed enough to assimilate such evidence.

12-1. Equation of state from the partition function

As we fix our attention on the equation of state obeyed by imperfect gases we may return to a simpler model of the molecules making up the gas than we employed in the preceding chapter. That chapter has taught us that for the purpose at hand the internal degrees of freedom of the molecules can be ignored. The individual molecules can therefore be characterized by their three coordinates and their three components of momentum. The question whether the continuum approximation is valid for their energy spectrum can be decided by some order of magnitude considerations of the molecular forces. It has been remarked in Section 11-5 that interatomic forces in molecules are such that a quantum treatment is usually necessary. In the special case of I_2, however, quantum theory supplies the sinh in the factor $\sinh (154.4/T)$ in formula (11.53), while classical theory would give just the argument alone; this means that for Fig. 11.9 classical theory would actually have been sufficient. For imperfect gases we deal with the van der Waals force. This force is very similar to the intermolecular force of Fig. 11.7, except that the energy scale is down by roughly a factor 50.* Now, the criterion whether the continuum approximation is valid depends on the magnitude of hf/kT with $2\pi f = (\kappa/m)^{1/2}$. Here κ is some sort of spring constant, and m some sort of molecular mass. A reduction of the energy by a factor 50 would, with the dimensions remaining the same, reduce κ by a similar factor. This, in turn, would reduce f by a factor 7. Thus classical physics is favored by this estimate. Experience has shown that quantum features are relevant only for the light gases helium and hydrogen. These exceptions rather confirm the general impression that quantum phenomena are not essentially involved in the theoretical problem of imperfect gases.

The most straightforward approach to an equation of state is through the partition function F. Its form is written out in the denominator of (4.32), with a supplementary factor $1/N!$ entering because of the Gibbs paradox. It reads then

$$F(N) = \frac{1}{h^N N!} \int \int \int \cdots \int \exp\left[-\beta \left\{ \frac{1}{2m} (\mathbf{p}_1^2 + \mathbf{p}_2^2 + \mathbf{p}_3^2 + \cdots \mathbf{p}_N^2) \right. \right. \tag{12.01}$$

$$\left. \left. + \Phi(\mathbf{r}_1, \mathbf{r}_2, \ldots \mathbf{r}_N) \right\} \right] dp_{1x}\, dp_{1y}\, dp_{1z}\, dp_{2x} \ldots dp_{Nz}\, dx_1\, dy_1\, dz_1\, dx_2 \ldots dz_N$$

where Φ is the total potential energy of the system which depends on the positions $\mathbf{r}_1, \mathbf{r}_2, \ldots \mathbf{r}_N$ of the molecules. Thanks to the use of classical physics the kinetic energy part of (12.01) splits off from the potential part and can be computed explicitly. This feature has already been emphasized

---

* Y. K. Syrkin and M. E. Diatkina. *Structure of Molecules and the Chemical Bond* (translation). New York: Interscience Publishers, 1950.

in Chapter 4 in connection with the equipartition theorem. Therefore, F reduces to

$$F(N) = \frac{1}{N!} \left(\frac{2 \pi m k T}{h^2} \right)^{\frac{3}{2}N} Q(N) \tag{12.02}$$

with

$$Q(N) = \int \int \int \ldots \int \exp[- \beta \, \Phi(\mathbf{r}_1, \mathbf{r}_2, \ldots , \mathbf{r}_N)] \, dx_1 \, dy_1 \ldots dz_N \tag{12.03}$$

The problem of imperfect gases, and even liquids, is thereby made identical with the problem of computing $Q(N)$, the *configurational partition function* of N particles. Since it is the only part of F dependent on volume, formula (5.34) for the pressure simplifies to

$$p = k T \frac{\partial \ln Q(N)}{\partial V} \tag{12.04}$$

12-2. Equation of state from the virial theorem

As an alternative to the approach from the partition function we have the approach from the virial theorem of Chapter 2. The theorem (2.37) was used then to derive the equation of state for perfect gases; all that was necessary was to add to the theorem the assumption that the molecules do not interact with each other, so that the only interaction left was the one with the walls of the container. Thus, the dropping of this assumption will lead us back to the present problem. Actually, a little more than that is needed to get results in convenient form. We shall assume spherically symmetric molecules subject to pairwise forces which act along their line of centers and whose magnitude depends only on their mutual distance. We shall also assume these forces to obey the principle of superposition, that is, any pairwise force shall be unaltered by the presence of a third molecule. This superposition principle shall include the interactions with the walls of the container, which are assumed to have short range, as previously.

The formula we are about to investigate here in detail is (2.38), which reads

$$\tfrac{3}{2} N k T = \tfrac{3}{2} p V + \text{(Internal virial)} \tag{12.05}$$

Equation (12.05) represents a great advance in understanding, if not in ease of computation, over (12.04). The modification of the laws of Boyle and Charles by the intermolecular forces is immediately apparent as an additive term. Now we introduce the pairwise nature of the forces for further simplification. This yields

$$\text{Internal virial} = -\tfrac{1}{2} \sum_i \mathbf{r}_i \cdot \mathbf{F}_i$$

$$= -\tfrac{1}{2} \sum_i \mathbf{r}_i \cdot \sum_{j \neq i} \mathbf{F}_{ij}$$

$$= -\tfrac{1}{4} \sum_{i,j} (\mathbf{r}_i \cdot \mathbf{F}_{ij} + \mathbf{r}_j \cdot \mathbf{F}_{ji})$$

$$= -\tfrac{1}{4} \sum_{i,j} (\mathbf{r}_i - \mathbf{r}_j) \cdot \mathbf{F}_{ij}$$

$$= -\tfrac{1}{4} \sum_i \{ \sum_{j \neq i} (\mathbf{r}_i - \mathbf{r}_j) \cdot \mathbf{F}_{ij} \}$$

Here the fourth line was gained through Newton's third law. The last line seems at first sight a trivial restatement of the line preceding. The way of writing it is meant to indicate a change in interpretation. The molecule i, over which we sum last, might be any molecule; the curly bracket following represents a sum over all other molecules of quantities which refer to the molecule i: distance to the molecule i and force on the molecule i. When the time average is taken the molecule i will get into all possible situations and so will any other molecule, such as the molecule 1. We can therefore replace i by 1 and the sum over i by a factor N. This brings (12.05) into the form

$$p V = N k T + \tfrac{1}{6} N \overline{\sum_{j=2}^{N} (\mathbf{r}_1 - \mathbf{r}_j) \cdot \mathbf{F}_{1j}} \qquad (12.06)$$

By equation (12.06) knowledge of the equation of state needs only the knowledge of the *pair distribution* of a gas, that is, the arrangement of the remainder of the gas around a given molecule assumed at the origin. This is of course a quantity fluctuating in time; we want its time average. In a perfect gas two molecules are not correlated. The time average of the number of molecules per unit volume is therefore not affected by the fact that there is one at the origin; we could thus simply write ρ for this quantity, where ρ is the number density of the gas. A sum of the type occurring in (12.06) could then be given the form

$$\overline{\sum_{j=2} \varphi(|\mathbf{r}_1 - \mathbf{r}_j|)} = 4 \pi \rho \int_0^\infty \varphi(r) r^2 \, dr$$

where φ is any function depending only on the intermolecular distance, and r is a polar coordinate. For large values of r this estimate cannot be much in error. However, at short distances there are strong discrepancies. In particular, the probability is zero of finding another molecule at a position for which r equals zero or is smaller than twice a molecular radius. This deficit in density is often followed by a surplus at the so-called nearest neighbor distance. The position of this maximum is determined by the attractive forces in the system. Further oscillations follow which gradually dampen out. This state of affairs is rendered explicit by the introduction of a *pair correlation function* $g(r)$, which is a multiplicative factor reproducing these departures from randomness at a distance r. In favorable cases the function $g(r)$ can be measured directly by X-ray diffraction. Figure 12.1 shows the curve $g(r)$ as measured for liquid sodium at 100°

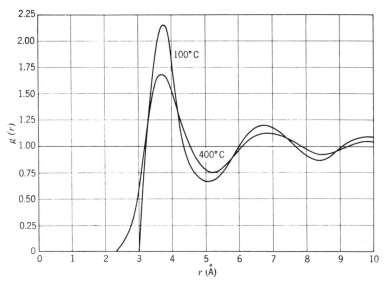

Fig. 12.1. Pair correlation function $g(r)$ for liquid sodium at 100° C and 400° C. The curves were obtained from X-ray data [Trimble and Gingrich, *Phys. Rev.*, **53**, 278 (1938)].

and 400° C. With the help of $g(r)$ we can make the preceding equation accurate by putting it into the form

$$\overline{\sum_{j=2}^{N} \varphi(|\mathbf{r}_1 - \mathbf{r}_j|)} = 4\pi \frac{N}{V} \int_0^\infty g(r)\,\varphi(r)\,r^2\,dr \tag{12.07}$$

The second term in (12.06) is of the form discussed here since, by assumption, F_{1j} is parallel to $\mathbf{r}_1 - \mathbf{r}_j$ and is determined by the magnitude of that quantity. The identity (12.07) can therefore be introduced into (12.06), yielding

$$pV = NkT + \tfrac{2}{3}\pi \frac{N^2}{V} \int_0^\infty g(r)\,F(r)\,r^3\,dr \tag{12.08}$$

Here the force $F(r)$ is counted positive if repulsive. The formula brings in only known quantities except for the pair correlation function $g(r)$. This function is at least in principle amenable to experimental determination. In addition, it can be made the subject of sound intuitive guesses, as we shall see below.

It may be mentioned in passing that the pair correlation function does in fact permit determination of all equilibrium properties of a gas. This is, for instance, the case for the internal energy U. By assumption, the potential energy $\Phi(\mathbf{r}_1, \mathbf{r}_2, \ldots, \mathbf{r}_N)$ defined in (12.01) is of the form

$$\Phi(\mathbf{r}_1, \mathbf{r}_2, \ldots, \mathbf{r}_N) = \tfrac{1}{2} \sum_{i,j} \phi(|\mathbf{r}_i - \mathbf{r}_j|) \qquad (12.09)$$

By the arguments previously given its time average can be written in the form

$$\Phi(\mathbf{r}_1, \mathbf{r}_2, \ldots, \mathbf{r}_N) = \tfrac{1}{2} N \sum_{j=2}^{N} \phi(|\mathbf{r}_1 - \mathbf{r}_j|)$$

This is of the form (12.07). The total energy therefore equals

$$U = \tfrac{3}{2} N k T + 2\pi \frac{N^2}{V} \int_0^\infty g(r)\,\phi(r)\,r^2\,dr \qquad (12.10)$$

With two state variables known for the gas all the others follow by suitable manipulations.

We conclude this section with a proof, given by Born and Green, that the two methods discussed for construction of the equation of state are in principle identical. To prove this we reduce expression (12.04) to (12.06) by transformation. We start by eliminating the size of the container from the limits of integration in (12.03). This is done through the substitution

$$\mathbf{r}_i = L\,\mathbf{x}_i$$

Here L is of the order of the linear dimensions of the container; the variables $\mathbf{x}_i$ can then remain fixed as the volume is varied. This means that

$$\frac{d \ln L}{d \ln V} = \frac{1}{3}$$

and that

$$Q(N) = L^{3N} \int \int \int \ldots \int \exp[-\beta\,\Phi(L\mathbf{x}_1, L\mathbf{x}_2, \ldots)]\,d\mathbf{x}_1 \ldots d\mathbf{x}_N$$

The differentiation (12.04) can now be carried out by the intermediate step of differention with respect to L. It yields

$$p = \frac{NkT}{V} - \frac{\int \int \int \ldots \int \left(\sum_i \mathbf{x}_i \cdot \dfrac{\partial \Phi}{\partial \mathbf{r}_i} \right) \exp[-\beta\Phi]\,d\mathbf{x}_1 \ldots d\mathbf{x}_N}{\int \int \int \ldots \int \exp[-\beta\,\Phi]\,d\mathbf{x}_1 \ldots d\mathbf{x}_N} \cdot \frac{1}{3}\,\frac{1}{V}$$

$$(12.11)$$

The factor L in the numerator just goes to replace $\mathbf{x}_i$ in the virial expression by $L\,\mathbf{x}_i = \mathbf{r}_i$. Then $d\mathbf{x}_i$ can be replaced in the numerator and denominator by $d\mathbf{r}_i$, and the factor $1/3V$ is required by the virial theorem. The main novel feature here is that the average of the virial appearing is a partition function average, while we introduced it as a time average originally. It is a basic assumption of statistical mechanics that these two types of averages are equivalent. The proof given shows incidentally that the conventional statistical definition of $g(r)$ is

$$g(r_{12}) = \frac{V^2 \int\int\int \ldots \int \exp[-\beta\Phi]\,d\mathbf{r}_3, d\mathbf{r}_4, \ldots, d\mathbf{r}_N}{\int\int\int \ldots \int \exp[-\beta\,\Phi]\,d\mathbf{r}_1\,d\mathbf{r}_2\,d\mathbf{r}_3 \ldots d\mathbf{r}_N} \tag{12.12}$$

The concept thus presupposes a preliminary process of averaging over $N-2$ molecules before we can consider the relative arrangement of the two molecules singled out. Thus $g(r_{12})$, which is an extremely direct concept when considered as a time average, becomes very complicated when considered as a spatial or ensemble average.

12-3. Approximate results from the virial theorem; van der Waals' equation

Formula (12.08) derived from the virial theorem is less easily adapted to permit a successive sequence of approximations to the equation of state than the more symmetric procedure making use of the partition function. It has, on the other hand, a direct intuitive content which the partition function lacks. It is therefore particularly suited to the task of producing quickly a good, but not quite accurate, rendition of the equation of state. The natural step in such a task is to assume that $g(r)$ is statistically self-consistent. By this, one means that it is itself a Boltzmann exponential in the potential $\phi(r)$ whose average it determines in (12.10). This assumption receives some support from the fact that the term $\exp[-\beta\,\phi(r_{12})]$ is the only factor which can be taken out of the integral sign in (12.12). The assumption is of course basically a low-density assumption: it would be valid if no third body were present to interfere when molecules 1 and 2 are in close proximity. We thus write down the approximate ansatz

$$g(r) \approx \text{constant} \cdot \exp[-\beta\,\phi(r)]$$

The constant can be determined by consideration of the limiting case of large r. If $\phi(r)$ is a function of the type shown in Fig. 11.7 it tends to zero for large r; under the same conditions $g(r)$ approaches unity. The constant is therefore 1, and we have

$$g(r) \approx \exp[-\beta \phi(r)] \qquad (12.13)$$

Insertion of this assumption into (12.08) yields an equation of state of the form

$$p V = N k T - \tfrac{2}{3} \pi N \rho \int_0^\infty \frac{d\phi}{dr} \exp[-\beta \phi(r)] \, r^3 \, dr$$

The integral above may be integrated by parts with the integrated out part, $(1/\beta)\{1 - \exp[-\beta \phi(r)]\}$, being chosen to preserve convergence at infinity. The result is

$$p V = N k T + 2\pi \frac{N^2 k T}{V} \int_0^\infty \{1 - \exp[-\beta \phi(r)]\} \, r^2 \, dr \quad (12.14)$$

The equation of state can of course be written out for a variety of inter-molecular potentials $\phi(r)$. We shall make no attempt here to choose the best one, but shall tie in with earlier work by taking a potential having a finite-sized hard core with a weak attractive interaction following at a larger distance. This means that we assume

$$\phi(r) = +\infty \quad \text{if} \quad 0 < r < d \qquad (12.15a)$$

$$-k T \ll \phi(r) < 0 \quad \text{if} \quad r > d \qquad (12.15b)$$

$$\phi(r) \to 0 \quad \text{if} \quad r \to \infty \qquad (12.15c)$$

The convergence under (12.15c) must be sufficiently rapid to ensure convergence of the integral in (12.14). This need specifically excludes Coulomb forces from consideration: the gas being studied must not consist of ions and electrons.

With the interatomic potential obeying (12.15) the integral in (12.14) breaks up into two parts, one from 0 to d, and one from d to ∞. In the latter part the curly bracket can be approximated by $\beta \phi$, because of the estimate (12.15b). The result is

$$\frac{p V}{N k T} = 1 + \frac{2 \pi N d^3}{3 V} + \frac{2 \pi N}{V k T} \int_d^\infty \phi(r) \, r^2 \, dr \qquad (12.16)$$

The substitutions

$$-a = 2 \pi \mathscr{A}^2 \int_d^\infty \phi(r) r^2 \, dr \qquad (12.17a)$$

and

$$b = \tfrac{2}{3} \pi \mathscr{A} \, d^3 \qquad (12.17b)$$

bring it into the form

$$\frac{p + \dfrac{a}{V^2}}{1 + \dfrac{b}{V}} = RT \tag{12.18}$$

This is almost, but not quite, the van der Waals equation. To get it exactly we must make the substitution

$$\frac{1}{1 + \dfrac{b}{V}} \to 1 - \frac{b}{V}$$

whereupon equation (1.11) results. Although the step is harmless for large values of V/b, it is rather arbitrary as an analytical procedure. It is this last step which brings in the critical and condensation phenomena discussed below. These phenomena are absent from either (12.14) or (12.18). The latter is, however, closely related to the van der Waals equation. It is the beginning of its virial expansion, as is seen when we write it in the form

$$pV = RT\left\{1 + \frac{b}{V} - \frac{a/RT}{V}\right\} \tag{12.19}$$

It thus agrees with the van der Waals equation in the value of the second virial coefficient and reproduces the pressure correction of that equation exactly. And what is even more to the point, the nature of the coefficients a and b is just as van der Waals predicted. The quantity a is a constant which takes into account the long-range attraction of the molecules, while b takes into account their hard core repulsion; it is a low multiple of the close-packed volume of the molecules. Thus the van der Waals equation is based on a considerable amount of intuitive insight which can partially be justified.†

12-4. The Joule–Thomson effect

In later parts of this chapter we shall discuss as best we can the critical phenomenon as something typical of real gases. We shall see that the phenomenon is not easy to handle, because by the time we approach the

† An even more far-reaching justification of the van der Waals equation was carried out by Kac, Uhlenbeck, and Hemmer in the one-dimensional case [*J. Math. Phys.* **4,** 216 and 229 (1963)]. To get the result one must let the range of the attraction tend to infinity.

critical point, phenomena akin to condensation occur, and the gaseous structure has already broken down in a serious way. A phenomenon which occurs at much lower pressure and thus is a better indicator of nonideal conditions from the point of view of computation is the Joule–Thomson effect. A schematic picture of the phenomenon is shown in Fig. 12.2.

Fig. 12.2. Schematic picture of the experimental arrangement for the Joule–Thomson phenomenon.

A gas flows continuously through a pipe having a constriction which is so narrow that the gaseous density is permanently lower beyond as compared to ahead of the constriction. To understand what is going on we assume the gas to be sealed off by two pistons as shown in Fig. 12.3. The piston

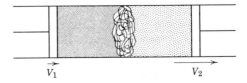

Fig. 12.3. Explanation of the Joule–Thomson effect by the insertion of imaginary pistons.

at the left pushes in slowly at such a rate that the density to the left of the constriction remains constant. Similarly, the piston to the right moves out at a slightly higher speed so as to keep the density beyond the constriction constant at a lower value. When a certain quantity of gas passes through the constriction the work done by the first piston is $p_1 V_1$, where p_1 is the pressure ahead of the constriction and V_1 is the initial volume of the quantity of gas considered. Simultaneously work is done on the second piston; its amount is $p_2 V_2$, where p_2 is the pressure beyond the constriction and V_2 the final volume of the quantity of gas. If the process takes place adiabatically we get from the first law

$$U_2 - U_1 = p_1 V_1 - p_2 V_2 \qquad (12.20)$$

The process is obviously most easily dealt with if the enthalpy concept is introduced as defined in (7.11). We simply have

$$H_1 = H_2 \qquad (12.21)$$

and the question arises whether the gas heats or cools upon decompression at constant enthalpy. For perfect gases the temperature remains constant, for in that case both U and $p\,V$ are functions of temperature only, and the constancy of H implies the constancy of T. The gaseous imperfections just discussed are going to modify that, and we shall follow up the modification. We may start out with equation (7.12), which, at constant enthalpy, yields S as a function of p. For S itself we write

$$dS = \frac{C_p}{T}\,dT + \left(\frac{\partial S}{\partial p}\right)_T dp$$

The unknown derivative $(\partial S/\partial p)_T$ we get by cross differentiation of (7.14). Thereby (7.12) becomes transformed into

$$dH = C_p\,dT + \left\{ V - T\left(\frac{\partial V}{\partial T}\right)_p \right\} dp \qquad (12.22)$$

The temperature in the Joule–Thomson experiment goes with or against the pressure, depending on whether the sign of the curly bracket is negative or positive. The technical importance of the effect lies in the possibility of cooling a gas by passing it through constrictions. The negative sign is therefore the technologically interesting one.

It is a straightforward matter to work out the quantities in (12.22) for a gas obeying (12.19). The result is

$$\left(\frac{\partial T}{\partial p}\right)_H = -\frac{1}{\mathscr{C}_p}\,\frac{\mathscr{V}(R\,T\,b - 2\,a)}{R\,T\,\mathscr{V} + 2\,R\,T\,b - 2\,a} \qquad (12.23)$$

The sign of the denominator is always fixed by its first term, which is positive. The sign of the whole expression is thus the opposite of that of the numerator. The numerator is also positive if the temperature is high enough, which makes the effect technically useless. However, there is a temperature, the *inversion temperature*, below which the numerator is negative and the effect has the technically desired positive sign. The inversion temperature is obtained by annulling the numerator of (12.23). It yields

$$T_{\text{inv}} = \frac{2\,a}{R\,b} \qquad (12.24)$$

If we remember from Problem 18, Chapter 1, that the critical temperature has the same dimensional form, but has a factor $\frac{8}{27}$ instead of a factor 2, we see that the Joule–Thomson effect is a good indicator of a departure from perfection in a gas before the imperfection has become serious. Table 12–1 shows the comparison of the result (12.24) with experimentally observed inversion temperatures. Agreement is good, in view of the crude model used.

TABLE 12-1. Theoretical computations of the inversion temperature for some gases from the second virial coefficient and comparison with experiment.

Gas	a [atmosphere (liter)^2]	b (liters)	Inversion Temperature Theoretical ($^\circ$K)	Experimental ($^\circ$K)
H_2	0.1975	0.02096	230	202
He	0.0216	0.01400	38	$\sim$25–60
N_2	1.3445	0.05046	648	621
Ar	1.2907	0.03931	800	723
CO_2	5.0065	0.10476	1165	$\sim$1500

Note. The constants a and b are from Hirschfelder *et al.*; the experimental inversion temperatures from Zemansky.

The Joule–Thomson effect is the standard method of liquefying the common gases. Our analysis shows that it can be used to cool and finally liquefy air instead of the more obvious but clumsier method of adiabatic expansion in a cylinder by mechanical motion of a piston. The table shows that hydrogen and helium cannot be liquefied in this manner, starting from room temperature, because their inversion temperatures are too low. For several decades this was a real obstacle to their liquefaction. However, the table shows that the inversion temperature of hydrogen is above the boiling point of nitrogen, and the inversion point of helium above the boiling point of hydrogen. The Joule–Thomson effect can therefore be used to liquefy these gases if suitable precooling arrangements are made.

12-5. Ursell–Mayer expansion of the partition function; diagram summation

As has been explained earlier, the entire problem of determining the thermal properties of gases and liquids can be said to depend on the determination of the configurational partition function $Q(N)$, defined by (12.03). If we insert in this the assumption (12.09) of pairwise forces we get

$$Q(N) = \int \int \int \ldots \int \exp\left[-\beta \sum_{i<j} \phi(|\mathbf{r}_i - \mathbf{r}_j|)\right] d\mathbf{r}_1 \, d\mathbf{r}_2 \ldots d\mathbf{r}_N \quad (12.25)$$

An interlocked product of terms is a very difficult thing to handle. Ursell first found a useful guiding principle to deal with (12.25). It is based on the substitution

$$f_{ij} = \exp[-\beta\,\phi(|r_i - r_j|)] - 1 \qquad (12.26)$$

With this substitution $Q(N)$ becomes

$$Q(N) = \int\int\int\cdots\int \prod_{i<j}(1+f_{ij})\,dr_i\,d\mathbf{r}_2\ldots d\mathbf{r}_N \qquad (12.27)$$

The transformation from (12.25) to (12.27) appears at first sight a formality. However, the form (12.27) allows the introduction of summation techniques into the problem. In order to make them plausible f_{ij} should be small. It is in fact small when $\phi(|\mathbf{r}_i - \mathbf{r}_j|)$ is small, but has the value -1 in the "hard core" region. Thus it is not perfect for the purpose contemplated.

The introduction of the functions f_{ij} allows an ordering of $Q(N)$ according to an entirely new principle. As our first term in (12.27) we take the one containing no f_{ij}'s at all. The integrand is then 1 and the integral V^N. This is the value of $Q(N)$ in the perfect gas limit. Next we take terms having just one f_{ij}, or one "cluster" of two molecules. We then continue with terms having two, three, four factors f_{ij}. We can associate with each of those terms a "cluster diagram." These diagrams represent the molecules by dots and the factors f_{ij} by lines joining these dots. Every possible cluster diagram of lines linking N labeled points will be in one-to-one correspondence to one integral in the Ursell–Mayer expansion. It is convenient to follow modern usage and to employ the cluster diagram as a shorthand notation for the corresponding integral itself. We shall write down here the Ursell–Mayer expansions for the cases of three and four particles as examples. For three particles we have

$$Q(3) = \int\int\int\cdots\int(1+f_{23})(1+f_{31})(1+f_{12})\,d\mathbf{r}_1\,d\mathbf{r}_2\,d\mathbf{r}_3$$
$$= \therefore + 3\,.\diagdown + 3\diagup\diagdown + \triangle \qquad (12.28)$$

and for four particles

$$Q(4) = \int\int\int\int\cdots\int \prod_{i>j=1}^{4}(1+f_{ij})\,d\mathbf{r}_1\,d\mathbf{r}_2\,d\mathbf{r}_3\,d\mathbf{r}_4$$
$$= \vdots\vdots + 6\,\vdots\_ + 12\,\lfloor\dot{} + 3\,\overline{\underline{}}$$
$$+ 4\,\diagup\!\!\!\!\diagdown + 12\,\sqcup + 4\,\diagdown\!\!\!\!\diagup + 3\,\square$$
$$+ 12\,\diagdown\!\!\!\square + 6\,\boxtimes\!\!\! + \boxtimes \qquad (12.29)$$

Here we have already grouped equivalent diagrams together and written a multiplying factor for them; for it is only the topology of these diagrams that matters, that is, topologically equivalent diagrams yield the same integral. Advantage is taken of this in the factor which tells how often it is present. In this count each dot is considered a distinct entity. Diagrams handled in this way are said to have "labeled" corners.

Further progress in the diagram summation of the configurational partition function is based on the role which the connected diagrams play in the theory. A cluster of N points is a diagram of N points or vertices all members of which are linked. The first picture in Fig. 12.4 is such a cluster. As a comparison the second picture in Fig. 12.4 shows a disconnected diagram of nine points which consists of three clusters, a cluster

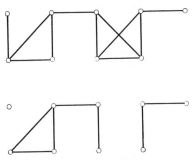

Fig. 12.4. Example of a cluster of nine points and a disconnected diagram of nine points having three clusters.

of one, a cluster of three, and a cluster of five vertices. A glance at the integral (12.27) shows that an integral represented by a disconnected diagram decomposes into a product of integrals, each going with a cluster. Among the integrals associated with nine-vertex diagrams the first one of Fig. 12.4 will be an integral which has never shown up for a lesser number of vertices, while the second will be just a product of integrals previously encountered. It is therefore the cluster type diagram which brings in the new "information" as we expand the partition function to include diagrams of more and more vertices, while disconnected diagrams fail to yield new "information" about an internally coupled system. Now if we disregard for the moment diagrams having isolated points, there is no disconnected diagram in the example (12.28); in the example (12.29) there is just one, and it occurs only three times. However, if we take a large N the situation is entirely different. To return to the example just discussed, among N particles the diagram $=$ occurs with a factor $\frac{1}{8} N (N-1) (N-2) (N-3)$, while the cluster diagram $\llcorner$, which also has two factors f_{ij}, has only a factor $\frac{1}{2} N (N-1) (N-2)$.

This analysis has two consequences. First, it seems to destroy the possibility of expanding $Q(N)$ according to the number of f_{ij} factors in an

Ursell–Mayer expansion, because the large diagrams have also large powers of N as factors. Now N may very well be 10^{20}. No reasonable decrease of the integrals can compensate for such factors. Second, even among diagrams having the same number of links the irrelevant disconnected diagrams have factors outweighing the ones containing clusters, which are new for the number of links considered. Thus even among diagrams with equal numbers of links the disconnected ones outweigh the clusters. Yet it is the clusters which contain the new information.

12-6. Mayer's cluster expansion theorem

It was known intuitively for some time by many workers that the large powers of N occurring in diagram expansions for many-body systems do not produce true divergencies. The factors come in because $Q(N)$ is basically an Nth power (it equals V^N for the perfect gas) and expansion of an Nth power is apt to produce factors containing powers of N in many of the terms. There developed therefore a semiempirical method of removing these factors by formal extraction of the Nth root. This was a tedious and unsystematic procedure. Mayer was able in the present instance to replace this procedure by the following theorem:

Let $G(N)$ be any labeled diagram of N vertices and $G_1(N)$ be any diagram in which the N members form a single cluster. Let $W(G)$ be an arbitrary weight factor associated with G, except for two features:

(a) It shall depend on the topology of the diagram but not on its labeling.

(b) The weight of a disconnected diagram shall equal the product of the weights of the clusters composing it.

Then the following algebraic identity holds:

$$1 + \sum_{N=1}^{\infty} \frac{x^N}{N!} \sum_{G(N)} W[G(N)] = \exp\left\{ \sum_{M=1}^{\infty} \frac{x^M}{M!} \sum_{G_1(M)} W[G_1(M)] \right\} \quad (12.30)$$

Here $\sum_{G}$ is a summation over all diagrams subsumed under G.

Proof. Denote by $G_m(N)$ any diagram of N vertices which decomposes into m clusters, and by $\sum_{G_m(N)} W[G_m(N)]$ the sum of the weights over this subclass of diagrams. We can construct this subclass by creating m collections of all connected cluster diagrams which have less than $N + 1$ vertices and by picking one cluster in each collection. The restriction that the sum of the vertices equal N must of course be maintained by deliberate choice. For each such choice the weight equals

$$W = \frac{N!}{N_1! \, N_2! \ldots N_m!} W[G_1(N_1)] \, W[G_1(N_2)] \ldots W[G_1(N_m)]$$

The result follows from the product assumption with the factorial arising from the labeling: we can exchange the labeling between collections. By making all possible choices in the m collections, we cover all members of $G_m(N)$ a number of times which is $m!$. This factor arises because the m collections are identical and any particular choice can therefore be made in as many ways as one can interchange collections. We have therefore the result

$$\sum_{G_m(N)} W[G_m(N)] = \frac{1}{m!} \sum_{\substack{N_1\ N_2 \\ N_1+N_2+N_3+\ldots+N_m=N}} \cdots \sum_{N_m} \frac{N!}{N_1!\, N_2!\ldots N_m!}$$

$$\times \sum_{G_1(N_1)} W[G_1(N_1)] \sum_{G_1(N_2)} W[G_1(N_2)] \ldots \sum_{G_1(N_m)} W[G_1(N_m)] \quad (12.31)$$

From (12.31) theorem (12.30) follows by straightforward algebra:

$$1 + \sum_{N=1}^{\infty} \frac{x^N}{N!} \sum_{G(N)} W[G(N)] = 1 + \sum_{N=1}^{\infty} \frac{x^N}{N!} \sum_{m=1}^{N} \sum_{G_m(N)} W[G_m(N)]$$

$$= 1 + \sum_{N=1}^{\infty} x^N \sum_{m=1}^{N} \frac{1}{m!} \sum_{\substack{N_1\ N_2 \\ N_1+N_2+\ldots N_m=N}} \cdots \sum_{N_m} \prod_{N_i} \frac{1}{N_i!} \sum_{G_1(N_i)} W[G_1(N_i)]$$

At this stage x^N can be brought inside the product sign as it equals $\prod_{N_i} x^{N_i}$. The sum over N can then be dropped, simply by summing over m to infinity and removing the restriction on the N_i. This means that each sum runs independently from 1 to infinity. The sums are thereby rendered completely independent of each other, and the sum of products becomes a product of sums. We thus get

$$1 + \sum_{N=1}^{\infty} \frac{x^N}{N!} \sum_{G(N)} W[G(N)] = 1 + \sum_{m=1}^{\infty} \frac{1}{m!} \prod_{i=1}^{m} \sum_{N_i=1}^{\infty} \frac{x^{N_i}}{N_i!} \sum_{G_1(N_i)} W[G_1(N_i)]$$

The product of m factors is m times the same factor which makes it an mth power. So we conclude finally

$$1 + \sum_{N=1}^{\infty} \frac{x^N}{N!} \sum_{G(N)} W[G(N)] = 1 + \sum_{m=1}^{\infty} \frac{1}{m!} \left\{ \sum_{m=1}^{\infty} \frac{x^M}{M!} \sum_{G_1(M_i)} W[G_1(M)] \right\}^m$$

$$= \exp\left\{ \sum_{M=1}^{\infty} \frac{x^M}{M!} \sum_{G_1(M)} W[G_1(M)] \right\}$$

which is the result stated in (12.30).

Mayer's theorem eliminates in a rigorous way all disconnected diagrams from the problem of imperfect gases and disposes of the spurious factors

N in its cluster expansion. Its main disadvantage is that it is associated with the Ursell–Mayer expansion; the truncation of the sum when passing from the left to the right of (12.30) does as a rule render it impossible to express the result in closed form after the transformation is completed.

As we go back over the proof of Mayer's theorem we realize that relation (12.31) is the heart of the proof, and in particular the factorials occurring in it. It is hard to believe that the factorials can be so cooperative. Perhaps the reader will gain some confidence if he follows the author in checking the relation numerically in a simple case. Let us take $N = 4$, $m = 2$. The left-hand side is then already worked out in (12.29); it is the third, fourth, and seventh term of that expansion. Thus (12.31) predicts that

$$12 \; \llcorner\!\cdot \;\; + 3 \; \boldsymbol{\boxminus} \;\; + 4 \; \boldsymbol{\triangle} \;\; = \frac{1}{2} \sum_{N=1}^{3} \frac{4!}{N!(4-N)!}$$
$$\sum_{G_1(N)} W[G_1(N)] \sum_{G_1(4-N)} W[G_1(4-N)]$$

$N = 1$ has only one cluster diagram, the isolated point; $N = 2$ has only one, $\cdot\;\cdot$; $N = 3$ has two, namely, $\llcorner$ and $\boldsymbol{\triangle}$; the first has a factor three because there are three nonequivalent ways to label the corners 1, 2, 3. Hence we get

$$12 \; \llcorner\!\cdot \;\; + 3 \; \boldsymbol{\boxminus} \;\; + 4 \; \boldsymbol{\triangle} \;\; = 2(\cdot)\{3 \; \llcorner \;\; + \boldsymbol{\triangle}\} + 3(-\!\!-)(-\!\!-)$$
$$+ 2\{3 \; \llcorner \;\; + \boldsymbol{\triangle}\}(\cdot)$$

The identity of the two sides is obvious.

12-7. Mayer's formulation of the equation of state of imperfect gases

In order to apply Mayer's theorem to imperfect gases it is necessary to make use of the grand partition function. It equals, according to (8.13) and (12.02),

$$\mathscr{F} = 1 + \sum_{N=1}^{\infty} \frac{1}{N!} z^N Q(N) \tag{12.32}$$

with the so-called *fugacity* z equal to

$$z = \left(\frac{2 \pi m k T}{h^2} \right)^{3/2} \exp \beta \mu \tag{12.33}$$

Now, according to (12.27) and the subsequent discussion, the configurational partition function $Q(N)$ is of the form

$$Q(N) = \sum_{G(N)} W[G(N)] \tag{12.34}$$

where $\sum\limits_{G(N)}$ goes over all labeled diagrams of N vertices, and $W[G(N)]$ is the integral of a product of Ursell–Mayer functions represented by the diagram. Theorem (12.30) is therefore applicable to (12.32) and yields

$$\mathscr{F} = \exp\left\{\sum_{M=1}^{\infty} \frac{z^M}{M!} \sum_{G_1(M)} W[G_1(M)]\right\} \tag{12.35}$$

This formula has already been written down for the case of the perfect gas. It is the combination of (8.59) and (8.50). The exponent has in that case a single term, the cluster of one, and

$$W[G_1(1)] = V$$

The dependence on volume remains in fact fairly simple, even for the imperfect gas: each linked cluster integral produces the factor V just once for the coordinate of the center of mass of the cluster. The cluster integral $W[G_1(M)]$ therefore takes the form

$$W[G_1(M)] = V\, w[G_1(M)] \tag{12.36}$$

where $w[G_1(M)]$ is a function of the parameters of the force law and the temperature, but not the volume. Therefore $\mathscr{F}$ takes the form

$$\mathscr{F} = \exp\left\{V \sum_{M=1}^{\infty} \frac{z^M}{M!} \sum_{G_1(M)} w[G_1(M)]\right\} \tag{12.37}$$

with

$$w[G_1(1)] = 1 \tag{12.38}$$

The form (12.37) is one to which both (8.08) and (8.15) are applied easily, yielding

$$\frac{p}{kT} = \sum_{M=1}^{\infty} \frac{z^M}{M!} \sum_{G_1(M)} w[G_1(M)] \tag{12.39}$$

and

$$\frac{N}{V} = \sum_{M=1}^{\infty} \frac{z^M}{(M-1)!} \sum_{G_1(M)} w[G_1(M)] \tag{12.40}$$

Elimination of z from this simultaneous pair of equations yields the equation of state.

The advantage of the equation pair (12.39) and (12.40) over other formulations is that it is exact and at the same time good for use in approximations. These approximations can be improved indefinitely by calculation of more and more complicated cluster integrals. It is also a perfectly straightforward matter, when two quantities are given as power series in the same variable without a constant term, to write one variable as a power series in terms of the other with predictable coefficients. If we do this in the form of expressing the quantity (12.39) in powers of the quantity (12.40) we get the virial expansion with cluster integrals as coefficients. We shall not follow up this practical aspect very far. If we stop the two series at the pair cluster diagram we obtain

$$\frac{p}{kT} = z + \tfrac{1}{2} z^2 I + 0(z^3) \tag{12.41}$$

and

$$\frac{N}{V} =' z + z^2 I + 0(z^3) \tag{12.42}$$

where I is the pair diagram of the interaction (12.26), stripped of a factor V by (12.36). This means that

$$I = -4\pi \int_0^\infty \{1 - \exp[-\beta\phi(r)]\} r^2 \, dr \tag{12.43}$$

The result of the elimination process comes out to be

$$\frac{p}{kT} = \frac{N}{V} - \frac{1}{2} I \frac{N^2}{V^2} + 0\left(\frac{N^3}{V^3}\right) \tag{12.44}$$

This result is identical with the result (12.14) obtained from the virial theorem. The present formalism is, however, capable of being continued to higher powers without any essential difficulty. This was not the case for the result (12.14). The Mayer method represents therefore a very fundamental progress in the theory of imperfect gases.

Before this subject is dropped it should be mentioned that Ono and Kilpatrick‡, §, ‖ have shown that the Mayer cluster expansion theorem holds under much more general conditions than was originally supposed. Neither the restriction to pairwise forces, nor the one to classical mechanics, is necessary. The identity can be traced to relation (8.58), which expresses the grand partition function $\mathscr{F}$ in terms of the canonical partition functions $F(N)$. It is therefore a result of wider significance than appears from the proof given here.

12-8. Phase equilibrium between liquid and gas; critical phenomenon

The gaseous and liquid phases of a substance stand in a closer mutual relation than phases do in general. For the phase boundary between them

‡ S. Ono. *J. Chem. Phys.* **19**, 504 (1951).
§ J. E. Kilpatrick. *J. Chem. Phys.* **21**, 274 (1953).
‖ T. L. Hill. *Statistical Mechanics.* New York: McGraw-Hill, 1956, pp. 134 and 135.

when plotted in a pressure–temperature plane terminates at a point at which distinction of the two phases becomes impossible. In Fig. 12.5 a schematic phase diagram of a substance is plotted. It has three regions: solid, liquid, and gaseous. The triple point P_t is the point at which the three phases coexist, in accordance with the phase rule. A second point P_c is marked, at which the boundary line between the two fluid phases ceases to exist. This created very early the general notion that there is perhaps no distinction of principle between the two phases, and an equation

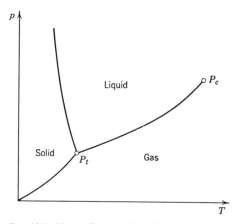

Fig. 12.5. Phase diagram of a simple substance (schematic). P_t = triple point; P_c = critical point.

might be found describing them both. Van der Waals definitely had this ambition when formulating his equation. We shall take as a basis of discussion the Dietrici equation introduced in Problem 18, Chapter 1. It yields the same second virial coefficient as the van der Waals equation and avoids the awkward problem of negative pressures. The equation reads

$$p\left(V-b\right) = R\,T \exp\left[-\frac{a}{R\,T\,V}\right] \qquad (12.45)$$

Some of the isotherms derived from this equation are shown in Fig. 12.6. They show three distinct types of behavior. There is a critical isotherm having a point at which both $(\partial p/\partial V)_T$ and $(\partial^2 p/\partial V^2)_T$ vanish simultaneously. This is the critical point which was introduced in the problems of Chapter 1. At that point the bulk modulus of the substance is zero. For temperatures larger than T_c the slope of every isotherm is always negative; for very large values of T the perfect gas condition is approached. For temperatures smaller than the critical temperature the isotherms have two regions of negative slope, separated by a central region of positive slope. It is quite obvious that this central portion can have no physical existence because a positive slope represents a physical instability. It does therefore have to be rejected, but there is some question what it is to be replaced by.

The basic defect of a single analytic equation of state in this region is

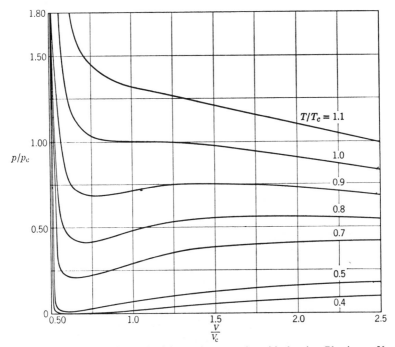

Fig. 12.6. Isotherms of the Dietrici equation near the critical point. Plot in a *p*-V diagram.

that it assumes a uniform density. Actually this region is a two-phase region with two entirely different densities in different portions of the sample. As discussed in Chapter 7, in such a region the pressure is actually independent of the volume and is controlled by the temperature only, as worked out in (7.46). Thus the true isotherm consists of two sections of negative slope, joined by a straight line of horizontal slope. Along this line the structure of each phase remains constant as represented by the end points, and only the relative abundance of the two phases changes.

Maxwell solved the problem of how the horizontal line should be laid in plots of the type shown in Fig. 12.6. He specified that the horizontal line must be laid across the tipped-over S-shaped curve in such a way that the areas to the right and to the left of the straight line are equal. This construction is shown in Fig. 12.7. The areas which are to be equal are cross-hatched.

We shall give not Maxwell's argument but rather a modernized version of it which is logically clearer. According to (7.43), two coexistent phases of the same pure substance must have the same chemical potential or

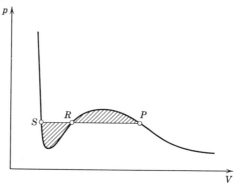

Fig. 12.7. Maxwell construction for the location of the
equilibrium vapor pressure.

Gibbs free energy g per molecule. The variation of this quantity along an isotherm is described, from (7.14), as

$$\int dg = \int v \, dp \qquad (12.46)$$

where v is the volume per molecule. The geometrical meaning of (12.46) is exhibited in Fig. 12.8. Let PQR be a piece of conventional isotherm. The difference in chemical potential between the points P and Q is then the area cross-hatched on the figure with the sign such that Q has a higher chemical potential than P. This surplus remains as the point R is reached, the surplus at R being the area bounded by the horizontal line PR and the curve PQR. The point R on the curve is thus immediately shown as unstable in comparison to a phase mixture having the same mean molecular volume. The curve can now be followed beyond R, where it yields negative increments of g. Finally if the Maxwell construction has been followed the horizontal line PR intersects the curve a third time at S as shown in Fig. 12.7, and the total increment of g from P to S vanishes. The states represented by P and S form therefore the proper coexistent phases, and the chemical potential along the horizontal line is equal to that of the end points, and smaller than that on the curved isotherm having the same volume v. Thus the phase mixture is the stable form.

There is an obvious objection to the Maxwell construction, namely, that it makes use of incorrect pieces of isotherm in which the pressure increases with volume. Such points cannot even be local minima of the Gibbs free energy, and thus an argument making use of such pieces of curve is subject to doubt. In the case of equation (12.45), where the entire equation of state is summed up in the same analytic equation, this objection can be laid to rest. The free energy g is a state variable, which means that its value is determined entirely by the point in the phase diagram.

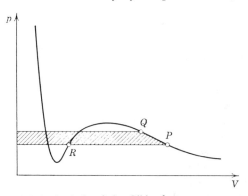

Fig. 12.8. Analysis of the Gibbs free energy on a pressure-volume plot.

We can therefore compute the free energy by following a path passing around the outside of P_c in the plot of Fig. 12.5. All intermediate points are then proper stable points, and the free energy obtained by integrating (7.14) must come out correctly. By the analyticity argument for g and its property as a state variable, the difference between the values at P and S in Fig. 12.7 must come out the same as it does by the Maxwell construction. Thus the Maxwell construction is justified. Of course, if the equation of state is pieced together from different analytic approximations for different parts of the p–V plane, the argument against unphysical equations regains its full force and the Maxwell construction is no longer necessarily correct.

A. RECOMMENDED PROBLEMS

1. In Section 12-4, Joule–Thomson inversion is discussed as if it occurred at a fixed temperature. Actually it occurs along a curve in the T–p plane or the T–V plane. Find the equation of this curve for the van der Waals gas.

2. Acquire facility with Mayer diagram notation by writing down the diagram expression for $Q(5)$, following the models (12.28) and (12.29).

B. GENERAL PROBLEMS

3. When the equation of state for a gas is modified by the inclusion of a second virial coefficient

$$p = \frac{RT}{V}\left(1 + \frac{B(T)}{V}\right)$$

the expression for the chemical potential μ is also changed. Find the new expression.

4. A Simon helium liquefier consists essentially of a thermally insulated vessel containing gaseous helium at high pressure and $10°$ K (above the critical point of helium). From this vessel the helium is allowed to escape slowly until the pressure is 1 atm, and the temperature $4.2°$ K, the boiling point of helium. Assuming that the insulation is perfect and the heat capacity of the vessel negligible compared to that of the gas, and that gaseous helium is a perfect gas, calculate what the initial pressure must be, if the vessel is to be entirely filled with liquid at the end.

5. It was pointed out that the identity (12.31) forms the core of Mayer's cluster theorem. Verify the identity for $N = 5$, $m = 2$.

6. It was pointed out that relations (12.41) and (12.42) for pressure and density are the beginning of power series developments in z. Continue the developments to z^4, and derive from them the equation of state (12.44) to the fourth virial coefficient.

13

Thermal properties of crystals

This chapter deals essentially with the explanation of the origin of the heat capacity of solids, and crystals in particular. It will be seen that the heat capacity arises from the vibration of the atoms about their equilibrium positions. The analysis of these vibrations is facilitated by the symmetry of the crystal and by the relation which these vibrations have to the phenomenon of transmission of sound.

13-1. Relation between the vibration spectrum and the heat capacity of solids

Solids are like molecules, structures which are geometrically fixed. The atoms or ions compressing them are approximately at rest. The electrons, on the other hand, are in rapid motion, but in wave functions which are fixed by the presence of the heavy nuclei; the phenomenon of excitation of electrons is interesting, but sufficiently rare that it can be neglected in a first study of thermal properties of solids. We are then left to consider a framework of atoms held in static equilibrium by the forces within the solid.

It was pointed out in Chapter 4 that this system is among the easiest in physics to analyze if one is satisfied with a semiquantitative explanation. A general theorem of mechanics tells us that if one has a system of N

mass points in stable equilibrium under their mutual forces, then the total potential energy can be developed into a power series in their displacements. This power series will be devoid of linear terms because of the equilibrium conditions. The first nontrivial terms are thus the quadratic terms. If the development is broken off at this point (which is legitimate for small enough amplitudes of vibration), then the potential energy appears as a homogeneous quadratic form in the displacements. This quadratic form, together with the quadratic form representing the kinetic energy, may be analyzed by vibration theory, a well understood chapter of mechanics. The theory asserts that a linear transformation always exists which transforms simultaneously the kinetic energy K and the potential energy Φ to a sum of squares as follows:

$$K = \tfrac{1}{2} \sum_i \frac{dX_i^2}{dt} \tag{13.01}$$

$$\Phi = \tfrac{1}{2} \sum_i \omega_i^2 \, X_i^2 \tag{13.02}$$

The variables X_i are called *normal coordinates*, and ω_i is the *eigenfrequency* associated with the normal coordinate X_i. The number of terms in the sums (13.01) and (13.02) is equal to three times the number of atoms making up the solid.

If we combine this analysis with the equipartition theorem in the form (4.46), we are immediately led to the law of Dulong and Petit for atomic crystals. The corresponding generalization for molecular crystals would ascribe to a mole of solid a heat capacity equal to

$$\mathscr{C} = 3 \, n \, R \tag{13.03}$$

where n is the number of atoms comprising a molecule.

There are two discrepancies from (13.03) to be noted. First, at low temperature $\mathscr{C}$ falls below (13.03) and finally approaches zero at absolute zero, as required by the Third Law. The semiquantitative explanation of this phenomenon is due to Einstein. He showed that the harmonic oscillator has a dormancy phenomenon as regards its thermal properties, which sets in when $k\,T$ becomes small compared to the quantum steps. To facilitate analysis he assumed that all frequencies are identical. The curves for the energy and heat capacity are then the ones shown in Fig. 4.3 and Fig. 4.4. A second discrepancy to be noted is that even at room temperature the proposed law (13.03) is well obeyed only for atomic crystals, that is, for the case $n = 1$. In other cases the heat capacity tends to be smaller, but gently rising. The reason for this lies in the feature that interatomic forces in molecules are particularly stiff and therefore of high frequency. The dormancy thus sets in at relatively elevated temperatures.

Einstein's idea of taking all frequencies the same is therefore not even approximately valid for molecular crystals. It is possible to take care of these difficulties by a number of Einstein terms, each for a particular type of interatomic vibration. It is more profitable to skip this stage and to pass directly to the kind of formalism suggested by (13.01) and (13.02). It consists of introducing a *frequency distribution function* $g(f)$ which gives the number of modes per unit frequency range. It must obey the relation implicit in (13.03), namely,

$$\int_0^\infty g(f)\,df = 3\,n\,N \tag{13.04}$$

where N is the number of molecules in the solid.

The introduction of $g(f)$ permits immediate adaptation of the formulas of Chapter 4 to the present purpose. The decomposition (13.01) and (13.02) into normal modes permits decomposition of the partition function (5.01) into a set of factors of the type (4.54). The result is best written in a logarithmic form as follows

$$\ln F = \int_0^\infty \ln\left(\tfrac{1}{2}\operatorname{cosech}\tfrac{1}{2}\,\beta\,h\,f\right)g(f)\,df \tag{13.05}$$

It is to be noted that the factor $1/N!$ introduced in (8.48) is *not* to be employed in connection with (13.05). For this factor was placed there to correct for a specific defect which the partition function would have without it: namely, that it accepts interchange of two identical particles as a possible operation leading to a new state. This defect is not present here. The partition function deals with a quantized wave field for which the problem of "exchange of particles" does not arise. Furthermore, as far as interchange of the atoms or electrons of the crystal is concerned, the normal mode approximation does not include them as possible operations. Thus no correction of the partition function is necessary here.

The internal energy U of the crystal may be obtained either directly from (4.55), or from the partition function (13.05) with the help of (5.02). Either procedure yields the result

$$U = \int_0^\infty \tfrac{1}{2}\,h f \coth \tfrac{1}{2}\beta h f\, g(f)\,df \tag{13.06}$$

From (13.05) and (13.06) the entropy follows by (5.09) as

$$\frac{S}{k} = \int_0^\infty \left\{\ln\left(2\operatorname{cosech}\tfrac{1}{2}\,x\right) + \tfrac{1}{2}\,x\coth\tfrac{1}{2}\,x\right\}g(f)\,df \tag{13.07}$$

where

$$x = \frac{h f}{k T} \tag{13.08}$$

Finally we get the heat capacity most directly from

$$\frac{C}{k} = \beta^2 \frac{d^2 \ln F}{d\beta^2} \qquad (13.09)$$

or from (4.56). The result is with (13.08)

$$\frac{C}{k} = \int_0^\infty \tfrac{1}{4}\, x^2 \operatorname{cosech}^2 \tfrac{1}{2}x \ \ g(f) \ df \qquad (13.10)$$

The Dulong–Petit limit arises in (13.10) when x is small for *all* frequencies. The part of the integrand in (13.10) preceding $g(f)$ then equals unity, and (13.04) can be applied to the remainder, yielding

$$C = 3\, n\, N\, k \qquad (13.11)$$

The formula is rarely valid in this extreme form.

13-2. Vibrational bands of crystals; models in one dimension

The purpose of the two examples to follow is to familiarize the reader with the idea that symmetry permits a complete analysis of the vibration spectrum of a crystal in wave number space. The problem is thus uniquely one of transformation from a wave number scale to a frequency scale. Those to whom this idea is familiar should pass up this section and continue directly with Section 13-3.

The simplest model of a vibrating crystal consists of a linear array of equal masses M connected by equal massless springs of spring constant k (Fig. 13.1). Let us investigate the possible energy states associated with

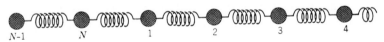

Fig. 13.1. One-dimensional array of equal masses and springs; this is the simplest model of a vibrational band.

the longitudinal motions of such an array. For simplicity, the first mass is assumed linked up with the last one. This is entirely to avoid special consideration of the end members. It stands to reason that the disposal of the end members has no profound influence upon the modes of motion if the number N of masses is assumed large. The problem is thus one of N degrees of freedom, with coordinates $x_0, x_1, x_2, x_3, \ldots x_{N-1}$, which measure the displacement of each mass from its equilibrium position. It is sometimes convenient to look at this same problem as if the array were infinite, but subject to a periodic boundary condition*

$$x_{n+N} = x_n \tag{13.12}$$

The kinetic energy K of the system so defined equals

$$K = \tfrac{1}{2} M \sum_{\nu=0}^{N-1} \left(\frac{dx_\nu}{dt}\right)^2 \tag{13.13}$$

and the potential energy Φ

$$\Phi = \tfrac{1}{2} k \sum_{\nu=0}^{N-1} (x_\nu - x_{\nu+1})^2 \tag{13.14}$$

Before transforming (13.13) and (13.14) to normal coordinates we shall look for a moment at the classical equations of motion which they engender. The equation is the same for each index n and reads

$$M \frac{d^2 x_n}{dt^2} = k \, (x_{n+1} + x_{n-1} - 2x_n) \tag{13.15}$$

The right-hand side is the second difference of x with respect to n, which for long waves can be approximated by the second derivative. Equation (13.15) then becomes

$$M \frac{\partial^2 x}{\partial t^2} = k \frac{\partial^2 x}{\partial n^2} \tag{13.16}$$

which is D'Alembert's equation. The wave phenomenon which is picked up here is the propagation of sound. It is relatively easy to verify that the constant in (13.16) is indeed the square of the velocity of sound for the system.

For waves whose phase varies significantly from one mass to its neighbor the analysis leading to (13.16) is not correct. We must instead go through with the transformation to normal coordinates. Actually, in the present instance the modes are entirely determined by group theory, owing to the symmetry of the array. The element of symmetry which enters is the elementary translation

$$x_n \rightarrow x_{n+1} \tag{13.17}$$

We shall not use any group theoretical language in this book, but simply take advantage of the simplified algebra which (13.17) puts at our disposal.

We shall carry out the transformation to normal coordinates by two

---

* This boundary condition is sometimes called the Born–von Karman boundary condition.

successive unitary transformations. The first arises from the symmetry element (13.17) and the second from time reversal. The first transformation reads

$$x_n = \frac{1}{\sqrt{N}} \sum_{\mu=0}^{n-1} e^{-2\pi i \mu n/N} X_\mu \tag{13.18}$$

The transformation (13.18) dominates all work on crystals to such an extent that some comments are in order concerning it. It is unitary because

$$\frac{1}{N} \sum_{\mu=0}^{N-1} e^{2\pi i(n-n')\mu/N} \cong \delta_{n,n'} \tag{13.19}$$

Here the sign $\cong$ is used to indicate that the Kronecker δ is also 1 if n and n' differ by a multiple of N. Indeed both (13.18) and (13.19) have the indeterminacy (13.12) built in. The same is true for the result of the inversion formula of (13.18), namely,

$$X_m = \frac{1}{\sqrt{N}} \sum_{\nu=0}^{N-1} e^{2\pi i m \nu/N} x_\nu \tag{13.20}$$

X_m also obeys the identity

$$X_{m+N} = X_m \tag{13.21}$$

which means that the finite trigonometric transformation employed here yields automatic closure if one lets the index run over more than N successive integers. It also means that the summations in (13.18), (13.19), or (13.20) could be extended over any other N consecutive integers and yield the same answers. The geometric meaning of the transformation is also of the greatest importance in many different fields of solid state physics. It furnishes the passage from the localized description (exemplified here by x_n) to the wave description (exemplified here by X_m). The wave type coordinate has the property of being multiplied only by a factor when the translational symmetry operation (13.17) is applied. In the present instance it must therefore be pretty close to being a normal coordinate. Indeed, if we substitute (13.18) into (13.13) and (13.14) we find through repeated use of (13.19)

$$K = \tfrac{1}{2} M \sum_{\mu=0}^{N-1} \frac{dX_\mu}{dt} \frac{dX_{-\mu}}{dt} \tag{13.22}$$

and

$$\Phi = \tfrac{1}{2} M \sum_{\mu=0}^{N-1} \omega_\mu^2 X_\mu X_{-\mu} \tag{13.23}$$

where we set

$$\omega_\mu = 2\pi f_\mu = 2 \left(\frac{k}{M}\right)^{1/2} \sin \frac{\pi |\mu|}{N} \tag{13.24}$$

The mode $\mu = 0$ is already a normal mode; it has zero frequency.

Because of the identity (13.21) the mode $\mu = \frac{1}{2} N$ also is already a normal mode if present (which is the case for even N); it has the highest possible frequency

$$\omega_{\max} = 2 \left(\frac{k}{M}\right)^{\frac{1}{2}} \tag{13.25}$$

For the other modes a further transformation must be carried out which transforms to standing waves. It reads

$$X_m = \frac{1}{\sqrt{2}} (A_m + i B_m) \tag{13.26a}$$

$$X_{-m} = \frac{1}{\sqrt{2}} (A_m - i B_m) \tag{13.26b}$$

$$0 < m < \tfrac{1}{2} N \tag{13.26c}$$

The combination of (13.18) and (13.26) is the real orthogonal transformation to normal coordinates which we have been looking for. The kinetic and potential energies now take the form (for even N)

$$K = \tfrac{1}{2} M \left(\frac{dX_0}{dt}\right)^2 + \tfrac{1}{2} M \left(\frac{dX_{\frac{1}{2}N}}{dt}\right)^2 + \tfrac{1}{2} M \sum_{\mu=1}^{\frac{1}{2}N-1} \left[\left(\frac{dA_\mu}{dt}\right)^2 + \left(\frac{dB_\mu}{dt}\right)^2\right] \tag{13.27}$$

$$\Phi = \tfrac{1}{2} M \omega_{\max}^2 X_{\frac{1}{2}N} + \tfrac{1}{2} M \sum_{\mu=1}^{\frac{1}{2}N-1} \omega_\mu^2 [A_\mu^2 + B_\mu^2] \tag{13.28}$$

With the results (13.27) and (13.28) the system shown in Fig. 13.1 is reduced to a set of harmonic oscillators with a frequency spectrum given

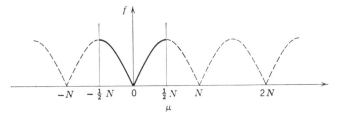

Fig. 13.2. Plot of frequency versus wave number for the model of Fig. 13.1. The information outside the part shown in full outline is redundant.

by (13.24). A plot of the frequency versus wave number relation is shown in Fig. 13.2. Some aspects of this curve are of wide significance for solids in general. First, at very low frequency, frequency and wave number are proportional to each other, as required by the theory of sound. Then a departure from proportionality sets in; this departure becomes significant when the wave length becomes comparable to the interparticle distance.

Finally, the spectrum breaks off altogether, owing to the discrete character of the vibrating units. For if one considers wave lengths smaller than twice the interparticle distance, one finds a phase difference larger than π between neighbors; such a phase difference can just as well be expressed as a number absolutely smaller than π.

Quantization of the hamiltonian composed of (13.27) and (13.28) is immediate. The quantum states of the system are

$$\varepsilon(p_0, n_1, n_2, \ldots n_{N-1}) = \frac{p_0^2}{2NM} + \sum_{\mu=1}^{N-1} (n_\mu + \tfrac{1}{2}) \hbar \, \omega_\mu \qquad (13.29)$$

From this elaborate analysis only two things need be retained for the theory of thermal properties. First, it follows from (13.26) and (13.18) that the normal modes are plane waves which are uniformly spaced in wave number to a maximum wave number for which the phase difference between successive masses is π. Second, equation (13.24) provides a relation between frequency and wave number which permits determination of $g(f)$. By the preceding analysis the average of any function of the frequency f equals

$$\langle \phi(f) \rangle = \frac{1}{N} \left[\phi(0) + \phi(f_{\max}) + 2 \sum_{\mu=1}^{\frac{1}{2}N-1} \phi(f_\mu) \right]$$

Equating the sum to an integral over μ and transforming to f as the variable of integration, we get

$$\langle \phi(f) \rangle = \frac{2}{\pi} \int_0^{f_{\max}} \phi(f) \, \frac{df}{(f_{\max}^2 - f^2)^{1/2}} \qquad (13.30)$$

which means that $g(f)$ equals

$$g(f) = \frac{\frac{2}{\pi} N}{(f_{\max}^2 - f^2)^{1/2}} \qquad (13.31)$$

The curve is shown in Fig. 13.3. The essential feature to retain for future references is not its exact form but the fact that the frequency spectrum has a lower limit zero and a finite upper limit, and that it forms a quasi-continuum between these limits.

We shall briefly discuss a second one-dimensional model which differs from the model of Fig. 13.1 only in that heavy balls of mass M alternate with light balls of mass m. Such a structure forms a reasonable model for ionic crystals such as sodium chloride or even atomic crystals such as diamond or silicon, which have two atoms per primitive cell. The mathematics is very similar to the one used in the first example and can be supplied by the reader. We pass immediately to the result which is significant for statistical purposes, namely, the frequency versus wave number relation. It reads

$$f^2 = \frac{k(m+M) \pm [k^2(m+M)^2 - 4k^2 m M \sin^2 \pi p/N]^{\frac{1}{2}}}{4\pi^2 m M} \quad (13.32)$$

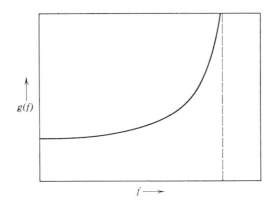

Fig. 13.3. Frequency density versus frequency for the model of Fig. 13.1.

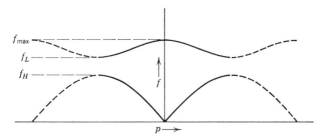

Fig. 13.4. Plot of frequency versus wave number for the model of Fig. 13.1 with alternating masses. The two frequency bands are called "acoustical" and "optical."

where p is an integer. A plot of this function is shown in Fig. 13.4. It is seen that the frequency spectrum consists now of two branches which are associated with the two signs of the square root in (13.32). $p = 0$ yields the smallest frequency, which is zero, and the largest one, which equals

$$f^2_{\max} = \frac{k}{2\pi^2}\left(\frac{1}{M}+\frac{1}{m}\right) \quad (13.33)$$

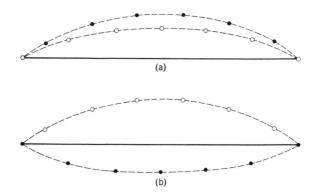

Fig. 13.5. Nature of the vibration in the acoustical (a) and optical (b) branches of the vibration spectrum. The open circles represent the lighter masses, the full circles the heavier ones.

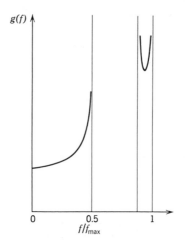

Fig. 13.6. Frequency density versus frequency for the model of Fig. 13.1 with masses alternating in the ratio 3:1.

The lowest frequency of the upper branch equals

$$f_L^2 = \frac{k}{2\pi^2 m} \tag{13.34}$$

and the highest frequency of the lower branch equals

$$f_H^2 = \frac{k}{2\pi^2 M} \tag{13.35}$$

The nature of the two branches is shown in Fig. 13.5. In the lower or "acoustical" branch the two masses vibrate in phase; its low-frequency portion is correctly represented as sound, as previously. The upper or "optical" branch has the two masses vibrate out of phase. Its name comes from the alkali halides in which the two masses also have opposite charges.

For statistical purposes only the quantity $g(f)$ resulting from (13.32) need be retained. The result is Fig. 13.6. It is seen that structural complications in the crystal also produce complications in the function $g(f)$.

13-3. Vibrational bands of crystals; general theory

It is well known that crystals are objects exhibiting a high degree of regularity. This regularity can be analyzed accurately according to the elements of symmetry involved and varies from case to case. There is one type which transcends by far all others in physical importance, and that is the symmetry with respect to small translational displacements. The vectors representing these displacements are called *lattice vectors*. Any lattice vector **r** can be made up out of three independent elementary translations, represented by three *basis vectors* **a**, **b**, **c**, thus

$$\mathbf{r} = l\,\mathbf{a} + m\,\mathbf{b} + n\,\mathbf{c} \tag{13.36}$$

where

$$l, m, n = \text{integer} \tag{13.37}$$

The lattice of the points formed by (13.36) and (13.37) is called the *Bravais lattice* of the crystal. The result of tracing it out with the help of the basis vectors **a**, **b**, **c** is to subdivide space into a number of parallelepipeds as shown in Fig. 13.7. The content of one such parallelepiped is identical to that of any other. The parallelepiped is called the *primitive cell* of the crystal. Exploration of the content of one such cell is equivalent to exploration of the entire crystal. This is most conveniently done by adopting (13.36) but letting l, m, and n vary within the range

$$0 \leq l, m, n \leq 1 \tag{13.38}$$

and to analyze the contents of such a parallelepiped. The result of such an analysis, which is carried out by X-rays, is a set of positions (l, m, n) for the atoms of the crystal and a set of density contours in the cell for the rapidly moving electrons. Neither of these data is affected much by temperature. The primary influence of temperature takes place as a rule through thermal expansion which slowly changes the size and shape of the primitive cell. Ignoring this effect, we arrive then at the picture of a lattice with respect to which the atoms have fixed equilibrium positions. For simplicity we shall limit the discussion to simple monatomic crystals in which the lattice shown in Fig. 13.7 is formed by the atoms themselves. Equations

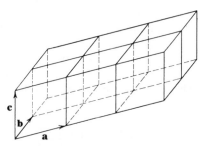

Fig. 13.7. Lattice structure traced out by
the three basis vectors of a crystal.

(13.36) and (13.37) then label the position of the atoms. To avoid irrelevant complications we shall assume the crystal as a whole to be a parallelepiped with sides parallel to the primitive cell. The sides of the crystal can then be described as $L\,\mathbf{a}$, $M\,\mathbf{b}$, and $N\,\mathbf{c}$, with L, M, N large integers.

Let us define by $\mathbf{x}_{l,m,n}$ the displacement from equilibrium of the atom at the position l, m, n. We assume for simplicity the Born–von Karman closure

$$\mathbf{x}_{l+L,m,n} = \mathbf{x}_{l,m,n} \tag{13.39}$$

and we employ classical mechanics, because we require only the determination of the frequencies of a set of harmonic oscillators. With these assumptions we arrive at a generalization of (13.15) of the form

$$\frac{d^2 x^i_{l,m,n}}{dt^2} = \sum_{\lambda=0}^{L-1} \sum_{\mu=0}^{M-1} \sum_{\nu=0}^{N-1} \sum_{k=1}^{3} \kappa^{ik}_{\lambda\mu\nu}\, x^k_{l+\lambda,m+\mu,n+\nu} \tag{13.40a}$$

where

$$i = 1, 2, 3$$
$$l = 0, 1, 2, 3, \ldots, L-1$$
$$m = 0.1, 2, 3, \ldots, M-1$$
$$n = 0.1, 2, 3, \ldots, N-1 \tag{13.40b}$$

Here i and k label the three coordinate axes; l, m, n, λ, μ, ν label the three cell numbers of each primitive cell. $\kappa_{\lambda\,\mu\,\nu}^{ik}$ gives the elastic coupling (divided by a mass) of the i coordinate of the atom l, m, n to the k coordinate of the atom $l+\lambda$, $m+\mu$, $n+\nu$. The crystalline symmetry enters into the expression because the coupling constant depends only on the difference of the cell numbers, not on the cell numbers themselves. This feature permits introduction of plane waves as normal coordinates. Generalizing (13.20), we introduce

$$X^i_{l\star,m\star,n\star} = \frac{1}{\sqrt{LMN}} \sum_{l,m,n} \exp\left[2\,\pi\,i\left(\frac{l\,l^\star}{L} + \frac{m\,m^\star}{M} + \frac{n\,n^\star}{N}\right)\right] x^i_{l,m,n} \quad (13.41)$$

The equation of motion for this coordinate results by multiplication of (13.40) with the above exponential and addition. We find

$$\frac{d^2 X^i_{l\star,m\star,n\star}}{dt^2} = \sum_{k=1}^{3} X^k_{l\star,m\star,n\star} \sum_{\lambda,\mu,\nu} \kappa^{ik}_{\lambda,\mu,\nu} \exp\left[2\,\pi\,i\left(-\frac{l^\star\lambda}{L} - \frac{m^\star\mu}{M} - \frac{n^\star\nu}{N}\right)\right] \quad (13.42)$$

Observe that only three normal coordinates appear on the right. Indeed, if we let i run from 1 to 3, we have a closed system of three equations in three unknowns with all the complications tucked away into the nine coefficients on the right; in other words, we can write down for each value of the parameters $l^\star$, $m^\star$, $n^\star$ an equation system of the form

$$\frac{d^2 X^{(1)}}{dt^2} = \sigma^{(11)} X^{(1)} + \sigma^{(12)} X^{(2)} + \sigma^{(13)} X^{(3)}$$

$$\frac{d^2 X^{(2)}}{dt^2} = \sigma^{(21)} X^{(1)} + \sigma^{(22)} X^{(2)} + \sigma^{(23)} X^{(3)}$$

$$\frac{d^2 X^{(3)}}{dt^2} = \sigma^{(31)} X^{(1)} + \sigma^{(32(} X^{(2)} + \sigma^{(33)} X^{(3)}$$

The system yields three circular frequencies ω which can be written as roots of a secular equation

$$\begin{vmatrix} \sigma^{(11)} + \omega^2 & \sigma^{(12)} & \sigma^{(13)} \\ \sigma^{(21)} & \sigma^{(22)} + \omega^2 & \sigma^{(23)} \\ \sigma^{(31)} & \sigma^{(32)} & \sigma^{(33)} + \omega^2 \end{vmatrix} = 0 \quad (13.43)$$

The reason we get three frequencies is that elastic waves have three states of polarization. Generally, the cubic equation (13.43) must be solved to find the correct polarizations for a given wave length. However, in crystals of high symmetry, the equation will factor for propagation in certain special directions. We shall not discuss fully how to obtain the nine coefficients $\sigma^{(ik)}$. It is more important here to clarify the wave character of the (almost) normal coordinates (13.41). To see what they stand for we invert their defining relation. This amounts to applying the inversion (13.20) to three sets of indices, which is rendered easy because of (13.19). We get

$$x(l, m, n) = \frac{1}{\sqrt{LMN}} \sum_{l^\star=0}^{L-1} \sum_{m^\star=0}^{M-1} \sum_{n^\star=0}^{N-1}$$

$$\exp\left[2 \pi i \left(-\frac{l\,l^\star}{L} - \frac{m\,m^\star}{M} - \frac{n\,n^\star}{N}\right)\right] X(l^\star, m^\star, n^\star) \quad (13.44)$$

This equation shows that a particular $X(l^\star, m^\star, n^\star)$ represents a plane wave of displacements x_{lmn}^i except that real and imaginary parts must be sorted out as in (13.26). Being a plane wave, it must have a wave vector $\mathbf{k}$. This wave vector is best represented in *reciprocal space* or $\mathbf{k}$ space. This reciprocal space is generated by the *reciprocal lattice vectors* $\mathbf{a}^\star$, $\mathbf{b}^\star$, $\mathbf{c}^\star$ of the basis vectors $\mathbf{a}$, $\mathbf{b}$, $\mathbf{c}$. They are defined by

$$\begin{array}{lll}
\mathbf{a}^\star \cdot \mathbf{a} = 1 & \mathbf{a}^\star \cdot \mathbf{b} = 0 & \mathbf{a}^\star \cdot \mathbf{c} = 0 \\
\mathbf{b}^\star \cdot \mathbf{a} = 0 & \mathbf{b}^\star \cdot \mathbf{b} = 1 & \mathbf{b}^\star \cdot \mathbf{c} = 0 \\
\mathbf{c}^\star \cdot \mathbf{a} = 0 & \mathbf{c}^\star \cdot \mathbf{b} = 0 & \mathbf{c}^\star \cdot \mathbf{c} = 1
\end{array} \quad (13.45)$$

or explicitly

$$\mathbf{a}^\star = \frac{\mathbf{b} \times \mathbf{c}}{\mathbf{a} \cdot \mathbf{b} \times \mathbf{c}} \qquad \mathbf{b}^\star = \frac{\mathbf{c} \times \mathbf{a}}{\mathbf{a} \cdot \mathbf{b} \times \mathbf{c}} \qquad \mathbf{c}^\star = \frac{\mathbf{a} \times \mathbf{b}}{\mathbf{a} \cdot \mathbf{b} \times \mathbf{c}} \quad (13.46)$$

In terms of the reciprocal lattice vectors $\mathbf{k}(l^\star, m^\star, n^\star)$ takes the form

$$\mathbf{k}(l^\star, m^\star, n^\star) = 2 \pi \left[\frac{l^\star \mathbf{a}^\star}{L} + \frac{m^\star \mathbf{b}^\star}{M} + \frac{n^\star \mathbf{c}^\star}{N}\right] \quad (13.47)$$

One may verify indeed that multiplication of (13.47) with (13.36) yields the exponential occurring in (13.44), as required. Since $\mathbf{k}$ is actually defined through that exponential it has the indeterminacies already noted in the one-dimensional examples. In every way $\mathbf{k}(l^\star + L, m^\star, n^\star)$ is equivalent to $\mathbf{k}(l^\star, m^\star, n^\star)$. This means that reciprocal space is also divided into equivalent primitive cells like the crystalline space and that the basis vectors of the cell are $2 \pi \mathbf{a}^\star$, $2 \pi \mathbf{b}^\star$, and $2 \pi \mathbf{c}^\star$. All physical quantities, and the frequency in particular, are periodic functions in reciprocal space, and they are completely described if they are given in one such primitive cell.

It is customary to take the primitive cell in reciprocal space not as a parallelepiped of sides $2 \pi \mathbf{a}^\star$, $2 \pi \mathbf{b}^\star$, and $2 \pi \mathbf{c}^\star$, but as a polyhedron around the origin of the space which is so constructed that each wave vector has the least possible length among the possibilities which are inherent in its indeterminacy. This polyhedron is called the *first Brillouin zone*. The vibration spectrum of a solid thus appears as a number of bands defined over the first Brillouin zone. The number of bands equals 3 n, where n is the number of atoms per primitive cell. In simple lattice crystals, for which $n = 1$, the three bands can be roughly identified with one longitudinal and two transverse acoustic waves for every wave length and every direction. In crystals with two atoms per cell one can similarly label the six bands with semiquantitative accuracy as longitudinal acoustic (1), transverse acoustic (2), longitudinal optic (1), transverse optic (2).

It is important for the following discussion to realize that the simple proportionality between wave number and frequency valid for long wave length persists in the general framework of the equations (13.40). The two illustrations, Fig. 13.3 and Fig. 13.4, have made this highly plausible. In (13.40) this expresses itself in the two identities on the coefficients

$$\sum_{\lambda=0}^{L-1} \sum_{\mu=0}^{M-1} \sum_{\nu=0}^{N-1} \kappa_{\lambda\mu\nu}^{ik} = 0 \qquad (13.48)$$

and

$$\kappa_{-\lambda,-\mu,-\nu}^{ik} = \kappa_{\lambda\mu\nu}^{ik} \qquad (13.49)$$

The first relation is a consequence of the Gallilean invariance of our problem, that is, that only differences between displacements can be responsible for forces, and the second that a simple lattice always has inversion symmetry.

Suppose we now introduce the two constraints into the equation of motion (13.42). The terms in λ, μ, ν and $-\lambda$, $-\mu$, $-\nu$ then combine to give cosines because of (13.49), and a 1 can be subtracted from the cosines because of (13.48). As a result the coefficients in (13.42) take the form

$$\sum_{\lambda,\mu,\nu} \kappa_{\lambda,\mu,\nu}^{ik} \left\{ \cos\left(\frac{l^\star \lambda}{L} + \frac{m^\star \mu}{M} + \frac{n^\star \nu}{N}\right) - 1 \right\}$$

For small $l^\star$, $m^\star$, $n^\star$ the curly bracket above is proportional to the square of these numbers. This means, because of (13.47), proportionality to the magnitude of k^2 at fixed angle. This dependence transfers to (13.43) and makes each of the three roots ω of the determinant proportional to $|\mathbf{k}|$ at fixed angle. This is the essence of the concept of a velocity of propagation, or rather three velocities of propagation. The theory permits of course a quite capricious dependence of each of these velocities on the direction of propagation and the polarization, but the essence of the elastic wave concept is saved for the discrete model. A similar reasoning can be carried out for the three acoustical branches in a more complicated crystal.

13-4. Debye theory of the heat capacity of solids

The foregoing discussion contains within itself the possibility of a complete spectral analysis of the vibration spectrum of a crystal, including ways to determine the function $g(f)$ entering into the final thermal results (13.05), (13.06), and (13.10). Physics has gone through two stages concerning such calculations. In the early years, when facilities for numerical computation were small, transposition from a density in k space to a density along a frequency axis was a considerable undertaking, since no analytic procedures are known to carry out the transformation. Later, when numerical computers were more common, and tentative frequency versus k plots became available, it became clear that thermal data are not very suitable to check against the theory, because they are relatively insensitive to the exact form of the $g(f)$ used in their computation. Neutron diffraction is in fact the tool to use in a proper investigation. Consequently, while it was believed at first that a good $g(f)$ might be difficult to compute, it is now felt that it is not worth while to compute it well just to account for thermal properties. Contributory to this outlook is the fact that a good approximate theory of solid heat capacities is available which is in essence sufficient for an understanding of thermal phenomena in solids.

The Debye theory of solid heat capacities is based on the recognition that the vibrations discussed in this chapter are identical with solid elastic waves if the wave lengths are long. Starting from this basis, Debye took the step of extrapolating the density formulas resulting from the theory of sound to the entire vibration spectrum. We ended up in the last section working out this basis. For long waves there must be three velocities of propagation c_1, c_2, and c_3 which link the frequencies and k in a relation of the form

$$\omega_i = c_i(\vartheta, \varphi) \, |\mathbf{k}| \tag{13.50a}$$

where

$$i = 1, 2, 3 \tag{13.50b}$$

The sums we must carry out in statistics are generally of the form

$$\sum_{i=1}^{3} \sum_{l^\star, m^\star, n^\star} F(f)$$

where $l^\star$, $m^\star$, $n^\star$ are linked to $\mathbf{k}$ by (13.47). An elementary transformation of crystal physics permits expressing this sum as an integral in k space as follows:

$$\sum_{i=1}^{3} \sum_{l^\star, m^\star, n^\star} F(f) = \frac{V}{8\,\pi^3} \int \int \int \sum_{i=1}^{3} F(f) \, d\mathbf{k} \tag{13.51}$$

Introducing (13.50) into this, we get in polar coordinates

$$\sum_{i=1}^{3} \sum_{l\star,m\star,n\star} F(f) = V \int F(f)\, f^2\, df \int\int \sum_{i=1}^{3} \frac{1}{c_i^3(\vartheta, \varphi)}\, d\Omega$$

which means that $g(f)$ equals

$$g(f) = V f^2 \int\int \left\{ \frac{1}{c_1^3} + \frac{1}{c_2^3} + \frac{1}{c_3^3} \right\} d\Omega \qquad (13.52)$$

The integral over the solid angle $d\Omega$ is over all directions of the wave propagation vector and does not involve the frequency. The result is equivalent to the result (10.26) for light except that light cannot be longitudinally polarized and that its velocity of propagation is a constant under the circumstances discussed in Chapter 10.

Whenever the formula (13.52) is adequate for discussing solid vibrations the Debye theory will give results similar to those for light. In particular there will be a "Stefan–Boltzmann" law according to which the total energy is proportional to T^4 and the heat capacity to T^3. This means that formulas (10.07), (10.08), and (10.32) for the total energy can be taken over, with the modification of replacing the constant (10.26) by (13.52). This yields

$$U = \frac{\pi^4}{15} \frac{V\, k^4}{h^3} \int\int \left\{ \frac{1}{c_1^3} + \frac{1}{c_2^3} + \frac{1}{c_3^3} \right\} d\Omega \cdot T^4 \qquad (13.53)$$

This equation leads to the "Debye T^3 law" for the heat capacity of solids which is in fact well verified for most atomic crystals at moderately low temperatures. The constant in the formula depends only on the crystalline elastic constants, and thus can be computed a priori from mechanical data. We shall make the comparison between elastic and thermal data a little further on.

Before proceeding to compare (13.53) with experiment we must remember that it is only a limiting law for low temperature because (13.52) is only a limiting formula for low frequencies. Very simply, formula (13.52) violates the integral relation (13.04), which states that the total number of vibrational modes is finite. Debye's theory consists in reconciling these two aspects of the vibration spectrum of solids in the simplest possible way, namely, by letting $g(f)$ vary as the square of f until the total number $3\,N$ is reached, and then breaking the spectrum off. A comparison of the Debye spectrum with the correctly calculated spectrum of tungsten is shown in Fig. 13.8. The rendition of detail is only mediocre, except at the low-frequency end.

The cut-off frequency F introduced by Debye is customarily expressed in temperature units through the relation

$$h F = k \Theta \tag{13.54}$$

where Θ is called the *Debye temperature*. If we assume that $g(f)$ is proportional to f^2 up to the cut-off point, then we can determine $g(f)$ from (13.04) alone. We find

$$g(f) = \frac{9 N h^3}{k^3 \Theta^3} f^2 \quad \text{for} \quad f < \frac{k \Theta}{h} \tag{13.55a}$$

and

$$g(f) = 0 \quad \text{for} \quad f > \frac{k \Theta}{h} \tag{13.55b}$$

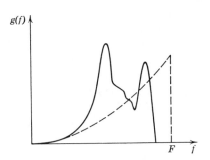

Fig. 13.8. Frequency spectrum of tungsten, and comparison with the Debye approximation (after R. Becker).

We can now employ (13.55) to get expressions for thermal quantities. Substitution into (13.05) yields for the partition function

$$\ln F = 9 N \left(\frac{T}{\Theta}\right)^3 \int_0^{\Theta/T} \ln \left(\tfrac{1}{2} \operatorname{cosech} \tfrac{1}{2} x\right) x^2 \, dx \tag{13.56a}$$

or by an integration by parts removing the logarithm

$$\ln F = - 3 N \ln \left(2 \sinh \frac{1}{2} \frac{\Theta}{T}\right) + \tfrac{3}{2} N \left(\frac{T}{\Theta}\right)^3 \int_0^{\Theta/T} \coth \tfrac{1}{2} x \, x^3 \, dx \tag{13.56b}$$

Similarly substitution into (13.06) yields the energy and into (13.10) the heat capacity

$$U = \frac{9}{2} N k \frac{T^4}{\Theta^3} \int_0^{\Theta/T} \coth \tfrac{1}{2} x \, x^3 \, dx \tag{13.57}$$

and

$$\mathscr{C} = \frac{9}{4} R \left(\frac{T}{\Theta}\right)^3 \int_0^{\Theta/T} \frac{x^4 \, dx}{\sinh^2 \tfrac{1}{2} x} \tag{13.58a}$$

or through an integration by parts

$$\frac{\mathscr{C}}{R} = -\frac{9}{2}\frac{\Theta}{T}\coth\frac{1}{2}\frac{\Theta}{T} + 18\left(\frac{T}{\Theta}\right)^3\int_0^{\Theta/T}\coth\tfrac{1}{2}\,x\,x^3\,dx \quad (13.58b)$$

The Debye theory is apt to be particularly poor for the zero-point energy which is contained in (13.56) and (13.57). The reason is of course that the high frequencies make the most substantial contributions to this energy, and the Debye theory is a low-frequency theory. It is therefore desirable to separate it from the remainder and find perhaps independent ways to determine its value. The necessary transformation formulas are

$$\ln\left(2\sinh\tfrac{1}{2}\,x\right) = \tfrac{1}{2}\,x + \ln\left(1 - e^{-x}\right)$$

and

$$\coth\tfrac{1}{2}\,x = 1 + \frac{2}{e^x - 1}$$

Let us first make this transformation on the energy (13.57). It becomes

$$U = \tfrac{9}{8}\,N\,k\,\Theta + 3\,N\,k\,T\,D\!\left(\frac{\Theta}{T}\right)$$

where we have introduced the Debye function $D(y)$ by the formula

$$D(y) = \frac{3}{y^3}\int_0^y\frac{x^3\,dx}{e^x - 1} \quad (13.59)$$

The Debye function has the limiting properties[†]

$$D(0) = 1 \quad (13.60a)$$

and

$$D(y) \approx \frac{\pi^4}{5\,y^3} \text{ for } y \text{ large} \quad (13.60b)$$

For the reasons just stated the value obtained for the zero point energy should not be taken literally. We write therefore U in the form

$$U = U(0) + 3\,N\,k\,T\,D\!\left(\frac{\Theta}{T}\right) \quad (13.61)$$

If we apply the same reduction to (13.56b) we get

$$\ln F = -\beta\,U(0) - 3\,N\ln\left(1 - e^{-\Theta/T}\right) + N\,D\!\left(\frac{\Theta}{T}\right) \quad (13.62)$$

[†] An excellent table of the Debye function and related functions is in an article of J. A. Beattie, *Math. and Phys.* **6**, 1, (1916).

Finally we may also transform (13.58b) and find

$$\frac{\mathscr{C}}{R} = 12\ D\!\left(\frac{\Theta}{T}\right) - 9\ \frac{\Theta/T}{e^{\Theta/T} - 1} \tag{13.63}$$

It is seen that all thermodynamic functions involve only the one higher transcendental $D(y)$ defined in (13.59). The limiting values (13.60) easily outline the limiting behavior of the expressions at high and low temperatures. Since $D(y)$ is only a function of one variable, tables are available to find its value for intermediate values of y.

What makes the Debye theory of solid heat capacities so remarkable is

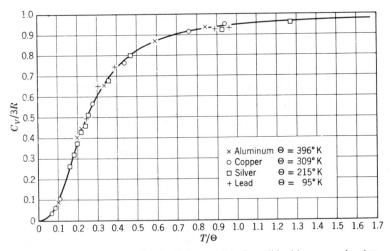

Fig. 13.9. Specific heat curve for the Debye model of a solid with measured points for aluminum, copper, silver, and lead. The abscissa scale contains the Debye temperature Θ, and is adjusted for optimum fit.

that it yields a molar heat capacity for monatomic solids which is a universal function of T/Θ. This may be seen from either (13.58) or (13.63). It should therefore be possible to bring all these curves to coincidence by a simple change in the temperature scale. This law is remarkably well verified for actual monatomic solids even though the Debye temperature Θ varies widely. In Fig. 13.9 are shown data for aluminum, copper, silver, and lead, together with the theoretical curve. It is seen that the Debye temperatures for those materials vary widely. But once the scale adjustment on the temperature is made, the heat capacity curves agree well with each other and with the Debye prediction.

The universal functions of T/Θ which the Debye theory yields for $\mathscr{U}/R\,T$ and $\mathscr{C}_v/R$ contain simple limiting laws which are worth considering in greater detail. These laws hold respectively in the temperature regions in which T is very much less and very much greater than Θ.

To derive the low-temperature formula we observe that if

$$T \ll \Theta$$

the function $D(y)$ behaves as shown in (13.60b). If we insert this into (13.61) we get

$$U = U(0) + \tfrac{3}{5} \pi^4 N k \frac{T^4}{\Theta^3} \tag{13.64}$$

Except for the zero-point energy term this formula is identical with (13.53), which is accurate and not based on a model. The model parameter Θ is thus in principle computable from mechanical data. This comparison is made in Table 13-1, using elastic data at room temperature and also at

TABLE 13-1. Values of the Debye temperature Θ obtained from thermal and elastic measurements (after Kittel)

	Fe	Al	Cu	Pb	Ag
Thermal value	453	398	315	88	215
Elastic value (room temp.)	461	402	332	73	214
Elastic value (abs. zero)		488	344		235

low temperature. The agreement is generally good. Strangely enough, the agreement is better if room-temperature elastic data are used rather than low-temperature ones. This is one of the indications that there is a fortuitous element in the excellent fit of the Debye theory. Differentiation of (13.64) yields the low-temperature limiting law for the specific heat:

$$\frac{\mathscr{C}}{R} = \tfrac{12}{5} \pi^4 \left(\frac{T}{\Theta}\right)^3 \tag{13.65}$$

The formula ought to be rigorously correct at low temperature for the properly chosen value of Θ.

To derive an expression of the heat capacity when

$$T \gg \Theta$$

we start from (13.57) and observe that, now, the variable x remains small. The coth can therefore be expanded

$$\coth \tfrac{1}{2} x = \frac{2}{x} + \tfrac{1}{6} x + 0(x^3)$$

yielding

$$U = 3\,N\,k\,T + \frac{3}{20}\,N\,k\,\frac{\Theta^2}{T} + 0\left(\frac{N\,k\,\Theta^4}{T^3}\right) \tag{13.66}$$

and

$$\frac{\mathscr{C}}{R} = 3 - \frac{3}{20}\left(\frac{\Theta}{T}\right)^2 + \ldots \tag{13.67}$$

These formulas have not the same standing as the low-temperature ones because the approximation neglects essential features of the frequency spectrum. The standard way to check on the applicability of the Debye

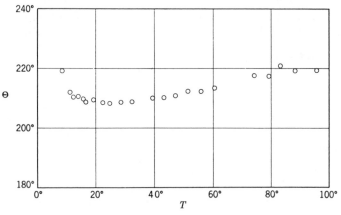

Fig. 13.10. Empirical Θ versus T curve for silver, obtained from a point-by-point fit of the specific heat to (13.63), with Θ as an adjustable parameter (after Hill).

theory is to measure $\mathscr{C}/R$ as a function of T and to adjust for each point the value of Θ so that (13.63) produces an exact fit. The result is shown for silver in Fig. 13.10. In view of the fact that the theory is admittedly approximate the constancy of Θ is truly remarkable.

Another indication of the undeservedly good fit of the Debye theory is the fact that apparent improvements of the theory tend to make the fit worse. It would be natural, for instance, to define separate Debye temperatures for longitudinal and transverse waves since the former are considerably stiffer. This would bring into the picture much higher frequencies which destroy the reasonable agreement with real frequency distributions exhibited in Fig. 13.8. The Debye theory is thus a very satisfactory theory of solid heat capacities which cannot stand theoretical tampering, and whose excellent fit with experiment is not fully understood.

13-5. Vapor pressure of solids

The general thermodynamic theory of phase transitions was studied in Chapter 7. The differential equation of Clausius and Clapeyron governing the change of vapor pressure with temperature was derived, and the generalization to many phases was sketched. In this treatment phases were accepted as such without question. We have now the opportunity to study one such phase change in detail, namely, the process of sublimation of crystals. For we are now in possession of detailed knowledge of the thermodynamic functions of either phase. We have thus the unusual opportunity to write down the explicit form of the vapor pressure equation—the pressure–volume relation which is valid when a substance is present simultaneously in crystalline and in gaseous form.

Discussion will be limited to monatomic substances crystallizing in a simple lattice: copper, argon, sodium, etc. The equation to be implemented is (7.42), the equality of the chemical potential in the two phases. We shall write down the chemical potential for either phase and then gain the desired equation by setting them equal.

For the gaseous phase we are given a particle partition function f in terms of which the chemical potential μ_g is directly given by (11.07)

$$- \beta \mu_g = \ln \frac{f}{N} \qquad (13.68)$$

Here f is given by (11.01) and (11.06) so that we have

$$- \beta \mu_g = \ln \left\{ \left(\frac{2 \pi m k T}{h^2} \right)^{3/2} \frac{k T}{p} \right\} + \ln f_{\text{int}} \qquad (13.69)$$

Since the gas is atomic by assumption, f_{int} can only consist of electronic and nuclear spin multiplicity factors. Presumably, these factors are also present in the solid phase, although the behavior of the electronic spin has to be watched. If they are present on both sides they will cancel out of the result, and we shall treat them in this manner. Thus only the first term of (13.69) need be retained.

For the solid phase we are given a system partition function F whose relation to the thermodynamic functions is given by (7.10). The Gibbs free energy which results from this equals

$$- \beta G = \ln F - \beta p V_s$$

where V_s is the volume of the solid phase. We now go to units per molecule, using (7.27),

$$- \beta \mu_s = \frac{1}{N} \ln F - \frac{p V_s}{N k T} \qquad (13.70)$$

Before inserting some expression into this we must remember to employ the same energy reference for the solid and the gas, since a formula such as (13.62) refers to the solid ground state. The energy difference is always such as to stabilize the solid phase since the gas has a much higher entropy. The simplest procedure is to call the quantity $U(0)$ in (13.62) $-\Lambda$, and let Λ be the energy of sublimation minus the zero point energy of the solid. The lower case letter λ shall have the same meaning referred to an individual atom. Then (13.70) becomes

$$- \beta \mu_s = \beta \lambda - 3 \ln (1 - e^{-\Theta/T}) + D\left(\frac{\Theta}{T}\right) - \frac{p \mathcal{V}_s}{R T} \qquad (13.71)$$

where $\mathcal{V}_s$ is the molar volume of the solid phase.
   Equating (13.69) and (13.71), we get

$$\ln p = \ln \left(\frac{2 \pi m}{h^2}\right)^{3/2} + \ln (k T)^{5/2} - \frac{\lambda}{k T}$$

$$+ 3 \ln (1 - e^{-\Theta/T}) - D\left(\frac{\Theta}{T}\right) + \frac{p \mathcal{V}_s}{R T} \qquad (13.72)$$

For very small pressures the pressure term on the right can be neglected and the equation solved for p. We find then

$$p = \frac{(2 \pi m)^{3/2}}{h^3} (k T)^{5/2} (1 - e^{-\Theta/T})^3 \exp\left[-\frac{\lambda}{k T} - D\left(\frac{\Theta}{T}\right)\right] \qquad (13.73)$$

The front factors which have the dimension of a pressure determine the order of magnitude of p: 10^{10} dynes/cm² or 10^4 atm which is a substantial number. The dimensional factor is usually reduced by the Boltzmann factor $\exp[-\beta \lambda]$ because the binding energy λ in the solid is many times $k T$. The remaining solid state factors are of the order unity and contribute little to the final result.
   It is amusing and instructive to derive the Clausius–Clapeyron equation from relation (13.72). As a first step we may assume the solid to be rigid. $\mathcal{V}_s$ is then a constant independent of pressure and temperature. The same is then also true for Θ and Λ. Under these conditions, differentiation of (13.72) with respect to T yields

$$\frac{1}{p} \frac{dp}{dT} = \frac{5/2}{T} + \frac{\lambda}{k T^2} - \frac{3}{T} D\left(\frac{\Theta}{T}\right) + \frac{\mathcal{V}_s}{R T} \frac{dp}{dT} - \frac{p \mathcal{V}_s}{R T^2} \qquad (13.74)$$

Now on the left $1/p = \mathscr{V}_g/RT$. Furthermore the heat of reaction is the difference of the molar enthalpies, thus,

$$\mathscr{H} = \mathscr{A}\,\lambda + \tfrac{5}{2}\,R\,T - 3\,R\,T\,D\!\left(\frac{\Theta}{T}\right) - p\,\mathscr{V}_s \tag{13.75}$$

Multiply the equation with RT and collect terms in dp/dT on the left. The result is

$$(\mathscr{V}_g - \mathscr{V}_s)\frac{dp}{dT} = \frac{1}{T}\,\mathscr{H} \tag{13.76}$$

which is the desired equation.

We now lift the restriction of a constant $\mathscr{V}_s$. As we are working in the harmonic approximation it is not logical to assume it to be a function of T. We therefore take it to be a function of pressure only. The same is then true for Θ and Λ. If we now repeat the previous differentiation, three terms are added to (13.74). Since (13.74) is correct as it stands, these three terms must cancel mutually. We thus get the following equation of state for the solid:

$$p\mathscr{V}_s - \frac{d\Lambda(\mathscr{V}_s)}{d\ln\mathscr{V}_s} + [\mathscr{U}_s(T,\mathscr{V}_s) - \mathscr{U}_s(0,\mathscr{V}_s)]\frac{d\ln\Theta}{d\ln\mathscr{V}_s} = 0 \tag{13.77}$$

This is essentially the Grüneisen equation of state as presented in Problem 22, Chapter 1. The Grüneisen constant Γ controls the variation of the Debye temperature with pressure according to the formula

$$\Gamma = -\frac{d\ln\Theta}{d\ln\mathscr{V}_s} \tag{13.78}$$

The Grüneisen equation of state is thus seen to be a consequence of the Debye theory of solid heat capacities. The only supplementary assumption made by Grüneisen is that the Γ defined in (13.78) is a constant. Thermodynamics alone would allow it to be an arbitrary function of $\mathscr{V}_s$.

A. RECOMMENDED PROBLEMS

1. The Debye theory is based on relation (13.50), valid at low temperature. For spin waves, on the other hand, the corresponding relation is

$$\omega \propto k^2$$

How does the energy of the spin system depend on temperature when the temperature is low enough?

2. Suppose you are given the following data about the vapor pressure of silver:

Temperature, °K	1219	1121	1040
Vapor pressure, mm Hg	10^{-3}	10^{-4}	10^{-5}

Make from these data three independent determinations of the energy of sublimation of silver at absolute zero, being given that its Debye temperature is 220°.

B. GENERAL PROBLEMS

3. Prove from the second model of Section 13-2 that there is a qualitative distinction between acoustical and optical modes, in that the phase difference between the two kinds of vibrating atoms is zero in the first instance, and π in the second.

4. Calculate frequencies and wave characteristics for a problem similar to the ones in 13–2, namely, one in which all masses are equal, but the springs are alternately weak and strong. Prove that one can again distinguish an acoustical and an optical branch with phase shifts of zero and π respectively, provided phases are measured at the symmetry centers of the structure rather than at the location of the masses.

5. Establish the validity of (13.51), which converts any sum over frequencies into an integral. Do *not* assume the axes of the crystal to be rectangular.

6. Suppose we interpret (13.61) as a relation valid at all pressures and temperatures by taking $U(0)$ and Θ to be given functions of the volume only. Abbreviate the relation to read

$$U(T, V) = U(0) + U'$$

Derive the equation of state which results from these assumptions. What additional assumption is needed to get the Grüneisen equation of state written down in Problem 22, Chapter 1?

14

Statistics of conduction electrons in solids

It was pointed out in the preceding chapter that most electrons in atoms, molecules, or solids are in the mechanical condition which is normal for condensed matter generally, namely, a state in which mechanical behavior is only weakly coupled to temperature. One should therefore expect that there are few properties of electrons in condensed matter which are of statistical interest. For their temperature behavior can only produce small corrections to a theory which either deals with the electrons without reference to temperature, or else deals with thermal phenomena without involving the electrons.

The picture just given needs some retouching because of the phenomenon of electric conduction. If we pass an electric current through a wire it is indeed true that only a small fraction of the total charge in the wire is moving at all, and that the carriers acquire a drift velocity in the field which is small compared to their natural velocity. However, the resulting electric current is nevertheless important to us, and electronics of solids acquires its importance from this circumstance. From this same point of view, a chapter on electrons in solids as part of equilibrium statistics is somewhat unsatisfactory. The phenomenon of principal interest—namely, electric conduction—is not an equilibrium phenomenon and cannot be treated with the methods now at our disposal. We shall take care of this defect as best we can here by presenting a short intuitive picture of electric conduction before entering into the main topic.

In most cases when a solid is capable of carrying current, the current is carried by electrons. The electrons which carry current are loosely bound members of some outermost shell of an atom. When an electric field **E** of reasonable size is applied, these electrons acquire a drift velocity $\langle \mathbf{v} \rangle$ in the field direction which is proportional to the field, thus

$$\langle \mathbf{v} \rangle = - \mu \, \mathbf{E} \tag{14.01}$$

The constant of proportionality μ is called the *mobility* of these electrons. Sometimes, there are several groups of electrons, each having its own mobility. In crystalline materials having strong anisotropy it may be essential to treat μ as a tensor of rank 2. We pass from (14.01) to *Ohm's law* by introducing the electric current density **i** through the formula

$$\mathbf{i} = n \, e \, \langle \mathbf{v} \rangle \tag{14.02}$$

where n is the number density of the electrons under consideration and e the electronic charge. Combination of (14.02) with (14.01) yields for the electric conductivity

$$\sigma = \frac{\mathbf{i}}{\mathbf{E}} = n \, e \, \mu \tag{14.03}$$

where σ is of course also a tensor if μ is.

The connection of electrical conductivity to equilibrium statistics rests on equation (14.03). The actual conductivity of a material is the product of two factors: the number n of mobile electrons per unit volume; and their mobility μ, which is a speed per unit field. The second factor cannot be derived from equilibrium statistics, but the first one can. It also so happens that the first factor varies over many more orders of magnitude (10^{18}) than the second (10^5). Thus conductivity can be discussed in a qualitative way by examination of the density of current-carrying electrons. This is a subject of equilibrium statistical physics.

14-1. The distinction of metals and insulators in Fermi statistics

The theory of electrons in solids has a base very similar to the theory of electrons in atoms. The concept of one-electron states is found to be approximately valid, and electrons can then be placed into these states according to the principles of Fermi statistics discussed in Chapter 9. The main distinction is that in atoms the states form a discrete set, while in crystals we deal with bands. Correspondingly, we can handle Fermi statistics of electrons in atoms as if we were at the absolute zero of temperature (as shown in Fig. 9.1), while in solids there are more refined features which are associated with Fig. 9.2. For the present qualitative discussion, Fig. 9.1 will suffice. According to it, all energy levels below the Fermi energy μ are filled, and above that energy, all are empty.

The reason for having the energy states for electrons in crystals occur in bands is the same as that which makes the interatomic vibrations occur in bands: the translational periodicity of the crystal lattice. This periodicity makes a $\mathbf{k}$ vector appear in the electronic wave functions as it does in the crystal vibration spectrum. This can be quite easily seen. Suppose we are given a Schroedinger equation for a one-electron wave function of the form

$$-\frac{\hbar^2}{2m}\nabla^2\Psi + V(\mathbf{x})\Psi = E\Psi \tag{14.04}$$

then we have by definition the identity for $V(\mathbf{x})$

$$V(\mathbf{x}+\mathbf{r}) = V(\mathbf{x}) \tag{14.05}$$

with $\mathbf{r}$ given by (13.36) and (13.37). Suppose now we have a solution $\Psi(\mathbf{x})$ of (14.04) going with a particular value of the energy E. Then other solutions of the same equation are

$$\Psi(\mathbf{x}+\mathbf{a}),\quad \Psi(\mathbf{x}+2\,\mathbf{a}),\quad \Psi(\mathbf{x}+\mathbf{a}+\mathbf{b})$$

etc. It is slightly preposterous to assume these 10^{20} solutions all linearly independent, but the actual structure of Ψ is not directly derivable from such a negative argument. It is better to subject the functions $\Psi(\mathbf{x}+l\,\mathbf{a}+m\,\mathbf{b}+n\,\mathbf{c})$ to the transformation (13.41) whose fundamental importance to crystal physics was emphasized in connection with (13.18):

$$b(\mathbf{x};l^\star,m^\star,n^\star)$$
$$=\frac{1}{\sqrt{LMN}}\sum_{l,m,n}\exp\left[2\pi i\left(\frac{ll^\star}{L}+\frac{mm^\star}{M}+\frac{nn^\star}{N}\right)\right]\Psi(\mathbf{x}-l\mathbf{a}-m\mathbf{b}-n\mathbf{c}) \tag{14.06a}$$

With the help of the definition (13.47) this can be put into the more condensed form

$$b(\mathbf{x};\mathbf{k}) = \frac{1}{\sqrt{LMN}}\sum_{\rho}\exp[i\,\mathbf{k}\cdot\mathbf{\rho}]\,\Psi(\mathbf{x}-\mathbf{\rho}) \tag{14.06b}$$

Naturally, most of the sums (14.06) are zero, but not all of them: for the invertibility of the unitary transformation (14.06) was proved earlier, and therefore some b's cannot vanish. In the normal case there is only one* $\mathbf{k}$ for which the sum does not vanish. The right-hand side of (14.06)

---

* More exactly two. Because of the time inversion symmetry of the Schroedinger equation the energies for the states $\mathbf{k}$ and $-\mathbf{k}$ are always equal.

then displays the structure of the wave function on the left. It is a plane wave of wave vector $\mathbf{k}$, except that it may have structural features within the primitive cell. It is called a *Bloch function*. We may read off from (14.06) the symmetry properties of Bloch functions

$$b(\mathbf{x} + \mathbf{r}; \mathbf{k}) = \exp(i\,\mathbf{k}\cdot\mathbf{r})\,b(\mathbf{x}; \mathbf{k}) \qquad (14.07)$$

if $\mathbf{r}$ obeys (13.36) and (13.37). Furthermore

$$b(\mathbf{x}; \mathbf{k} + 2\,\pi\,\mathbf{r}^\star) = b(\mathbf{x}; \mathbf{k}) \qquad (14.08)$$

if $\mathbf{r}^\star$ is a vector in the reciprocal lattice, that is, a linear combination of $\mathbf{a}^\star$, $\mathbf{b}^\star$, $\mathbf{c}^\star$ with integral coefficients. This second property was already pointed out in connection with (13.47). Finally from the reality of the wave equation (14.04) the conjugate complex function of $b(\mathbf{x}; \mathbf{k})$ must also be a solution; it has the wave vector $-\mathbf{k}$:

$$b(\mathbf{x}; -\mathbf{k}) = b^\star(\mathbf{x}; \mathbf{k}) \qquad (14.09)$$

In the present context we are interested not so much in the wave functions as in the energies associated with them. As $\mathbf{k}$ varies, the energy varies with it and becomes itself a function of $\mathbf{k}$

$$E = W(\mathbf{k}) \qquad (14.10)$$

$W(\mathbf{k})$ will have the symmetries of the wave function, namely,

$$W(\mathbf{k} + 2\,\pi\,\mathbf{r}^\star) = W(\mathbf{k}) \qquad (14.11)$$

and

$$W(-\mathbf{k}) = W(\mathbf{k}) \qquad (14.12)$$

Equation (14.11) shows that it is sufficient to trace out the energies in the first Brillouin zone, and that within this zone the energy levels form a number of bands; outside this zone the energies are determined by their property of being periodic in reciprocal space. In addition, by (14.12), the energy function has inversion symmetry within the first Brillouin zone.

From the count (14.06a) it is plausible that the number of states in a band equals the number of primitive cells of the crystal. Thus, if we think for the moment of a simple lattice type monatomic crystal, each atom contributes one state to a band. The band is thus the logical equivalent in the crystal of the atomic state, and is sometimes given the atomic designation. As an example we may take the much-studied $3s$ band of sodium metal. The relation of that band to the atomic $3s$ state can be brought out by an artificial calculation, in which the sodium atoms are brought together from infinite separation to the actual crystalline spacing by diminishing gradually the lattice parameter. The result is a picture as shown in Fig. 14.1: the discrete level widens out into a band of levels. In favorable cases the identity of the band with the atomic state is maintained. However, this is not always the case. A very important case for the following discussion

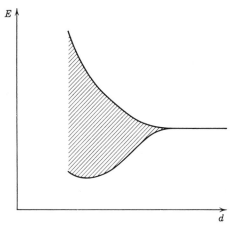

Fig. 14.1. Change of an energy band with inter-atomic distance (schematic).

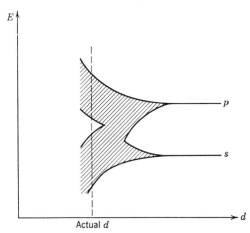

Fig. 14.2. Merging and reseparating of s and p bands in column 4 elements. At large distance the band states number 1:3, at small distance 2:2, and are respectively attractive and repulsive sp^3 hybrids (schematic).

is the rearrangement of the s and p orbitals in diamond, silicon, and germanium. As the lattice parameter is diminished they merge and then reseparate again as shown in Fig. 14.2. However, before they merge the bands take respectively two and six electrons, and after they separate each of the two bands takes four, thus allowing the lower band to be filled and the upper to be empty in the case of atoms having four valence electrons.

This brings us naturally to the distinction of metals and insulators as it results from band structure. Since the electrons are subject to the Pauli principle and have spin $\frac{1}{2}$, not more than two electrons can be placed into any orbital. The filling of the states proceeds from states of low energy up, until all electrons are accommodated. This produces of course the well known periodic table of elements. The feature which interests us here particularly is whether such a filling produces a full band or, on the contrary, a partially filled band. This has an important bearing on the possibility of the electrons carrying current. If the well-known principles about group velocity of wave phenomena in general are applied to De Broglie waves, we find that electrons in Bloch states have a velocity associated with them which is given by

$$\mathbf{v} = \frac{1}{\hbar} \frac{\partial W(\mathbf{k})}{\partial \mathbf{k}} \tag{14.13}$$

Now according to (14.12) the states $\mathbf{k}$ and $-\mathbf{k}$ always occur in degenerate pairs of equal energy. Equation (14.13) shows then that they have opposite velocity. This means that a full band carries no current, nor can it be made to carry current by application of a perturbation such as a weak field. If a current is to be carried, an energy gap has to be bridged to the next higher band. This cannot be accomplished with a very weak perturbation. On the other hand, if a band is only partially filled, then equations (14.12) and (14.13) still yield a zero current for such a system if it is at rest, because the velocities cancel in pairs. But a small perturbation has now states available whose separation is infinitesimal from the filled states; it can alter thereby this distribution, and a state of finite current can thus be produced. Such materials are naturally electric conductors. One sees from this that metals must be considered solids having partially filled energy bands. The classical example is the alkali metals whose s band outside their noble gas core is half filled. Insulators, on the other hand, are solids which have all their bands either completely empty or completely filled, with a forbidden gap separating the highest filled from the lowest empty band.

14-2. Semiconductors: electrons and holes

The division of solids into metals and insulators is based on the simplistic picture of the Fermi distribution shown in Fig. 9.1, valid at the absolute zero of temperature. When this picture applies, only the electrons in the direct neighborhood of the Fermi energy are able to respond to small forces. Thus, if the Fermi level μ is within a band of allowed energies, the material is a metal. If, on the contrary, it is in a forbidden zone, it is an insulator. It is clear that temperature, which produces an electron distribution of the type shown in Fig. 9.2, must soften up this distinction. The softening up is naturally more pronounced for insulators, for insulators should carry zero current according to the crude picture, and zero is never for long an answer in experimental physics.

Figure 14.3 shows the basic principle underlying the behavior of an

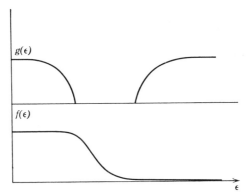

Fig. 14.3. Representation of a semiconductor, showing separate plots of $g(\varepsilon)$ and the "Fermi factor" $f(\varepsilon)$ against ε. According to (9.14) their product is the number of electrons.

intrinsic semiconductor. It is a modification of the insulator concept. The density of states $g(\varepsilon)$, defined in connection with (9.14), has indeed an energy range in which it is zero, and the Fermi energy μ falls in this range. Thus the substance would be an insulator if the Fermi distribution function were a step function. But expression (9.13) is a step function only for $\beta = \infty$. For any finite β, the function shows tails reaching into regions of higher and lower energy in which $g(\varepsilon)$ does not vanish. It is immediately obvious that the tail of the distribution reaching into higher energy must produce carriers of current, for we are now faced with a small number of electrons in an otherwise empty band. They can respond to small forces and thus follow an applied field. It is not quite as obvious, but just as true, that the tail reaching into lower energies must produce carriers of current. For we said originally that a full band cannot carry any current; the tail of the Fermi distribution reaching into that band prevents it from being full; consequently this tail must be also productive of electrical conductivity, just as the tail reaching into higher energies.

There is a far-reaching equivalence between excess electrons in an almost empty band and defect electrons ("holes") in an almost full band. This equivalence has a dynamical and a statistical aspect. The dynamical aspect is such that the defect electron is equivalent to a band electron of reversed charge, for which the energy versus $\mathbf{k}$ plot must be read upside down ("bubble in water" picture). The reason for this is that, for most purposes, the hamiltonian acting on a full band is equivalent to a null result, or more precisely, a fixed constant independent of small disturbances. If now a state in such a band is left empty, the diagonal matrix element of the hamiltonian for this state must be *subtracted* from this fixed constant in order to yield the first-order approximation to the energy of the many-body system. This will, in the first place, reverse the sign of a relation such as (14.10), making it read

$$E = - W(\mathbf{k}) \tag{14.14}$$

In addition, it will also reverse the sign of perturbing potentials, thus in effect giving the "hole" a reversed sign in its dynamical response to electric fields. It can therefore be treated with good accuracy as a positive charge, from the point of view of its dynamical response to forces.

From the point of view of this book, the *statistical* equivalence of holes to positive charges is of major importance. We shall see now that this is an effect specific to the Fermi distribution which would not arise if one of the two other statistics were valid. To verify this point we start out with (9.14), which states that the number of electrons per unit energy range equals

$$n(\varepsilon) = \frac{g(\varepsilon)}{\exp \beta(\varepsilon - \mu) + 1} \tag{14.15}$$

Suppose we now reverse the question and ask for the number $g(\varepsilon) - n(\varepsilon)$, which is the number of electrons *missing* per unit energy range. We find

$$g(\varepsilon) - n(\varepsilon) = \frac{g(\varepsilon)}{\exp \beta(\mu - \varepsilon) + 1} \tag{14.16}$$

This is manifestly the same formula as (14.15) except that the energy scale is turned upside down. Thus there is also no *statistical distinction* between an electron a certain distance above the Fermi level and a hole the same distance below it, except for a reversal in the sign of the energy.

It often happens in semiconductors that the Fermi level μ is sufficiently far removed from either the highest filled band (usually called *valence band*) and/or the lowest empty band (usually called *conduction band*) that the Boltzmann limits of the above formulas can be employed. This means for (14.15)

$$n(\varepsilon) = g(\varepsilon) \exp \beta (\mu - \varepsilon) \qquad (14.17)$$

This formula is of course well known to us. It was written out in (8.65). Whenever (14.15) and (14.17) are equivalent, the electron gas is said to be nondegenerate. The condition for nondegeneracy of electrons in the conduction band of a semiconductor is that the bottom of the band be removed from the Fermi level by an energy difference of several times $k\,T$. The same phenomenon can occur for holes. If the top of the valence band is separated from the Fermi level by several times $k\,T$, (14.16) becomes

$$g(\varepsilon) - n(\varepsilon) = g(\varepsilon) \exp \beta (\varepsilon - \mu) \qquad (14.18)$$

Equation (14.18) represents a "nondegenerate" gas of holes reaching with its Boltzmann tail downward into the filled bands. In using (14.18) one must of course know when to stop the electron-hole analogy. It works fine in dealing with electrical responses of the electron system. But it cannot be used in dealing with questions involving the energy of formation of the substance or the equation of state. The analysis of Chapter 9 for the pressure of a degenerate electron gas may not be quite right for electrons in a band. It does show, however, that an almost filled band of electrons brings into play pressures which are orders of magnitude higher than what (14.17) or (14.18) could produce.

When Fermi energy μ was first introduced into physics it was naturally thought of as a material constant. It still is in many materials, particularly metals. But recent decades have witnessed the phenomenon of a variable μ, often manipulated artificially for purposes of design. This should not be surprising. In Boltzmann statistics μ has no standing and may be replaced at will by the partition function, using (8.66). Even in quantum statistics it is in principle a quantity to be eliminated by a relation such as (9.15) which specifies the total number of particles. Thus, if the total number of electrons can be changed in a material, the value of μ will be changed at the same time.

An early picture of a variable μ is shown in Fig. 14.4. It was thought with some reason that the band structure of the outer electrons varies little for the transition metals from titanium to zinc. Thus their electronic structure could be visualized by a progressively higher position of the Fermi level in the band system as shown. A variant of this idea is the idea of alloying neighboring elements in the periodic table and treating the electronic properties as due to a shift in the Fermi level. These notions have had only moderate success. In order to create a substantial shift in the

Fermi level of a metal, a substantial fraction of impurity has to be admixed. This impurity will do many other things besides communicating its electron surplus or deficit to the common pool. The idea is therefore sound, but only good for qualitative reasoning.

The idea of manipulating the Fermi level really opened up when it was transferred from metals to semiconductors. Semiconductors have their Fermi level at an energy for which $g(\varepsilon) = 0$. Therefore a very small

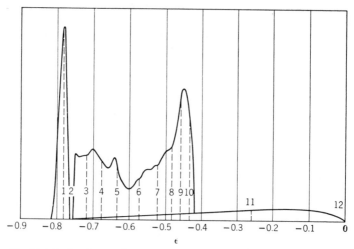

Fig. 14.4. Early picture of a variation of Fermi level with atomic number in the transition metals. The numbered vertical dotted lines indicate the position of the Fermi level for a number of outer shell electrons per atom as marked (after Slater).

admixture of a material with an electron surplus (or deficit) will swamp a distribution such as the one shown in Fig. 14.3. In order to accommodate a substantial number of electrons the Fermi level will have to be moved all the way from the middle of the forbidden band close to or even into the conduction band. The number of electrons accommodated thereby is small, but the shift in μ is substantial. The most successful field of application of this notion is for the column four semiconductors silicon and germanium. The shift of the Fermi level toward the conduction band is accomplished by admixture of pentavalent impurities, called *donor impurities*, such as phosphorus or arsenic. These substances enter the host lattice substitutionally, making the usual four-coordinated valence bonds. In the process of doing this they find themselves with a surplus electron which either is loosely bound to the impurity (which carries an extra charge) or helps to fill the conduction band of the host lattice.

Even the bound state is very close to the conduction band minimum in energy, and thus the Fermi level (which has to lie at the position at which the occupation probability is $\frac{1}{2}$) gets squeezed between this impurity state and the conduction band minimum. In a very similar way trivalent impurities called *acceptors*, such as boron or gallium, can be introduced to depress the Fermi level to a position near the valence band. Either the bonds to its neighbors are short of an electron, which means a "bound hole," or the bonding electron is made up from neighboring atoms, which means a "free hole" in the valence band. The Fermi level is thereby forced into the neighborhood of the valence band.

In addition to these equilibrium shifts of the Fermi level by "doping" of semiconductor material, temporary and local shifts are produced in practice by injection of electric carriers. This is the base of operation of transistors and related devices. This shift is again outside the scope of the equilibrium theory developed here, but has of course great practical interest.

The notion of a μ which is easily changed by varying the external conditions suggests the possibility of finding among the preceding equations some which are not sensitive to the position of μ. Such an equation is the mass action law between electrons and holes, which is valid when neither carrier is degenerate. If we denote by n the density of electrons in the conduction band, and by p the density of holes in the valence band, we have from (14.17)

$$n = \exp[\beta \mu] \frac{1}{V} \int \exp[-\beta \varepsilon] \, g_n(\varepsilon) \, d\varepsilon \qquad (14.19)$$

and from (14.18)

$$p = \exp[-\beta \mu] \frac{1}{V} \int \exp[\beta \varepsilon] \, g_p(\varepsilon) \, d\varepsilon \qquad (14.20)$$

Multiplication of (14.19) and (14.20) yields

$$n \cdot p = \text{constant} \qquad (14.21)$$

The constant is a function of temperature, but no longer of μ. It is the product of the two integrals shown. Its primary feature is the energy difference between the two bands in a Boltzmann exponent. We can therefore write

$$\text{Constant} = \exp[-\beta G] \times \text{Density of states factor} \qquad (14.22)$$

where G is the band gap between the valence and conduction bands. Actually the mass action law expresses in another way what we have so far expressed in terms of a variable Fermi level, namely, that a large number of electrons in the conduction band makes holes in the valence band very scarce and vice versa.

A very neat way of putting (14.21) is to write

$$n\,p = n_i^2 \qquad (14.23a)$$

where n_i^2 is the carrier density of intrinsic material. The equation is reminiscent of the relation of H^+ and OH^- ions in water, as written in equation (17.26), p. 375. It is a relatively simple matter to list n_i for a given substance at a given temperature. We find at 300° K

$$n_i = 1.4 \times 10^{10}\,\text{cm}^{-3} \text{ for silicon} \qquad (14.23b)$$

and

$$n_i = 2.4 \times 10^{13}\,\text{cm}^{-3} \text{ for germanium} \qquad (14.23c)$$

Semiconductors which acquire their negative carriers by a pentavalent or donor impurity are said to be *n* type, while those which have acceptor impurities and conduct through holes are called *p* type. Most semiconductors at room temperature are either *n* or *p* type. Their conductivity rises exponentially with temperature because their carriers have to be activated from the "bound" to the "free" state. The activation energy is small—about $\frac{1}{10}$ eV—but noticeable. Only at high temperature (about 1000°) does the Fermi distribution become so flat that donors and acceptors no longer play any role as providers of charge carriers. The semiconductor then becomes intrinsic, the Fermi level settles in the middle of the forbidden band, and the number of electrons and holes becomes substantially equal. Once *n* and *p* are equal, the square root of the equilibrium constant in (14.21) actually equals their number. It follows then from (14.22) that the activation energy for this type of conduction is *one half* of the band gap *G*; this point is occasionally overlooked in qualitative arguments. A silicon or germanium sample which is to be an intrinsic semiconductor at room temperature demands high-purity chemistry in its manufacture: impurity control to the order 10^{-10}; it is then naturally also a high-resistivity sample, so that resistivity is quoted in the trade as a measure of purity. Even so, most so-called high-purity samples are compensated, that is, have an equal number of donor and acceptor impurities. The donors then give their surplus electron to the equally numerous acceptors, and neither the valence nor the conduction band has an appreciable number of electric carriers.

It should be mentioned in closing that there is an effect involving the use of a simultaneous electric and magnetic field to measure directly the number n of electric carriers. The effect is called the *Hall effect*. It permits the splitting apart of n and μ in the easily measured quantity σ shown in (14.03). The theory of the Hall effect will be found in Section 20-4.

14-3. Theory of thermionic emission

When a metal is heated to a sufficiently high temperature, a certain number of electrons are emitted from the metal surface and can be collected as thermionic current. The picture that we make ourselves of this current is quite simple. The metal has electrons in a Fermi distribution whose Fermi level lies below the "vacuum level," by which we mean the level for which the electron is at rest when outside the metal. The difference between

Fig. 14.5. Picture of a metal as a "lake of electrons." The letters W_0 and W indicate two alternative definitions of the work function. W_0 is the potential difference between metal and vacuum, but W, the difference between the vacuum and Fermi levels, has more experimental importance.

the Fermi level and the vacuum level is the most convenient definition of the work function; it is indicated as W in Fig. 14.5. Herring and Nichols[†] have shown that, even though the quantity W is obviously a scalar, it is not an absolute constant, even if the metal is clean; it depends, on the contrary, on the crystal face exposed. The basic reason for this unexpected complication is that if two crystal faces are simultaneously exposed one will acquire an electric dipole layer as compared to the other. Removal of these layers thus produces different values of W for different faces. Large shifts in W can, in fact, be produced artificially by the deposition of chemicals, but cleaning of the metal does not remove these shifts entirely. Fortunately, in work of moderate precision, these uncertainties in the value of W can often be ignored and W treated as a constant for the crystal under consideration. For typical metals the quantity is of the order of 4 volts.

† C. Herring and M. H. Nichols. *Revs. Modern Phys.* **21,** 185 (1949).

Thermionic emission is explained in terms of Fig. 14.5 by the Maxwellian type tail in the velocity distribution of the electrons, which reaches higher than the vacuum level. Electrons located in this tail have enough energy to escape from the metal. From here on, the problem could be treated as a vapor pressure problem, except that in practice such a pressure is not allowed to develop. A field sufficient to snatch up all emerging electrons is usually applied, and the calculation has to be slightly deflected to yield a saturation current instead of a vapor pressure.

Since we deal with electrons having energies considerably above the mean, it is not too bad a picture to treat the electrons as free with their natural mass. This apparently old-fashioned viewpoint has had considerable success in recent years for elements in the first three columns of the periodic table. We start out by computing the number of electrons in an element of volume of phase space inside the metal. To do this we begin with (14.17), which is the correct limiting relation of (14.15). We combine this with formula (4.22) and multiply with 2 for the two values of spin. The number of electrons per unit volume having their momentum within dp'_x, dp'_y, and dp'_z when in the metal thus equals

$$n_1(p'_x, p'_y, p'_z)\, dp'_x\, dp'_y\, dp'_z = \frac{2}{h^3}\, dp'_x\, dp'_y\, dp'_z \exp \beta \left[\mu - \frac{1}{2m}\, (p'^2_x + p'^2_y + p'^2_z) \right]$$

(14.24)

Since the experimental arrangement is usually such that all electrons escaping are collected, we must inject here an argument of kinetic theory. For an area A perpendicular to the z direction the number striking the area per second within the same limits of momentum as n_1 is the density times the area times the perpendicular speed, or

$$n_2\, dp'_x\, dp'_y\, dp'_z = n_1\, A\, \frac{p'_z}{m}\, dp'_x\, dp'_y\, dp'_z \qquad (14.25)$$

We shall not be fussy here but shall simply assume that all such electrons escape. After the escape they find themselves in vacuum, having climbed a hill sloped in the z direction and inclined toward the metal. The z momentum is thereby reduced, and we have new momenta p_x, p_y, p_z given by

$$p_x = p'_x$$
$$p_y = p'_y$$
$$\frac{p_z^2}{2\,m} = \frac{p'^2_z}{2\,m} - W_0 \qquad (14.26)$$

The quantity W_0 is shown in Fig. 14.5. Transformation to the new momenta is simple. We find

$$\frac{\partial(p'_x, p'_y, p'_z)}{\partial(p_x, p_y, p_z)} = \frac{p_z}{p'_z}$$

The amount of charge escaping per unit area in unit time within the restricted momentum range becomes therefore from (14.25)

$$i_z \, dp_x \, dp_y \, dp_z = \frac{2\,e}{h^3} \frac{p_z}{m} \exp\left[\beta\left(\mu - W_0 - \frac{1}{2\,m}(p_x^2 + p_y^2 + p_z^2)\right)\right] dp_x \, dp_y \, dp_z \tag{14.27}$$

Establishment of the total current density is now simply a matter of integration over all values of p_x and p_y, and all positive values of p_z. At the same time we set, in accordance with Fig. 14.5,

$$W = W_0 - \mu \tag{14.28}$$

We find thus

$$i = \frac{4\,\pi\,e\,m\,k^2\,T^2}{h^3} \exp[-\beta\,W] \tag{14.29}$$

This is the so-called *Richardson–Dushman* equation. If the law is written in the form

$$i = A\,T^2 \exp[-\beta\,W] \tag{14.30a}$$

then the derivation given suggests that A is a universal constant which equals

$$A = 120 \text{ amps/cm}^2 \text{ degree}^2 \tag{14.30b}$$

The verification of this factor is not very good. One can first argue that a reflection coefficient should have been added at the interface, but this should not change the result by much. More serious is the possible variation of W with temperature; such a dependence is likely to arise from a phenomenon such as thermal expansion. Experimental data are usually analyzed by making a semilog plot of the saturation current against $1/T$. The slope of the plot yields W with fair precision. The intercept is then supposed to yield the factor. This result will be confused if we have a temperature variation of W with T of the form

$$W = P + Q\,T$$

We then get a new "constant" in the Richardson equation which equals

$$A \exp\left[- \frac{Q}{k} \right]$$

A value of Q only slightly larger than k will obviously confuse the results. It is therefore felt that experimental evidence against the value of A is not conclusive.

14-4. Degeneracy and non-degeneracy: electronic heat capacity in metals

At the present time, when engineers manipulate the Fermi level of their materials to suit their needs, a proof that electrons obey the Fermi distribution law may seem superfluous. However, as late as 1920, electrons in a metal were thought of as a gas obeying Boltzmann statistics. It is therefore worth while to bring out the qualitative features which forced this profound change from classical statistics upon the scientific community.

Shortly after the discovery of the electron, Drude found that if we assume the interior of metals to be filled with a gas of free electrons the conductivity of metals is explained. The number of free electrons needed equals about the number of valence electrons, and their mean free path has to be about 10^{-5} cm. Today we consider this result of Boltzmann statistics an accident. It arises partly from the fact that an electric field accelerates an electron according to the law

$$\frac{d\mathbf{k}}{dt} = - \frac{e}{\hbar} \mathbf{E}$$

The Pauli principle is no hindrance to this acceleration process because the $\mathbf{k}$ value being occupied by one electron is simultaneously vacated by another. Therefore, an entire partially filled band can be accelerated by a field, the Pauli principle notwithstanding. The randomizing collisions, on the other hand, are inhibited by the Pauli principle because the state scattered into is likely to be occupied. This explains the long mean free path. It is natural, however, that the early workers simply observed that the kinetic theory of gases yielded a perfectly straightforward explanation of Ohm's law without any essential modification. A detailed discussion of these kinetic questions will be found in Chapter 20. The discussion of this chapter will be limited to problems connected with equilibrium statistics.

Drude's explanation of conductivity did in fact pose a serious problem for equilibrium statistical mechanics. For, although he found that his perfect gas of electrons had the right transport properties, it did not seem to have the heat capacity required by his picture. We have explained the heat capacity of solids in Chapter 13 without any reference to whether they are metals or insulators. Yet a value $\frac{3}{2} R$ for the molar heat capacity of a perfect gas is almost inescapable, and this is 50% of the Dulong–Petit value for a monovalent substance. The physicists of 1920 felt that this difference ought to be noticeable. Yet actually there is no such difference between metals and insulators as regards either the law of Dulong and Petit or the theory of Debye. The free electrons which seem so copious in electrical experiments simply cannot be found when we look for their thermal effects.

It was Sommerfeld who first suggested that perhaps conduction electrons were a gas, but a gas obeying a strange statistics because of the Pauli exclusion principle. This principle forces a high zero-point energy on a gas if the gas is degenerate. Consequently, the increase of energy with temperature comes out to be abnormally low. We shall now verify Sommerfeld's prediction in detail.

We verified in Section 9-3 that if the valence electrons of a metal do form something like a gas, the gas is definitely in the condition we described as degenerate. We now want to calculate for such a system of particles the effect which is most directly indicative of their degenerate condition, namely, their heat capacity. Let us introduce the Fermi function $F(x)$ as

$$F(x) = \frac{1}{e^x + 1} \tag{14.31}$$

Then the total number N of conduction electrons equals

$$N = \int_{-\infty}^{+\infty} g(\varepsilon)\, F[\beta\,(\varepsilon - \mu)]\, d\varepsilon \tag{14.32}$$

The total energy U of these electrons is

$$U = \int_{-\infty}^{+\infty} \varepsilon\, g(\varepsilon)\, F[\beta\,(\varepsilon - \mu)]\, d\varepsilon \tag{14.33}$$

The quantity to be computed is the heat capacity of the electron gas, that is, the temperature derivative of (14.33) or

$$\frac{C}{k} = -\beta^2 \left(\frac{\partial U}{\partial \beta} \right)_N$$

In taking the derivative we must assume μ to be a function of β, such that the integral (14.32) for N remains constant. This means that we have

$$\frac{C}{k} = -\beta^2 \left\{ \left(\frac{\partial U}{\partial \beta}\right)_\mu - \frac{\left(\frac{\partial U}{\partial \mu}\right)_\beta \left(\frac{\partial N}{\partial \beta}\right)_\mu}{\left(\frac{\partial N}{\partial \mu}\right)_\beta} \right\} \qquad (14.34)$$

With (14.32) and (14.33) this becomes

$$\frac{C}{k} = -\beta^2 \int_{-\infty}^{+\infty} \varepsilon (\varepsilon - \mu) g(\varepsilon) F_x'[\beta (\varepsilon - \mu)] \, d\varepsilon$$

$$+ \beta^2 \frac{\displaystyle\int_{-\infty}^{+\infty} \varepsilon g(\varepsilon) F_x'[\beta (\varepsilon - \mu)] \, d\varepsilon \int_{-\infty}^{+\infty} (\varepsilon - \mu) g(\varepsilon) F_x' [\beta (\varepsilon - \mu)] \, d\varepsilon}{\displaystyle\int_{-\infty}^{+\infty} g(\varepsilon) F_x'[\beta (\varepsilon - \mu)] \, d\varepsilon}$$

where F_x' is the derivative of the function (14.31). To simplify this we observe that if the factors ε in the first term and in the first integral of the second term are replaced by $\varepsilon - \mu$, equal contributions are added and subtracted. We shall make this substitution and simultaneously adopt $\beta (\varepsilon - \mu) = x$ as the variable of integration. The above expression for C then becomes

$$\frac{C}{k} = -kT \int_{-\infty}^{+\infty} g(\mu + xkT) \, x^2 \, F_x' \, dx + kT \frac{\left[\displaystyle\int_{-\infty}^{+\infty} g(\mu + xkT) \, x \, F_x' \, dx \right]^2}{\displaystyle\int_{-\infty}^{+\infty} g(\mu + xkT) \, F_x' \, dx}$$

Because of the presence of F_x' in all integrals the contributions to C come exclusively from the neighborhood of the Fermi level. Expansion of $g(\mu + xkT)$ in powers of x yields automatically an expansion of the heat capacity in powers of kT/μ. If the sample is degenerate this quantity is small. If it is very small, as it is in metals, we should expect a fair approximation to the heat capacity by the first term only, which takes $g(\varepsilon)$ equal to $g(\mu)$. In this approximation the numerator of the second term vanishes by symmetry, and we get

$$\frac{C}{k} = kT g(\mu) \left[-\int_{-\infty}^{+\infty} x^2 \, F_x' \, dx \right]$$

The integral in brackets is easily calculated. First we observe that, from

(14.31), the integrand is even. Hence the integral equals twice the integral from 0 to ∞. An integration by parts then shows it to be equal to

$$4 \int_0^\infty x\, F(x)\, dx = 4 \int_0^\infty x\, dx \{e^{-x} - e^{-2x} + e^{-3x} - \ldots\} = \frac{\pi^2}{3}$$

The electronic heat capacity is therefore equal to

$$\frac{C}{k} = \frac{\pi^2}{3} g(\mu)\, k\, T \tag{14.35}$$

Formula (14.35) is not limited to the free electron model, but is generally valid for degenerate electrons. For a numerical estimate of the formula it is convenient to employ the free electron model again, perhaps with an effective mass correction. For metals, absolute zero can be assumed for the temperature in estimating $g(\mu)$. Formula (9.16) then applies, and the integral (14.32) can be evaluated as

$$N = \int_0^\mu g(\varepsilon)\, d\varepsilon$$

or

$$N = \tfrac{2}{3} g(\mu)\, \mu \tag{14.36}$$

Substitution of (14.36) into (14.35) yields

$$\frac{C}{N\,k} = \tfrac{1}{2} \pi^2 \frac{k\,T}{\mu} \tag{14.37}$$

In reducing this to molar units, a decision must be made about the number of conduction electrons per atom. This number is usually taken as the valence v, which is a rather arbitrary choice in the case of metals having a complicated band structure. However, the choice works often enough to be worth while. Then (14.37) becomes

$$\frac{\mathscr{C}}{R} = \tfrac{1}{2} \pi^2 v \frac{k\,T}{\mu} \tag{14.38}$$

This formula makes clearly evident the fact that the electronic contribution to the specific heat of metals is very small. For it is a formula in the standard dimensional unit R containing on the right the number kT/μ; this number was estimated earlier as being of the order 10^{-3}.

There is more to the electronic heat capacity than just the fact that it is small. The theory also shows that it should be proportional to the first power of the temperature. This result is a direct consequence of the Fermi distribution law and therefore worth checking against experiment. Rather surprisingly, the evidence for the law is found in the low-temperature region even though the term approaches zero as T approaches zero. The reason is that the phenomenon is in competition with the Debye form of heat capacity, which varies as T^3. This Debye contribution thus dies away faster as the temperature is lowered than the electronic term; the latter is therefore observed most easily in the low-temperature range.

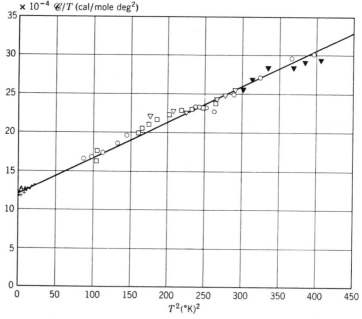

Fig. 14.6. Experimental heat capacity data for iron. The plot shows $\mathscr{C}/T$ versus T^2. It permits simultaneous examination of the electronic and lattice terms, which appear respectively as the intercept and the slope of a straight line (after Keesom and Kurrelmeyer).

A beautiful way to see both contributions was developed by the Leiden school and is shown in Fig. 14.6 for the case of iron. On the abscissa is plotted T^2, and on the ordinate $\mathscr{C}/T$. The advantage of such a plot is that if the heat capacity has the form

$$C = A T + B T^3$$

this type of plot will yield a straight line. The figure shows that this is clearly the case, and values for A and B can be read off as the intercept and the slope of the line, respectively. The slope determines a Debye Θ by (13.65). The intercept yields primarily a $g(\mu)$ as shown in (14.35). It is customary, partly for historical reasons, to transform this into a mass by using (14.38) and (9.23) for an interpretation, replacing m by an effective mass $m^\star$. We know now that very few bands resemble free electron parabolas, so that this usage might seem outdated. However, the thermal data yield but a single number about conduction electrons, and this number may just as well be expressed as a mass as in any other form. To distinguish it from other "masses" we call it the *density of states mass* of the electrons in the metal. Table 14–1 gives a list of typical values which result from this procedure. It is seen that the values of $m^\star$ obtained do not depart much from the free electron mass m; this is a satisfactory outcome. Iron, for which the data are given in Fig. 14.6, is out of line. One finds the same situation for transition metals in general. The empirical values of $m^\star$ come out high, which simply means, because of (9.23), that the empirical μ in (14.38) is low. Very little of this can be blamed on an incorrect assignment of the valence v because it enters into $m^\star$ only through its cube root. The error is of course in the passage from (14.35) to (14.38), which is entirely incorrect in the case of transition metals with unfilled inner shells. Figure 14.4 gives a rough view of the density of states function for such metals, and it is clear that the association of a "density of states mass" with the d band is not likely to lead to significant numbers. The picture does explain, however, that metals like iron have a very high $g(\mu)$ because of their unfilled shells and this, in turn yields, by (14.35), a very high linear term in the heat capacity.

TABLE 14–1. Density of states masses m^* for conduction electrons in metals as obtained from the linear term in the heat capacity (after Kittel, except for iron)

Metal	Valence (assigned)	$\dfrac{m^\star}{m}$
Na	1	0.6
Cu	1	1.5
Ag	1	0.95
Be	2	0.46
Mg	2	1.33
Zn	2	0.85
Fe	2	7.95

14-5. "Doped" semiconductors: n–p junctions

Having outlined the evidence for Fermi statistics for electrons, we now proceed to indicate how the Fermi level, and thus the electron and hole densities, are subject to external manipulation in semiconductors. It is clear to start with that in a perfectly pure material the Fermi level has a fixed position typical of the substance. The condition which determines μ in such a case is that the number of electrons in the conduction band and the number of holes in the valence band be equal. This means equality of expressions (14.15) and (14.16). These expressions are of such a structure that, regardless of the factors $g(\varepsilon)$ in the two bands, the Fermi factors at the band edges must be equal when the temperature becomes small. For it is seen that these factors assume infinitesimally small values as the temperature approaches zero and that their ratio can be kept finite only if the energy denominators are equal. This means that

$$\exp \beta(\varepsilon_c - \mu) + 1 \approx \exp \beta(\mu - \varepsilon_v) + 1$$

where ε_c is the energy at the bottom of the conduction band, and ε_v the energy at the top of the valence band. This solves to

$$\mu = \tfrac{1}{2}(\varepsilon_c + \varepsilon_v) \tag{14.39}$$

in the limit of $\beta = \infty$. If we do not quite pass to the limit but accept a parabolic band shape for either band (in the valence band, of course, a parabola facing downward), then we can employ (9.16) for either band, with a suitable effective mass replacing m. Both the electrons and the holes can be considered as nondegenerate in such a problem, and we get

$$2\left(\frac{2\pi m_c}{h^2 \beta}\right)^{3\!/\!2} \exp \beta(\mu - \varepsilon_c) = 2\left(\frac{2\pi m_v}{h^2 \beta}\right)^{3\!/\!2} \exp \beta(\varepsilon_v - \mu) \tag{14.40}$$

which yields

$$\exp \beta(2\mu - \varepsilon_c - \varepsilon_v) = \left(\frac{m_v}{m_c}\right)^{3\!/\!2} \tag{14.41}$$

This means that at finite temperature μ shifts a little toward the band with the smaller effective mass.

In practice the "intrinsic" condition (14.41) is only realized at quite elevated temperatures for most semiconductors. At room temperature the distribution of carriers is controlled by the impurities in the sample. Suppose we deal with an n-type semiconductor which has a set of donor impurities. Let the extra electrons of these impurities be in a bound state close to the bottom of the conduction band, separated from the band by an energy difference D as shown in Fig. 14.7. Equation (14.40) is then to be replaced by an equilibrium condition between the donor states and the conduction band, requiring that the number of holes in the donor states be equal to the number of electrons in the band. This condition reads

$$\frac{2 \left(\dfrac{2 \pi m_c}{h^2 \beta}\right)^{3/2}}{\exp \beta(\varepsilon_c - \mu) + 1} = \frac{d}{\exp \beta(\mu - \varepsilon_D) + 1} \qquad (14.42)$$

Here d is the density of the donor impurities, and ε_D the energy of the electrons in the bound donor state. Equation (14.42) works out differently

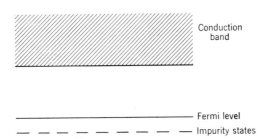

Conduction band

Fermi level

Impurity states

Fig. 14.7. Impurity states and conduction band in an n-type semiconductor. Equilibrium is controlled by the requirement that the number of electrons in the band equal the number of holes in the impurity states.

in practice from (14.40) because usually the dimensional factor on the left is of the order 10^{21} cm^{-3}, while d is more like 10^{15}. There is then no question of degeneracy on the left, and the 1 can be left off. For extremely small temperature μ will still lie halfway between ε_c and ε_D, but the energy difference D is often not very large, perhaps 0.1 eV. So the condition will soon arise that even for $\mu = \varepsilon_D$ the left side is larger than the right. Equilibrium will now be reached by μ falling below ε_D. The donor centers become almost fully ionized, and the exponential in the denominator on the right can be discarded. Equilibrium is then obtained from the condition

$$\exp \beta(\varepsilon_c - \mu) = \frac{2 \left(\dfrac{2 \pi m_c k T}{h^2}\right)^{3/2}}{d} \qquad (14.43)$$

As the temperature rises μ will drop more and more; finally it will land in the middle of the forbidden band. The semiconductor will cease to be "extrinsic," and equilibrium between electrons and holes in the two bands will become the controlling circumstance. Figure 14.8 shows a diagram for germanium which gives the location of the Fermi level within the forbidden band as a function of the carrier density and the temperature.

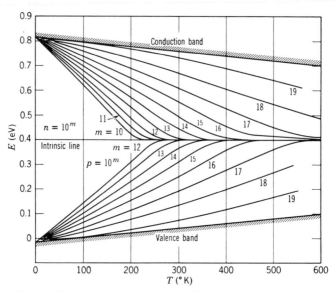

Fig. 14.8. Variation of the Fermi level with temperature in germanium
for various carrier concentrations (after Jonscher).

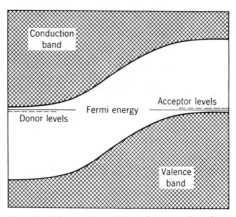

Fig. 14.9. Distorted band shape in an *n–p* junction.
On the two sides the Fermi level is "pinned" to
the valence and conduction bands, respectively. A
redistribution of charge allows the two parts to
join smoothly with the Fermi level remaining
horizontal.

The shift toward the intrinsic line at high temperature is clearly marked. The graph also shows that for very low temperature and high density the Fermi level penetrates the bands slightly and degeneracy sets in. In that case the argument about the size of the two dimensional factors in (14.42) is reversed. This is a rather exceptional circumstance.

An *n–p* junction is a boundary surface in a single piece of semiconductor at which a section doped with donors meets a section doped with acceptors. We shall discuss not its properties as a circuit element, but only its structure when no current is drawn. The junction is then in thermal equilibrium. The salient feature of this structure is the distorted band shape shown in Fig. 14.9. We shall only examine the simplest model of such a junction,

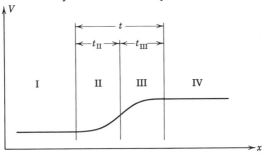

Fig. 14.10. Four-region solution for the abrupt junction at absolute zero. The curve shown is the potential for a negative charge as function of position. It consists of horizontal portions in the regions I and IV, and parabolic segments in the regions II and III.

the so-called *abrupt junction* in which material having constant donor density d is in direct contact with material having constant acceptor density a.

For reasons of mathematical convenience we shall first treat the junction at absolute zero. Carriers can then exist only very close to the Fermi level. The central portion of the junction which has the Fermi level crossing over from the conduction to the valence band is thus carrier-free and acts as an insulator separating n and p-type material (*depletion layer*). If we combine this insight with the abrupt junction model we find the specimen divided into four zones: n-type neutral, n-type carrier free, p-type carrier free, and p-type neutral. These four regions are labelled I, II, III, and IV in Fig. 14.10. Since the two central zones have no carriers, the charge of the ionized impurities remains uncompensated. The potential, obeying Poisson's equation, is thus bent into parabolic arcs whose curvature depends on the impurity density. In the outer regions, on the other hand, the number of carriers and ionized impurities is equal, the sample neutral, and the potential a horizontal straight line.

Discussing the four regions one by one, we start with the n-type neutral region I. Its potential is constant. In the n-type carrier-free region II, the Fermi level is below the donor impurity level and all levels are ionized, carrying a positive charge. The potential energy for negative charges thus obeys the equation

$$\frac{d^2 V_{II}}{dx^2} = \frac{4 \pi e d}{\kappa} \tag{14.44}$$

where κ is the dielectric constant of the semiconductor. Suppose we call the thickness of the layer t_{II} and place the origin of coordinates and the reference zero for the potential at the junction. Equation (14.44) must then be solved subject to the boundary conditions that the potential vanish at the border of II and III, which is at $x = 0$, and the field vanish at the border of I and II, which is at $x = -t_{II}$. This yields

$$V_{II} = \frac{4 \pi e d}{\kappa} \{ t_{II} \, x + \tfrac{1}{2} \, x^2 \} \tag{14.45a}$$

The same reasoning yields for region III

$$V_{III} = \frac{4 \pi e a}{\kappa} \{ t_{III} \, x - \tfrac{1}{2} \, x^2 \} \tag{14.45b}$$

The boundary condition at all border points is that the fields and potentials be continuous. The solutions (14.45) do this everywhere except for the matching of slopes at $x = 0$. This condition reads

$$\frac{t_{II}}{a} = \frac{t_{III}}{d} = \frac{t}{a + d} \tag{14.46}$$

Here we have introduced the letter t for the total thickness of layers II and III combined. This is the layer devoid of carriers or the depletion layer. It is characterized by high resistivity in experiments involving electric currents; t may therefore be properly called the thickness of the junction. The distribution of the depletion layer between the n and the p side is inversely as the impurity density; the layer is thus primarily on the side with the smaller impurity concentration. The potential shift is such as to keep the line for μ horizontal. The potential must therefore raise the acceptor level at $x = t_{III}$ to the same height as the donor level at $x = -t_{II}$. This potential drop is approximately equal to, but less than, the energy gap G; the donor and acceptor binding energies must be subtracted.

We denote the potential drop by g to indicate its resemblance to G. We have then from (14.45)

$$g = \frac{2 \pi e}{\kappa} (d\, t_{\mathrm{II}}^2 + a\, t_{\mathrm{III}}^2)$$

Substitution of t_{II} and t_{III} from (14.46) yields for this finally

$$g = \frac{2 \pi e}{\kappa} \frac{a\, d}{a + d}\, t^2 \qquad (14.47)$$

The equation gives the junction thickness t in terms of the doping densities a and d and the potential drop g. A characteristic quantity in the junction is the maximum field, which, from (14.45) and (14.46), comes out as

$$E_{\max} = \frac{4 \pi e}{\kappa} \frac{a\, d}{a + d}\, t \qquad (14.48)$$

As an example of the order of magnitude of the quantities entering into these formulas we may take an n–p junction in germanium. g will be about 0.7 volt, $\kappa = 16$, and common density values for a and d will be 10^{17} cm^{-3}. The resultant thickness of the junction then comes out from (14.47) as 1.6×10^{-5} cm, and the maximum field from (14.48) as 10^5 volt/cm. There is a technical advantage in making thin junctions because of transit time considerations and the possibility of electron tunneling through the junction. Such junctions must be made through high doping densities, as this is the only variable at the disposal of the experimenter in equation (14.47).

Having formed a reasonable picture of a junction by approximating the Fermi distribution function by a step function, we want to repeat the calculation at a finite temperature. The calculation is of interest because we deal with a Fermi distribution of the electrons in a potential which they partially set up themselves through their charge. We have thus a self-consistency problem. A look at Fig. 14.10 must make us expect that the sharp distinction between regions I and II will become blurred, and similarly between III and IV. In order to avoid all unnecessary complications we shall assume the junction symmetric, that is, the two halves of the junction shall be mirror images of each other. We then need only recompute the potential distribution in regions I and II.

We shall assume the donor impurities fully ionized; this is an assumption which is usually verified at room temperature, as was explained in connection with equation (14.43). The densities of the electrons and holes are then necessarily also nondegenerate as their energies are even further removed from the Fermi level. The densities are then listed in (14.40), with the following modifications:

(a) m_c and m_v must be taken as equal if the junction is to be truly symmetric about the midpoint.

(b) ε_c and ε_v are functions of position because the electric potential is not a constant. They are instead given by

$$\varepsilon_c = \tfrac{1}{2} e\,G + e\,V$$

and

$$\varepsilon_v = -\tfrac{1}{2} e\,G + e\,V$$

Here G is the band gap in potential units, and V the electric potential for negative charges. The expressions in (14.40) then become

$$n = 2 \left(\frac{2\pi m^\star}{h^2\,\beta} \right)^{3/2} \exp \beta(\mu - \tfrac{1}{2}eG - eV)$$

and

$$p = 2 \left(\frac{2\pi m^\star}{h^2\,\beta} \right)^{3/2} \exp \beta(-\tfrac{1}{2}eG + eV - \mu)$$

We can simplify this with the help of (14.23) to

$$n = n_i \exp \beta(\mu - eV) \tag{14.49a}$$

and

$$p = n_i \exp \beta(eV - \mu) \tag{14.49b}$$

Restricting ourselves now to regions I and II of an abrupt junction, we get Poisson's equation in the form

$$\frac{d^2V}{dx^2} = \frac{4\pi e}{\kappa}\,(d + p - n) \tag{14.50}$$

This replaces $V = $ constant in region I, and equation (14.44) in region II. It is convenient at this point to introduce the dimensionless variable

$$\Psi = \beta\,(e\,V - \mu) \tag{14.51}$$

With the help of (14.51) and (14.49), (14.50) becomes

$$\frac{d^2\Psi}{dx^2} = \frac{4\pi \beta e^2}{\kappa}\,(d + 2\,n_i \sinh \Psi) \tag{14.52}$$

The equation is to be solved between $x = -\infty$ and $x = 0$. At $x = 0$ the boundary condition is $\Psi = 0$. At $x = -\infty$ the right-hand side of (14.52) must vanish so that

$$\sinh \Psi(-\infty) = -\frac{d}{2 n_i} \tag{14.53}$$

Between these limits Ψ is negative throughout. With the help of (14.53) we can bring (14.52) into the form

$$\frac{d^2\Psi}{dx^2} = \frac{4\pi\beta e^2 d}{\kappa} \left\{ 1 - \frac{\sinh \Psi(x)}{\sinh \Psi(-\infty)} \right\} \tag{14.54}$$

The differential equation (14.54) can be given a simpler form if the interest is restricted to junctions having a potential drop large compared to kT/e. Because of (14.51) this means that

$$|\Psi(-\infty)| \gg 1$$

and

$$\sinh \Psi(-\infty) = -\tfrac{1}{2} e^{-\Psi(-\infty)} \ll -1$$

The term containing $\sinh \Psi$ in (14.54) is then entirely negligible unless $|\sinh \Psi(x)|$ is also large. We commit thus a negligible error if we set

$$\sinh \Psi(x) \approx -\tfrac{1}{2} e^{-\Psi(x)}$$

With the abbreviation

$$\Psi(x) - \Psi(-\infty) = \Phi(x) > 0 \tag{14.55}$$

(14.54) becomes then

$$\frac{d^2\Phi}{dx^2} = \frac{4\pi\beta e^2 d}{\kappa} \{1 - e^{-\Phi}\} \tag{14.56}$$

The equation can be integrated once by multiplication with $d\Phi/dx$. The boundary condition that $d\Phi/dx$ vanish at $-\infty$ determines the constant of integration. The result is

$$\frac{1}{2}\left(\frac{d\Phi}{dx}\right)^2 = \frac{4\pi\beta e^2 d}{\kappa} \{\Phi + e^{-\Phi} - 1\} \tag{14.57}$$

Either one of the two equations (14.56) or (14.57) is a universal equation for the positive function Φ, except for a scale factor in the length. The length L which enters is

$$L = \left(\frac{\kappa k T}{4\pi e^2 d}\right)^{1/2} \tag{14.58}$$

It was first introduced by Debye and Hückel in the theory of electrolytes. The length is therefore called the *Debye length*. It is not directly associated with the junction thickness, for the thickness depends on the magnitude to which Φ rises. L has rather to do with the rate at which Φ falls to zero at a large distance from the junction. Equations (14.56) and (14.57) give a universal relation which determines this rate in terms of the Debye length L. The two length parameters t and L have this in common, that they vary both inversely as the square root of the density of impurities. Therefore, the entire scale of the junction varies in the same manner.

One thing equation (14.57) can be used for without further integration is to give the maximum field at the junction which occurs at $x = 0$. The equation gives it as a function of the potential drop g. From (14.51) we have

$$E = \frac{1}{e\,\beta} \frac{d\Phi}{dx} \tag{14.59}$$

and because of the symmetry assumed for the two halves of the junction

$$\Phi(0) = \tfrac{1}{2}\,e\,\beta\,g \tag{14.60}$$

Here g is of the same order as the band gap G but is apt to be smaller, as Fig. 14.8 has illustrated. Substitution of (14.59) and (14.60) into (14.57) and neglect of the small exponential term yields then

$$E_{\max}^2 = \frac{4\,\pi\,e\,d}{\kappa}\left(g - \frac{2\,k\,T}{e}\right) \tag{14.61}$$

The same formula, without the subtracted term $2kT/e$, results from the analysis at absolute zero by elimination of t from (14.47) and (14.48). Equation (14.61) is therefore a broad indication that the junction proper is not greatly modified by temperature.

The preliminary result (14.61) reduces the interest in the second integration of Poisson's equation. It actually can be carried out, starting from (14.57), by a simple quadrature. It yields a universal curve if one takes the scaled potential Φ as ordinate and the scaled distance x/L as abscissa, with the scaling factors given by $\beta\,e$ and the Debye length (14.58). In addition a sliding origin is implicit for the abscissa because the curve rises until the maximum field is reached. The field depends on the potential drop g as shown in (14.61). Explicitly the integration of (14.57) takes the form

$$\frac{x}{L} = \frac{1}{\sqrt{2}} \int^{\Phi} \frac{d\varphi}{\{\varphi - 1 + \exp(-\varphi)\}^{\frac{1}{2}}} \tag{14.62}$$

The curve is plotted in Fig. 14.11. One simple feature of the curve which can be seen directly is the asymptotic behavior of the potential at large distance from the junction. Then x/L is large and negative and Φ is

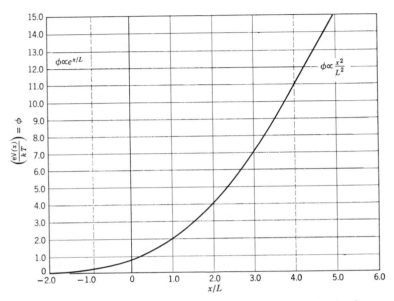

Fig. 14.11. Potential versus distance curve as one approaches a junction from bulk material. The curve is universal if eV/kT is plotted against x/L. The termination point on the right depends on the voltage reached. Limiting laws are indicated in the extreme right and left portions of the figure.

sufficiently small that the denominator can be expanded to the first nonvanishing term. We find

$$\Phi \approx \Phi_0 \exp(x/L) \tag{14.63}$$

This clears up the significance of the parameter L defined in (14.58). It is associated not so much with phenomena within the junction as with the damping of a potential disturbance at a large distance from the disturbance. The picture which one can associate with the Debye length L is that the thermal agitation of the mobile carriers prevents perfect screening of a perturbing potential by the charge cloud of density $e\,d$. Thus an exponential tail type of extension is created for any local potential disturbance. The logarithmic decrement of the disturbance in space equals the Debye length L.

A. RECOMMENDED PROBLEMS

1. Derive equation (14.13) for the expectation value of the velocity of a Bloch electron by using quantum mechanics rather than an argument based on wave packets. In other words, show that the expectation value of the operator $-i\dfrac{\hbar}{m}\dfrac{\partial}{\partial x}$ for a Bloch function is $\dfrac{1}{\hbar}\dfrac{\partial W}{\partial k_x}$.

2. The following facts are known for solutions of the one-dimensional Schroedinger equation with a periodic potential $V(z)$ of period 1:

(a) Bands do not overlap in energy, and can be labeled $j = 1, 2, 3, 4, \ldots$ in order of increasing energy.

(b) The energy band functions $W_j(k)$ are alternately monotonically increasing and monotonically decreasing in the interval $\langle 0, \pi \rangle$.

(c) The energy extrema provide a natural definition of a dynamic effective mass at the energy extrema; the definition is adequate for an almost empty or almost full band.

Substantiate the claim (c), assuming the other two. Proceed from this to investigate solutions for the two-dimensional Schroedinger equation with the potential $V(x)+V(y)$. Find the location of the energy extrema and the associated dynamic "masses," and study the problem of band overlap. Is there a circumstance which might naturally produce isotropic masses?

B. GENERAL PROBLEMS

3. An ingot of germanium is formed by melting together 100 gm of germanium and 3.22×10^{-6} gm of antimony. Compute the conductivity of the ingot (atomic weight of Ge 72.60, atomic weight of Sb 121.76, density of Ge 5.46 gm/cm³, mobility of electrons in Ge 3600 cm²/volt sec).

4. Assume that one adds to the ingot of Problem 3 0.78×10^{-6} gm of gallium, atomic weight 69.72, in addition to the antimony. What is the new conductivity (mobility of holes in Ge 1700 cm²/volt sec)?

5. Suppose that the lowest energy band function of Problem 2 is reasonably well approximated by the formula

$$W = -W_0 \cos k$$

Construct and describe the contours of constant energy in the $k_x - k_y$ plane for the corresponding two-dimensional problem.

6. Calculate the modification which the density of states expression

(9.16) undergoes if, as is the case in n-type germanium or silicon, the dependence of energy on $\mathbf{k}$ has the form

$$W = \frac{\hbar^2}{2} \left\{ \frac{k_1^2}{m_1} + \frac{k_2^2}{m_2} + \frac{k_3^2}{m_3} \right\}$$

The indices 1, 2, 3 refer to three mutually perpendicular directions in $\mathbf{k}$ space.

7. In analogy to Problem 6, Chapter 10, a number of two-level "atoms" are sunk into an assembly of fermions, inducing transitions between states which differ by their own energy splitting. Show that, if we assume Boltzmann distribution for the atoms and equal transition probability in the two directions, the Fermi–Dirac distribution follows for the fermions. Account must be taken of the fact that fermions make transitions from occupied to unoccupied states only.

15

Statistics of magnetism

In electromagnetic theory, magnetism usually appears as a secondary phenomenon associated with changing electric fields. This approach does not do justice to the subject. In modern physics magnetic moments arise directly in association with angular momentum. This means in many cases that magnetic moments have to be treated as given constants of atoms and subatomic particles. These moments have an interesting thermal behavior which has inspired some of the deepest investigations of statistical mechanics. As a result of this effort we possess a wealth of statistical understanding concerning magnetic phenomena which is surpassed only by our understanding of near-perfect gases.

There is a magnetization phenomenon which does not involve temperature and to which, therefore, the decoupling argument of Chapter 5 is applicable, namely, *diamagnetism*. In atoms and molecules not having an intrinsic moment, application of a magnetic field produces a precessional motion of the electrons by the laws of induction. The moment due to this precession is opposed to the applied field, and the resultant susceptibility is therefore negative. It is also independent of temperature. An extreme form of diamagnetism is the *Meissner effect* of superconductors. Superconductors have an induced moment opposing an external field, which is so strong as to cancel completely the magnetic induction in their interior. There are some interesting thermodynamic problems associated with superconductivity; they are associated with the equilibrium between the superconducting and the normal phase of a substance. The internal structure of superconductors may be treated without reference to temperature. Thus, diamagnetism and superconductivity have no further interest here, except that they compete with the magnetic phenomena which we are about to study.

15-1. Paramagnetism of isolated atoms and ions

Paramagnetism is the response to a magnetic field of particles which have a permanent moment and are free to orient that moment in space. In the absence of a field this orientation will be random, but when a field is applied there will be a torque tending to align the moment with the field. This torque will be opposed by thermal agitation. Thus, a positive susceptibility results which is usually strongly temperature dependent. This feature makes the subject of interest here.

Langevin was the first to grasp the essence of paramagnetic behavior. Suppose a number of identical localized freely orientable magnetic dipoles of moment μ are subject to a magnetic field H. Each one acquires then a potential energy ϕ equal to

$$\phi = - \mu \, H \cos \vartheta \tag{15.01}$$

where ϑ is the angle between the dipole and the field. If we denote by x the combination

$$x = \beta \, \mu \, H \tag{15.02}$$

then the probability for the angle lying between ϑ and $\vartheta + d\vartheta$ equals

$$P(\vartheta) \sin \vartheta \, d\vartheta = \frac{\exp(x \cos \vartheta) \sin \vartheta \, d\vartheta}{\displaystyle\int_0^\mu \exp(x \cos \vartheta) \sin \vartheta \, d\vartheta} \tag{15.03}$$

The quantity usually wanted is the total moment M along the field, that is,

$$M = N \int_0^\pi \mu \cos \vartheta \, P(\vartheta) \sin \vartheta \, d\vartheta$$

which comes out to be

$$M = N \mu \, L(\beta \, \mu \, H) \tag{15.04}$$

where $L(x)$ is called the *Langevin function*. It equals

$$L(x) = \coth x - \frac{1}{x} \tag{15.05}$$

A plot of this function is shown in Fig. 15.1. The magnetization I is defined as the magnetic moment per unit volume so that we get from (15.05)

$$I = \frac{N}{V}\, \mu\, L(x) \qquad\qquad (15.06)$$

If x is small we can replace coth x in (15.05) by its power series, which equals

$$\coth x = \frac{1}{x} + \tfrac{1}{3}\, x - \tfrac{1}{45}\, x^3 + \ldots$$

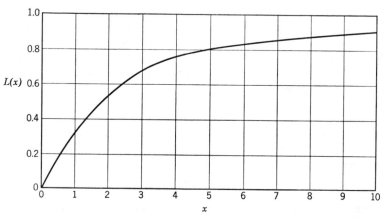

Fig. 15.1. Langevin curve for the magnetic moment of a freely orientable dipole. The abscissa contains the combination H/T.

The leading term cancels so that I becomes proportional to H. The constant of proportionality is called the *paramagnetic susceptibility* χ. We have then

$$I = \chi\, H \qquad\qquad (15.07)$$

and

$$\chi = \frac{N\, \mu^2}{3\, V\, k\, T} \qquad\qquad (15.08)$$

The susceptibility is seen to vary inversely as the temperature. This feature is called *Curie's law*, and the constant in (15.08) having the dimension of a temperature the *Curie constant*. Figure 15.2 shows a plot of the susceptibility of a copper salt, plotted against $1/T$. We get a linear curve going through the origin. Curie's law is therefore obeyed. From the plot we read off a Curie constant C' equal to

$$C' = 0.002 \text{ deg}$$

Salts containing paramagnetic ions often obey Curie's law very well because the assumptions going into the theory are justified to a high degree. The law is in fact so reliable in many cases that certain salts are used as thermometric substances at low temperature.

The thermodynamics of (15.06), namely, of a magnetization I which

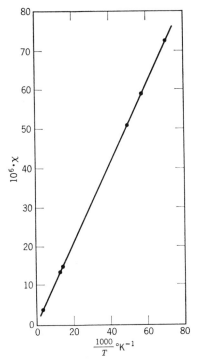

Fig. 15.2. Plot of susceptibility versus reciprocal temperature for powdered copper-potassium sulfate hexahydrate (after Hupse).

depends only on H/T was exploited in earlier chapters, notably Chapters 5 and 6. We defined substances behaving in this manner as "perfect paramagnets," and showed that their internal energy depends on temperature only. The work done in magnetizing a paramagnetic sample thus goes primarily into diminishing the entropy. To do it at constant temperature an amount of heat equal to the work done must be simultaneously withdrawn. Under adiabatic conditions the sample will tend to heat up; the reverse process, cooling through demagnetization, is of practical importance in low-temperature physics.

The main modification of the Langevin theory through quantum mechanics lies not in the modification of the statistics. Boltzmann statistics remains correct because the atoms carrying the magnetic moments are localized. This localization makes them distinguishable. What modern quantum mechanics contributes is the insight that the angle between the magnetic moment and the field is quantized. The quantization is linked with that of angular momentum. A fairly straightforward calculation of elementary electromagnetism shows that if a magnetic moment μ arises exclusively from circulating electrons (orbital paramagnetism) then the conversion factor from an angular momentum A to the associated magnetic moment μ equals

$$\mu = - \frac{e}{2\,m\,c}\,A \tag{15.09}$$

The factor in (15.09) is called the *gyromagnetic ratio* and can be determined directly by precision experiments on rotating metal objects. Now A is quantized in units of $\hbar$. Therefore μ is also quantized. The quantum equals

$$\mu_{\mathrm{B}} = \frac{e\,\hbar}{2\,m\,c} \tag{15.10a}$$

or

$$\mu_{\mathrm{B}} = 0.927 \times 10^{-20}\ \text{esu} \tag{15.10b}$$

μ_{B} is called the *Bohr magneton*.

If the relation (15.09) generally connected electronic angular momenta and magnetic moments, then all magnetic moments of atoms and molecules would be integral multiples of one half of a Bohr magneton. This simple result is not correct because of the electronic spin anomaly. According to this anomaly, a full Bohr magneton is associated with the spin of the electron even though the spin is only $\frac{1}{2}$. One expresses this more complicated situation by writing

$$\mu = g\,\mu_{\mathrm{B}}\,j \tag{15.11}$$

where j is the angular momentum quantum number, and g is a pure number, the *g factor*. If the angular momentum is entirely orbital, g equals 1, if entirely arising from electron spins, g equals 2. Most angular momenta are of mixed origin, and g may therefore be another number, even one larger than 2 or smaller than 1. In solids, crystalline symmetry conditions often enforce angular wave functions of the type $\sin m\varphi$ or $\cos m\varphi$, rather than $e^{\mathrm{i}m\varphi}$ and $e^{-\mathrm{i}m\varphi}$ as in atoms. The expectation value of the orbital angular momentum is then zero instead of $\pm\,m\,\hbar$. An orbital angular momentum of this type is called *quenched*. As a consequence of orbital quenching, spin is often the only angular momentum left, and g's close to 2 are a common occurrence.

Because of the quantization of angular momentum the continuous distribution of orientations of the moment is to be replaced in a quantum calculation by $2j+1$ discrete orientations, characterized by the $2j+1$ possible values of the magnetic quantum number m. The axis of quantization in the problem is the applied field $\mathbf{H}$. From (15.11) we see that the moment along this axis for a state of magnetic quantum number m equals

$$\mu = g \, \mu_\mathrm{B} \, m \tag{15.12}$$

where $m = -j, \ -j+1, \ldots, j-1, j$. The extreme values of this component are reached for $m = \pm j$ and equal the value (15.11), apart from a sign. If we adopt again the abbreviation (15.02), with μ having the meaning (15.11), the probability for the quantum number m being realized equals

$$P_j(m) = \frac{\exp\left[x \, \dfrac{m}{j}\right]}{f_j(x)} \tag{15.13}$$

where the partition function $f_j(x)$ equals

$$f_j(x) = \sum_{m=-j}^{j} \exp\left[\frac{m}{j} x\right] = \frac{\sinh\left\{1 + \dfrac{1}{2j}\right\} x}{\sinh \dfrac{1}{2j} x} \tag{15.14}$$

The total moment M along the field then results from (15.11), (15.12), and (15.13) as

$$M = N \mu \, \frac{\displaystyle\sum_{m=-j}^{j} \frac{m}{j} \exp\left[\frac{m}{j} x\right]}{\displaystyle\sum_{m=-j}^{j} \exp\left[\frac{m}{j} x\right]}$$

This is again an expression of the form (15.04), with a new Langevin function $L_j(x)$ equal to the above fraction. We evaluate the fraction by observing that its numerator is the derivative with respect to x of the denominator. We therefore get from (15.14) in the place of (15.05)

$$L_j(x) = \left(1 + \frac{1}{2j}\right) \coth \left(1 + \frac{1}{2j}\right) x \; - \; \frac{1}{2j} \coth \frac{1}{2j} x \tag{15.15}$$

It is seen that the traditional Langevin function (15.05) is the limit of $L_j(x)$ for j going to infinity. The other limiting case, $j = \frac{1}{2}$, which arises for a single electron spin, is also worthy of note. It gives simply

$$L_{1_2}(x) = \tanh x \tag{15.16}$$

In the low field limit (15.07) and (15.15) lead to a slightly altered value of the paramagnetic susceptibility χ, namely,

$$\chi_j = \tfrac{1}{3}\left(1 + \frac{1}{j}\right)\frac{N\mu^2}{VkT} \tag{15.17}$$

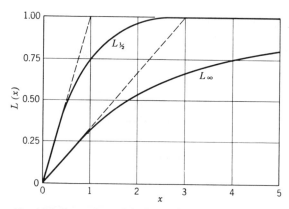

Fig. 15.3. Comparison of the Langevin curves for $j = \frac{1}{2}$ and $j = \infty$. The initial slopes are marked by dashed straight lines.

for $j = \frac{1}{2}$ this reduces simply to

$$\chi_{1_2} = \frac{N\mu^2}{VkT} \tag{15.18}$$

Figure 15.3 shows a plot of the Langevin curves (15.16) and (15.05) for comparison. The resemblance is more important than the differences. Both curves start out linearly for small x and approach 1 asymptotically when x gets large. They both have the combination $\mu H/kT$ as the argument. The initial slopes differ by a factor 3. Thus the novelty of quantization produces no essential change in the theory. It is obvious from the formulas that all other functions $L_j(x)$ are intermediate between the two functions shown.

The thermal manifestation of paramagnetism is best seen by analyzing its contribution to the heat capacity. According to (7.17), the heat capacity at constant field is the temperature derivative at constant field of the magnetic enthalpy $E = U - M H$. It was shown in (5.53) that, for paramagnetic materials, U has no connection with magnetism; the magnetic contribution arises therefore entirely from the second term. For simplicity we shall take $j = \frac{1}{2}$. From (15.04) and (15.16) we obtain then for the heat capacity C_H

$$C_H = N k x^2 \operatorname{sech}^2 x \tag{15.19}$$

where the abbreviation (15.02) has again been employed.

The behavior of C_H according to (15.19) is shown in Fig. 15.4 in a C

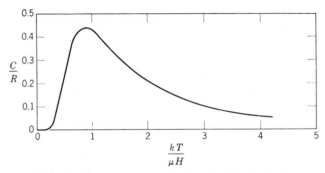

Fig. 15.4. Heat capacity of a specimen having localized paramagnetism; plot of the magnetic contribution versus temperature at a nonvanishing field.

versus T plot. C_H equals zero for both high and low temperatures and comes to a peak at an intermediate value. This type of heat capacity curve is typical for many ordering processes: a specific heat hump, clustered around a definite temperature. There is a qualitative difference between this type and those previously encountered which leveled off at some equipartition value. This time the energy spectrum is limited on the high energy side, and the entire passage from nearly complete order to nearly complete disorder occupies only a limited temperature range. The total change in magnetic enthalpy equals

$$\Delta E = \int_0^\infty C_H(T) \, dT = N \mu H \tag{15.20}$$

This result is obtained by a simple reversal of the preceding differentiation. The total change in entropy equals

$$\Delta S = \int_0^\infty C_H(T) \frac{dT}{T}$$

which, from (15.02) and (15.19), can also be written as

$$\Delta S = N k \int_0^\infty x \, \text{sech}^2 \, x \, dx$$

This becomes by an integration by parts

$$\Delta S = N k \int_0^\infty (1 - \tanh x) \, dx$$

$$= N k \lim_{x = \infty} \{x - \ln \cosh x\}$$

and finally reduces to

$$\Delta S = N k \ln 2 \tag{15.21}$$

The interpretation of (15.21) comes from (5.08). At low temperature S equals zero; at high temperature it equals k times the logarithm of the multiplicity. For spin $\frac{1}{2}$ the multiplicity is 2^N, which is just the result (15.21).

There is a second aspect of (15.19) which needs some discussion. The heat capacity depends only on x, which contains H and T in the combination H/T. The temperature range of the hump in the heat capacity can therefore be shifted at will by an adjustment of the field. More precisely, the entropy of the spin system described here can be increased by the amount (15.21) by the simple device of demagnetizing a magnetically saturated specimen. If this process is carried out adiabatically, a corresponding decrease in temperature will be enforced to keep the total entropy constant. If it is carried out in thermal contact with a second body, heat will be extracted from that body. This procedure is the standard cooling method at temperatures so low that pumping on liquid helium is no longer an efficient cooling process.

15-2. Pauli paramagnetism

The preceding calculation has to be modified if we are not dealing with localized magnetic moments. The typical case is the alkali metals, whose valence electrons have a spin and a moment, and are also carriers of electric current. Boltzmann statistics is then not applicable, and we must repeat the calculation for a continuum of levels subject to Fermi statistics. We know from previous experience that Fermi statistics is worth studying only when we have degeneracy ,and that the electrons in metals are in fact in a degenerate condition. We shall therefore limit ourselves here to the study of the spin paramagnetism of conduction electrons at absolute zero, remembering in the meantime that the results might be modified somewhat at finite temperature.

Suppose we draw a simple kind of density of states versus energy curve as shown in Fig. 15.5. At absolute zero the energy levels will be filled up to the Fermi energy ε_F. If these electrons have a spin $\frac{1}{2}$ and a magnetic moment, and if a magnetic field is present, two versions of the density

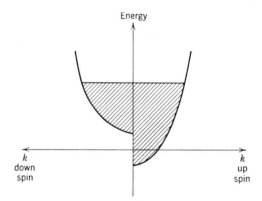

Fig. 15.5. Plot of energy versus density of states for the two spin orientations in the case of Pauli paramagnetism at absolute zero. The levels in the shaded regions are occupied.

of states curve have to be drawn: one displaced upward by μH; the other downward by μH. Instead of a simple $g(\varepsilon)$ such as (9.16) we now have two such functions: one for each orientation of spin. The total magnetic moment of the electronic assembly then equals

$$M = \tfrac{1}{2} \mu \left\{ \int_0^{\varepsilon_F + \mu H} g(\varepsilon)\,d\varepsilon - \int_0^{\varepsilon_F - \mu H} g(\varepsilon)\,d\varepsilon \right\} \tag{15.22}$$

Usually μH is very small compared to ε_F. The resultant integration reduces therefore to a simple multiplication and yields

$$M = \mu^2 H g(\varepsilon_F) \tag{15.23}$$

This yields a susceptibility which equals

$$\chi = \frac{\mu^2 g(\varepsilon_F)}{V} \tag{15.24}$$

We may estimate the susceptibility (15.24) by taking for $g(\varepsilon)$ the formula (9.16). The result is a spin susceptibility of the order 10^{-5}, which is this time not dependent on temperature. Unfortunately the formula is not easily checked quantitatively because the actual susceptibility of a metal is made up of a number of contributions among which the above is only one. It is true, however, that alkali metals have a positive susceptibility independent of temperature which is of the order of magnitude (15.24).

15-3. Ferromagnetism; internal field model

A phenomenon which was extremely fruitful for the development of statistical mechanics is ferromagnetism. We call a material ferromagnetic

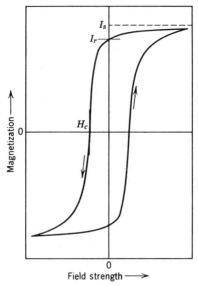

Fig. 15.6. Empirical magnetization curve of a ferromagnetic material; the curve is irreversible, and is roughly characterized by the saturation magnetization I_s, the remanence I_r, and the coercive field H_c.

if it is capable of having a magnetic moment in the absence of an applied field. Ferromagnetism is not very widespread among substances in general; the majority of ferromagnets known are alloys or compounds of the elements iron, nickel, cobalt. If a ferromagnet is heated it loses its characteristic properties at a definite temperature, the so-called Curie temperature T_c. Above the Curie point, ferromagnetic materials are paramagnetic, with an enhanced value of the paramagnetic susceptibility. Figure 15.6

shows a typical plot of the magnetization as a function of an applied magnetic field for a ferromagnetic material (magnetization curve). It is seen that we deal with a phenomenon showing hysteresis; the magnetization is not a unique function of the field but depends upon past history. In particular, the *remanence* I_r, that is, the magnetization at zero field, depends on the direction in which the specimen was last saturated. Similarly, a finite opposing field, the *coercive field* H_c, is necessary to bring the magnetization to zero. A characteristic aspect is the saturation of the magnetization at a value I_s as the field grows. This aspect is reminiscent of the Langevin curves shown in Fig. 15.3. However, paramagnetism saturates only at very high fields (10,000 oersted) or very low temperatures, while in ferromagnetism the phenomenon occurs in easily accessible ranges of temperature and field. The one element which is truly reminiscent of Langevin's theory is the value of the saturation magnetization, which is essentially the same for paramagnets and ferromagnets.

The interpretation of the mechanism of ferromagnetism has long been handicapped by the obvious nonequilibrium aspect of the phenomenon as commonly observed. The phenomenon as depicted in Fig. 15.6 is nonconservative; in fact, by equation (1.27), the area of the magnetization curve gives directly the amount of energy converted into heat in one hysteresis cycle. The phenomenon contains therefore frictional aspects which are not capable of a direct atomic explanation. Pierre Weiss* took the fundamental step necessary for an understanding by postulating that the atomic nature of ferromagnetism is obscured to us by a structure involving relatively large magnetized blocks, called *domains*. A single domain was supposed to have a magnetization curve of the type shown in Fig. 15.7; the magnetization appears as a definite function of temperature and field. Thermodynamics and statistics are therefore applicable. What distinguishes the curve from the Langevin curve of Fig. 15.1 is the *spontaneous magnetization* which the material possesses at zero field. According to Weiss, those aspects of Fig. 15.6 not reproduced in Fig. 15.7 were to be explained by the assumption of interlocking domains which are spontaneously magnetized but are not free to follow external changes reversibly. This hypothesis of Weiss has been brilliantly verified in the last few years. We shall not try to discuss this aspect here, but shall refer the reader to the appropriate literature.† The atomic aspect of ferromagnetism, represented by Fig. 15.7, is in itself a very interesting problem.

The original hypothesis of Weiss to explain Fig. 15.7 was the hypothesis of an *internal field* proportional to the magnetization. At first sight this

---

* P. Weiss. *J. phys.* **6**, 667 (1907).

† C. Kittel and J. K. Galt. "Ferromagnetic domain theory," *Solid State Phys.* **3**, 437 (1956).

seems a reasonable hypothesis because such an effect is known from electro-dynamics. Let us therefore modify equation (15.06) by making on the right-hand side the substitution

$$H \rightarrow H + \alpha I \qquad (15.25)$$

with α an undetermined number. We get then for the case $j = \frac{1}{2}$, with (15.02) and (15.16)

$$I = \frac{N}{V} \mu \tanh \beta\mu(H+\alpha I) \qquad (15.26)$$

The equation of state (15.26) connecting the thermodynamic variables I, H, T, is in qualitative agreement with the facts about ferromagnetism.

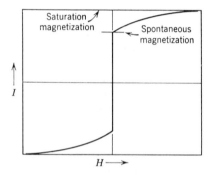

Fig. 15.7. Magnetization curve of a single magnetic domain; the curve is reversible, and differs from the Langevin curves mainly through the presence of a spontaneous magnetization at zero magnetic field.

We shall consider first the fact about spontaneous magnetization. Setting H in (15.26) equal to 0 and solving for T, we get

$$kT = \frac{\alpha\,\mu\,I}{\text{arctanh}\left(\dfrac{I\,V}{N\,\mu}\right)} \qquad (15.27)$$

The equation yields a T for every I between 0 and $N\mu/V$. When I is almost equal to $N\mu/V$ the denominator is very large and T is correspondingly small. Thereupon as I decreases T increases. However, T does not increase indefinitely because the inverse hyperbolic tangent is always larger than its argument (see Fig. 15.3). We therefore find

$$kT < \frac{\alpha\,\mu\,I}{I\,V/N\,\mu} = \frac{\alpha\,N\,\mu^2}{V} = k\,T_c \qquad (15.28)$$

When the inequality does not hold, the only solution of (15.27) is $I = 0$. The temperature $T = T_c$ separates therefore a high-temperature region, in which the specimen is paramagnetic, from a low-temperature region, in which it is ferromagnetic. This transition temperature is the Curie temperature discussed earlier. Figure 15.8 shows the spontaneous magnetiza-

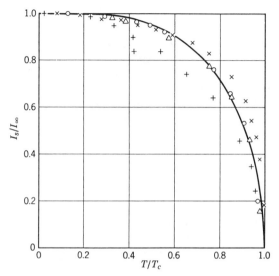

Fig. 15.8. Plot of spontaneous magnetization versus temperature; theoretical two-level internal field curve with experimental points for iron ($\times$), nickel ($\circ$), cobalt ($\triangle$), and magnetite ($+$) (after R. Becker).

tion versus temperature curve predicted by (15.27) together with empirical results for iron, nickel, cobalt, and magnetite; it is seen that the agreement is close. A particularly interesting aspect of it is the way the magnetization approaches zero near the Curie point. To see this in detail from (15.27), we may assume $I V / N \mu$ small and expand the arctanh, thus

$$k T \approx \frac{\alpha \mu I}{\dfrac{I V}{N \mu} + \dfrac{1}{3} \dfrac{I V^3}{N \mu}}$$

or, with the definition (15.28),

$$I \approx \frac{N \mu}{V} \sqrt{3} \left(\frac{T_c}{T} - 1\right)^{\frac{1}{2}} \tag{15.29}$$

This is an approach to zero with infinite slope. In addition to the spontaneous magnetization, the internal field theory also predicts correctly the susceptibility of ferromagnetic materials above the Curie temperature. In the paramagnetic range I is small compared to its saturation value, and therefore the tanh in (15.26) is small and can be set equal to its argument; the resultant equation for I is, with (15.28),

$$I = \frac{N \mu^2 H}{V k (T - T_c)} \tag{15.30}$$

This equation is the *Curie–Weiss law* for the paramagnetic susceptibility of ferromagnetic materials. It differs from the previously derived Curie law only in the replacement of T by $T - T_c$. The Curie–Weiss law is also in good agreement with experiment.

It is interesting that such a successful theory as the internal field theory shows its insufficiency only when the experimental facts are used to estimate α. We get from (15.28)

$$\alpha = \frac{k T_c V}{N \mu^2} \approx \frac{10^{-16} \times 10^3 \times 1}{10^{23} \times 10^{-40}} = 10^4 \tag{15.31}$$

This is in contradiction to electrodynamics, which predicts α's in (15.25) at most equal to 4π. Or, in other words, the observed Curie temperatures, which lie in the neighborhood of $1000°$, are much too high if it is assumed that the atomic magnetic dipoles couple through their magnetic interaction.‡

15-4. Ferromagnetism; Ising model

The discovery of the correct nature of the ferromagnetic coupling force came as a by-product of quantum mechanics. Dirac§ showed that the electronic spin and the Pauli exclusion principle combine in such a way as to produce between the spins of two neighboring electrons a coupling of the form

$$\phi = J \, \boldsymbol{\sigma}_1 \cdot \boldsymbol{\sigma}_2 \tag{15.32}$$

‡ It is interesting that in the related phenomenon of ferroelectricity the conclusion goes the other way. Electric moments are about two orders of magnitude larger (in Gaussian units), and ferroelectric Curie points generally lower than their magnetic analogues. The field constant α thus comes out to be of reasonable order; the internal field hypothesis is therefore right in this case, at least in its basic idea. Of course, the coupling of electric dipoles through their electrostatic interaction has also a local aspect, which is rather more complicated than the ferromagnetic coupling discussed here.

§ P. A. M. Dirac. *Proc. Roy. Soc. A* **123**, 714 (1929). See also any textbook on quantum mechanics.

Here σ_1 and σ_2 are the two spins and J is a function of distance called the *exchange energy* or *exchange coupling*. It can be of either sign. If the wave functions of the two electrons interpenetrate substantially, the exchange energy is of the same order as the electrostatic interaction, but at larger distances it falls off exponentially as the wave functions themselves. We may therefore think of J as being of electrostatic order, but only acting between close neighbors. Heisenberg‖ was the first to realize that this exchange energy J, if of negative sign, offers a natural explanation for the phenomenon of ferromagnetism. For a local coupling of spins of the form (15.32), strung throughout a crystalline material, will tend to align spins to form a large resultant; in atoms, this effect is well known as *Hund's rule* and arises from just this cause.

We have today only partial answers concerning a set of spins coupled by exchange coupling. For some purposes approximate answers may be obtained by the use of approximate methods. There is one question of principle, however, which needs a straightforward answer, and that is whether the spontaneous magnetization and the Curie point phenomenon, as exhibited in Fig. 15.8, are in fact consequences of the interaction (15.32). A transition of the Curie point type, in which a system passes through a continuity of intermediate states, yet exhibits a discontinuity in some temperature derivatives at a temperature T_c, is called a *second-order phase transition*. The question whether a coupling between neighbors, strung over large distances, can entail such second-order phase changes is a question of principle which statistics must answer by yes or no. This is the reason for the intense interest in this type of system on the part of workers in statistical mechanics.

At the present time, the question raised in the preceding paragraph cannot be answered mathematically for the full exchange interaction (15.32). There is, however, a simplified form of the exchange interaction, the so-called *Ising model*, for which far-reaching answers have been obtained. We shall therefore turn to this model for further discussion.

The Ising model is essentially a truncated exchange hamiltonian. It replaces $\sigma_{1x}\sigma_{2x} + \sigma_{1y}\sigma_{2y} + \sigma_{1z}\sigma_{2z}$ by $\sigma_{1z}\sigma_{2z}$ only; this is qualitatively plausible because, if the quantization is along the z direction, only this part is diagonal, and the expectation values of the other operators are zero. In addition, in its simpler aspects, the Ising problem needs no quantum mechanics for its discussion, because all variables of the truncated hamiltonian commute. The quantum variable σ_z thus becomes a classical scalar σ capable of two values, $+1$ and -1. A network of such scalar spins is

‖ W. Heisenberg. *Z. Physik* **49**, 619 (1928).

thus assumed to stretch through the crystal, forming a coupled system. The magnetic enthalpy of the model is thus taken as

$$E = -\tfrac{1}{2} \sum_{i,k} J_{ik}\, \sigma_i\, \sigma_k - \mu\, H \sum_i \sigma_i \qquad (15.33)$$

where each summation runs over all spins. The magnitude of the interaction constants J_{ik} is almost always chosen so as to leave only adjacent pairs in the first sum. In fact, in some forms of the model there is just one coupling constant linking all neighbors. Expression (15.33) then takes the form

$$E = -J \sum_{\langle i,k \rangle} \sigma_i\, \sigma_k - \mu\, H \sum_i \sigma_i \qquad (15.34)$$

Here the summation $\sum\limits_{\langle i,k \rangle}$ runs over all pairs of direct neighbors. The suppression of the x and y components of spin in the Ising model produces consequences which one must keep in mind to interpret the results of the model reasonably. In the first place, it deprives magnetism of its angular momentum aspect and thus falsifies its dynamics. It is, furthermore, incorrect at low temperature, as we shall see later. It appears, however, that there is no essential defect in the model at and above the Curie point, where the statistical count of states assumes preponderant importance; this count is right in the model. We may therefore presume that the cooperative aspect of an exchange coupled network is similar to the one resulting from the truncated expressions (15.33) and (15.34).

The first thing to do in a discussion of the Ising model is to simplify it still further so as to yield an approximate derivation of the Weiss theory. In this approximate derivation we treat the plus spins and minus spins as two chemical species. Let N_+ be the number of plus spins and N_- the number of minus spins, and define their concentrations by the customary relations

$$c_+ = \frac{N_+}{N_+ + N_-} = \frac{N_+}{N} \qquad (15.35a)$$

$$c_- = \frac{N_-}{N} = 1 - c_+ \qquad (15.35b)$$

If we consider these concentrations fixed but the distribution of the spins random, then the entropy of the arrangement is given by equation (5.66) for the entropy of mixture:

$$S = -N k\, (c_+ \ln c_+ + c_- \ln c_-) \qquad (15.36)$$

To estimate the energy, we limit ourselves to the case for which there is only one nearest neighbor interaction J. The assumption of a random arrangement of spins implies then that the relative abundances of $+\,+$, $-\,-$, and $+\,-$ neighbor pairs have their random values, that is,

are as c_+^2, c_-^2 and $2c_+ c_-$, respectively. If we insert this into the energy expression (15.34) we get

$$E = -\tfrac{1}{2} z N J (c_+^2 + c_-^2 - 2\, c_+\, c_-) - N \mu H (c_+ - c_-) \quad (15.37)$$

Here z is the number of nearest neighbors of any given spin (two for a linear array, four for a two-dimensional square net, six for a simple cubic lattice, eight for a body-centered cubic lattice, etc.). To simplify further, let us introduce the total magnetic moment M as a variable through the substitutions

$$c_+ = \tfrac{1}{2} \left(1 + \frac{M}{N \mu} \right) \quad (15.38a)$$

$$c_- = \tfrac{1}{2} \left(1 - \frac{M}{N \mu} \right) \quad (15.38b)$$

We then get for E

$$E = -\frac{1}{2} \frac{z J M^2}{N \mu^2} - M H \quad (15.39)$$

and for S, from (15.36),

$$S = N k \left\{ \ln 2 - \tfrac{1}{2} \left(1 - \frac{M}{N \mu} \right) \ln \left(1 - \frac{M}{N \mu} \right) - \tfrac{1}{2} \left(1 + \frac{M}{N \mu} \right) \ln \left(1 + \frac{M}{N \mu} \right) \right\}$$

$$(15.40)$$

Now we introduce the free energy Φ at constant field

$$\Phi = U - T S - H M = E - T S \quad (15.41)$$

It is the free energy entering on the right-hand side of (8.38). By (8.39), it has the property of being a minimum for fixed temperature and fixed magnetic field. If, therefore, we succeed in approximating this quantity for a given physical situation, and if we find the approximation contains one or more undetermined parameters describing the internal state of the system, then we must dispose of them in accordance with the Second Law, and make Φ a minimum with respect to these parameters. In the present instance, we have one such parameter M; we must dispose of it by demanding that

$$\left(\frac{\partial \Phi}{\partial M} \right)_{H,T} = 0 \quad (15.42)$$

This yields, after some rearrangement of terms,

$$\frac{M}{N \mu} = \tanh \frac{\mu}{k T} \left(H + \frac{z J M}{N \mu^2} \right) \quad (15.43)$$

We recognize in (15.43) equation (15.26) previously derived, with a physical interpretation for the "internal field constant" α:

$$\alpha = \frac{z\,J}{N\,\mu^2/V} \tag{15.44}$$

Since N/V is the number of spins per unit volume, it is of the order $1/r^3$, where r is the distance between neighboring spins; the denominator of (15.44) is thus of the order of the magnetic interaction energy μ^2/r^3 of two neighboring spins, which is roughly 10^{-16} erg. The numerator, on the other hand, is the exchange coupling, which is more like an electrostatic quantity; this means that it may rise toward 10^{-12} erg. The large magnitude of α is thus explained, and the "internal field" idea reduced to a more rational set of assumptions. With the reduction of α to fundamental constants, quantities derived from it are also so reduced. We shall note here only the new expression for the Curie temperature T_c which replaces (15.28), namely,

$$\frac{z\,J}{k\,T_c} = 1 \tag{15.45}$$

According to the Weiss internal field theory, the Curie point transition is characterized, not only by the magnetic effects just discussed, but also by a characteristic thermal behavior. This thermal behavior is worth discussing here because it is indicative of cooperative phenomena in general. Viewed from this point of view, paramagnetism and ferromagnetism deal with the same ordering process, first without, then with, cooperative action. The modification of the thermal behavior as a consequence of this action is then of basic interest. Part of this modification can be predicted from equation (7.44), which is valid for structural transformation in general. The coupling makes the enthalpy change larger, while the entropy change stays the same. The transformation is thereby shifted into a higher temperature range. However, this is not the most striking change. To see the modification we eliminate M between (15.39) and (15.43), setting $H = 0$. The result is

$$\frac{k\,T}{J} = z \frac{\sqrt{-\dfrac{2E}{zNJ}}}{\operatorname{arctanh}\sqrt{-\dfrac{2E}{zNJ}}} \tag{15.46}$$

Differentiation of T with respect to E yields the heat capacity curve shown in Fig. 15.9. It is seen that the curve has a discontinuity at the Curie point: the specific heat rises to a maximum and thereupon goes discontinuously to zero. Yet, at the same time, the change is not as abrupt as for normal phase changes, where two forms of different material properties coexist at one temperature. In other words, we have reproduced the features of a second-order phase change.

Having verified that a certain type of approximate treatment of the

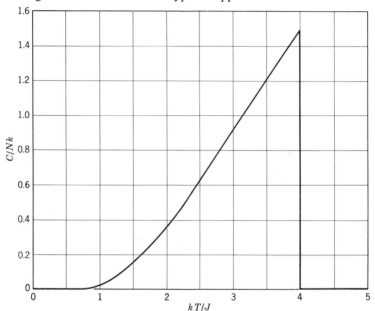

Fig. 15.9. Heat capacity at constant field versus temperature in the internal field theory of ferromagnetism; comparison of this curve with Fig. 15.4 shows the effect of a cooperative coupling.

Ising model will yield the Weiss internal field theory and the second-order phase change associated with it, we must go on to verify whether these results are a consequence of the model or of the approximations used in evaluating its properties. For many years this distinction was not properly appreciated. It is true that we deal with approximations in either case: in one case we deal with a model which is only an approximate image of reality; in the other we derive a result which is only an approximate consequence of the theory. However, there is a world of difference between the two cases. In the approximate evaluation of a theory we simply make mistakes; that is, we depart from logic. We hope the departure is small. But when we get striking results, such as those shown in Figs. 15.8 and 15.9, then it is possible that they are a consequence of our mistakes;

the results are then worthless and prove nothing. If, on the other hand, we adopt an internally consistent model which admittedly does not correspond exactly to anything in nature, and if we evaluate conscientiously the properties of that model, striking results of the type mentioned are of value. We know then that there is an interacting system which exhibits second-order phase changes according to the known laws of physics, and that perhaps experimentally observed systems are not too far removed from our model system.

It is this kind of reason which has led physicists to investigate seriously the Ising model. Second-order phase changes are tricky: by equation (4.19), erratic changes in the specific heat imply erratic changes in the fluctuations of the system. Even today the nature of these fluctuations near the critical temperature is poorly understood. But the Ising model has at least shown that second-order phase changes can arise as a consequence of the known laws of physics. This is the reason why this model is worth studying in detail.

Careful investigation of the Ising model means construction of its partition function. Since we want to investigate its properties in the presence of a magnetic field, the partition function will be of the type $F^\star$ discussed in Section 8-2. Anticipating this later need, we wrote down the necessary relations at that time. The "energy" in the Boltzmann exponent is the quantity (8.37), which we usually call magnetic enthalpy for clarity. Fortunately it is also the energy discussed in (15.33) and (15.34); it differs from U in that it also contains the potential energy in the applied field H. For this type of energy the partition function $F^\star$ equals

$$F^\star = \sum_{\sigma_i = \pm 1} \exp\left[\beta J \sum_{\langle i,k \rangle} \sigma_i \sigma_k + \beta \mu H \sum_i \sigma_i \right] \tag{15.47}$$

Here $\sum_i$ goes over all spins, $\sum_{\langle i,k \rangle}$ over all pairs of direct neighbors, and $\sum_{\sigma_i = \pm 1}$ over the 2^N combinations ± 1 of the N spins. We verify that we get from $F^\star$ by differentiation

$$E = -\left(\frac{\partial \ln F^\star}{\partial \beta} \right)_H \tag{15.48}$$

and

$$M = \frac{1}{\beta} \left(\frac{\partial \ln F^\star}{\partial H} \right)_\beta \tag{15.49}$$

These relations were already written down in (8.39). It was shown in (8.38) that the partition function $F^\star$ is associated with the "magnetic free energy" $A - \mathbf{H} \cdot \mathbf{M}$.

The study of $F^\star$ for cases of physical interest is an extremely hard problem. Its difficulty resides in the fact that, if one wishes to evaluate it for the purpose of investigating phase transitions, one cannot assume the number of cooperating units to be finite. For in this latter case (15.47) consists of a finite sum of terms without temperature singularity; this makes the sum likewise nonsingular. All interesting results are therefore obtained by examination of the Nth root of $F^\star$ in the limit that N goes to infinity. The difficulties of a limiting process are therefore superimposed upon the combinatorial difficulty of evaluating (15.47).

Some further thought along the same lines shows that even an infinite Ising model cannot show ferromagnetism as long as it is infinite in one dimension only. For ferromagnetism means a spontaneous magnetization, which in turn means long-range order among the spins without an applied

Fig. 15.10. Destruction of order in a linear array of spins; one single reverse coupling is sufficient to destroy long-range order.

field. Suppose that we have such an ordered arrangement of N Ising spins with all spins parallel. Then we can upset it by introducing a single break in it, that is, introducing the energy $2J$ at one point as shown in Fig. 15.10. Since this break can be made at any of N positions (where $N + 1$ is the number of members of the linear chain), the entropy gain is $k \ln N$. The free energy change is therefore

$$\Delta\Phi = 2J - kT \ln N$$

Since we have to consider $\Delta\Phi$ in the limit of infinite N, there is no temperature, however small, sufficient to prevent destruction of long-range order, because Φ is always lower for the disordered state.

The same argument was developed by Peierls[¶] to prove the contrary for a square net of Ising spins: such a net is ferromagnetic. For this purpose an estimate must be made of the number of ways in which a border of L segments separating $+$ and $-$ spins can be laid in a square net. Such a border is shown in Fig. 15.11. The number of ways in which such a border can be laid determines the entropy, and the length of the border determines the energy. The result of the investigation is that the entropy, as well as the energy, varies as the first power of the length of such a border. A temperature can therefore be found which is sufficiently low so that the

¶ R. Peierls. *Proc. Cambridge Phil. Soc.* **32**, 477 (1936).

energy overpowers the entropy in the expression $E - TS$. The stable state is then one not containing such a border if the temperature is sufficiently low.

The chapter following this one gives the derivation of the system partition function for the rectangular Ising net without magnetic field, that is, a two-dimensional array of spins having the value ± 1 and having an interaction $- J_1 \sigma \sigma'$ between neighbors in the rows and an interaction $- J_2 \sigma \sigma'$ between neighbors in the columns. There are several derivations known, but they all make use of advanced algebraic methods which have

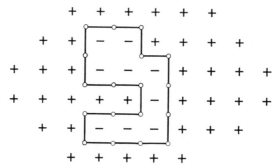

Fig. 15.11. Introduction of reverse orientation into the square Ising net: it is now necessary to maintain a border of high energy.

no easily discerned relationship to physical reasoning. We shall therefore separate the algebra from the physical discussion and take over the results (16.62), (16.63), and (16.67) from the next chapter. To avoid extraneous complications we shall restrict the detailed discussion to the case $J_1 = J_2$, with a few observations concerning the more general case.

As worked out in Chapter 16, the square net of Ising spins has a partition function F which equals

$$F = f^N \tag{15.50}$$

where N is the total number of spins, and f is given by

$$\ln f = \ln (2 \cosh 2\beta J) + \frac{1}{\pi} \int_0^{\frac{1}{2}\pi} \ln \left\{ \frac{1 + (1 - \kappa^2 \sin^2 \omega)^{1/2}}{2} \right\} d\omega \tag{15.51}$$

with

$$\kappa = \frac{2 \sinh 2 \beta J}{\cosh^2 2 \beta J} \tag{15.52}$$

The parameter κ is zero at small temperature, rising to a maximum of unity at a temperature at which

$$\sinh \frac{2J}{kT_c} = 1 \qquad (15.53a)$$

or

$$\frac{J}{kT_c} = 0.4407 \qquad (15.53b)$$

Then κ returns to zero at high temperature. This means that the integral in (15.51) vanishes for both very low and very high temperatures. The rough outline of the behavior of f is therefore contained in the first term; however the second term has an effect at intermediate temperatures.

The best way to investigate the result more closely is to compute the energy E by (15.48). If we maintain the distinction between the two terms in (15.51) we get E in the form

$$E = -2NJ \tanh 2\beta J - NJ \frac{\sinh^2 2\beta J - 1}{\sinh 2\beta J \cosh 2\beta J} \left\{ \frac{2}{\pi} K(\kappa) - 1 \right\} \quad (15.54)$$

where $K(\kappa)$ is the complete elliptic integral of the first kind. At first sight the statement that the first term gives roughly the behavior of the system seems even more appropriate for E than it was for f. For now the second term vanishes three times: at high and at low temperature (because of the vanishing of the curly bracket), and at the critical point defined by (15.53) [because of the vanishing factor $(\sinh^2 2\beta J - 1)$]. This rough result is not unexpected. The first term in (15.54) is the result (16.14) for a one-dimensional Ising chain, with an obvious doubling of the exchange interaction J for the two dimensions. However, the influence of the second term is now more apparent. $K(\kappa)$, the complete elliptic integral of the first kind, has a logarithmic infinity at the critical point for which $\kappa = 1$. Consequently the entire second term behaves as $(T-T_c) \ln |T-T_c|$ in the immediate neighborhood of T_c. The slope of the energy curve is therefore infinite at $T = T_c$, and the heat capacity is also infinite. Away from the critical point the second term has a typically "cooperative" influence. When the first term is strongly negative the correction is also negative; on the other hand, when the first term approaches zero the correction is positive and brings the result still closer to zero. Figure 15.12 shows the two energy versus temperature curves in heavy and dotted outline.

This is perhaps the right moment to refer back to the internal field approximation, which gave for the Curie point the result (15.45). In the language of equation (15.53) this means for the square net

$$\frac{J}{k\,T_c} = 0.25$$

This is not in very good agreement with (15.53). Even more disappointing is the energy versus temperature curve as a whole, which is shown for comparison in Fig. 15.12 as the curve in light outline. In addition to

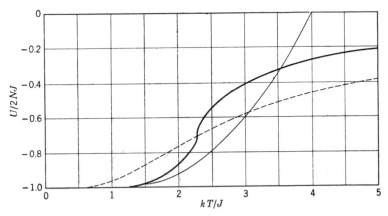

Fig. 15.12. Energy versus temperature for the Ising square net. The exact curve is shown in heavy outline, the "internal field" curve in light outline, and the equivalent one-dimensional curve dashed.

misplacing the Curie point the internal field theory misinterprets its significance. It yields zero energy, that is, complete randomness, at the Curie temperature, while in reality 70 % of the total coupling is still present. This is due to short-range alignment of neighboring spins. This alignment predominantly determines the energy, even if the sample as a whole has no resultant spin. The same discrepancy is shown in Fig. 15.13 in a comparison of the two heat capacity curves.

The behavior of the two-dimensional Ising model as a function of the magnetic field is not known at this time, the mathematics having been found intractable. We have therefore no good evidence about the validity of the Curie–Weiss law (15.30) in a theoretically consistent model. Yang** did succeed in computing the spontaneous magnetization. It is zero above the Curie point and below it has the value

$$M = N\,\mu \left[\frac{\cosh^2 2\beta J}{\sinh^4 2\beta J}\,(\sinh^2 2\beta J - 1)\right]^{\frac{1}{8}} \tag{15.55}$$

** C. M. Yang. *Phys. Rev.* **85**, 809 (1952).

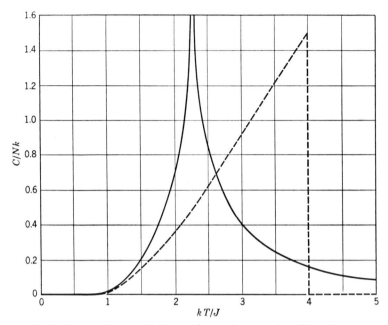

Fig. 15.13. Exact and internal field plots for the heat capacity of the Ising square net versus temperature.

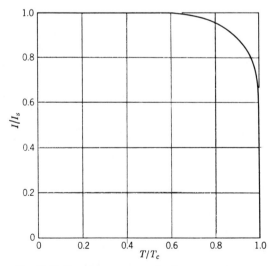

Fig. 15.14. Spontaneous magnetization versus temperature for the Ising square net.

So the magnetization does indeed vanish at the Curie point with vertical slope, as expected; but in detail, discrepancies with the internal field model are again strong. As seen in Fig. 15.14, the $\frac{1}{8}$ power in (15.55) leads to an extremely sharp drop of the magnetic moment very close to the Curie point. This is a result which does not agree with either the internal field theory or most experimental results, as illustrated in Fig. 15.8. We shall have occasion to discuss this deficiency when dealing with the theory of spin waves.

By presenting only the internal field model and exact results for the Ising model we have of course not given an adequate picture of the field. Very sophisticated approximations can be devised which make the numerical disagreement between the approximation and the exact theory very small. The three-dimensional Ising model has not been solved analytically, and there is little hope that it ever will be. We must therefore draw the best possible conclusion from the available evidence for three dimensions. The evidence is that the specific heat is also infinite at the Curie point, perhaps with a law which is not logarithmic. Also the power law in the drop-off of the spontaneous magnetization at the Curie point seems to be more than 1/8, which is encouraging. At present, hope for further progress is almost entirely based on the powerful numerical computing techniques which are now becoming available to us.

*15-5. Spin wave theory of magnetization

If we consider how difficult the problem is of making reliable statistical computations at or near the Curie point, we would be pretty foolhardy to use a formula such as (15.53) for an empirical determination of the exchange coupling in an actual material. The situation is in this respect very similar to the case for gases: the critical point is too severe a disturbance of the perfect gas to teach us much about intermolcular forces. We must look for a weaker disturbance. In Chapter 12 we turned to the Joule–Thomson effect or the second virial coefficient. In ferromagnets we find this kind of effect in the low-temperature behavior of the spontaneous magnetization.

In an almost saturated sample, the only excitation recognized by the Ising model is the reversal of an individual spin with the breaking of the corresponding bonds. In modern parlance we would call this a particle excitation. These excitations are expensive energy-wise and lead to a decrease of the spontaneous magnetization with increasing temperature as $1 - \exp[-\alpha/T]$. Bloch[††] first showed that a localized system of spins

†† F. Bloch. *Z. Physik* **61**, 206 (1930).

coupled by the Heisenberg interaction (15.32) has very much cheaper collective modes of excitation available. These modes are called *spin waves*. They lead to a more rapid departure from saturation for the spontaneous magnetization as the temperature is raised.

The interaction energy to be studied is the untruncated form of (15.34), namely,

$$E = -\frac{1}{s^2} J \sum_{\langle i,k \rangle} \mathbf{s}_i \cdot \mathbf{s}_j - g\,\mu_\beta H \sum_j s_j^z \qquad (15.56)$$

Here $\mathbf{s}_j$ is, in units of $\hbar$, the angular momentum operator of the moment at the location j, and s the maximum possible value of s_j^z. This magnitude will be assumed the same at all locations. The coefficient of the second term is directly taken over from (15.12). The factor $1/s^2$ in the first term needs some explaining. It is meant to make the interaction between two neighbors $-J$ when the spins are parallel. It implies that the Ising value is really the maximum value for the first sum, and not some value perhaps three times as great. That this is so can be seen from the following little Landé type argument. Define the resultant of two neighboring spins as j, that is,

$$\mathbf{j} = \mathbf{s}_1 + \mathbf{s}_2$$

Square the relation and isolate $\mathbf{s}_1 \cdot \mathbf{s}_2$. You get

$$\mathbf{s}_1 \cdot \mathbf{s}_2 = \tfrac{1}{2} j(j+1) - s(s+1)$$

The maximum value of j is $2s$. Hence

$$(\mathbf{s}_1 \cdot \mathbf{s}_2)_{\max} = s^2 = (s^z)_{\max}^2$$

while

$$(\mathbf{s}_1 \cdot \mathbf{s}_1)_{\max} = s(s+1)$$

This is the result implied in (15.56). For $s = \tfrac{1}{2}$ the value on the first line is indeed only one third of that on the second line. Generally, for any value of the quantum number s, the maximum possible value of the interaction is that given by the Ising model, as stated.

The commutation relations of the operators s_j are as usual

$$\mathbf{s}_j \times \mathbf{s}_j = -i\,\mathbf{s}_j \qquad (15.57)$$

One may verify that relation (15.57) holds in the same form for any vector sum of operators from different sites.

The first thing to verify for the interaction (15.56) is that if we define the total angular momentum $\mathbf{S}$ by

$$\mathbf{S} = \sum_j \mathbf{s}_j \qquad (15.58)$$

then both $\mathbf{S}^2$ and S^z commute with E. For the field term in E the proposition is trivial, as our statement is just the normal basis of all angular momentum theory. For the exchange term the easiest way to proceed is to show that all three components of $\mathbf{S}$ commute with it; this may be even further extended and simplified by saying that $\mathbf{S}$ commutes with every single term in the exchange hamiltonian. The relevant calculation is thereby reduced to

$$
\begin{aligned}
S^y\,(\mathbf{s}_l \cdot \mathbf{s}_m) - (\mathbf{s}_l \cdot \mathbf{s}_m)\,S^y &= (s_l^y + s_m^y)(\mathbf{s}_l \cdot \mathbf{s}_m) - (\mathbf{s}_l \cdot \mathbf{s}_m)(s_l^y + s_m^y) \\
&= (s_l^y s_l^x - s_l^x s_l^y)\,s_m^x + (s_l^y s_l^z - s_l^z s_l^y)\,s_m^z \\
&\quad + s_l^x\,(s_m^y s_m^x - s_m^x s_m^y) + s_l^z\,(s_m^y s_m^z - s_m^z s_m^y) \\
&= +\,i\,s_l^z s_m^x - i\,s_l^x s_m^z + i\,s_l^x s_m^z - i\,s_l^z s_m^x = 0
\end{aligned}
$$

Thus the quantities

$$
\mathbf{S}^2 = S\,(S+1)
$$

and

$$
S^z = M_s
$$

are good quantum numbers for the interaction (15.56). In the state of saturation we have

$$
S_{\max} = (M_s)_{\max} = N\,s \tag{15.59}
$$

Our next job is to expand the energy (15.56) about the ground state (15.59). For this purpose we introduce raising and lowering operators for the individual spins, as follows:

$$
s_j^+ = \frac{1}{\sqrt{2}}\,(s_j^x - i\,s_j^y) \tag{15.60a}
$$

$$
s_j^- = \frac{1}{\sqrt{2}}\,(s_j^x + i\,s_j^y) \tag{15.60b}
$$

Multiplication of the two operators yields

$$
s^-\,s^+ = \tfrac{1}{2}\,\{(s^x)^2 + (s^y)^2 + i\,(s^y\,s^x - s^x\,s^y)\}
$$

With (15.57) this reduces to

$$
s_j^-\,s_j^+ = \tfrac{1}{2}\,(s - s_j^z)\,(s + s_j^z + 1) \tag{15.61}
$$

We want to solve this equation for s^z near saturation and are thus tempted to write

$$
s \approx s_j^z \gg 1
$$

which brings (15.61) into the form

$$
s_j^-\,s_j^+ = s\,(s - s_j^z) \tag{15.62}
$$

The argument for (15.62) sounds like a poor excuse. However, there are further arguments for it:

(a) For $s = \frac{1}{2}$ the formula is exact.

(b) For any s the formula is equivalent to (15.61) for $s_j^z = s$ and $s_j^z = s - 1$. Thus, we must only assume that multiple spin excitation at one lattice site is improbable if we wish to substitute (15.62) for (15.61). This sets a natural limit to the validity of spin wave theory, unless $s = \frac{1}{2}$. Accepting (15.62) in this sense, we solve it for s_j^z and get

$$s_j^z = s - \frac{1}{s} s_j^- s_j^+ \tag{15.63}$$

For s_j^x and s_j^y we find obvious linear expressions from (15.60). A short calculation then yields for the exchange dot product

$$\mathbf{s}_l \cdot \mathbf{s}_j = s^2 - (s_l^- - s_j^-)(s_l^+ - s_j^+) + \frac{1}{s^2} s_l^- s_l^+ s_j^- s_j^+$$

As in most other solid state wave approximations we drop the quartic term and write

$$\mathbf{s}_l \cdot \mathbf{s}_j = s^2 - (s_l^- - s_j^-)(s_l^+ - s_j^+) \tag{15.64}$$

Substitution of (15.63) and (15.64) into (15.56) yields

$$E = -\tfrac{1}{2} N z J - g N s \mu_\beta H + \frac{1}{s} g \mu_\beta H \sum_j s_j^- s_j^+$$

$$+ \frac{1}{s^2} J \sum_{\langle l,j \rangle} (s_l^- - s_j^-)(s_l^+ - s_j^+) \tag{15.65}$$

To the energy (15.65) we apply the Born running wave transformation (13.41). To simplify notation we employ the concepts of lattice vector and wave vector as defined in (13.36) and (13.47). We may write then

$$\alpha(\mathbf{k}) = \frac{1}{\sqrt{N}} \sum_{\mathbf{r}} e^{i\mathbf{k}\cdot\mathbf{r}} s^-(\mathbf{r}) \tag{15.66a}$$

$$\beta(\mathbf{k}) = \frac{1}{\sqrt{N}} \sum_{\mathbf{r}} e^{-i\mathbf{k}\cdot\mathbf{r}} s^+(\mathbf{r}) \tag{15.66b}$$

The relations may be inverted to read

$$s^-(\mathbf{r}) = \frac{1}{\sqrt{N}} \sum_{\mathbf{k}} e^{-i\mathbf{k}\cdot\mathbf{r}} \alpha(\mathbf{k}) \tag{15.67a}$$

$$s^+(\mathbf{r}) = \frac{1}{\sqrt{N}} \sum_{\mathbf{k}} e^{+i\mathbf{k}\cdot\mathbf{r}} \beta(\mathbf{k}) \tag{15.67b}$$

We need the commutation relations of the new operators. We get from (15.66)

$$\alpha(\mathbf{k})\,\alpha(\mathbf{k}') - \alpha(\mathbf{k}')\,\alpha(\mathbf{k}) = 0 \qquad (15.68a)$$

$$\beta(\mathbf{k})\,\beta(\mathbf{k}') - \beta(\mathbf{k}')\,\beta(\mathbf{k}) = 0 \qquad (15.68b)$$

For the mixed commutator we first complete the calculation which led to (15.61) and find

$$s^-(\mathbf{r})\,s^+(\mathbf{r}) - s^+(\mathbf{r})\,s^-(\mathbf{r}) = -\,s^z(\mathbf{r})$$

This, in turn, leads to

$$\alpha(\mathbf{k})\,\beta(\mathbf{k}') - \beta(\mathbf{k}')\,\alpha(\mathbf{k}) = -\frac{1}{N}\sum_{\mathbf{r}} s^z(\mathbf{r})\exp i(\mathbf{k}-\mathbf{k}')\cdot\mathbf{r} \qquad (15.69)$$

The result (15.69) is rather cumbersome and must be simplified by a physical argument. The formula has a front factor $1/N$. To the extent that the s^z's are all alike, this factor cancels with the sum if $\mathbf{k}$ and $\mathbf{k}'$ are equal. This is the case for the leading term in formula (15.63) for s^z. Under all other conditions the sum yields a result of the order 1; the commutator is then of the order $1/N$ and thus negligible. One may therefore write to a good approximation

$$\alpha(\mathbf{k})\,\beta(\mathbf{k}') - \beta(\mathbf{k}')\,\alpha(\mathbf{k}) = -\,s\,\delta_{\mathbf{k},\mathbf{k}'} \qquad (15.70)$$

Together, (15.68) and (15.70) form the commutation rules of a set of bosons. One gets the eigenvalues by the usual argument. Let $\alpha\,\beta$ have the eigenvalues t, then $\beta\,\alpha$ has the eigenvalues $t + s$. Now multiply (15.70), for $\mathbf{k} = \mathbf{k}'$ with α; you get

$$\alpha\,(\alpha\,\beta) = (\alpha\,\beta)\,\alpha - s\,\alpha$$

Take matrix elements between two eigenstates t and t' of $\alpha\,\beta$. It yields

$$\langle t|\alpha|t'\rangle\,\{t - t' - s\} = 0$$

Thus $\langle t|\alpha|t'\rangle$ is different from zero only if $t' = t - s$. Taking for granted that there is a least possible value $t_{\min}$, there results for this value $\langle t_{\min}|\alpha|t'\rangle = 0$ for all t' and

$$\langle t_{\min}|\alpha\,\beta|t_{\min}\rangle = 0$$

and hence generally

$$t = n\,s, \quad n = 0, 1, 2, \ldots$$

$$\alpha(\mathbf{k})\,\beta(\mathbf{k}) = n(\mathbf{k})\,s \qquad (15.71a)$$

with

$$n(\mathbf{k}) = 0, 1, 2, 3, \ldots \qquad (15.71b)$$

The next step in the eigenvalue analysis is to express the energy (15.65) in terms of $\alpha(\mathbf{k}) \, \beta(\mathbf{k})$. This is accomplished by equations (15.67). We find

$$\sum_{\mathbf{r}} s^-(\mathbf{r}) \, s^+(\mathbf{r}) = \sum_{\mathbf{k}} \alpha(\mathbf{k}) \, \beta(\mathbf{k}) \tag{15.72}$$

and

$$\sum_{\langle \mathbf{r}, \mathbf{r}' \rangle} \{ s^-(\mathbf{r}) - s^-(\mathbf{r}') \} \{ s^+(\mathbf{r}) - s^+(\mathbf{r}') \} = \sum_{\mathbf{k}} \alpha(\mathbf{k}) \, \beta(\mathbf{k}) \left\{ z - \sum_{\rho} \cos \mathbf{k} \cdot \boldsymbol{\rho} \right\} \tag{15.73}$$

Here $\boldsymbol{\rho}$ stands for all vectors leading from one spin to a nearest neighbor, and z is the number of nearest neighbors. Substitution of (15.71), (15.72), and (15.73) into (15.65) yields finally the energy spectrum

$$E(n(\mathbf{k}_1), n(\mathbf{k}_2), \ldots) = -\tfrac{1}{2} \, N z J - g \, N s \, \mu_B \, h$$

$$+ \, g \, \mu_B \, H \sum_{\mathbf{k}} n(\mathbf{k}) + \frac{1}{s} J \sum_{\mathbf{k}} n(\mathbf{k}) \left\{ z - \sum_{\rho} \cos \mathbf{k} \cdot \boldsymbol{\rho} \right\} \tag{15.74}$$

The first line gives the ground state energy, and the second line the possible excitations in terms of the "magnon" quantum numbers $n(\mathbf{k})$. It is seen that, while the excitation energy contributed by the field proceeds always in full steps of $g \, \mu_B \, H$, the exchange contribution becomes very small for long waves because of the approximate vanishing of the curly bracket.

Before doing statistics on (15.74) it is useful to present one more piece of interpretation. According to the discussion preceding (15.59), S^2 and S^z are exact quantum numbers for the system. The spin wave states must therefore be classifiable according to these two numbers. For the quantum number M_s this is a simple matter because consultation of (15.56) shows it to be simply the multiplier of $- g \, \mu_B \, H$ in (15.74). Hence we have

$$M_s = N s - \sum_{\mathbf{k}} n(\mathbf{k}) \tag{15.75}$$

The computation of the total spin requires a little more work. By (15.60) and (15.66) we have

$$S^x = \sum_j s_j^x = \frac{1}{\sqrt{2}} \sum_j (s_j^- + s_j^+) = (\tfrac{1}{2} N)^{1/2} [\alpha(0) + \beta(0)]$$

$$S^y = \sum_j s_j^y = \frac{1}{i\sqrt{2}} \sum_j (s_j^- - s_j^+) = \frac{1}{i} (\tfrac{1}{2} N)^{1/2} [\alpha(0) - \beta(0)]$$

and therefore

$$(S^x)^2 + (S^y)^2 = N \, [\alpha(0) \, \beta(0) + \beta(0) \, \alpha(0)]$$

By (15.70) the second product equals

$$\beta \alpha = \alpha \beta + s$$

which yields with (15.71)

$$(S^x)^2 + (S^y)^2 = N s \, [2 \, n(0) + 1]$$

This combines with (15.75) into

$$\mathbf{S}^2 = N s \, [2 \, n(0) + 1] + \left[N s - \sum_{\mathbf{k}} n(\mathbf{k}) \right]^2 \qquad (15.76)$$

To a very high order in $1/Ns$, equation (15.76) is equivalent to

$$\mathbf{S}^2 = \left[N s - \sum_{\mathbf{k} \neq \mathbf{0}} n(\mathbf{k}) \right] \left[N s - \sum_{\mathbf{k} \neq \mathbf{0}} n(\mathbf{k}) + 1 \right]$$

which means that

$$\mathbf{S}^2 = S \, (S+1) \qquad (15.77a)$$

with

$$S = N s - \sum_{\mathbf{k} \neq \mathbf{0}} n(\mathbf{k}) \qquad (15.77b)$$

Thus we have the result that all magnons reduce M_s. The magnons of finite wave number also reduce S by the same amount, but the magnon of zero wave number leaves S the same and produces precession of the $\mathbf{S}$ vector about the field axis. This is in line with the feature of (15.74) that a zero wave number magnon does not affect the total exchange energy. In spite of the approximate derivation (15.77) must be an exact result, for both sides are exactly defined integers or half integers and thus cannot differ from each other by a small amount.

After this rather extensive mechanical preparation, we are prepared to do statistics on (15.74) with the particular idea of accounting for the magnetization versus temperature curve of ferromagnets. Denote by $K(\mathbf{k})$ the multiplier of $n(\mathbf{k})$ in (15.74), that is, set

$$K(\mathbf{k}) = \frac{1}{s} J \left\{ z - \sum_{\rho} \cos \mathbf{k} \cdot \mathbf{\rho} \right\} + g \, \mu_B \, H \qquad (15.78)$$

The system partition function

$$F^\star = \sum_{n(\mathbf{k}_1)=0}^{\infty} \sum_{n(\mathbf{k}_2)=0}^{\infty} \cdots \sum_{n(\mathbf{k}_N)=0}^{\infty} \exp[- \, \beta \, E(n(\mathbf{k}_1), n(\mathbf{k}_2) \ldots)] \qquad (15.79)$$

then factors as is usual for decoupled modes. The result is

$$F^\star = \exp[\tfrac{1}{2} \, \beta \, N z J + \beta \, g \, N s \, \mu_B \, H] \prod_{\mathbf{k}} \frac{1}{1 - \exp[- \, \beta \, K(\mathbf{k})]} \qquad (15.80)$$

The magnetic moment along the field is obtained from $F^{\star}$ by (15.49). We find

$$M = g\,N\,s\,\mu_B - \sum_k \frac{g\,\mu_B}{\exp\,[\beta\,K(\mathbf{k})] - 1} \qquad (15.81)$$

To make a usable equation out of this, the sum has to be transformed into an integral according to the general rule (13.51):

$$\sum_k \approx \frac{N\,\omega}{8\,\pi^3} \iiint d\mathbf{k}$$

Here the integral is over the first Brillouin zone, ω is the volume per paramagnetic atom, and N is the number of such atoms in the sample. If we are satisfied to compute the spontaneous magnetization only, we can discard the term depending on H in (15.78). For the remainder, we must realize that the main contribution to the result arises from long waves. We can therefore expand $K(\mathbf{k})$ in powers of $\mathbf{k}$ and retain only the quadratic term. If the material is also cubic, then the quadratic term cannot depend on direction and we get

$$z - \sum_\rho \cos \mathbf{k} \cdot \boldsymbol{\rho} = \tfrac{1}{2} \sum_\rho (\mathbf{k} \cdot \boldsymbol{\rho})^2 = \tfrac{1}{6}\, z\, k^2\, \rho^2$$

Equation (15.81) then becomes

$$M = g\,N\,s\,\mu_B - \frac{g\,\mu_B\,N}{8\,\pi^3} \iiint \frac{d\mathbf{k}}{\exp[\beta J z k^2 \rho^2 / 6s] - 1}$$

The integral converges automatically for large wave numbers and can be extended over all space. The integral which comes in is then

$$\int_0^\infty \frac{x^2\,dx}{\exp x^2 - 1} = \int_0^\infty x^2\,dx\,\{e^{-x^2} + e^{-2x^2} + e^{-3x^2} + \cdot\cdot\cdot\}$$

$$= \frac{\sqrt{\pi}}{4}\,\zeta(\tfrac{3}{2})$$

$$= \frac{\sqrt{\pi}}{4}\,2.612$$

With this result we get

$$M = g\,N\,s\,\mu_B \left\{ 1 - \left(\frac{3}{2\,\pi}\right)^{3/2} 2.612\,\frac{\omega}{\rho^3} \left(\frac{k\,T}{z\,J}\right)^{3/2} \right\} \qquad (15.82)$$

This is the Bloch $T^{3/2}$ law for the decrease of magnetization with increasing temperature. It is well obeyed by many salts. The law contains no other unknown number except J. Experimental magnetization curves can therefore be used to determine J reliably.

A. RECOMMENDED PROBLEMS

1. Some organic molecules have a triplet excited state not far above the singlet ground state. Calculate their paramagnetic susceptibility in terms of the triplet moment μ and the triplet–singlet splitting Δ.

2. Antiferromagnetism is a phenomenon akin to ferromagnetism. The simplest kind of antiferromagnet consists of two equivalent antiparallel sublattices A and B such that members of A have only nearest neighbors in B and vice versa. Show that the internal field theory of this type of antiferromagnetism yields a formula like the Curie–Weiss law (15.30), except that $T-T_c$ is replaced by $T+T_c$. T_c is the antiferromagnetic transition point (Néel's law).

B. GENERAL PROBLEMS

3. In manganous fluorosilicate hexahydrate the manganous complex has $j = \frac{5}{2}$ and $g = 2$. The density is 1.9 gm/cm^3, and the molecular weight 305. Find the Curie constant of the material.

4. Compute the magnetic contribution to the heat capacity of a paramagnetic material if the original expression (15.05) is used for $L(x)$ instead of $L_{\frac{1}{2}}(x)$. Compare the behavior of your expression with (15.19), particularly at low temperature. Comment!

5. For $H = 0$, $I = 0$ is always a solution of (15.26). On what grounds can the solution be rejected below the Curie point?

6. Show from the Second Law that if a material obeys the laws of Curie–Weiss or Néel the entropy decomposes into a purely thermal term plus a term depending on magnetization only.

7. Show that the observation in Problem 6 also applies to the ferromagnetic equation of state (15.26). What is the expression for the entropy?

8. Below the antiferromagnetic Curie point (Néel point) the susceptibility of an antiferromagnet drops again. Show that in the internal field theory of Problem 2 the rate of increase of the susceptibility with temperature immediately below the Néel point is double the rate of decrease immediately above. In the derivation it must be assumed that the applied field is parallel to the antiferromagnetic orientation (this is as a rule not true in experimental work).

9. A superconductor is a perfect diamagnet for which $\mathbf{B} = 0$, or $\mathbf{I} = -\mathbf{H}/4\pi$. Superconductivity can be destroyed by a field as well as by elevated temperature, so that the superconducting and the normal phases are separated by a curve in the H–T plane. Show that along that curve we have for the latent heat L of transformation

$$L = -\frac{VT}{4\pi} H \frac{dH}{dT}$$

To what conclusions does this lead about the transformation curve generally and at the λ point in particular?

10. Use the formalism of Section 15-5 to compute the energy of the ferromagnetic spin system at low temperature. Comment on the order of magnitude of the answer.

★16

Mathematical analysis of the Ising model

The most successful elaboration of technique in statistical mechanics exists in connection with the Ising model. As these techniques are rather far removed from either the basic starting assumptions or the final results, we anticipated the physical content in the preceding chapter and present here the techniques. We do this because we believe it unwise to suppress successful techniques when discussing a subject in mathematical physics, for in the last analysis a subject grows with the techniques available to handle it. In the present instance the physical content of the techniques is not well understood, particularly as they relate to second-order phase transitions. However, the physical information must be buried somewhere in the formalism and will perhaps be found sometime in the future.

16-1. Eigenvalue method for periodic nearest neighbor systems

A wide class of problems in statistical mechanics has benefited from the possibility of being transformed into the form of an eigenvalue problem. Let there be a number of units identical in structure which are lined up as beads on a string, and let them be numbered 1, 2, 3, 4, . . . N. Let the state of each unit be described by a variable x_1, x_2, x_3, . . . x_N. These variables may be discrete or continuous and may in fact stand for a set of variables. However, we do have in mind the idea of potential energy, which means we assume the variables to be classical and statistically separable from their conjugate momenta. Into this type of one-dimensional system, we now introduce the postulate of nearest neighbor interaction; it means that we assume each unit to interact with its two direct neighbors. only. This situation is symbolized by a set of crosses in Fig. 16.1 which

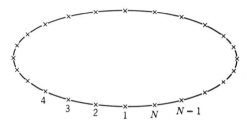

Fig. 16.1. A chain of N cooperating units.

interact according to the lines shown. Let the interaction between two neighbors be $V(x, y)$. Then the probability for a given state of the system is proportional to the Boltzmann exponential

$$\exp[-\beta\{V(x_1, x_2) + V(x_2, x_3) + \cdots + V(x_N, x_1)\}] \quad (16.01)$$

from which the partition function is formed by summation (or integration)

$$F = \sum_{x_1} \sum_{x_2} \cdots \sum_{x_N} \exp[-\beta\{V(x_1, x_2) + V(x_2, x_3) \cdots + V(x_N, x_1)\}] \quad (16.02)$$

A reasoning from probability calculus*,† leads one to associate the following eigenvalue problem with the summation written out in (16.02):

$$\sum_y \exp[-\beta\, V(y, z)]\, a(y) = \lambda\, a(z) \quad (16.03)$$

λ will have a number of different eigenvalues λ_ν. To each there belongs one eigenvector a_ν if multiple values are counted as often as they arise. The orthogonality relation for the a's is well known:

$$\sum_y a_\mu(y)\, a_\nu(y) = \delta_{\mu\nu} \quad (16.04)$$

* H. A. Kramers and G. H. Wannier. *Phys. Rev.* **60**, 252 (1941).
† Edwin Lassettre and John P. Howe. *J. Chem. Phys.* **9**, 747 and 801 (1941).

The formula developing the kernel in terms of the eigenvectors is also well known for integral equations:

$$\exp[-\beta V(y, z)] = \sum_\nu \lambda_\nu \, a_\nu(y) \, a_\nu(z) \tag{16.05}$$

It obviously must hold for matrices too because the two sides of (16.05) have identical eigenvectors and eigenvalues.

If we substitute (16.05) into (16.02) the summations over $x_1, x_2 \ldots x_N$ can be carried out explicitly with the help of (16.04). The result is

$$F = \sum_\nu \lambda_\nu^N \tag{16.06}$$

Relation (16.06) becomes particularly useful in the case when the number N of cooperating units is very large. We may then neglect all but the largest eigenvalue λ_1 and write

$$F \approx \lambda_1^N \tag{16.07}$$

This result shows that solution of the eigenvalue problem (16.03) is equivalent to carrying out the statistical summation. We shall therefore refer to the expression $\exp[-\beta V(y, z)]$ as the *resolving kernel* of the statistical problem.

Not only the largest eigenvalue λ_1, but also the associated eigenvector $a_1(x)$, has physical significance. If we stop short in our summation over x_i leading to (16.06) just before performing the summation over the last variable, then we are left with $a_1^2(x)$, which, by (16.01), is the probability distribution for the variable x. More complicated probabilities can be constructed in a similar manner. If the maximum eigenvalue λ has several eigenvectors, an indeterminacy in the associated probabilities arises which is of profound significance in the study of long-range order.‡ We shall not follow up these subsidiary topics but restrict ourselves to the study of the thermodynamic properties under the simplest possible conditions.

16-2. One-dimensional Ising model

In the case of the Ising model the variable x is the spin σ which is restricted to just the two values -1 and $+1$. The interaction $V(x, y)$ is therefore a matrix. In the case of a linear array of spins forming a closed ring, as shown in Fig. 16.1, the resolving kernel is just a 2×2 matrix. The interaction $V(x, y)$ reads in this case

$$V = -J \sigma \sigma' - \tfrac{1}{2} \mu H (\sigma + \sigma') \tag{16.08}$$

‡ J. Ashkin and W. E. Lamb. *Phys. Rev.* **64**, 159 (1943).

Here J is the exchange coupling constant, and μ the magnetic moment. If we introduce the abbreviations

$$K = \beta J \tag{16.09}$$

and

$$C = \beta \mu H \tag{16.10}$$

we find

$$\exp[-\beta V] = \begin{pmatrix} e^{K+C} & e^{-K} \\ e^{-K} & e^{K-C} \end{pmatrix} \tag{16.11}$$

and the associated eigenvalue problem (16.03) takes the form

$$\begin{pmatrix} e^{K+C} & e^{-K} \\ e^{-K} & e^{K-C} \end{pmatrix} \begin{pmatrix} \alpha \\ \beta \end{pmatrix} = \lambda \begin{pmatrix} \alpha \\ \beta \end{pmatrix} \tag{16.12}$$

The larger eigenvalue λ_1 equals

$$\lambda_1 = e^K \cos C + (e^{2K} \sinh^2 C + e^{-2K})^{1/2} \tag{16.13}$$

The results of this calculation have little intrinsic interest. The purpose of the entire effort is the study of second-order phase transitions and the related phenomenon of spontaneous magnetization. It has been shown in the preceding chapter that no one-dimensional system of the type proposed here can show such effects. We must therefore treat the problem as an exercise which is of interest for its method rather than its results. The results for the model are in line with the expectations. From (15.48), (16.07), and (16.13) we get for the energy E in the absence of a field

$$E = -NJ \tanh K \tag{16.14}$$

This is a smooth increase from $-NJ$ to 0 as the temperature rises. Further, we get the magnetic moment from (15.49) as

$$M = N\mu \frac{\sinh C}{(\sinh^2 C + e^{-4K})^{1/2}} \tag{16.15}$$

The formula represents an enhanced paramagnetism. M vanishes when H vanishes, but the low field susceptibility has an enhancement factor e^{2K} over the corresponding formula (15.08) for completely uncoupled magnetic moments. The $1/T$ rise in that formula for low temperature is in fact swamped by the much steeper temperature dependence of this enhancement factor.

16-3. Solution of the two-dimensional Ising model by abstract algebra

The adaptation of the two-dimensional Ising model to the eigenvalue method is illustrated in Fig. 16.2. The spins are assumed lined up on a cylinder, and the unit which was symbolized by a cross in Fig. 16.1, has now become a tier of N spins circling the mantle of the cylinder. The "coordinate" x of the N spins is capable of 2^N discrete values. Correspondingly the eigenvalue equation (16.03) will now be a matrix equation

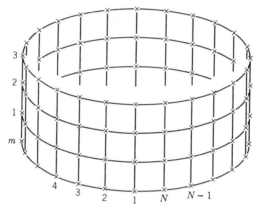

Fig. 16.2. Adaptation of the two-dimensional Ising net to the eigenvalue method.

with a matrix having 2^N lines and 2^N columns. The method to be described has been invented originally by Onsager;§ the actual form presented here is due to Schultz, Mattis, and Lieb.‖ The essence of the procedure consists in decomposing the large matrix into a direct product of N 2×2 commuting matrices. The eigenvalues are thereby reduced to a product of N factors. In order to perform this calculation the original matrix must undergo a number of transformations all of which are known from other applications, but which together form a procedure which is often considered obscure. In order to help dissipate this obscurity these transformations will be listed immediately with a short characterization. Thereafter, under the same headings, the calculation will be carried out in detail.

(*a*) Onsager transformation to Pauli spin matrices. The way in which spin matrices enter into the discussion is already exhibited in (16.11) in one dimension. In two dimensions there are more spins, but the idea is the same.

§ L. Onsager. *Phys. Rev.* **65,** 117 (1944).
‖ L. Schultz, D. Mattis, and E. Lieb. *Rev. Mod. Phys.* **36,** 856 (1964).

(b) Jordan–Wigner transformation to fermion operators. The transformation of independent spin operators with their mixed commutation–anticommutation rules into a set with anticommutation rules throughout has been known for many years. It is very convenient for the subsequent step because the fermion commutation rules are preserved under unitary transformations.

(c) Born transformation to running waves. Few solid state problems can escape this classical transformation, which was first introduced in (13.18) and (13.20), and generalized to three dimensions in (13.41). The original spin operators do not allow this transformation, but the new fermion operators do.

(d) As is usual in other fields, the original problem is now decoupled into separate problems, one for each wave. The decoupling appears here as a decoupling into a direct product. This is in line with the fact that the partition function for decoupled systems factors into a product, not a sum. The decoupling of the waves k and $-k$ is not complete, which is a phenomenon already observed in (13.22) and (13.23). This last reduction merges with the determination of the eigenvalues.

(a) We shall only discuss the problem in the absence of a magnetic field. By (15.48) this means also that we cannot determine the spontaneous magnetization. The resolving kernel which goes with a tier as shown in Fig. 16.2 reads then

$$\exp[-\beta V] = \exp\left[K_1 \sum_{\nu=1}^{N} \sigma_\nu \sigma'_\nu + K_2 \sum_{\nu=1}^{N} \sigma_\nu \sigma_{\nu+1}\right] \quad (16.16)$$

Here two interactions J_1 and J_2 have been defined, the first going from tier to tier and the second along a tier; a K has been defined for either coupling according to (16.09). Expression (16.16) is not a symmetric interaction $V(x, y)$ as was assumed in Section 16-1, but the eigenvalues are the same. The symmetric form would have had for the second sum: $\frac{1}{2}\sum_\nu(\sigma_\nu\sigma_{\nu+1}+\sigma'_\nu\sigma'_{\nu+1})$. An eigenvector of the form $\exp\left[-\frac{1}{2}\sum_\nu \sigma_\nu \sigma_{\nu+1}\right] a(\sigma)$ will correct that. The reduction to spin operators is very simple. We simply read σ^z for σ, which reduces the second factor. The reduction of the first factor was accomplished in (16.11) and reads in terms of Pauli spin matrices

$$e^{K_1} + e^{-K_1} \sigma^x$$

To simplify this we introduce the so-called dual transformation

$$\tanh K_1^\star = e^{-2K_1} \quad (16.17a)$$

or symmetrically

$$\sinh 2 K_1 \sinh 2 K_1^\star = 1 \quad (16.17b)$$

which brings the above factor into the form

$$(2 \sinh 2 K_1)^{1/2} \exp[K_1^\star \sigma^x]$$

The kernel (16.16) becomes then

$$\exp[-\beta V] = (2 \sinh 2 K_1)^{1/2 N} \exp\left[K_1^\star \sum_\nu \sigma_\nu^x\right] \exp\left[K_2 \sum_\nu \sigma_\nu^z \sigma_{\nu+1}^z\right]$$

It is desirable to symmetrize this kernel, leaving the more complicated term in the middle. This is accomplished by a similarity transformation with $V_1^{1/2}$, where V_1 is defined as

$$V_1 = \exp\left[K_1^\star \sum_{\nu=1}^N \sigma_\nu^x\right] \tag{16.18}$$

It brings the resolving kernel $\mathscr{K}$ into the final form

$$\mathscr{K} = V_1^{-1/2} \exp[-\beta V] V_1^{1/2} = (2 \sinh 2 K_1)^{1/2 N} V_1^{1/2} V_2 V_1^{1/2} \tag{16.19}$$

where we defined further

$$V_2 = \exp\left[K_2 \sum_{\nu=1}^N \sigma_\nu^z \sigma_\nu^{z+1}\right] \tag{16.20}$$

(b) The spin-raising and spin-lowering operators σ^+ and σ^- are defined through

$$(\sigma^+)^2 = (\sigma^-)^2 = 0 \tag{16.21a}$$

$$\sigma^+ \sigma^- + \sigma^- \sigma^+ = 1 \tag{16.21b}$$

They are here best related to the two components of σ through

$$\sigma^z = \sigma^+ + \sigma^- \tag{16.22}$$

$$\sigma^x = \sigma^+ \sigma^- - \sigma^- \sigma^+ \tag{16.23}$$

This leads to the subsidiary relations

$$\sigma^+ \sigma^z + \sigma^z \sigma^+ = \sigma^- \sigma^z + \sigma^z \sigma^- = 1 \tag{16.24}$$

and

$$\sigma^+ \sigma^x + \sigma^x \sigma^+ = \sigma^- \sigma^x + \sigma^x \sigma^- = 0 \tag{16.25}$$

All these relations have the form of anticommutators. Between different sites, however, the anticommutators have no special properties, and the commutators are zero. This unsymmetric state of affairs is remedied by the transformation of Jordan and Wigner[¶]

$$C_m^+ = \sigma_m^+ \prod_{\nu=1}^{m-1} \sigma_\nu^x \tag{16.26a}$$

$$C_m = \sigma_m^- \prod_{\nu=1}^{m-1} \sigma_\nu^x \tag{16.26b}$$

[¶] P. Jordan and E. Wigner. *Z. Physik* **47**, 631 (1928).

This converts commutation into an anticommutation between different sites, as one can verify by making use of (16.25). Yet at the same time, the relations (16.21) are not disturbed. As a result, the C's all anticommute with each other and even with themselves, except for the one non-vanishing anticommutator

$$C_m^+ C_m + C_m C_m^+ = 1 \qquad (16.27)$$

In terms of the fermion basis operators C and C^+, the operator V_1 defined in (16.18) becomes with (16.23)

$$V_1 = \exp\left[K_1^\star \sum_{\nu=1}^{N} (C_\nu^+ C_\nu - C_\nu C_\nu^+) \right] \qquad (16.28)$$

The operator V_2 is a little more complicated to reproduce. The combination is

$$(C_m - C_m^+)(C_{m+1} + C_{m+1}^+) = \sigma_m^z \sigma_{m+1}^z \qquad (16.29)$$

We prove this by inserting the definitions (16.26), remembering that the σ's on different sites commute and that $(\sigma^x)^2 = 1$. There remains then an expression

$$(\sigma_m^- - \sigma_m^+)(\sigma_m^+ \sigma_m^- - \sigma_m^- \sigma_m^+)(\sigma_{m+1}^- + \sigma_{m+1}^+)$$

Here expression (16.23) has been inserted for σ_m^x. The result (16.29) follows from this by the use of (16.21) and finally (16.22). From (16.20) and (16.29), we get therefore for V_2

$$V_2 = \exp\left[K_2 \sum_{\nu=1}^{N} (C_\nu - C_\nu^+)(C_{\nu+1} + C_{\nu+1}^+) \right] \qquad (16.30)$$

The derivation as given is incorrect for cyclic closure when $\nu = N$, $\nu + 1 = 1$. It is possible to set this term right by a good deal of extra labor. We shall ignore this point here. One incorrect term at $\nu = N$ produces simply a "seam" in the cylinder of Fig. 16.2. It is well known that such end effects do not influence physical results; in fact, different closure conditions have been used occasionally** in the past without any material effect on the final answer. It should also be observed that the resultant problem is formally symmetric along a tier. This formal symmetry must correspond to an actual symmetry, which we are unable to understand at this time.

(c) The advantage of the C's over the σ's is that they obey fermion commutation rules. These rules remain unaffected by a unitary transformation of the C's among themselves if only the C^+'s are subjected to the

** H. A. Kramers and G. H. Wannier. *Phys. Rev.* **60**, 263 (1941).

conjugate complex transformation. The transformation we want to apply is of course (13.18)

$$C_n = \frac{e^{i\pi/4}}{\sqrt{N}} \sum_{\mu=0}^{N-1} \exp[-2\pi i n\mu/N] \, \eta_\mu \qquad (16.31a)$$

and

$$C_n^+ = \frac{e^{-i\pi/4}}{\sqrt{N}} \sum_{\mu=0}^{N-1} \exp[2\pi i n\mu/N] \, \eta_n^+ \qquad (16.31b)$$

The factor $\exp(i\pi/4)$ is introduced to ensure later on real coefficients of all terms in V_2. Substitution of (16.31) into (16.28) and (16.30) yields

$$V_1 = \exp\left[K_1^\star \sum_{\mu=-\frac12 N}^{\frac12 N -1} (\eta_\mu^+ \, \eta_\mu - \eta_\mu \, \eta_\mu^+) \right] \qquad (16.32)$$

$$V_2 = \exp\left[K_2 \sum_{\mu=-\frac12 N}^{\frac12 N -1} \{ i \, e^{2\pi i\mu/N} \, \eta_\mu \, \eta_{-\mu} + i \, e^{-2\pi i\mu/N} \, \eta_\mu^+ \, \eta_{-\mu}^+ \right.$$
$$\left. + e^{2\pi i\mu/N} \, \eta_\mu \, \eta_\mu^+ - e^{-2\pi i\mu/N} \, \eta_\mu^+ \, \eta_\mu \} \right] \qquad (16.33)$$

The modes μ and $-\mu$ are coupled as is usual in such problems. Apart from that, the contributions from different modes commute, being bilinear forms of anti-commuting operators. The exponential of the sum therefore becomes a product of exponentials. We set therefore

$$V_1 = \prod_{\mu=0}^{\frac12 N} v_1(\mu) \qquad (16.34a)$$

$$V_2 = \prod_{\mu=0}^{\frac12 N} v_2(\mu) \qquad (16.34b)$$

and get directly from (16.32) and (16.33)

$$v_1(0) = \exp K_1^\star (\eta_0^+ \, \eta_0 - \eta_0 \, \eta_0^+) \qquad (16.35)$$

$$v_1(\tfrac12 N) = \exp K_1^\star (\eta_{\frac12 N}^+ \, \eta_{\frac12 N} - \eta_{\frac12 N} \, \eta_{\frac12 N}^+) \qquad (16.36)$$

$$v_2(0) = \exp K_2(\eta_0 \, \eta_0^+ - \eta_0^+ \, \eta_0) \qquad (16.37)$$

$$v_2(\tfrac12 N) = \exp K_2(\eta_{\frac12 N}^+ \, \eta_{\frac12 N} - \eta_{\frac12 N} \, \eta_{\frac12 N}^+) \qquad (16.38)$$

The other modes need a combination of μ and $-\mu$. We get from (16.32)

$$v_1(\mu) = \exp\{2 \, K_1^\star \, [\eta_\mu^+ \, \eta_\mu - \eta_{-\mu} \, \eta_{-\mu}^+]\} \qquad (16.39)$$

and finally from (16.33)

$$v_2(\mu) = \exp\left\{2 K_2\left[-\cos\frac{2\pi\mu}{N}\ (\eta_\mu^+\ \eta_\mu - \eta_{-\mu}\ \eta_{-\mu}^+)\right.\right.$$
$$\left.\left. + \sin\frac{2\pi\mu}{N}\ (\eta_\mu^+\ \eta_{-\mu}^+ + \eta_{-\mu}\ \eta_\mu)\right]\right\} \quad (16.40)$$

With the relations (16.34)–(16.40) the resolvent matrix (16.19) takes the form

$$\mathcal{K} = (2 \sinh 2\ K_1)^{\frac{1}{2}N} \prod_{\mu=0}^{\frac{1}{2}N} v_1^{\frac{1}{2}}(\mu)\ v_2(\mu)\ v_1^{\frac{1}{2}}(\mu) \quad (16.41)$$

Each μ factor in this product commutes with any other μ factor. The problem is thereby decomposed into a direct product of independent matrix problems. Any product of individual eigenvalues will be an eigenvalue of the resolvent matrix. We shall endeavor to write these eigenvalues as exponentials so that the logarithm of the total eigenvalue will appear as the sum of exponents. The sum over modes can then be converted into an integral in the usual way.

(d) Denote the eigenvalues of the μth factor in (16.41) by $\exp \varepsilon(\mu)$. We then get immediately $\varepsilon(0)$ and $\varepsilon(\frac{1}{2}N)$ by observing that these expressions involve only the operator

$$\mathcal{O} = \eta\ \eta^+ - \eta^+\ \eta$$

for which

$$\mathcal{O}^2 = 1$$

We obtain therefore

$$\varepsilon(0) = \pm\ (K_1^\star - K_2) \quad (16.42a)$$

and

$$\varepsilon(\tfrac{1}{2}N) = \pm\ (K_1^\star + K_2) \quad (16.42b)$$

Since the final answer will be in the form of an integral these results for two individual modes will have no influence on the answer. The important thing is the handling of the product for general μ. Let us introduce the abbreviations

$$\frac{2\ \pi\ \mu}{N} = q \quad (16.43)$$

$$\eta_\mu^+\ \eta_\mu - \eta_{-\mu}\ \eta_{-\mu}^+ = R \quad (16.44a)$$

$$\eta_\mu^+\ \eta_{-\mu}^+ + \eta_{-\mu}\ \eta_\mu = S \quad (16.44b)$$

where the summation index μ is temporarily suppressed. As the η's obey fermion commutation rules, the two operators R and S can be seen to obey the algebraic relations

$$R S + S R = 0 \tag{16.45}$$

$$R^3 - R = 0 \tag{16.46}$$

$$S^3 - S = 0 \tag{16.47}$$

$$R^2 - S^2 = 0 \tag{16.48}$$

All except the first of these rules are also verified for linear combinations of R and S having "rotated" form. In other words, it follows from (16.45) to (16.48) that

$$(R \cos \phi + S \sin \phi)^2 = R^2 \tag{16.49}$$

and

$$(R \cos \phi + S \sin \phi)^3 = R \cos \phi + S \sin \phi \tag{16.50}$$

Equations (16.46) and (16.50) show that the square brackets occurring in expressions (16.39) and (16.40) for $v_1(\mu)$ and $v_2(\mu)$ have only the eigenvalues 1, 0, and -1, and equation (16.49) shows that they share the eigenvectors for the eigenvalue 0. Then v_1 and v_2 both have the value 1, and $\varepsilon(\mu) = 0$. For the remaining eigenvalues we may set

$$R^2 = S^2 = 1 \tag{16.51}$$

and (16.45) remains as the only other nontrivial relation. For such operators the exponential form of (16.39) and (16.40) can be made linear. The relevant factor in (16.41) becomes then

$$v_1^{\frac{1}{2}} v_2 v_1^{\frac{1}{2}} = \{\cosh K_1^* + R \sinh K_1^*\} \{\cosh 2 K_2 +$$
$$\sinh 2 K_2 (-R \cos q + S \sin q)\} \{\cosh K_1^* + R \sinh K_1^*\}$$

Multiplying out this expression, we get

$$v_1^{\frac{1}{2}} v_2 v_1^{\frac{1}{2}} = \cosh 2 K_1^* \cosh 2 K_2 - \sinh 2 K_1^* \sinh 2 K_2 \cos q$$
$$+ R \{\sinh 2 K_1^* \cosh 2 K_2 - \cosh 2 K_1^* \sinh 2 K_2 \cos q\}$$
$$+ S \sinh 2 K_2 \sin q \tag{16.52}$$

It turns out that the sum of the squares of the coefficients of R and S, subtracted from the constant, equals 1. We are therefore allowed to set

$$v_1^{\frac{1}{2}} v_2 v_1^{\frac{1}{2}} = \cosh \varepsilon - R \sinh \varepsilon \cos \varphi + S \sinh \varepsilon \sin \varphi \tag{16.53}$$

with

$$\cosh \varepsilon = \cosh 2 K_1^* \cosh 2 K_2 - \sinh 2 K_1^* \sinh 2 K_2 \cos q \tag{16.54}$$

and

$$\tan \varphi = \frac{\sinh 2 K_2 \sin q}{\cosh 2 K_1^* \sinh 2 K_2 \cos q - \sinh 2 K_1^* \cosh 2 K_2} \tag{16.55}$$

Equation (16.53) means that the exponential form of the operators can be restored, yielding

$$v_1^{1/2} \, v_2 \, v_1^{1/2} = \exp[\, \varepsilon \, (- \, R \cos \varphi + S \sin \varphi)] \qquad (16.56)$$

The eigenvalues of (16.56) are $e^{\pm \varepsilon}$, with ε given by (16.54).

16-4. Analytic reduction of the results for the two-dimensional Ising model

By the preceding analysis all eigenvalues of the resolving kernel (16.41) are obtained in the form

$$\lambda = (2 \sinh 2 \, K_1)^{1/2 N} \, \exp \sum_{\mu=0}^{1/2 N} \varepsilon(\mu) \qquad (16.57)$$

where $\varepsilon(0)$ and $\varepsilon(\tfrac{1}{2} N)$ are given by (16.42), and the remaining quantities $\varepsilon(\mu)$ are either zero or are one of the two roots of the equation

$$\cosh \varepsilon(\mu) = \cosh 2 \, K_1^{*} \, \cosh 2 \, K_2 - \sinh 2 \, K_1^{*} \, \sinh 2 \, K_2 \, \cos \frac{2\pi\mu}{N} \quad (16.58)$$

It is seen that $\varepsilon(0)$ and $\varepsilon(\tfrac{1}{2} N)$ also fit this formula except that they are only half the size; this is to be expected as they arise from one mode only, while (16.58) is the contribution of two modes. Every possible combination of ε's forms a root of the resolving kernel, but the largest one alone determines the thermodynamic properties, by (16.07). This means taking the larger root of (16.58) every time. Proceeding to the limit $N = \infty$ is not difficult for that root. We employ (16.43) and get

$$\lambda_{\max} = (2 \sinh 2 \, K_1)^{\frac{1}{2} N} \, \exp\!\left\{ \frac{N}{2\pi} \int_0^{\pi} \varepsilon(q) \, dq \right\} \qquad (16.59)$$

The expression appears to be unsymmetric in the two coupling parameters K_1 and K_2. A symmetric form results if we transform the integral into a double integral with the help of the two identities

$$\varepsilon = \frac{1}{\pi} \int_0^{\pi} \ln \, (2 \cosh \varepsilon - 2 \cos p) \, dp \qquad (16.60)$$

and

$$(2 \sinh 2 \, K_1)^{\frac{1}{2} N} = \exp\!\left\{ \frac{N}{2 \, \pi^2} \int_0^{\pi} dp \int_0^{\pi} dq \, \ln \, (2 \sinh 2 \, K_1) \right\}$$

The relation (16.17b) then permits elimination of K_1^{*} if expression (16.54) is inserted into (16.60). The result is an expression having symmetric form, namely,

$$\lambda_{max} = \exp \left\{ \frac{N}{2\pi^2} \int_0^\pi dp \int_0^\pi dq \ln [4 \cosh 2 K_1 \cosh 2 K_2 \right.$$
$$\left. - 4 \sinh 2 K_1 \cos p - 4 \sinh 2 K_2 \cos q] \right\} \quad (16.61)$$

Because of (16.07) the above result yields a system partition function F of the form

$$F = f^{NN'} \quad (16.62)$$

where N' is the number of tiers in the cylinder of Fig. 16.2, and N the number of spins in the tier. Therefore $N N'$ is the total number of spins, and f is something like an "equivalent particle partition function" for the problem. From (16.61) and (16.07) we get for f

$$\ln f = \ln 2 + \frac{1}{2\pi^2} \int_0^\pi \int_0^\pi dp \, dq \ln [\cosh 2 K_1 \cosh 2 K_2$$
$$- \sinh 2 K_1 \cos p - \sinh 2 K_2 \cos q] \quad (16.63)$$

The energies derived from (16.63) can in all cases be expressed in terms of elliptic functions. We shall carry out this reduction here only for the square net for which

$$K_1 = K_2 = K \quad (16.64)$$

A simple transformation then brings (16.63) in the form

$$\ln f = \ln (2 \cosh 2 K) + \frac{1}{2\pi^2} \int_0^\pi \int_0^\pi d\omega_1 \, d\omega_2 \ln \{1 - \kappa \cos \omega_1 \cos \omega_2\}$$
$$(16.65)$$

Here we have defined

$$\kappa = \frac{2 \sinh 2 K}{\cosh^2 2 K} = \frac{2 \sinh 2 K^*}{\cosh^2 2 K^*} \quad (16.66)$$

We then use the identity (16.60) in reverse and get

$$\ln f = \ln (2 \cosh 2 K) + \frac{1}{\pi} \int_0^{\frac{1}{2}\pi} \ln \left\{ \frac{1 + (1 - \kappa^2 \sin^2 \omega)^{\frac{1}{2}}}{2} \right\} d\omega \quad (16.67)$$

The separation of (16.67) into two terms is rather interesting. The "dual transformation" (16.17) maps high temperatures on low temperatures and vice versa. This is most easily seen in the form (16.17b). This form also shows that one particular point K_c is mapped on itself. It is the point for which

$$\sinh 2 K_c = 1 \quad (16.68a)$$

or

$$K_c = 0.4407 \quad (16.68b)$$

If the temperature singularity associated with the Curie point were located anywhere else except at K_c, there would be two such singularities. But the Curie point, being the temperature at which the spontaneous magnetization goes to zero, must be single. Therefore K_c is the only possible location of the Curie point. Now the parameter κ introduced in (16.66) has the same value for the two temperatures related by the duality transformation. This is easily verified from (16.17) and indicated in (16.66). It equals zero at either extreme and reaches its maximum, unity, at the Curie point. Thus the split-up in (16.67) makes the integral the same on the two sides of the Curie point. The integrand is a well behaved function of κ, except when the square root under the logarithm vanishes. This happens precisely for $\kappa = 1$. Another way to see the same thing is to observe that the integral (16.65) can only be singular when the argument of the logarithm is allowed to vanish. This again is only possible for $\kappa = 1$. The reader is invited to extend this argument to the integral (16.63) for the Ising net with two interactions. He will find that the only possible singular point can arise if the condition

$$\sinh 2\, K_1 \sinh 2\, K_2 = 1 \qquad (16.69)$$

is obeyed. If it is desired to examine the nature of the singularity in detail, then it is easiest to differentiate (16.67) with respect to temperature and to examine the energy. This analysis is found in the preceding chapter.

There are no problems for this chapter.

17

Theory of dilute solutions

We have seen in the first part of this book that statistical mechanics makes certain predictions valid for all materials; these predictions, incorporated in the discipline of thermodynamics, do not form a complete theory. In particular, an equation of state is usually not available. This often means that thermodynamics has to be applied in combination with experimental information in numerical form. Even in the technology of steam, a relatively simple material, we are dependent on cumbersome steam tables for complete information.

In this second part we have essentially taken up selected topics for which statistical mechanics gives more than merely thermodynamic information. Broadly speaking, this extra information consists of something like an equation of state, derived on theoretical grounds. The value of the theory depends in practice on its formal simplicity, and on the accuracy with which it represents experimental facts. The theory of gases in the "perfect" range is an unqualified success on both counts. We should not leave the field of applications without dealing with the theory of dilute solutions. In this field certain simple general principles have been found by van't Hoff[*] and others which are in close parallelism to the properties of perfect gases.

[*] J. H. van't Hoff. Z. physik. Chem. **1,** 481 (1887).

We shall discuss these principles, together with a major refinement introduced by Debye and Hückel for electrolytes.†

17-1. Thermodynamic functions for dilute solutions

Suppose we consider a number of species I, dissolved in a solvent. Let us label these species by numbers $1, 2, 3, \ldots I$, and the respective numbers of dissolved molecules or ions by $N_1, N_2, \ldots N_I$; we denote further by N_0 the number of solvent molecules. If we now consider a simple property such as the volume of the total system, we can write for it

$$V = N_0\, v\left(T,p, \frac{N_1}{N_0}, \frac{N_2}{N_0}, \ldots \frac{N_I}{N_0}\right) \tag{17.01}$$

This equation assumes only the solution close enough to a perfect fluid so that the distinction of Section 7-2 between extensive and intensive variables is applicable. If all the solute molecules are absent this becomes simply

$$V = N_0\, v_0\,(T, p) \tag{17.02}$$

Obviously, the function (17.01) can be very complicated. To make any headway with it we must exploit the notion that the solution is dilute. We define as dilute a solution in which the solute molecules are so far apart that their mutual interaction is negligible. This implies necessarily that

$$\sum_{i=1}^{I} N_i \ll N_0 \tag{17.03}$$

Even beyond this it must mean for the vast majority of solute molecules that they interact not with each other, but only with the solvent. We can then ascribe to each of them a "sphere of influence," consisting of itself and solvent molecules only. Outside these spheres we have undisturbed solvent; inside we have solvent modified by one particular solute molecule. As a rule, one of these spheres does not overlap with any other sphere. The properties within each sphere may differ radically from the properties of the pure solvent, but there will be just N_i such spheres for the N_i molecules of species i. Denote by v_i the difference between the actual volume of the "sphere of influence" and the volume occupied by the solvent within the sphere if the solute molecule were absent (this could even be a negative number). Then we can write for the total volume

$$V = N_0\, v_0\,(T, p) + \sum_{i=1}^{I} N_i\, v_i(T, p) \tag{17.04}$$

† P. Debye and E. Hückel. *Physik. Z.* **24**, 185 (1923).

The same analysis can be repeated for the internal energy instead of the volume: u_i is then the difference between the internal energy of a "sphere of influence" and that of the same sphere with the solute removed. In other words we may write

$$U = N_0 u_0 (T, p) + \sum_{i=1}^{I} N_i u_i (T, p) \tag{17.05}$$

It would be quite incorrect to repeat this argument in the same form for the entropy. Entropy measures statistical multiplicity and cannot be assumed additive in this type of situation. We therefore follow the safer course of computing the entropy from the Second Law in the form (5.45). Substitution into this of (17.04) and (17.05) yields

$$dS = \frac{1}{T} \sum_{i=0}^{I} N_i \{du_i + p\, dv_i\} \tag{17.06}$$

This integrates to an expression of the form

$$S = \sum_{i=0}^{I} N_i s_i(T, p) + C(N_0, N_1, \ldots N_I) \tag{17.07}$$

where s_i is the result of the integration

$$s_i = \int \frac{1}{T} \{du_i + p\, dv_i\} \tag{17.08}$$

The reason (17.07) has an additive constant depending on the N's is that (17.06) is a differential equation which holds only if the numbers N_i are kept fixed. The way to determine it is to lower the value of p and raise the value of T so that the entire solution, including the solutes, vaporizes. Our system will then approximate a mixture of perfect gases, and the entropy will be given by (5.72). This expression is seen to be of the form (17.07): the first two terms depend on T and p and are linear in the number of molecules present, while the third term is nonlinear but independent of T and p. We must identify it with our constant $C(N_0, N_1, \ldots N_I)$ and write

$$C(N_0, N_1, \ldots N_I) = -k \sum_{i=0}^{I} N_i \ln \frac{N_i}{N_0 + N_1 + \cdots + N_I}$$

We can simplify this somewhat if we keep the order-of-magnitude relation

(17.03) in mind. For the zeroth term we get then, keeping only small terms of the first order,

$$N_0 \ln \frac{N_0}{N_0 + N_1 + \cdots + N_I}$$

$$= -N_0 \ln \left(1 + \frac{N_1 + N_2 + \cdots + N_I}{N_0}\right) \approx -(N_1 + N_2 + \cdots + N_I)$$

and for the others

$$N_i \ln \frac{N_i}{N_0 + N_1 + \cdots + N_I} \approx N_i \ln \frac{N_i}{N_0}, \quad i > 0$$

The integration constant in (17.07) becomes therefore to a good approximation

$$C(N_0, N_1, \ldots N_I) = -k \sum_{i=1}^{I} N_i \left\{\ln \frac{N_i}{N_0} - 1\right\} \tag{17.09}$$

and the entropy becomes

$$S = \sum_{i=0}^{I} N_i s_i (T, p) - k \sum_{i=1}^{I} N_i \left\{\ln \frac{N_i}{N_0} - 1\right\} \tag{17.10}$$

Combination of (17.04), (17.05), and (17.10) yields the other thermodynamic functions if desired. Our interest here is in the Gibbs free energy and the chemical potentials derivable therefrom. If we introduce the abbreviation

$$g_i = u_i + p v_i - T s_i \tag{17.11}$$

we get for G, using the definition (7.13),

$$G = \sum_{i=0}^{I} N_i g_i + k T \sum_{i=1}^{I} N_i \left\{\ln \frac{N_i}{N_0} - 1\right\} \tag{17.12}$$

It was shown in (7.29) that the chemical potential of the species i follows from (17.12) by differentiation of G with respect to N_i at constant pressure and temperature. This yields

$$\mu_0 = g_0 - \frac{k T}{N_0} \sum_{i=1}^{I} N_i \tag{17.13}$$

and

$$\mu_i = g_i + k T \ln \frac{N_i}{N_0}, \quad i > 0 \tag{17.14}$$

For either the solvent or the solutes the modification of μ due to the solution phenomenon has a relatively simple form. The consequences of

(17.13) for the solvent will occupy us in Section 17-2, and the consequences of (17.14) in Section 17-3.

17-2. Osmotic pressure and other modifications of solvent properties by the solutes

We shall now enumerate the modifications in the properties of the solvent which arise from the presence of the solutes. The modifications are summed up in (17.13); this equation has a very simple structure. The first term, g_0, is the chemical potential which the solvent would have if it were pure; the second term represents a reduction $\Delta\mu_0$, which gives the solution extra stability because of the presence of the dissolved molecules. One observes that $\Delta\mu_0$ contains only absolute constants and can thus be computed without difficulty. The summation shows that the solutes enter only through the total number of dissolved particles, whether they be molecules, atoms, or ions. Experimental physical chemistry makes use of this fact; by measuring $\Delta\mu_0$ and knowing the total amount of solute (if there is just one kind), the number (and thereby quite often the nature) of the individual dissolved particles can be identified. The various effects differ just in the method by which $\Delta\mu_0$ is determined.

The effect which has given rise to the most colorful language, and to an easily remembered formula, arises if it is possible to construct a semi-permeable membrane of the type discussed in Section 5-6. The membrane must have the property of being permeable to solvent molecules but not to any of the solute particles. If pure solvent is placed on one side of such a membrane and a solution on the other, the solvent can reach equilibrium, while establishment of equilibrium for the solute molecules is inhibited. Therefore μ_0 must be the same on the two sides. Since the solute particles reduce μ_0, the solvent molecules will tend to migrate away from the pure side, other things being equal. A schematic arrangement for observing this is shown in Fig. 17.1. The semipermeable membrane is placed at the bottom of a U-tube, whose two arms contain pure solvent and solution, respectively. Free migration of solvent molecules will set in until a pressure head P is built up on the solution side which cancels its extra stability. This pressure head is called the *osmotic pressure* of the solution. Its magnitude is read off from (17.11) and (17.13): it must be such as to make μ_0 the same on the two sides. This yields

$$P v_0 = \left(\sum_{i=1}^{I} N_i \right) \frac{k T}{N_0}$$

By (17.03) and (17.04), $N_0 v_0$ equals the volume of the solution up to first-order correction terms. The equation can therefore be given the form

$$P V = \left(\sum_{i=1}^{I} N_i \right) k T \qquad (17.15)$$

Formally, this is the same as (2.43) for a perfect gas mixture. We therefore have the following theorem:

The osmotic pressure P of a dilute solution may be computed as if the

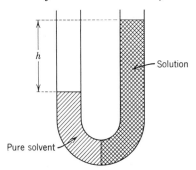

Fig. 17.1. Equilibrium across a semi-permeable membrane of a solvent between its pure form and a solution. The overpressure needed on the solvent side to maintain equilibrium is called the *osmotic pressure P.*

solute particles formed a perfect gas having the temperature of, and occupying the volume of, the solution.

The nature of the above effect is easily remembered and has been confirmed experimentally. For precision work on a routine basis it is preferable not to have to rely on the working of a semipermeable membrane. Measurement of other effects is therefore preferred. These effects are *reduction of the vapor pressure, increase of the boiling point,* and *depression of the freezing point.* The formulas for these effects are based on the assumption (not always correct) that the solutes are unable to enter the gaseous or the solid phase of the solvent; thereby, pure solvent can be placed in equilibrium with the solution across a phase boundary. The relevant formulas can all be obtained by modifying slightly the derivation of (7.45). If the second phase is denoted by an asterisk, and if one thinks of the solution as being formed gradually through the addition of solute, then relation (7.43b) governs the process

$$d\mu^\star = d\mu_0$$

By (17.13), μ_0 has acquired new degrees of freedom owing to the presence of the solute molecules. Expression (7.45) thus gets modified to

$$-(s^\star - s_0)\, dT + (v^\star - v_0)\, dp = -d\left(\frac{kT}{N_0} \sum_{i=1}^{I} N_i\right) \qquad (17.16)$$

Since the changes in temperature and pressure arising from osmotic effects are usually small, the differential form (17.16) is adequate to estimate finite temperature and pressure shifts. For the case of the vapor pressure reduction, the temperature is assumed constant, and one gets

$$\Delta p = -\frac{kT \displaystyle\sum_{i=1}^{I} N_i}{N_0 \; v^\star - v_0} \qquad (17.17a)$$

or in more practical units

$$\Delta p = -\frac{RT}{\mathscr{V}^\star - \mathscr{V}_0} \sum_{i=1}^{I} \left(\frac{N_i}{N_0}\right) \qquad (17.17b)$$

If the liquid molar volume is negligible compared to the gaseous one, and if the latter phase can be treated as a perfect gas, (17.17) can be given the form known as *Raoult's law:*

$$\frac{\Delta p}{p} = -\sum_{i=1}^{I} \left(\frac{N_i}{N_0}\right) \qquad (17.18)$$

The other two effects are concerned with temperature shifts. One gets them by holding p constant in (17.16) and varying T. This yields

$$\Delta T = \frac{kT \displaystyle\sum_{i=1}^{I} N_i}{N_0 \; s^\star - s_0} \qquad (17.19a)$$

or in molar units

$$\Delta T = \frac{RT^2}{\mathscr{H}_0^\star} \sum_{i=1}^{I} \left(\frac{N_i}{N_0}\right) \qquad (17.19b)$$

Here $\mathscr{H}_0^\star$ is the heat of transformation from the solvent phase into the other one present, assumed pure. For the shift of the boiling point one gets thus a positive ΔT, and for the shift of the freezing point a negative ΔT.

In view of the internal consistency of statistical theory it should be possible to derive (17.17) also from the arrangement shown in Fig. 17.1. For once the pure solvent is in equilibrium with the solution across a semipermeable membrane, it must also be in equilibrium across the atmosphere of solvent vapor which rests over either part of the U-tube. Otherwise a circulation movement would set in, violating the First Law. If the difference in height of the two levels is h, the osmotic pressure P in the arrangement equals

$$P = (\rho_0 - \rho^\star) g h$$

where $\rho^\star$ is the density of the vapor. The difference in the vapor pressure, Δp, on the other hand, equals

$$\Delta p = \rho^\star g h$$

so that

$$\Delta p = \frac{\rho^\star}{\rho_0 - \rho^\star} P$$

$$= \frac{v_0}{v^\star - v_0} P$$

$$= \frac{V}{N_0} \frac{P}{v^\star - v_0}$$

Substitution into this of (17.15) yields

$$\Delta p = \frac{\sum_{i=1}^{I} N_i}{N_0} \frac{k T}{v^\star - v_0}$$

in agreement with (17.17a).

We shall conclude this section with some indication of the numerical aspect of these results. We shall compute the freezing point depression produced by a *normal* aqueous solution, that is, a solution of 1 mole in 1 liter of water. The formula to use is (17.19b). The numbers are

$$\frac{N_1}{N_0} = \frac{18}{1000}$$

$$R = 1.98 \text{ cal/deg mole} \qquad \mathscr{H}_0^\star = -80 \times 18 \text{ cal/mole}$$

$$T = 273.2° \text{ K}$$

Multiplying this out, we get

$$\Delta T = -1.85°$$

We see that we deal with a substantial effect which is easily accessible to measurement.

17-3. Behavior of solutes in dilute solutions; analogy to perfect gases

The striking thing about properties of solutes in dilute solutions is their resemblance to perfect gases. The reason is of course that the interaction of individual particles with each other is minimal in either case. The fact that, in solutions, the particles are intensively interacting with the solvent creates the major difference. It means that some results are modified or have to be reinterpreted. The concept of osmotic pressure was in this latter category. We now come to some gas-like aspects of solute behavior.

Formula (17.14) for the chemical potential of a solute shows a striking similarity to formula (11.44) for the chemical potential of a component of a gas mixture. In either case there is a first term which is not dependent on the abundance of any of the components; this term depends on pressure and temperature. It is followed by a second term not dependent on temperature and proportional to the logarithm of the abundance. Because of the difference between the two cases the first term is not as easily calculable for solutions as it is for gases, for we have the properties of the solvent as a medium entering here.

We shall now show that (17.14) makes the *law of mass action* valid in dilute solutions. Let there be among the I solute molecules a chemical reaction possible of the form

$$\sum_{i=1}^{I} v_i((i)) = 0 \tag{17.20a}$$

such as, for instance, the dissociation reaction of silver chloride

$$Ag^+ + Cl^- - AgCl = 0 \tag{17.20b}$$

The general analysis of Section 7-4 then tells us immediately that there is a corresponding relation between chemical potentials, namely,

$$\sum_{i=1}^{I} v_i \mu_i = 0 \tag{17.21a}$$

which in the above example means that

$$\mu_{Ag} + \mu_{Cl^-} - \mu_{AgCl} = 0 \tag{17.21b}$$

The possible separation of (17.14) into two types of terms allows a similar separation in (17.21). If we define

$$[i] = \frac{N_i}{N_0} \tag{17.22}$$

we get immediately

$$\prod_{i=1}^{I} [i]^{v_i} = f(T, p) \tag{17.23a}$$

which, in our example of silver chloride, reduces to

$$\frac{[\text{Ag}^+] \, [\text{Cl}^-]}{[\text{AgCl}]} = f(T, p) \qquad (17.23\text{b})$$

The fact that the equilibrium constant $f(T, p)$ is not easily calculable is a disadvantage as compared to the case of gases. In practice, this disadvantage is compensated by an advantage: the constant may usually be treated as independent of pressure. The reason for this is that $f(T, p)$ deals with properties of the solvent only, and the usual experimental pressures are not sufficient to modify its structure significantly.

The law of mass action is of much greater value for solutions, particularly aqueous solutions, than it is for gases. The number of reactions which we know to proceed freely to a measurable equilibrium in water is very much greater than the number of such reactions known in gases, and the technique of making measurements is much easier. The equilibrium indicated above for silver chloride, for instance, is crucial in the analytical procedure for measuring the concentration of chlorides. The essential feature to add to (17.23b) is that [AgCl] is pinned to a permanently low value because of low solubility, so that, in practice, the product of the silver and chlorine concentrations is a constant. Thus the chlorine ion drives the silver out of solution, and one can be titrated by the other with great precision.

A particularly interesting and important reaction in water is

$$\text{H}^+ + \text{OH}^- = \text{H}_2\text{O} \qquad (17.24)$$

the dissociation reaction of water into hydrogen and hydroxyl ions. The corresponding relation between the chemical potentials reads

$$\mu_{\text{H}^+} + \mu_{\text{OH}^-} = \mu_{\text{H}_2\text{O}} \qquad (17.25)$$

Since H_2O is the solvent, its chemical potential is given by (17.13). This means to a first approximation that $\mu_{\text{H}_2\text{O}}$ is a constant which depends only on temperature and slightly on pressure. For the two ions, (17.14) is applicable. If we isolate the two logarithmic terms, (17.25) takes the form

$$[\text{H}^+] \, [\text{OH}^-] = g(T) \qquad (17.26\text{a})$$

The practical unit for measuring concentration is moles per liter, which is $1000/18$ times the natural unit (17.22). In these units we have at room temperature

$$g(293°) = 10^{-14} \qquad (17.26\text{b})$$

In other words, the concentration of either ion in pure water is 10^{-7} mole/liter. Equation (17.26) remains valid in the presence of other solutes. If one of these consists partly of hydrogen ions, as is the case with acids, the quantity $[H^+]$ will be raised, and $[OH^-]$ will drop correspondingly. The negative logarithm to the base 10 of $[H^+]$ is the commonly accepted measure of acidity, the so-called pH of a solution. It is higher than 7 for alkaline solutions and lower than 7 for acids.

Equation (17.14) can also be used to get a partial result for the relative distribution of a solute between two solvents A and B. The basic equation is (7.42), that is,

$$\mu_i^A = \mu_i^B \tag{17.27}$$

If we isolate again the logarithmic terms, we get from this

$$\frac{[i(A)]}{[i(B)]} = h(T) \tag{17.28}$$

or in words:

When a solute in the dilute range is in equilibrium between two different solvents, the ratio of its concentration in the two solvents is a function of temperature only.

Finally, equation (17.27), in combination with (11.44), yields a simple result for dissolved gases. We have already remarked that the equilibrium constants in liquids are in practice independent of pressure. The reason is that pressures used commonly in the laboratory are several orders of magnitude too small to produce structurally significant changes in liquids. In the gaseous phase, on the other hand, μ has the form (11.44). This expression contains the pressure and concentration only through the combination

$$k\,T \ln\,(c_i\,p) = k\,T \ln p_i \tag{17.29}$$

where p_i is the partial pressure of the particular gas component. If this term and the concentration term for the solution are isolated from the rest in (17.27), there results an equation of the form

$$\frac{[i]}{p_i} = j(T) \tag{17.30}$$

or in words:

The concentration of a dissolved gas in a liquid is, at fixed temperature, proportional to the partial pressure of the gas over the liquid (Henry's law).

More important than the case of dissolved gases is that of a dissolved substance in equilibrium with its pure crystalline form. The arrangement defines the concept of *solubility*. The methods of this chapter lead to certain predictions about solubility provided even the saturated solution is dilute. The predictions will be taken up in Problems 3, 6, and 7 at the end of the chapter.

17-4. Theory of strong electrolytes

It is natural to assume that expressions such as (17.04) and (17.05) form the beginning of an expansion of the general function (17.01) in powers of the concentration of solutes. If this is the case, the logarithmic term (17.09) is the only term not fitting into a series expansion. Furthermore, expressions (17.13) and (17.14) are then also the beginning of such power series, the first correction term to (17.13) being quadratic, and to (17.14) linear in the solute concentration. All the solute effects of the solvent thus come out to be limiting laws, with a higher power correction entering at normal concentrations. By the argument given, any percentage correction to the chemical potential shift, to the osmotic pressure, and to the temperature shifts (17.19) would start out as a correction linear in the concentration if the concentration is small.

It seems that, for solutions of molecular material, these formal considerations do apply. But the early work of Faraday, Arrhenius, etc., has shown that salts composed of strong acids and strong bases go into solution as individual ions. The osmotic effects are thereby doubled or tripled as compared to a molecular theory. In addition, it is found that the limiting behavior just described is invalid for those cases. In 1916, Bjerrum‡ took measurements of the freezing point depression of an aqueous solution of potassium chloride and found the points shown in Fig. 17.2. The departure from the limiting law (17.19) turns out not to go linearly with the concentration but to vary as its square root. The theory of this phenomenon, due to Debye and Hückel,§ will now be given.

First of all it is not very surprising that solutions of electrolytes show a departure from the "ideal" behavior of solutions indicated in the early parts of this chapter. The justification of this behavior was, after all, based on the idea of "spheres of influence" of individual solute molecules. Between ions, the acting force is the Coulomb force whose long range creates notorious difficulties in all fields of physics. Under these circumstances we must consider ourselves fortunate that we can salvage the equations written down in the first three sections for the case of electrolytes. It is only in the departure from these results at moderate concentration that the effect of the Coulomb forces enters. The effect is briefly that every ion tends to attract preferably ions of opposite charge; each ion rests thus in a negative potential which modifies in the first place the internal energy, and thereafter all other thermodynamic potentials. The effect is such as to reduce the osmotic pressure, and hence all other modifications of solvent properties.

‡ N. Bjerrum. *Skand. Naturforsk.* **16**, 229 (1916).

§ P. Debye and E. Hückel. *Phys. Z.* **24**, 185 (1923).

We can immediately write down a formal expression for the electrostatic energy $U^\star$ of a solution of ions. It equals

$$U^\star = \tfrac{1}{2} e \sum_{i=1}^{I} N_i z_i \phi_i \qquad (17.31)$$

Here z_i is the electrochemical valence of the ion ($+1$ for K^+, $+2$ for Ca^{++}, -1 for Cl^-, etc.), and ϕ_i the mean electrostatic potential experienced by

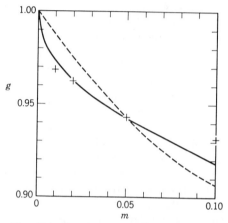

Fig. 17.2. Experimental points showing the ratio of experimental and theoretical freezing point depressions for a solution of potassium chloride as function of concentration. A linear plot (dotted line) does not fit the data but a square root plot (solid line) does (after Bjerrum).

an ion of species i owing to the presence of the other ions. The problem is of course the determination of this quantity ϕ_i.

To find ϕ_i, we assume an ion of species i placed at the origin of coordinates, and investigate the potential φ in the neighborhood of that ion. The number density n_j of the other ions in such a potential will be

$$n_j = \frac{N_j}{V} \exp[- \beta e z_j \varphi] \qquad (17.32)$$

The Boltzmann exponential is self-explanatory. The factor is determined so that n_j has its equilibrium value at a large distance from the chosen ion; it involves the assumption $\varphi = 0$ in this limit. We now close the argument by the self-consistency postulate employed in connection with

(12.13): the potential φ seen by the ions shall be the one which they themselves produce. Poisson's equation is then valid, and we have

$$\nabla^2 \varphi = -\frac{4\pi e}{\kappa} \sum_j z_j n_j \qquad (17.33)$$

Here κ is the dielectric constant of the solvent. Combination of (17.32) and (17.33) yields then

$$\nabla^2 \varphi = -\frac{4\pi e}{\kappa V} \sum_j z_j N_j \exp[-\beta e z_j \varphi] \qquad (17.34)$$

This equation was written down once before as (14.56), determining the potential distribution in the neighborhood of an n–p junction. At that time we had only two species of particles to consider, electrons and holes, and solved the equation only in one dimension. Now, we may have many species and need a spherically symmetric solution. However, we have the advantage that the potential may be assumed weak compared to kT, a point we shall check at the end of the reasoning. The exponential can then be expanded, yielding

$$\nabla^2 \varphi = -\frac{4\pi e}{\kappa V} \sum_j z_j N_j + \frac{4\pi \beta e^2}{\kappa V} \left(\sum_j z_j^2 N_j \right) \varphi$$

The first term vanishes because of the electrical neutrality of the solution. For the second term we introduce the *Debye length L*, generalizing the definition (14.58):

$$\frac{1}{L^2} = \frac{4\pi \beta e^2}{\kappa V} \sum_{j=1}^{I} z_j^2 N_j \qquad (17.35)$$

whereupon our differential equation becomes

$$\nabla^2 \varphi - \frac{1}{L^2} \varphi = 0 \qquad (17.36)$$

The spherically symmetric solution of (17.36) is

$$\varphi = \text{constant } \frac{e^{-r/L}}{r}$$

Around an ion of type i the constant must be $z_i e/\kappa$, so that we have

$$\varphi = \frac{z_i e}{\kappa} \frac{e^{-r/L}}{r} \qquad (17.37)$$

Expansion of the exponent finally yields

$$\varphi = \frac{z_i e}{\kappa r} - \frac{z_i e}{\kappa L} + 0(r) \qquad (17.38)$$

Equation (17.38) accomplishes the goal of finding the potential at the site of an ion of type i due to the other ions. For the first term is just the potential originating at the ion, and all omitted terms vanish for $r = 0$. The second term is therefore the desired potential ϕ_i:

$$\phi_i = -\frac{z_i e}{\kappa L} \tag{17.39}$$

Inserting this back into (17.31), we find for the total electrostatic energy $U^\star$

$$U^\star = -\frac{1}{2}\frac{e^2}{\kappa L} \sum_{i=1}^{I} N_i z_i^2 \tag{17.40}$$

The sum occurring in (17.40) is the same as the one occurring in the definition (17.35) of the Debye length. We can therefore eliminate one of the two quantities and write either

$$U^\star = -kT\frac{V}{8\pi L^3} \tag{17.41}$$

or

$$U^\star = -\frac{e^3}{\kappa^{3/2}}\left(\frac{\pi}{VkT}\right)^{1/2}\left(\sum_{j=1}^{I} z_j^2 N_j\right)^{3/2} \tag{17.42}$$

Before applying the result (17.42) we must verify that the potential of an ion is indeed weak compared to kT. This must be valid for the expression (17.39). We must therefore have

$$\frac{e^2}{\kappa L} \ll kT \tag{17.43}$$

This inequality can be twisted around to read that the Debye length must be very much larger than the mean distance between neighboring ions. This statement gives us a physical feel for the meaning of (17.43); it is not worth much in a numerical estimate because L contains itself the ionic density. For this latter purpose it is better to eliminate L with the help of (17.35). Let there be just one salt, consisting of $2N$ monovalent ions. The sum in (17.35) then equals $2N$. We isolate N/V, reading (17.43), as a condition for the magnitude of the concentration. It then takes the form

$$\frac{N}{V} \ll \frac{\kappa^3 k^3 T^3}{8\pi e^6} \tag{17.44a}$$

which means for water

$$\frac{N}{V} \ll \left(\frac{88.2 \times 1.38 \times 10^{-16} \times 300}{(4.8)^2 \times 10^{-20}}\right)^3 \frac{1}{8\pi} = 1.6 \times 10^{20}\ \text{cm}^{-3} \tag{17.44b}$$

This is not a very stringent condition, a normal solution (1 mole per liter) having a density of

$$\frac{N}{V} = 6 \times 10^{20} \text{ cm}^{-3}$$

However, the six orders of magnitude contributed by the dielectric constant of water are fairly essential for justifying the procedure.

The statistical consequences of the Debye–Hückel theory follow most directly if the correction to the Helmholtz free energy is computed from (7.06)

$$-\frac{A^\star}{T} = \int \frac{U^\star}{T^2}\, dT$$

Equation (17.42) yields then

$$A^\star = \tfrac{2}{3}\, U^\star$$

or

$$A^\star = -\frac{2}{3} \frac{e^3}{\kappa^{3/2}} \left(\frac{\pi}{V k T}\right)^{1/2} \left(\sum_{j=1}^{I} z_j^2 N_j\right)^{3/2} \tag{17.45}$$

On the assumption that $p\,V$ does not have any correction similar to (17.45), this expression is also the correction $G^\star$ for the Gibbs free energy. Combining this with (17.12), we get thus with (17.02)

$$G = k T \sum_{i=1}^{I} N_i \left\{\ln \frac{N_i}{N_0} - 1\right\} + \sum_{i=0}^{I} N_i g_i$$
$$- \frac{2}{3} \frac{e^3}{\kappa^{3/2}} \left(\frac{\pi}{N_0 v_0 k T}\right)^{1/2} \left(\sum_{i=1}^{I} z_i^2 N_i\right)^{3/2} + 0 \left(\frac{N_i^2}{N_0}\right) \tag{17.46}$$

It is seen that the effect of the electric charges on the ions inserts itself between the previously given terms and the naturally expected correction, which would vary as the square of the concentration.

We shall only investigate the effect of the electric charges on the properties of the solvent, that is, recompute μ_0. We get instead of (17.13)

$$\mu_0 = g_0 - \frac{k T}{N_0} \sum_{i=1}^{I} N_i + \frac{1}{3} \frac{e^3}{\kappa^{3/2}} \left(\frac{\pi}{v_0 k T}\right)^{1/2} \left(\sum_{i=1}^{I} z_i^2 \frac{N_i}{N_0}\right)^{3/2} + 0 \left(\frac{N_i^2}{N_0^2}\right) \tag{17.47}$$

It is seen that the Debye-Hückel term acts in the opposite direction to the classical solution term. However, it is smaller; for when we take their ratio, assuming again that we deal with a simple monovalent salt in solution, we get

$$\frac{\Delta \mu^\star}{\Delta \mu} = -\frac{1}{3} \left(\frac{N}{V} \frac{2 \pi e^6}{\kappa^3 k^3 T^3}\right)^{1/2} \tag{17.48}$$

The quantity in parenthesis is just the one which had to be small, by (17.44). Thus the Debye-Hückel theory is intrinsically limited to cases in which the departure from the classical behavior is small.

Experimental investigations usually give the factor g with which the

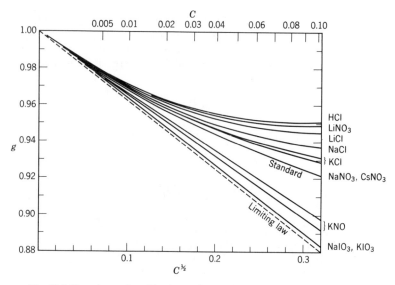

Fig. 17.3. Experimental verification of the Debye–Hückel theory. On ordinate is plotted the ratio of the actual to the "ideal" freezing point depression, on abscissa the square root of the concentration of various monovalent salts in mole/liter. The theory predicts the dashed straight line (after Guggenheim).

"ideal" osmotic pressure, freezing point depression, etc., have to be multiplied to get the actual values. We have thus from (17.48)

$$g = 1 - \frac{1}{3}\left(\frac{N}{V}\frac{2\pi e^6}{\kappa^3 k^3 T^3}\right)^{1/2} \qquad (17.49a)$$

This works out to be

$$g = 1 - \tfrac{1}{3}\, 1.03\, \sqrt{C} \qquad (17.49b)$$

where C is the concentration in moles per liter.

A plot comparing equation (17.49) with experimental data is shown in Fig. 17.3, where g is plotted on ordinate, and the square root of the concentration C on abscissa. The theoretical straight line is shown in heavy dashed outline. The experimental results are taken from freezing point depressions. It is seen that the theory predicts the limiting behavior correctly. Quantitative agreement, however, sets in only when the concentration is 1/100 molar or less. The theory is thus less precise than one might have predicted on the basis of the criterion (17.44). What this probably means is that the "normal" correction to the theory, which is linear in the concentration, also makes a noticeable contribution to the final result.

A. RECOMMENDED PROBLEMS

1. The triple point of water lies 0.0098° higher than the freezing point of water in air at atmospheric pressure. How many moles of air are dissolved in the water under these conditions?

2. Prove for a dilute solution the relation

$$\left(\frac{\partial(\beta\mu_i)}{\partial\beta}\right)_{N_j,\,p} = u_i + p\,v_i = h_i$$

$$\left(\frac{\partial(\beta\mu_i)}{\partial p}\right)_{N_j,\,T} = \beta\,v_i$$

3. Consider a system consisting of a crystalline substance and a saturated solution of this substance in a solvent. Let even this saturated solution be dilute in the sense of this chapter. Derive for these circumstances, using the results of Problem 2, a formula giving the rate at which the logarithm of the concentration varies with the temperature at fixed pressure.

B. GENERAL PROBLEMS

4. The coefficient $h(T)$ in equation (17.28) controlling the distribution of solute between two solvents is called the *distribution coefficient* for the solute between the two solvents. At room temperature, for iodine between carbon tetrachloride and water, the coefficient is 85. Suppose 1 gm of iodine is dissolved in water and (*a*) it is once shaken up with 1 liter of carbon tetrachloride, (*b*) it is twice shaken up with $\frac{1}{2}$ liter of carbon tetrachloride, (*c*) it is shaken repeatedly with small amounts of carbon tetrachloride which add up to 1 liter, so as to get the best extraction.

How much iodine remains behind in the water in each of the three cases?

5. The constant in equation (17.30) is called *Henry's constant*. Find the

law of variation of this constant with temperature, following the method of Problem 3.

6. At $25°$ C the solubility of orthorhombic sulfur in carbon tetrachloride is 0.84 gm/100 gm of solvent. Under these conditions the chemical potential of monoclinic sulfur is larger by 23 cal/mole.

(a) Find the solubility of monoclinic sulfur.

(b) Predict what will happen if you try to dissolve monoclinic sulfur up to the solubility limit.

7. Find the relation complementary to that of Problem 3 which gives the change of solubility with pressure.

8. Ferrous ions in water have a spin $S = 2$ and a g factor 2.05. In the presence of some diamagnetic chemicals they form an ion complex which is also diamagnetic. By what factor is the equilibrium constant between the bare Fe^{++} and the $(Fe^{++})_{complex}$ shifted by a magnetic field of 30000 oe at $300°$ K?

9. For a one hundredth normal solution of an electrolyte with monovalent ions, what is the radius of a sphere around a given ion which contains in the mean one half of a unit charge? How many ions are contained in such a sphere?

PART III

Kinetic theory, transport coefficients and fluctuations

*18

Kinetic justification of equilibrium statistics; Boltzmann transport equation

Equilibrium statistical mechanics is a very refined and sophisticated structure which yields results with a minimum of assumptions. It finds its most graceful application in the theory of perfect gases but is valid for much more complicated systems. The results for perfect gases, obtained by Maxwell, were originally derived, not by this means, but by a detailed molecular theory, the *kinetic theory of gases*. This theory takes the gas molecules to be particles in random motion which have occasional collisions. It was put into its final form by Boltzmann. The assumptions one has to make in using it are much more numerous than they are for equilibrium statistics. One might therefore suppose that kinetic theory is superseded as a discipline of physics. This is not the case; kinetic considerations are still necessary today whenever we discuss non-equilibrium situations. An effort is now proceeding at many places to reduce some of the very simplest nonequilibrium problems to equilibrium statistics. Some of these efforts will be dealt with in Chapter 23. Up to the present time these efforts have not succeeded in making detailed kinetic considerations unnecessary. Thus kinetic theory is still with us in the foreseeable future.

Kinetic theory itself has become much more manageable in recent years by the introduction of the concept of a *relaxation time*. We shall do all our applications, starting in the next chapter, with its help. Thus, readers interested in the applications of kinetic theory are invited to skip this chapter and pass immediately to Chapter 19. The present chapter will deal with a relatively accurate formulation of kinetic theory. This formulation is admittedly quite cumbersome. We shall not use it directly on any problems, but only indirectly as a means to justify and illuminate simpler methods. In the early parts of this chapter we shall use it to rederive the properties of the equilibrium state and to investigate the way this state is reached from other states. This investigation gives rise to some famous paradoxes which will be discussed fully. We shall then turn at the end of the chapter to the justification of the relaxation time approximation, as viewed from the more rigorous theory. This field of investigation is in its infancy, and we shall give the results as far as they are available at this time.

18-1. Derivation of the Boltzmann transport equation

Early workers in the kinetic theory of gases got qualitatively correct results by assuming that gas molecules move at random in different directions, but that all have the same speed. In the Boltzmann equation this primitive notion is replaced by the introduction of a probability distribution as to the position and velocity of the molecules of each species present. For simplicity, the following discussion will be limited to a gas consisting of a single species. We deal then with a single so-called *Boltzmann distribution function* $f(\mathbf{v}, \mathbf{r})$ which is taken to be a density in phase space. In other words, the quantity

$$f(\mathbf{v}, \mathbf{r})\, d\mathbf{v}\, d\mathbf{r} \qquad (18.01)$$

is the number of particles within the six-dimensional cube of sides dv^x, dv^y, dv^z, dr^x, dr^y, dr^z. The second assumption of early kinetic theory was that the gas molecules undergo only binary collisions. This assumption also enters into the Boltzmann equation. It has as a consequence that the Boltzmann equation can only be applied to dilute or moderately dense gases, for which the mean free path of the molecules is much larger than their size. Nobody has yet succeeded in developing a kinetic theory in which one molecule is at all times interacting with several other molecules.

In the Boltzmann theory the analysis of all possible binary encounters is treated by decomposing them into categories. A single category consists of encounters of all molecules having their velocity within $d\mathbf{u}$ with molecules having their velocity within $d\mathbf{v}$. To facilitate consideration of this category we adopt a coordinate system moving with velocity $\mathbf{u}$. The first kind of molecules then appears at rest while the second kind moves with a velocity $\mathbf{v} - \mathbf{u}$. To these moving molecules the stationary ones offer a total cross section $f(\mathbf{u}, \mathbf{r}) \, \sigma \, d\mathbf{u} \, d\mathbf{r}$ within a volume $d\mathbf{r}$, where σ is the cross section for an individual encounter. The total volume swept out in unit time by these targets is then $|\mathbf{v} - \mathbf{u}| f(\mathbf{u}, \mathbf{r}) \, \sigma \, d\mathbf{u} \, d\mathbf{r}$. Finally, the total number of encounters is the density of the moving molecules times the swept-out volume, which equals

Number of $\mathbf{u}$, $\mathbf{v}$ encounters per unit time and unit volume

$$= |\mathbf{v} - \mathbf{u}| \, \sigma f(\mathbf{u}, \mathbf{r}) f(\mathbf{v}, \mathbf{r}) \, d\mathbf{u} \, d\mathbf{v} \qquad (18.02)$$

In practice all encounters lead to the removal of a molecule from the element of volume of velocity space in which it was located originally. So integration of (18.02) over $\mathbf{u}$ yields the expression $-\left(\dfrac{df(\mathbf{v})}{dt}\right)_{\text{out}} d\mathbf{v}$. We have thus arrived at the equation

$$\left(\frac{df(\mathbf{v})}{dt}\right)_{\text{out}} = -f(\mathbf{v}) \int\int\int d\mathbf{u} \, |\mathbf{v} - \mathbf{u}| \, \sigma f(\mathbf{u}) \qquad (18.03)$$

Equation (18.03) is relatively simple. The integral has the dimension of a reciprocal time, which generally depends on the magnitude of $\mathbf{v}$. It is the *collision rate* for molecules of the type v, and its reciprocal is often called the *mean free time* between collisions. Dividing v by the collision rate yields a length, which is the *mean free path*. Equation (18.03) suggests an exponential rate of attrition of molecules of velocity $\mathbf{v}$. It can occasionally be used directly, namely, when we deal with a beam of velocity $\mathbf{v}$ and wish to analyze the attrition of the beam through collisions. Then $f(\mathbf{v})$ and $f(\mathbf{u})$ are actually different functions, the first referring to the beam, the second to the scattering gas.

From the point of view of kinetic theory (18.03) is, alas, incomplete. For, within a gas, the expression is of use only when taken together with another which gives the number of molecules scattered *into* the velocity range in the neighborhood of $\mathbf{v}$. Computation of this quantity requires a detailed analysis of the collision process. Suppose we consider a pair of molecules which end up with velocities $\mathbf{u}$, $\mathbf{v}$ after an encounter, and suppose we denote their velocities before the encounter by $\mathbf{u}'$, $\mathbf{v}'$. Then we have

$$\mathbf{u} + \mathbf{v} = \mathbf{u}' + \mathbf{v}' \qquad (18.04)$$

We shall assume all encounters to be elastic. Then the kinetic energy is conserved in a collision as well as the momentum, and we have in addition

$$u^2 + v^2 = u'^2 + v'^2 \qquad (18.05)$$

Equations (18.04) and (18.05) together have the important consequence that the difference of the velocities remains constant in absolute magnitude in any encounter. We bring this out by defining

$$\mathbf{w} = \mathbf{v} - \mathbf{u} \tag{18.06a}$$

$$\mathbf{w}' = \mathbf{v}' - \mathbf{u}' \tag{18.06b}$$

and writing down the relation

$$w = w' \tag{18.07}$$

This quantity w appears explicitly in (18.03) and is also contained implicitly in the cross section σ.

To avoid extraneous difficulties (which are very serious in practice) we shall assume the molecules to be point centers of force with the force acting along the line of centers; we shall also assume them to be free of internal structure. Each encounter is then individually symmetric with respect to initial and final conditions, and might just as well be assumed to proceed from $(\mathbf{u}, \mathbf{v})$ to $(\mathbf{u}', \mathbf{v}')$ as inversely. When this is the case an individual encounter is determined by the parameters of the impact. Classically, these parameters are the impact parameter proper, usually denoted by b, and an azimuth. The impact parameter b determines the angle of scattering χ uniquely. Quantum-mechanically we can only specify the angular momentum quantum number which produces a specific angular pattern. We take care of either viewpoint by resolving the cross section σ occurring in (18.03) into a *differential cross section $d\sigma$* by the well known procedure

$$d\sigma = d\sigma(\chi) = \frac{d\sigma}{d\chi} d\chi \tag{18.08}$$

It is convenient to combine an azimuth with χ and to employ the differential cross section σ' for scattering into unit solid angle. In other words, we set

$$\sigma'(\chi, w) = \frac{1}{2\pi \sin \chi} \frac{d\sigma(\chi)}{d\chi} \tag{18.09}$$

whereupon (18.08) becomes

$$d\sigma = \sigma' \sin \chi \, d\chi \, d\varepsilon = \sigma' \, d\Omega \tag{18.10}$$

Here ε is a suitably defined azimuth, and $d\Omega$ is the element of solid angle described by the vector $\mathbf{w}$ with respect to $\mathbf{w}'$ or vice versa. The advantage of having an element of solid angle appear in an integration is that the axis of the polar coordinate system representing it can always be shifted to suit the requirements of a particular calculation.

The definition (18.10) is suitable for putting (18.03) into a form adapted to inverse encounters. The expression now takes the form

$$\left(\frac{df(\mathbf{v})}{dt}\right)_{\text{out}} d\mathbf{v} = - f(\mathbf{v})\, d\mathbf{v} \int\int\int f(\mathbf{u})\, d\mathbf{u} \int\int w\, \sigma'\, d\Omega_{\mathbf{w}'} \quad (18.11)$$

where $\Omega_{\mathbf{w}'}$ is the solid angle described by $\mathbf{w}'$ with respect to $\mathbf{w}$. There is an intrinsic symmetry in the five integrations appearing here. This is brought out by taking a last step, namely, replacing $d\mathbf{u}$ by $d\mathbf{w}$ as variable of integration. By (18.06a) and the constancy of $\mathbf{v}$ this replacement proceeds with unit jacobian.* We get then, marking each solid angle for the vector whose orientation it describes,

$$\left(\frac{df(v)}{dt}\right)_{\text{out}} = - \int\int\int\int\int f(\mathbf{u}) f(\mathbf{v})\, \sigma' w^3\, dw\, d\Omega_{\mathbf{w}'}\, d\Omega_{\mathbf{w}'} \quad (18.12)$$

In other words, the five integrations are over the relative speed w which the initial and the final state have in common, and the angular variables of $\mathbf{w}$ and $\mathbf{w}'$ for which they are independent. All integrations refer to motion in the center of mass frame of reference. If the integrations in (18.12) are complemented by an integration over $\mathbf{v}$ as indicated in (18.11), we do in fact bring in integration over the motion of the center of mass. Any other velocity may be used in its place, such as $\mathbf{u}$, $\mathbf{u}'$, or $\mathbf{v}'$, and the jacobian which appears when we pass from one to the other is unity for symmetry reasons. In particular, a return to the form (18.11) shows that

$$d\mathbf{u}\, d\mathbf{v}\, d\Omega_{\mathbf{w}'} = d\mathbf{u}'\, d\mathbf{v}'\, d\Omega_{\mathbf{w}} \quad (18.13)$$

Relation (18.13) is of capital importance in the study of inverse encounters. By (18.02) we get the collisions ending up in $d\mathbf{u}$ and $d\mathbf{v}$ by taking among the events

$$w f(\mathbf{u}') f(\mathbf{v}')\, d\mathbf{u}'\, d\mathbf{v}'\, \sigma'(\chi, w)\, d\Omega_{\mathbf{w}'}$$

those particular ones ending up in $d\mathbf{u}$ and $d\mathbf{v}$. For the determination of the correct values of $\mathbf{u}'$, $\mathbf{v}'$, χ, and ε, a detailed analysis of the impacts is essential. This requires a knowledge of the law of force and varies from case to case. However, the relation (18.13) permits us to defer this task to a later stage and to transform the differentials without this detailed information. We get then

$$\left(\frac{df(\mathbf{v})}{dt}\right)_{\text{in}} d\mathbf{v} = d\mathbf{v} \int\int\int\int\int f(\mathbf{u}') f(\mathbf{v}')\, \sigma'\, w\, d\mathbf{u}\, d\Omega_{\mathbf{w}'} \quad (18.14)$$

---

* Differentials like $d\mathbf{u}$ or $d\Omega$ are always taken as intrinsically positive in this book.

Combination of (18.11) and (18.14) is immediate and yields

$$\frac{df(\mathbf{v})}{dt} = \int\int\int\int\int \{f(\mathbf{u}')f(\mathbf{v}') - f(\mathbf{u})f(\mathbf{v})\}\, \sigma'\, w\, d\mathbf{u}\, d\Omega_{\mathbf{w}'} \quad (18.15)$$

To bring the Boltzmann transport equation into its final form it is important to remember that the time derivative on the left is the hydro-dynamic total derivative in phase space. In explicit form, the equation becomes therefore

$$\frac{\partial f(\mathbf{v}, \mathbf{r}, t)}{\partial t} + \mathbf{v} \cdot \frac{\partial f(\mathbf{v}, \mathbf{r}, t)}{\partial \mathbf{r}} + \mathbf{a} \cdot \frac{\partial f(\mathbf{v}, \mathbf{r}, t)}{\partial \mathbf{v}} \quad (18.16)$$

$$= \int\int\int\int \{f(\mathbf{u}', \mathbf{r}, t)f(\mathbf{v}', \mathbf{r}, t) - f(\mathbf{u}, \mathbf{r}, t)f(\mathbf{v}, \mathbf{r}, t)\}\, \sigma'\, w\, d\mathbf{u}\, d\Omega_{\mathbf{w}'}$$

Here $\mathbf{a}$ is the acceleration of the gas particles arising from an external field. It is obvious that the hydrodynamic derivative on the left is much simpler than the right-hand side, which is a nonlinear integral operator; it is called the *Boltzmann collision operator*. We shall devote the end of this chapter to an effort to replace it by a more convenient expression which is roughly equivalent.

18-2. Equilibrium solutions of the Boltzmann equation; Maxwellian distribution

The simplest task which can be accomplished with (18.16) is to rederive the equilibrium properties of a gas. For this we assume that f depends neither on position nor on time, and that no external forces are present. The left-hand side of (18.16) then vanishes identically, and the equation reduces to the vanishing of the right-hand side. This is certainly accomplished if the curly bracket within the integral vanishes for all compatible values of the velocity. We shall defer the question whether this is a necessary condition to the next section, and solve the equation in this manner. Any solution obeys then the condition

$$f(\mathbf{u})f(\mathbf{v}) = f(\mathbf{u}')f(\mathbf{v}') \quad (18.17)$$

or

$$\ln f(\mathbf{u}) + \ln f(\mathbf{v}) = \ln f(\mathbf{u}') + \ln f(\mathbf{v}') \quad (18.18)$$

We define as an *additive invariant* of a binary encounter a quantity which is defined separately for each of the colliding particles, and which has the property that the sum of the two quantities is the same after the encounter as it was before. By (18.18) the unknown function $\ln f(\mathbf{v})$ is such an additive invariant. Other additive invariants were found previously in (18.04) and (18.05), namely, the three components of the velocity and its square. If it is possible to enumerate all additive invariants of an encounter, then $\ln f(\mathbf{v})$ is necessarily a linear combination of these. Now, in addition to the ones just named, any constant is trivially an additive invariant. But beyond this, there can be no other additive invariants in a two-body collision; for if there were more than four such invariants, the four vectors $\mathbf{u}$, $\mathbf{v}$, $\mathbf{u}'$, $\mathbf{v}'$ would have less than eight independent components. This would mean for equation (18.15), in which $\mathbf{v}$ is held fixed, fewer than five integrations. But such a reduction is clearly impossible, for all values of $\mathbf{u}$ for the second molecule are clearly possible and all angles of scattering. Thus our invariants yield the general solution of (18.18), which is of the form

$$\ln f(\mathbf{v}) = A + B\,v^x + C\,v^y + D\,v^z + E\,v^2 \qquad (18.19)$$

By a somewhat different disposal of the five constants we may put this into the form

$$f(\mathbf{v}) = a \exp[-\,b\,(\mathbf{v}-\mathbf{v}_0)^2] \qquad (18.20)$$

Three of the five constants, the three components of $\mathbf{v}_0$, are associated with the motion of the gas as a whole, for we have by symmetry from (18.20)

$$\int\int\int v^x f(\mathbf{v})\,d\mathbf{v} = v_0^x \int\int\int f(\mathbf{v})\,d\mathbf{v}$$

Such a motion is present if the container holding the gas is in motion, or in freely blowing large air masses. It can be eliminated from the distribution function on the basis of the Gallilean invariance of all physics; in other words, we can ask for the velocity distribution function in a frame of reference with respect to which the gas as a whole is at rest, and $\mathbf{v}_0$ equal to zero. Once this is done we have the Maxwellian distribution (4.25). The essential constants a and b in (18.20) are

$$b = \tfrac{1}{2}\,m\,\beta \qquad (18.21a)$$

and

$$a = n \left(\frac{m\,\beta}{2\,\pi}\right)^{3/2} \qquad (18.21b)$$

In other words, the essential irreducible constants of a gas in equilibrium are its temperature $T = 1/k\,\beta$ and its number density n. The Boltzmann transport equation allows no other constants. The reasoning of equilibrium statistics, which led to the same result, is thereby confirmed for gases by an entirely different method.

18-3. Boltzmann's *H*-theorem

We have solved the Boltzmann equation (18.16), and derived from it the Maxwellian distribution, by adopting the apparently more stringent relation (18.17). This relation is clearly sufficient for the solution of the equation. Boltzmann himself supplied the proof that it is necessary in his famous *H*-theorem, which we shall now explain.

In considering the possibility of other stationary solutions of Boltzmann's equation we shall assume without proof that they would be uniform in space in the absence of external forces. This assumption leaves only the first of the three terms on the left of (18.16). Physically, this assumption means that we limit our interest to the question of equilibration of the velocity distribution, and exclude temporarily other equilibration processes associated with some unbalance in space. For the resolution of this limited question we introduce the quantity H, which is defined as

$$H = \int \int \int d\mathbf{v} f(\mathbf{v}) \ln f(\mathbf{v}) \tag{18.22}$$

and investigate the variation of this quantity in time. The definition (18.22) immediately brings to mind the definition (5.13) of entropy to which the expression for H bears a close resemblance. One can in fact show that H *is* the entropy, apart from a negative multiplying factor and an additive constant. The present investigation therefore amounts to a study of the variation of the entropy in time under certain restrictive conditions.

We introduce now the derivative of H with respect to time

$$\frac{dH}{dt} = \frac{d}{dt} \int \int \int f(\mathbf{v}) \ln f(\mathbf{v}) \, d\mathbf{v}$$

$$= \int \int \int \frac{\partial}{\partial t} [f(\mathbf{v}) \ln f(\mathbf{v})] \, d\mathbf{v}$$

$$= \int \int \int \left[\frac{\partial f}{\partial t} \ln f + \frac{\partial f}{\partial t} \right] d\mathbf{v}$$

As we deal with a constant number of molecules, we must have that

$$\frac{dN}{dt} = V \frac{d}{dt} \int \int \int f \, d\mathbf{v} = V \int \int \int \frac{\partial f}{\partial t} \, d\mathbf{v} = 0$$

The second term in the expression for dH/dt therefore vanishes, and we are left with

$$\frac{dH}{dt} = \int \int \int \frac{\partial f}{\partial t} \ln f \, d\mathbf{v} \tag{18.23}$$

The quantity $\partial f/\partial t$ is now substituted from (18.16). Thereby dH/dt becomes

$$\frac{dH}{dt} = \int\int\int\int\int\int\int\int \{f(\mathbf{u}')f(\mathbf{v}') - f(\mathbf{u})f(\mathbf{v})\} \ln f(\mathbf{v}) \ \sigma' \ w \ d\mathbf{u} \ d\mathbf{v} \ d\Omega_{\mathbf{w}'}$$

(18.24)

This equation is now written again with some changes in the nomenclature of the variables of integration. The first one is to interchange $\mathbf{u}$ and $\mathbf{v}$, as well as $\mathbf{u}'$ and $\mathbf{v}'$. This only changes the logarithm, everything else being symmetric in $\mathbf{u}$ and $\mathbf{v}$. Within equation (18.24) this permutation is equivalent to the substitution

$$\ln f(\mathbf{v}) \to \tfrac{1}{2} \ln \left[f(\mathbf{u}) f(\mathbf{v}) \right]$$

(18.25)

We now want to exchange primed and unprimed variables. According to the observations preceding (18.08), this creates no problem mechanically; the switch in the integration variables is, in fact, exactly taken care of by the identity (18.13). The only change occurring beside the logarithm is a change in sign of the curly bracket in (18.24). We can incorporate this change into the logarithm term by generalizing (18.25) to read

$$\ln f(\mathbf{v}) \to \tfrac{1}{4}\{\ln \left[f(\mathbf{u}) f(\mathbf{v}) \right] - \ln \left[f(\mathbf{u}') f(\mathbf{v}') \right]\}$$

(18.26)

Equation (18.24) then becomes

$$\frac{dH}{dt} = -\frac{1}{4} \int\int\int\int\int\int\int\int \{f(\mathbf{u}) \, f(\mathbf{v}) - f(\mathbf{u}') \, f(\mathbf{v}')\}$$

(18.27)

$$\{\ln \left[f(\mathbf{u}) f(\mathbf{v}) \right] - \ln \left[f(\mathbf{u}') f(\mathbf{v}') \right]\} \ \sigma' \ w \ d\mathbf{u} \ d\mathbf{v} \ d\Omega_{\mathbf{w}'}$$

The interesting point about this calculation is that we have now expressed dH/dt as an integral containing the product of two factors indicated by the curly brackets. The first factor has the form $x - y$, the second $\ln x - \ln y$. With x and y intrinsically positive, these two quantities always have the same sign and vanish together. Since everything else in the integrand is positive, it follows then that the integral will always be positive and H will decrease. The distribution will therefore not be stationary. The only way to avoid this situation and to produce a stationary distribution is to have both curly brackets in (18.27) vanish. This produces just (18.17). Relation (18.17) is therefore not only sufficient, but also necessary for the existence of a velocity distribution which is stationary in time.

It is useful to indicate in detail the physical content of the *H*-theorem. It shows that a non-Maxwellian velocity distribution is converted into a Maxwellian one with an eventual body-drift by the collision operator in the Boltzmann equation; it also shows incidentally that the entropy is thereby increased. In addition it indicates the time taken by the equilibration process as being of the order of the mean free time between collisions. In this respect it goes beyond the Second Law as given in Chapter 5. In that chapter it was shown only that the equilibrium state is also the state of maximum entropy, but no indication was given of the time it might take for a given system to attain equilibrium. The reason that no answer containing the time can in general be expected is that these times differ enormously from case to case. The *H*-theorem gives us this time for a very special kind of equilibration in gases, namely, the establishment of a local Maxwellian distribution. It is important to remember that the theorem contains no indication about the rate at which any spatial unbalance might disappear. This remains, even for gases, a question to be investigated. A good part of kinetic theory deals in fact with this question. The most notable spatial differences are of density, temperature, body velocity, and composition. One can tentatively conclude from the *H*-theorem that the times involved in these equilibrations will probably be longer than the time taken to establish a local Maxwellian distribution. This tentative conclusion is in fact correct. The fastest equilibration is to be expected for differences in pressure, for equilibration is brought about here by an acting force. In the other cases a gradual propagation of equilibrium through collisions is necessary, leading to the slow processes of heat conduction, viscosity, and diffusion. These processes will form the subject matter of the next chapter.

18-4. Paradoxes associated with the Boltzmann transport equation; Kac ring model

We shall now discuss the paradoxes connected with the Boltzmann equation. At the outset of such a discussion it should be perfectly clear to the reader that there is not the slightest experimental evidence against the Boltzmann equation. It is true that in many situations it is only an approximate equation and has to be amended to take in neglected effects (triple collisions, molecular structure, quantum effects). It is possible, however, to adjust experimental conditions so that such corrections are not required; in such conditions the confirmation of the theory has always been perfect. It is also true that the derivation of exact results from the nonlinear integrodifferential equation (18.16) is almost a special science.† The literature abounds therefore with approximate results, approximately derived. These results are sometimes in error. However, more careful computation has always restored agreement with experiment. We must therefore start the discussion at the outset with the notion that we are dealing with an equation which is confirmed by experiment. This circumstance makes the paradoxes associated with the equation all the more puzzling and interesting.

The central paradox of the Boltzmann equation is that it turns a corner in scientific thought which no equation should ever be able to turn. The equation contains irreversibility, that is, it distinguishes between past and future. The appearance of the first derivative with respect to time in (18.16) already indicates this in a general way. The point is brought out quite clearly in the H-theorem, according to which the entropy increases in time. This result is obtained, starting from the laws of mechanics which are symmetric with respect to time reversal.‡ There is thus an obvious objection to the Boltzmann equation, the so-called *Umkehreinwand*. Any system obeying the laws of mechanics should also obey time reversal, and the Boltzmann equation does not. To this objection the research of Poincaré in mechanics has added a second, the so-called *Wiederkehreinwand*, the objection of recurrence in time. Poincaré showed that for every finite mechanical system a return to a state arbitrarily close to the initial state is achieved after a sufficiently long time. The time needed for such a return is called a Poincaré cycle. The Boltzmann equation does not allow such a return, even approximately. For it would require a decrease of the entropy back to something like the initial value; such a decrease is excluded by the H-theorem.

It is well to remember here that equilibrium statistical mechanics does not turn that corner from reversibility to irreversibility. Careless formulations of the Second Law often give that impression. Our proof of the Second Law from quantum mechanics only showed that the equilibrium state has a higher entropy than any other state. It thereby maximizes statistical multiplicity, which is sometimes called probability. The theorem so stated does not contain any indication about the development of a nonequilibrium system in time. An indirect reference to time is actually hidden in the notion of probability itself. It expresses itself in an observation such as "In a well shuffled deck of cards, the cards are not in a regular sequence." But any explicit computational detail respects the principles of mechanics.

---

† The standard reference work of this field is S. Chapman and T. G. Cowling, *The Mathematical Theory of Non-uniform Gases*, Cambridge University Press, 1952.

‡ This is also true of quantum mechanics, where the symmetry operation is $t \to -t$, $\Psi \to \Psi^*$. The latter substitution changes the sign of the expectation value of any velocity.

In particular, no H-theorem could possibly be derived from equilibrium statistics, for according to the rules for statistical weight discussed in Chapter 3 the region around $q_1, q_2, \ldots q_N, p_1, p_2, \ldots p_N$ has the same probability as the time-reversed region around $q_1, q_2, \ldots q_N, -p_1, -p_2, \ldots -p_N$. A unidirectional development in time can thus not occur in equilibrium statistics. It is true, however, that there is the implicit assumption that such unidirectional developments take place *prior* to the establishment of equilibrium. These developments are supposed to produce the equilibrium state as an end result. This end result is then investigated in equilibrium statistics.

This equivocal situation is radically altered by the Boltzmann transport equation. The equation exhibits for a special case how equilibrium arises in time from nonequilibrium, and it does this in a manner which is confirmed by experiment. It must therefore contain an ingredient which is not itself part of mechanics, which eliminates the time-reversal symmetry of mechanics, and which, in addition, is also true. What is the nature of this ingredient?

Although a great deal of reasoning power is needed to resolve this difficult question, we are not entirely dependent on abstract arguments, because some of the points at issue can be understood with the help of simple models. For the purpose at hand the Kac ring model §,‖ is excellent. The version presented here is slightly modified from the original presentations so as to be strictly reversible.

The model is shown in Fig. 18.1. A series of n points $P_1, P_2, P_3, \ldots P_n$ are arranged on a circle. Some of these points do, some do not, have a marker (indicated by check mark). There are m such markers. At the n halfway points between any two neighbouring points P_i and P_{i+1}, there are n balls, some black, some white. These balls are moving together in elementary steps; in one such elementary step, all balls move simultaneously one unit counterclockwise in the circle; those which sweep across a marker change color, the others do not. The problem is to predict the color of the balls after t steps.

The Kac ring model has the properties of a mechanical system which are essential here. First it is reversible; if the balls are stopped at any moment and rotated clockwise they will sweep across the markers in an inverted time sequence, and the pattern of colors will be the same as previously in the same positions. Second, it has a Poincaré cycle. After two complete revolutions of the system every ball will be at its starting position with its starting color.

In order to write down equations of motion for the system, let us denote

§ M. Kac. *Bull. Acad. Roy. Belg.* **42**, 356 (1956).

‖ M. Dresden. *Studies in Statistical Mechanics.* Amsterdam: North Holland, 1956, p. 303.

by $B(t)$ the number of black balls at time t and by $W(t)$ the number of white balls. We have obviously

$$B(t) + W(t) = n \tag{18.28}$$

We further have to distinguish the black and the white balls according to whether they have a marker ahead of them or not. We denote the numbers

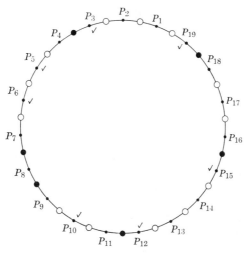

Fig. 18.1. Picture of the Kac ring model. Black or white balls move in single steps across points P_i; they change color if there is a marker ($\checkmark$) at the point they cross.

of the ones that do by $b(t)$ and $w(t)$. This definition entails a second "conservation law"

$$b(t) + w(t) = m \tag{18.29}$$

The numbers B and W at the time $t + 1$ can be predicted from the definitions given as being

$$B(t+1) = B(t) - b(t) + w(t) \tag{18.30a}$$

$$W(t+1) = W(t) - w(t) + b(t) \tag{18.30b}$$

The same equations are also valid for the time-reversed operation, except that $t+1$ must be replaced by $t-1$. Equations (18.30) are not complete and do not in general allow the determination of $B(t)$. We can solve the equations, however, by making the natural assumption that the color of the ball and the property of having a marker ahead of it are uncorrelated. This means that with the definition

$$\frac{m}{n} = \mu \tag{18.31}$$

we assume

$$\frac{b(t)}{B(t)} = \frac{w(t)}{W(t)} = \mu \tag{18.32}$$

Thereupon (18.30) becomes

$$B(t+1) - W(t+1) = (1-2\mu)\{B(t) - W(t)\} \tag{18.33}$$

and therefrom by recurrence

$$B(t) - W(t) = (1-2\mu)^t \{B(0) - W(0)\} \tag{18.34}$$

Equation (18.34) is a very interesting answer because it is both "right" and "wrong." The rightness of it is in its statistical implication. Since $1-2\mu$ is a number between -1 and $+1$, the difference in the number of the two kinds of balls decreases exponentially toward a random condition in which there is the same number of balls of either kind. This expresses our feeling of motion toward disorder in the same way as the Boltzmann equation does. However, the solution is wrong because it violates two mechanical features of the model, namely, of being reversible and of being periodic in time. We could save the first principle by saying that if the same assumption of no correlation is made for clockwise motion, the same equation (18.30) can be used, and we get

$$B(t) - W(t) = \frac{B(0) - W(0)}{(1 - 2\mu)^t} \tag{18.35}$$

which is time-reversed with respect to (18.34). However, the new equation cannot hold indefinitely because, finally, the left side of (18.35) will come out larger than n. Also the assumption which led to (18.35) is not the same as previously. The assumption is now that there is no correlation between the color of a ball and the fact whether the point P_i *just passed* had a marker. This new assumption may seem less plausible than the earlier one but cannot be refuted easily for a reversible system. Anyhow, what is clearly shown in the result is that the two assumptions of no correlation are mutually exclusive. If the assumption of no correlation is made before the marker points are passed, then it is invalid after, and vice versa. As to the second point, namely, the one of periodicity, it is clearly lost in (18.34) unless we started originally with $B(0) = W(0)$.

In order to clear up this point we shall write down proper equations of motion. We introduce the notation

$$\varepsilon_i = + 1 \text{ if there is no marker at } P_i \tag{18.36a}$$

$$\varepsilon_i = - 1 \text{ if there is a marker at } P_i \tag{18.36b}$$

Further

$$\eta_i(t) = + 1 \text{ if the ball at the position preceding } P_i \text{ is black at the time } t \tag{18.37a}$$

$$\eta_i(t) = - 1 \text{ if the ball at the position preceding } P_i \text{ is white at the time } t \tag{18.37b}$$

With the help of these definitions the model rules take the form

$$\eta_{j+1}(t+1) = \varepsilon_j \, \eta_j(t) \tag{18.38}$$

By iteration we thus get the solution to the mechanical problem

$$\eta_j(t) = \varepsilon_{j-1} \, \varepsilon_{j-2} \, \varepsilon_{j-3} \ldots \varepsilon_{j-t} \, \eta_{j-t}(0) \tag{18.39}$$

All indices here are to be read modulo n. The Poincaré cycle $t = 2\,n$ is then immediately read off from the equation. The quantity $B - W$ studied previously equals in terms of the definition (18.37)

$$B - W = \sum_{v=1}^{n} \eta_v \tag{18.40}$$

and the value of this quantity at the time t comes out to be

$$B(t) - W(t) = \sum_{v=1}^{n} \varepsilon_{v-1} \, \varepsilon_{v-2} \, \varepsilon_{v-3} \ldots \varepsilon_{v-t} \, \eta_{v-t}(0) \tag{18.41}$$

No further progress can be made beyond this point without a knowledge of the distribution of the markers. This distribution determines the mechanical properties of the system, and ultimately the results (18.39) and (18.41)

The question now comes up naturally: are we really vitally interested in the exact distribution of the markers? As the answer is clearly in the negative, we do well to embed our Kac ring model in an ensemble of other such models, following ideas outlined earlier. Let us suppose an ensemble of Kac rings, all of n balls, but differing in detail with respect to the distribution of the markers on the marker points. Let us further assume that we start every member of the ensemble with the same distribution of black and white balls at $t = 0$. If we further denote by pointed brackets an ensemble average, we find for the averaged form of (18.41) the expression

$$\langle B(t) - W(t) \rangle = \sum_{\nu=1}^{n} \langle \varepsilon_{\nu-1} \, \varepsilon_{\nu-2} \, \varepsilon_{\nu-3} \, . \, . \, . \, \varepsilon_{\nu-t} \rangle \, \eta_{\nu-t}(0)$$

Let us now assume in greater detail that the marker distribution is random. The ensemble average of any t consecutive factors ε_κ is then independent of the first factor $\varepsilon_{\nu-1}$; therefore the summation sign can be brought in front of the η's and evaluated by (18.40). The result is

$$\langle B(t) - W(t) \rangle = \langle \varepsilon_1 \, \varepsilon_2 \, \varepsilon_3 \, . \, . \, . \, \varepsilon_t \rangle \, \{B(0) - W(0)\} \qquad (18.42)$$

For any individual Kac ring the product in pointed brackets on the right-hand side of (18.42) equals ± 1. For our ensemble of Kac rings, however, we need the ensemble average of this quantity, which is comprised somewhere between the two.

For simple results it is best to have all members of the ensemble of Kac rings share the quantity μ defined in (18.31), not as a ratio of two fixed numbers, but as a probability, valid independently at each point P_i, that there is a marker at that point. If we start out by taking $t \leq n$, we find then that formula (2.02) applies to the present situation; it gives the probability of having s markers among t points. This means explicitly in the present notation

$$p(s, t) = \frac{t!}{s! \, (t-s)!} \, \mu^s \, (1-\mu)^{t-s}, \qquad 0 \leq t \leq n \qquad (18.43a)$$

When t is larger than n, but less than $2n$, a certain number of factors ε_κ occur twice and give 1. Then $p(s, t)$ becomes

$$p(s, t) = \frac{(2n-t)!}{s! \, (2n-t-s)!} \, \mu^s \, (1-\mu)^{2n-t-s}, \qquad n \leq t \leq 2n \qquad (18.43b)$$

Beyond $t = 2n$ the solution repeats periodically. Now our ensemble average is given by

$$\langle \varepsilon_1 \, \varepsilon_2 \, \varepsilon_3 \, . \, . \, . \, \varepsilon_t \rangle = \sum_{s=0}^{t} (-)^s \, p(s, t)$$

Substitution of (18.43) into this yields

$$\langle \varepsilon_1 \, \varepsilon_2 \, \varepsilon_3 \, . \, . \, . \, \varepsilon_t \rangle = (1 - 2\mu)^{n - |n-t|}$$

Finally, (18.42) becomes

$$\langle B(t) - W(t) \rangle = (1-2\mu)^{n - |n-t|} \, \{B(0) - W(0)\} \qquad (18.44)$$

For an initial period $t \leq n$ this is exactly (18.34). Thereafter, the time-reversed solution (18.35) takes over for a while. The long-range solution is a periodic oscillation, as required by the theorem of Poincaré. If, on the other hand, we think of n as being very large, say "beyond the range of human experience," then the difference $B - W$ dies away exponentially, as described in (18.34).

The derivation of the result (18.44) is very instructive because it shows clearly the mathematical origin of irreversibility. Irreversibility arises because our answers are not strictly mechanical: to get the expected result, averages and limits must be taken in a way which is extraneous to an individual mechanical system. In the Kac ring model the average is taken over the position of the markers, and the number n of members of the ring must be permitted to tend to infinity. These processes are of such a nature that the original unbalance of black and white balls could never have arisen in the first place, except by external action outside the rules of the model. In this respect the calculation follows very closely the experimental situation. We observe the damping of a galvanometer by first disturbing the instrument by an external impulse, and then watching the deflection decay according to its natural laws. Only very rarely do we have the chance to observe the natural growth of a fluctuation. Onsager has suggested that if we observed such a natural growth it would in the mean follow the time-inverted law. This conjecture is confirmed for the Kac ring model by the result (18.44).

The type of averaging and limit formation necessary to get the "statistical" answer (18.34) in the case of the Kac ring model cannot, as a rule, be duplicated for a mechanical system. Luttinger and Kohn¶ came very close to it when discussing the scattering of electrons by fixed impurities having random location within a solid. In the case of gases one would be hard put to find such a feature which can be randomized as readily. Gibbs has suggested that an averaging over initial conditions may perform that function. This does not work for the Kac ring model but may be effective in other cases. The correct view in the case of gases may very well be that statistics is never done on a mechanical system which is completely shielded from outside influences. Therefore, a hamiltonian does not describe the system completely, and stray interactions from the outside randomize certain features which would remain fixed according to the laws of mechanics.

Whatever the exact nature of the averaging process may be which yields a result such as (18.44), the original argument yielding the equivalent answer (18.34) is simpler and in some obscure way equivalent. The assumption is contained in equation (18.32), which states that before a ball meets a marker point there is no correlation between the color of the ball and the presence of the marker. The "time-reversed" derivation of (18.35)

¶ J. M. Luttinger and W. Kohn. *Phys. Rev.* **109**, 1892 (1958).

shows us that this means also that there is a definite correlation between the color of the ball and the presence of the marker *after they have met*. This feature has its equivalent in the derivation of the Boltzmann equation, starting right with (18.02). The feature is the *Stosszahlansatz*, the assumption that the velocity distribution of molecules about to have a collision is the same as that of molecules generally. This assumption is not valid after a collision if it is valid before. It brings in the correct type of irreversibility in a way which is plausible and natural, but which cannot be justified on purely mechanical grounds. Another point is connected with this one. While the final answers of kinetic theory can perhaps be laboriously justified for some sort of ensemble, they are always applied to an individual system. An *individual* galvanometer shows a certain type of damping, not an ensemble of similar galvanometers in a large collection. This aspect of statistics and kinetic theory is not explained by any of the preceding arguments, but needs the theory of fluctuations for an adequate justification.

18-5. Relaxation rate spectrum for Maxwellian molecules

In the last few years, experimentalists and theorists closely associated with experimental work have preferred to discuss kinetic phenomena in terms of the so-called relaxation time approximation. In this approximation the collision operator of the Boltzmann equation is replaced by an expression of the form

$$\frac{1}{\tau} \{ f(\mathbf{v}, \mathbf{r}) - f_0(\mathbf{v}, \mathbf{r}) \} \qquad (18.45)$$

Here f_0 is a local Maxwellian distribution, that is, a Maxwellian function having the same density, mean drift velocity, and energy as f. The quantity τ is sometimes made to depend on $\mathbf{v}$ as well as the local pressure and temperature. It is obvious that replacement of the right-hand side of (18.16) by (18.45) is a great simplification. For this reason we shall use it ourselves in the next two chapters.

It is widely supposed that the relaxation time approximation is a crude empirical device having no theoretical basis. This is not so. The Boltzmann operator is very naturally analyzed in terms of its relaxation rate spectrum, that is, in terms of particular solutions which decrease exponentially in time. That this might be so is hinted at in the "out" term of the collision operator. This term, introduced in (18.03), has indeed the same form as the first term in (18.45). However, the "in" term (18.14) does not resemble the second half of (18.45), so that the hint does not represent a completely valid argument.

It is well to precede the general discussion of this topic by an example for which the relaxation rate spectrum can be worked out in full detail. This is the example of molecules repelling each other as the inverse fifth power of the distance; such molecules are called Maxwellian. It is unfortunate that this type of force does not correspond to a situation occurring in nature. Attraction by this power law occurs between ions and polarizable atoms, but it is the repulsive case for which the theory is simple.

We shall accept without proof from classical mechanics that for any simple power law of force between two bodies the angular distribution of their scattering is independent of the velocity of encounter, provided it is analyzed in the center of mass frame of reference. There remains then the second question, why Maxwellian molecules are particularly simple to handle in kinetics. This is cleared up by a dimensional consideration concerning the cross section. An energy expression of the form

$$E = \tfrac{1}{2} m v^2 + \frac{\alpha}{r^4} \tag{18.46a}$$

contains two parameters, α/m and $E/m = \tfrac{1}{2} v_0^2$, out of which the scattering cross section must be made up. There is only one way to make from these quantities something which equals the square of a length, and that is by taking

$$\sigma = c \left(\frac{\alpha}{m}\right)^{1/2} \frac{1}{v} \tag{18.46b}$$

with c a pure number. Thus the cross section varies inversely as the speed. Now we see from (18.02) that the rate of binary encounters is governed by the product of the cross section and the relative speed. By (18.46b) this quantity is a constant. Having accepted earlier that the angular aspect of the scattering pattern is always the same, we now have the result that the rate of scattering into any angle is independent of the speed. The collision integral proper, which in (18.11) is made to precede the averaging over velocity, is thus independent of velocity. By this feature, velocity can be scaled entirely in terms of temperature; collision rates and temperature become independent phenomena, and kinetic problems can be solved at once for all temperatures. We shall take advantage of this feature by using a simplified notation for the velocity, setting

$$\tfrac{1}{2} m \beta v^2 \rightarrow v^2 \tag{18.47}$$

Equation (18.47) will be used in the rest of this section.

The relaxation of a non-Maxwellian distribution to Maxwellian form is governed by (18.16). It is clearly density dependent, as the equation is non-linear. If the departure from the Maxwellian character is weak we can linearize the equation by the substitution

$$f(\mathbf{v}, t) = n \pi^{-3/2} e^{-v^2} \{1 + g(\mathbf{v}, t)\} \qquad (18.48)$$

and the neglect of products of the small quantities g. Equation (18.16) becomes then

$$e^{-v^2} \frac{\partial g}{\partial t} = -\frac{n}{\pi^{3/2}} \iiiint e^{-(u^2+v^2)} \{g(\mathbf{u})+g(\mathbf{v})-g(\mathbf{u}')-g(\mathbf{v}')\} \, d\mathbf{u} \, [\sigma'w] \, d\Omega_{\mathbf{w}'} \qquad (18.49)$$

Here the two factors in the square bracket still have their dimension, together with n and t, while the remainder is rendered dimensionless in accordance with (18.47).

Equation (18.49) is a linear equation in the unknown function g, valid for small departures from equilibrium. It is meaningful to ask for the eigenvalues of the integral operator on the right. This is what is meant by eigenvalues of the collision operator in the case of gases. For dilute electrons or ions in a gas of much higher density, collisions occur generally with foreign objects; the Boltzmann equation is linear in the first place, and the linearization procedure is not necessary.

Before studying (18.49) generally, we first treat the completely soluble case of Maxwellian molecules. We can then employ (18.46b) on the two factors in square brackets and set

$$\sigma' w = \left(\frac{\alpha}{m}\right)^{1/2} \Gamma(\chi) \qquad (18.50)$$

where $\Gamma(\chi)$ is a dimensionless quantity independent of w; it is not necessarily normalizable as a probability because it may very well diverge for forward scattering. It is, however, *proportional* to the probability per unit solid angle for scattering by an angle χ. Expression (18.49) then becomes

$$e^{-v^2} \frac{\partial g}{\partial t} = -n \left(\frac{\alpha}{\pi^3 m}\right) \iiiint\!\!\int e^{-(v^2+u^2)}$$
$$\{g(\mathbf{u}) + g(\mathbf{v}) - g(\mathbf{u}') - g(\mathbf{v}')\} \, d\mathbf{u} \, \Gamma(\chi) \, d\Omega_{\mathbf{w}'} \qquad (18.51)$$

We associate with (18.51) the following eigenvalue problem

$$\frac{1}{\pi^{3/2}} \iiiint e^{-(v^2+u^2)} \{g_s(\mathbf{u})+g_s(\mathbf{v})-g_s(\mathbf{u}')-g_s(\mathbf{v}')\} \, d\mathbf{u} \, \Gamma(\chi) \, d\Omega_{\mathbf{w}'} = \lambda_s \, e^{-v^2} g_s(\mathbf{v}) \qquad (18.52)$$

and we are asking for the solutions g_s and the associated eigenvalues λ_s of this equation.

To start out with, there are five solutions of eigenvalue zero, namely,

$$g(\mathbf{v}) = 1 \tag{18.53a}$$

$$g(\mathbf{v}) = v_x \tag{18.53b}$$

$$g(\mathbf{v}) = v_y \tag{18.53c}$$

$$g(\mathbf{v}) = v_z \tag{18.53d}$$

$$g(\mathbf{v}) = v^2 \tag{18.53e}$$

These are the usual five conserved quantities. The five solutions are in fact also eigenfunctions of eigenvalue zero in the more general case (18.49). Eigenvalue zero means equilibrium and is of no further interest here. We wish to see the solutions having a nonzero relaxation rate.

It has been found that the solutions of (18.52) are the same as the polynomial solutions of the three-dimensional harmonic oscillator, separated in polar coordinates.** We shall give a proof of this now. First of all, the operator in (18.52) is a linear operator which is such that each of its four pieces is invariant under rotations. It follows then from group theory that the angular part of the eigenfunctions must be $Y_{lm}(\vartheta, \varphi)$, with degeneracy in the quantum number m. We can thus restrict ourselves to consideration of functions of the speed v, multiplied with $P_l(\cos \vartheta)$. Let us write the harmonic oscillator functions in the form

$$\Psi_l^s(\mathbf{v}) = e^{-\frac{1}{2}v^2} H_l^s(v) \, P_l(\cos \vartheta) \tag{18.54}$$

H_l^s are polynomials of degree $2s+l$ in v, having fixed parity and the lowest power equal to l. The first one, H_0^0, equals 1; the next, H_1^0 equals v. They obey the equation

$$-\nabla_v^2 \, \Psi_l^s(\mathbf{v}) + v^2 \, \Psi_l^s = (4s+2l+3) \, \Psi_l^s \tag{18.55}$$

They are orthogonal to each other when integrated over $d\mathbf{v}$. We now assert that the solutions of (18.52) are

$$g_l^s(\mathbf{v}) = H_l^s(v) \, P_l(\cos \vartheta) = e^{+\frac{1}{2}v^2} \Psi_l^s(\mathbf{v}) \tag{18.56}$$

These functions are orthogonal to each other in the presence of a weight factor e^{-v^2}. This implies in particular that the product $e^{-v^2} g$ is orthogonal to unity for all but the function g_0^0.

** C. S. Wang Chang and G. E. Uhlenbeck, unpublished. For a published version see L. Waldmann. *Handbuch der Physik*. **12**, 38 and 39.

Now if we take a function such as $e^{-\frac{1}{2}(v^2+u^2)} g_l^s(\mathbf{u})$ we see from (18.55) that it satisfies a six-dimensional harmonic oscillator equation of the form

$$\{-\nabla_v^2 - \nabla_u^2 + v^2 + u^2\}\,\phi = (4s+2l+6)\,\phi \tag{18.57}$$

It is also possible to show that the six-dimensional laplacian is preserved in passing from $(\mathbf{u}, \mathbf{v})$ to $(\mathbf{u}', \mathbf{v}')$. This is not so difficult as it seems. In passing from the left to the right of (18.13), one has one translation coordinate which is always the same; the remaining variables can be given the symmetric form (18.13), leaving just a pair of polar coordinates which must actually be transformed. The term $u^2 + v^2$ also transforms into $u'^2 + v'^2$ by (18.05). It follows that $e^{-\frac{1}{2}(v^2+u^2)} g_l^s(\mathbf{u}')$ also satisfies equation (18.57). It is therefore a linear combination of eigenfunctions having that eigenvalue. We thus get the development

$$g_l^s(\mathbf{u}') = \sum_{\substack{l',s',m' \\ l'',s''}} \kappa_{l,l',l'',m'}^{s,s',s''}\, H_{l'}^{s'}(v)\, Y_{l'}^{m'}(\vartheta, \varphi)\, H_{l''}^{s''}(u)\, Y_{l''}^{-m'}(\gamma, \psi) \tag{18.58}$$

with the restriction

$$l' + l'' + 2\,s' + 2\,s'' = l + 2\,s \tag{18.59}$$

The restriction on m' in the two spherical harmonics is well known from angular momentum composition. By (18.52) we must integrate $e^{-(v^2+u^2)} g_l^s(\mathbf{u}')$ over $d\mathbf{u}$. In this process all terms having l'' and s'' different from zero die. This leaves just the angular function $Y_{l'}^0$. It is again known from angular momentum composition that in such a case l' must equal l. Hence finally

$$e^{-(v^2+u^2)} g_l^s(\mathbf{u}') = \kappa_l^s\, e^{-(v^2+u^2)} g_l^s(\mathbf{v}) + \text{terms whose average over } \mathbf{u} \text{ is zero} \tag{18.60}$$

The property of $g_l^s(\mathbf{v})$ of solving (18.52) is thereby proved.

There remains the task of computing the coefficient κ_l^s. This task requires knowledge of the relations

$$u'^2 = F(\mathbf{u}, \mathbf{v}, \chi, \varepsilon) \tag{18.61}$$

and

$$u_z' = u' \cos \gamma = G(\mathbf{u}, \mathbf{v}, \chi, \varepsilon) \tag{18.62}$$

where χ and ε are the angles of scattering in the center of mass system. These relationships are written down in Section 21-1. For the purpose at hand we do not need to know them completely. For the polynomial surviving in the passage from (18.58) to (18.60) is the one of highest power in v. This means that we can get κ_l^s by comparing the coefficients of the highest powers of u' and v only. It follows that it is sufficient to find the coefficient of the highest power in v in (18.61) and (18.62). Since these relations are homogeneous in u and v we may get those coefficients by setting

$$\mathbf{u} \approx 0 \tag{18.63}$$

Equation (18.06) becomes then

$$\mathbf{v} = \mathbf{w} + 0(u) \tag{18.64}$$

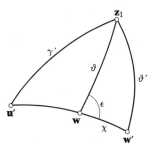

Fig. 18.2. Spherical triangles used in determining eigenvalues of the relaxation rate spectrum of Maxwellian molecules. In such a diagram, vectors show up through their piercing points, polar angles as sides, azimuths as angles.

We can further employ (18.04) in the form

$$\mathbf{u}' + \mathbf{v}' = \mathbf{v} + 0(u)$$

or with (18.64)

$$2\,\mathbf{u}' = \mathbf{v} - (\mathbf{v}' - \mathbf{u}') + 0(u)$$

which yields

$$2\,\mathbf{u}' = \mathbf{w} - \mathbf{w}' + 0(u) \tag{18.65}$$

We now first square (18.65) and get

$$4\,u'^2 = 2\,w^2 - 2\,w^2 \cos^2 \chi + 0(u)$$

which with (18.64) becomes

$$u' = v \sin \tfrac{1}{2}\chi + 0(u) \tag{18.66}$$

As a second step we introduce a polar z axis as shown in Fig. 18.2 and orient our velocity vectors about that axis to define our spherical harmonics.

We employ again (18.65) and the law of cosines for spherical triangles and get

$$2 u' \cos \gamma' = w \cos \vartheta - w \cos \vartheta' + 0(u)$$

or

$$2 u' \cos \gamma' = v \cos \vartheta - v \{\cos \vartheta \cos \chi + \sin \vartheta \sin \chi \cos \varepsilon\} + 0(u)$$

Combining terms and dividing by (18.66), we obtain from this

$$\cos \gamma' = \cos \vartheta \sin \tfrac{1}{2}\chi - \sin \vartheta \cos \tfrac{1}{2}\chi \cos \varepsilon + 0(u) \qquad (18.67)$$

From (18.67) we finally get by the composition law for spherical harmonics

$$P_l(\cos \gamma') = P_l(\cos \vartheta) P_l(\sin \tfrac{1}{2}\chi) + 2 \sum_{m=1}^{l} (-)^m \frac{(l-m)!}{(l+m)!}$$
$$P_l^m(\cos \vartheta) P_l^m(\sin \tfrac{1}{2}\chi) \cos m\varepsilon + 0(u)$$

Here ε is an azimuth of integration shown in Fig. 18.2 and has no further influence on the nature of the encounter. Integration over ε removes therefore all but the first term, and we are left with

$$P_l(\cos \gamma') \approx P_l(\cos \vartheta) P_l(\sin \tfrac{1}{2}\chi) + 0(u) \qquad (18.68)$$

Combination of (18.66) and (18.68) yields thus finally

$$u'^{l+2s} P_l(\cos \gamma') \approx v^{l+2s} \sin^{l+2s} \tfrac{1}{2}\chi P_l(\cos \vartheta) P_l(\sin \tfrac{1}{2}\chi) + 0(u) \quad (18.69)$$

The parameter in (18.60) equals therefore

$$\kappa_l^s = \sin^{l+2s} \tfrac{1}{2}\chi P_l(\sin \tfrac{1}{2}\chi) \qquad (18.70)$$

Repetition of this calculation for v' yields the same formula, with $\cos \tfrac{1}{2}\chi$ substituted for $\sin \tfrac{1}{2}\chi$.

Substitution of these results into (18.52) yields

$$\lambda_l^s = 2\pi \int_0^\pi \{1 + \delta_{l,0}\, \delta_{s,0} - \sin^{l+2s} \tfrac{1}{2}\chi P_l(\sin \tfrac{1}{2}\chi)$$

$$(18.71a)$$

$$- \cos^{l+2s} \tfrac{1}{2}\chi P_l(\cos \tfrac{1}{2}\chi)\} \Gamma(\chi) \sin \chi \, d\chi$$

and the associated relaxation rates R_l^s from (18.51)

$$R_l^s = n \left(\frac{\alpha}{m}\right)^{1/2} \lambda_l^s \qquad (18.71b)$$

Equation (18.71) yields the three zero relaxation rates listed in (18.53) and relates a number of the succeeding ones by trigonometric identities which the reader is invited to verify. In Table 18–1 these values are listed in units of η/p, where η is the viscosity and p the gas pressure. The slowest relaxation rate is $\frac{2}{3}$ on this scale; it will be the one occurring in most problems. Viscosity happens to produce an ellipsoidal distortion of the velocity distribution which belongs to $l = 2$; it decays at a slightly faster rate, namely 1.

TABLE 18-1. List of the first few relaxation rates R_l^s for the relaxation rate spectrum of Maxwellian molecules

l \ s	0	1	2	3
0	0	0	$\frac{2}{3}$	1
1	0	$\frac{2}{3}$	1	
2	1	$\frac{7}{6}$		
3	$\frac{3}{2}$			

Note. The units are η/p, where η is the viscosity, p the gas pressure.

To conclude, we see from the preceding analysis that if the interaction potential between two molecules varies inversely as the fourth power of the distance, and classical mechanics is applicable, the linearized Boltzmann equation can be analyzed completely in terms of particular solutions which decay exponentially in time. As these solutions form a complete set, any other solution is a linear combination of these particular solutions. When we now come to deal with a gas kinetic problem we can make use of this circumstance. The most important part of an actual solution is that part which has the slowest relaxation rate under the circumstances considered. To the extent that exclusive consideration of this slowest decay mode is justified, the relaxation time approximation (18.45) is also justified. To make the approximation work as well as possible the angular distortion of the velocity distribution in a particular application should be examined, so that the slowest decay rate going with the correct spherical harmonic can be selected.

18-6. Formal relaxation theory of the Boltzmann equation

For molecules which are not Maxwellian, but interact with each other according to some different central force law, the eigenvalue problem for the linearized Boltzmann equation can be defined, and certain indications given about the nature of the eigenfunctions. At the present time it is not known whether the solutions so obtained are complete. If the plausible hypothesis is made that they are, then the general solution of the linearized Boltzmann equation can be analyzed in terms of these eigenfunctions.

As our first step we define the linearized Boltzmann operator $\mathscr{C}(g)$ so as to separate the time from the other variables in (18.49):

$$e^{-v^2}\mathscr{C}(g) = \frac{1}{\pi^{3/2}} \iiiint e^{-(u^2+v^2)}\{g(\mathbf{u})+g(\mathbf{v})-g(\mathbf{u}')-g(\mathbf{v}')\}\,d\mathbf{u}\,[\sigma'w]\,d\Omega_{\mathbf{w}'}$$

(18.72)

We then take as the eigenvalue problem associated with the operator $\mathscr{C}(g)$ the equation

$$\mathscr{C}(g) = \kappa\,g \tag{18.73}$$

A certain amount of information about the solutions of (18.73) can be built up by elementary methods. We can start by observing that $e^{-v^2}\mathscr{C}$ is selfadjoint, that is,

$$\iiint h(\mathbf{v})\,e^{-v^2}\mathscr{C}(g)\,d\mathbf{v} = \iiint g(\mathbf{v})\,e^{-v^2}\mathscr{C}(h)\,d\mathbf{v} \tag{18.74}$$

This fact is proved by separate examination of the four terms in (18.72) and exchange of integration variables, as in the proof of the H-theorem. It then follows from (18.73) and (18.74) that

$$(\kappa - \kappa')\iiint g(\mathbf{v})\,g'(\mathbf{v})\,e^{-v^2}\,d\mathbf{v} = 0 \tag{18.75}$$

The functions $g(\mathbf{v})$ are thus orthogonal to each other in conjunction with the weight factor e^{-v^2}. Equation (18.75) excludes the presence of complex eigenvalues κ or eigenfunctions g; for if there were such an eigenvalue we would set $g'(\mathbf{v}) = g^{\star}(\mathbf{v})$, and find $\kappa^{\star} = \kappa$; g and $g^{\star}$ are then degenerate and their real and imaginary parts are separately eigenfunctions. By a refinement of this argument†† we can also exclude negative eigenvalues, for we can write (18.73) in the form

$$\kappa = \frac{\iiint g(\mathbf{v})\,\mathscr{C}(g)\,e^{-v^2}\,d\mathbf{v}}{\iiint g^2(\mathbf{v})\,e^{-v^2}\,d\mathbf{v}} \tag{18.76}$$

This expression can be handled according to the model of (18.25) and (18.26), through permutation of the variables of integration. The result analogous to (18.26) is that we can make in the numerator of (18.76) the substitution

$$g(\mathbf{v}) \to \tfrac{1}{4}\{g(\mathbf{u}) + g(\mathbf{v}) - g(\mathbf{u}') - g(\mathbf{v}')\} \tag{18.77}$$

†† G. E. Uhlenbeck and G. W. Ford. *Lectures in Statistical Mechanics.* Providence: American Mathematical Society, 1963, p. 82.

which, with (18.72), puts (18.76) into the form

$$\kappa = \frac{1}{4\,\pi^{3/2}} \frac{\iiiint\!\!\iiint e^{-(u^2+v^2)}\{g(\mathbf{u})+g(\mathbf{v})-g(\mathbf{u'})-g(\mathbf{v'})\}^2\,d\mathbf{u}\,d\mathbf{v}\,[\sigma'w]\,d\Omega_{\mathbf{w}'}}{\iiint e^{-v^2}\{g(\mathbf{v})\}^2\,d\mathbf{v}} \tag{18.78}$$

Expression (18.78) is manifestly positive or zero, as stated. The only functions having the eigenvalue zero are the ones listed in (18.53).

A further step which can conveniently be taken is to observe that the integral equation (18.73) must be separable in polar coordinates, with the angular functions being the spherical harmonics $Y_l^m(\vartheta, \phi)$. This follows because the operator $\mathscr{C}(g)$ is linear and invariant with respect to a rotation of the coordinate system. Group theory then imposes the functions Y_l^m, with degeneracy with respect to the index m. The functions g can therefore be indexed as $g_{l,m}^s(\mathbf{v})$ and the corresponding eigenvalues as κ_l^s. The index s is to indicate a set of radial functions.

The question whether the spectral index s is discrete or continuous is the one for which no complete answer is available at this time. For Maxwellian molecules s is discrete throughout. Ford and Schreiber[‡‡] have shown for the case of a gas of hard spheres that there is continuous spectrum, with at least one discrete nonzero eigenvalue separated from this continuum. That value goes with the angular index $l = 1$. Phillips,[§§] on the other hand, established by the Ritz method that the smallest nonzero eigenvalue for $l = 0$ is also bounded away from zero. If one follows him in the assumption that this is the smallest κ, then the slowest decay rate R for hard spheres comes out to be

$$R = 0.97\,\tfrac{128}{15}\,\pi^{1/2}\,n\,a^2\left(\frac{kT}{m}\right)^{1/2} \tag{18.79}$$

Here a is the hard sphere radius. To get a feeling for this number we may compare it with the mean free time τ between collisions, which we shall work out in Section 19-1. We anticipate this result, which comes out to be

$$\frac{1}{\tau} = \sqrt{2}\,\bar{v}\,n\,\sigma = 16\,\pi^{1/2}\,n\,a^2\left(\frac{kT}{m}\right)^{1/2} \tag{18.80}$$

We form the quantity $R\tau$ and find

$$R\tau = \tfrac{8}{15} \tag{18.81}$$

[‡‡] G. W. Ford and M. Schreiber. Private communication.
[§§] N. J. Phillips. *Proc. Phys. Soc.* **73**, 800 (1959).

The slowest relaxation rate is therefore smaller, but not so very much smaller, than $1/\tau$.

If we pass from the case of hard spheres to other force laws the information available becomes very much scantier. The general properties derived here remain valid for other central forces. Nothing is known, on the other hand, concerning the question whether the spectrum is continuous or discrete.

The analysis of the linearized Boltzmann equation in terms of exponential decay modes has a definite meaning in physical problems if the eigenfunctions so obtained form a complete set. This appears highly probable on physical grounds but is unproven mathematically. If the conjecture is accepted as true, the entire field gets a certain "finish" which appears plausible physically. It is then possible to expand an "arbitrary" function $e^{-v^2} G(\mathbf{v})$ in terms of these eigenfunctions, thus

$$e^{-v^2} G(\mathbf{v}) = \sum_{s,l,m} \gamma_s^{l,m} e^{-v^2} g_{l,m}^s (\mathbf{v}) \tag{18.82}$$

If the normalization

$$\int\int\int e^{-v^2} \{g_{l,m}^s(\mathbf{v})\}^2 \, d\mathbf{v} = 1 \tag{18.83}$$

is adopted, then, because of the orthogonality relation (18.75), $\gamma_s^{l,m}$ equals

$$\gamma_s^{l,m} = \int\int\int e^{-v^2} G(\mathbf{v}) g_{l,m}^s(\mathbf{v}) \, d\mathbf{v} \tag{18.84}$$

The Boltzmann equation (18.49) can now be solved completely in terms of its spectrum of relaxation rates. Comparison of (18.49), (18.72), and (18.73) shows that the eigenfunctions decay in time according to the scheme

$$g(\mathbf{v}, t) = g_{l,m}^s(\mathbf{v}) \exp[-n\kappa_l^s t] \tag{18.85}$$

If we now assume that the function $e^{-v^2} G(\mathbf{v})$ was the velocity distribution at the time $t = 0$, then we get from (18.82) and (18.85) the development of this function in time:

$$e^{-v^2} G(\mathbf{v}, t) = \sum_{s,l,m} \gamma_s^{l,m} e^{-v^2} g_{l,m}^s(\mathbf{v}) \exp[-n\kappa_l^s t] \tag{18.86}$$

It should be repeated in closing that the final result (18.86) is not mathematically proven at this time. Since the utility of the eigenvalue analysis (18.73) is finally based on a formula such as (18.86), the entire relaxation approach to the Boltzmann equation is still in doubt. However, its physical plausibility is extremely great. The relaxation ansatz (18.45) thereby also acquires a high degree of plausibility.

A. RECOMMENDED PROBLEMS

1. Compute the multiplying factor and the additive constant which convert the Boltzmann H into the entropy, as defined by (5.13).

2. Show from the Boltzmann transport equation (18.16) that the validity of the law of atmospheres (4.31) is not associated with any modification of the Maxwellian distribution.

B. GENERAL PROBLEMS

3. Suppose we are investigating the mutual scattering of two particles whose interaction varies inversely as a simple power of the distance. Show that if classical mechanics is applicable the following statements are true:

(a) The angular pattern of scattering in the center of mass system is independent of speed.

(b) The differential cross section varies as a simple power of the speed. Find that power.

4. Work out proofs of the four laws of spherical trigonometry. If we deal with a triangle of sides a, b, c (in units of the radius of the sphere), and angles α, β, γ, we have the following:

(a) First law of cosines (three sides, one angle).
$$\cos c = \cos a \cos b + \sin a \sin b \cos \gamma$$

(b) Law of sines (two sides, two opposing angles).
$$\sin a \sin \beta = \sin b \sin \alpha$$

(c) Law of cotangents (two sides, enclosed and adjoining angle).
$$\cotan a = \cotan b \cos \gamma + \cosec b \cotan \alpha \sin \gamma$$

(d) Second law of cosines (one side, three angles).
$$\cos \gamma = -\cos \alpha \cos \beta + \sin \alpha \sin \beta \cos c$$

5. Work out tables of the harmonic oscillator functions, as far as they occur in Table 18–1, taking the equation in the form (18.55). Verify the eigenvalues of the linearized Boltzmann operator appearing in that table, accepting formula (18.71).

19

Transport properties of gases

In the preceding chapter, the kinetic theory of gases was introduced and analyzed from a fundamental viewpoint. The theory was employed to throw light on questions which arise in statistical mechanics. In this chapter kinetic theory will be applied to problems which are outside equilibrium statistics. The mathematical methods employed will be simpler than those used earlier; for this reason, the text is written in such a way as to be independent of Chapter 18.

Since the kinetic theory of gases is a complete molecular theory, it is capable in principle of predicting experimental results under a wide variety of conditions. Many such situations have been analyzed in the last hundred years, and some of them were of great value in corroborating the basic assumptions of the theory. One may mention in this connection the escape of a gaseous beam through an opening in an oven, or the passage of gases through porous plugs. Nowadays, interest in formulas for experiments of this type is mostly incidental, in connection with experiments which have other purposes. The part of kinetic theory which has retained the general interest is the kinetic derivation of the transport coefficients of gases. These transport coefficients describe the reaction of a gas to mild but very common departures from thermal equilibrium. The most common ones are:

(a) *Viscosity*. If a fluid is subject to a gradient in the z direction of its

416

body velocity c which is perpendicular to the z direction, a shearing stress π is set up which is given by

$$\pi = \eta \frac{\partial c}{\partial z} \tag{19.01}$$

where η is the *coefficient of viscosity*.

(b) *Conduction of heat.* If a body is subject to a temperature gradient $\partial T/\partial z$ heat will flow down the gradient according to *Fourier's law*

$$h = -\lambda \frac{\partial T}{\partial z} \tag{19.02}$$

where h is an energy current density and λ is the *conductivity of heat*.

(c) *Diffusion.* If a density gradient $\partial n/\partial x$ exists for a substance within a body, and if the density gradient is balanced in such a way that there is no gradient of the total pressure, the substance will spread out according to *Fick's law*

$$i = -D \frac{\partial n}{\partial x} \tag{19.03}$$

Here i is a particle current density and D is the *coefficient of diffusion*.

(d) *Thermal diffusion.* If a gas mixture contains a temperature gradient $\partial T/\partial z$ a particle current will be set up for either species, owing to the temperature gradient. The current is expressed in the form

$$i = k\,D \frac{1}{T} \frac{\partial T}{\partial z} \tag{19.04}$$

Here k is called the thermal diffusion ratio and may have either sign. Combination of (19.03) and (19.04) allows for the possibility of a steady state concentration gradient in a gas if two ends are kept at different temperatures. The formula is

$$\frac{\partial n}{\partial z} = \frac{k}{T} \frac{\partial T}{\partial z} \tag{19.05}$$

The phenomenon of thermal diffusion is interesting because it depends for its bare existence on the law of force of the molecules. It vanishes exactly for Maxwellian molecules. For hard spheres and thus of course also for actual molecules, the sign of k is such that the heavier species tends to collect in the colder region. By itself, the effect is very small, of the order of 1%, but it can be enhanced in diffusion columns containing many units, such as the one shown in Fig. 19.1. Each chamber supports only a concentration gradient of 1 % or less, but the cool gas, enriched with heavy material, sinks on one side, while the light warm gas rises.

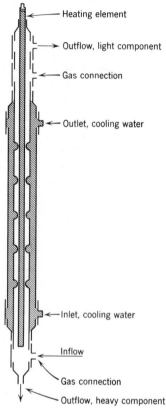

Fig. 19.1. Schematic picture of a diffusion column for isotope separation.

The effect thus can be multiplied so that it leads to a substantial enrichment at the top and bottom. In this form, the effect is good enough to permit separation of isotopes.

It is to be noticed that the empirical definitions (19.01) to (19.05) apply to a wider class of cases than just gases. In particular, they all apply to liquids and liquid solutions. Study of the case of gases thus serves a double purpose. First, it permits us to compute the simplest possible nonequilibrium quantities for a gas, and thereby yields information about the gas molecules. Second, it sets a precedent for nonequilibrium thermodynamics in general, giving us hope that equilibrium statistics can be extended.

19-1. Elementary theory of transport phenomena in gases

A very elementary approach to transport coefficients in gases can be obtained from the following simple reasoning.

Suppose we take a gas at equilibrium and divide it mentally by the insertion of an imaginary plane, as shown in Fig. 19.2. We can then ask

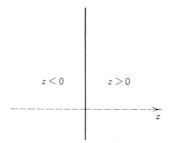

$z < 0$ $z > 0$

Fig. 19.2. Study of the flow properties of a Maxwellian gas by insertion of an imaginary dividing plane.

the question: how many molecules cross a unit surface of this plane in unit time? This question has a very simple answer. Let the plane be the plane $z = 0$ in a rectangular coordinate system. We then take the Maxwellian distribution, normalized as in (18.01),

$$f(\mathbf{v}, \mathbf{r}) = n \left(\frac{m\beta}{2\pi}\right)^{3/2} \exp[-\tfrac{1}{2}\beta m v^2] \tag{19.06}$$

and find the total particle current density across $z = 0$ arising from all molecules having $v^z > 0$. We get this by multiplying (19.06) with v^z and integrating over all velocities for which v^z is positive. This yields

$$j_+ = \int_{-\infty}^{+\infty} dv^x \int_{-\infty}^{+\infty} dv^y \int_0^\infty dv^z \, v^z f(\mathbf{v})$$

which comes out to be

$$j_+ = n \left(\frac{kT}{2\pi m}\right)^{1/2} \tag{19.07}$$

This number is very closely associated with the mean speed $\bar{v}$ worked out in Problem 3, Chapter 4. We find

$$j_+ = \tfrac{1}{4} n \bar{v} \tag{19.08}$$

There is another current j_- from the positive to the negative side of the z plane. In equilibrium it is of course exactly the same as (19.08). The elementary theory of transport coefficients assumes that the result (19.08) is not seriously modified by the nonequilibrium aspects of the actual situation, and proceeds to compute transport of various quantities across the z plane with its help.

The element which brings the kinetics into this simple approach is the concept of a *mean free path l*. For a particle moving among stationary obstacles of cross section σ and density n the mean free path is given by the well-known relation

$$l = \frac{1}{n\,\sigma} \tag{19.09}$$

In the present instance we have to be a little more careful because all objects are in motion. For a particle of fixed speed the best definition is to define l as its speed divided by the collision rate. It would be possible to average this quantity directly. Rather than do this we shall in the following define it as the mean speed divided by the mean collision rate; this takes into account that the collision rate determines gaseous properties. The collision rate $R(v)$ of a molecule of fixed speed v is simply the integral in (18.03), namely,

$$R(v) = \int\int\int d\mathbf{u}\ |\mathbf{v} - \mathbf{u}|\ \sigma f(\mathbf{u})$$

where $f(\mathbf{u})$ is a Maxwellian distribution. The average collision rate R is then

$$\bar{R} = \frac{\int\int\int\int\int\int d\mathbf{u}\, d\mathbf{v}\ |\mathbf{v} - \mathbf{u}|\ \sigma f(\mathbf{u})\, f(\mathbf{v})}{\int\int\int f(\mathbf{v})\, d\mathbf{v}} \tag{19.10}$$

In the case of hard spheres, which have a constant collision cross section, evaluation of (19.10) for a Maxwellian distribution is elementary but slightly tedious. We find

$$\bar{R} = \frac{4}{\pi} \left(\frac{k\,T}{m}\right)^{\!1/2} n\,\sigma = \sqrt{2}\,\bar{v}\, n\, \sigma \tag{19.11}$$

Now we have by definition

$$l = \frac{\bar{v}}{R}$$

and therefore

$$l = \frac{1}{\sqrt{2}\, n\, \sigma} \tag{19.12}$$

Dimensionally (19.12) and (19.09) are of the same type. In this sense the relation is valid for any law of force, excepting very long range forces such as the Coulomb force. This is especially true for that aspect of the two formulas which states that the product of the mean free path and the density of obstacles is a quantity which depends only on the individual encounters. The difference between (19.09) and (19.12) is specific for hard spheres. It expresses the fact that, since the molecules as well as their obstacles (which are other molecules), are in motion, the mean free path is decreased by a factor 0.707. That the factor is specific for the hard sphere model is seen by a consideration of Maxwellian molecules, as discussed in Chapter 18. In that case the scattering process is independent of the speed of encounter; there is then no difference between this speed being u or $|\mathbf{u} - \mathbf{v}|$; consequently, (19.09) and (19.12) must come out the same.

The way of estimating roughly the transport of a quantity across the z plane is to assume that for molecules making up the current j_+ the quantity is as it would be in equilibrium at a distance $z = -\kappa\, l$ with κ a number close to 1; for molecules making up the current j_-, on the other hand, it is as it would be at $z = +\kappa\, l$. This assumption produces a net flux of the quantity in question because it is varying along z.

Taking first the case of viscosity, we assume that the gas has a body velocity c_0 parallel to the z plane which changes slowly in the z direction. We are then interested in the transport of momentum across the plane; a rate of transport of momentum across a surface has the dimension of a force; in fact, it *is* a force. In the present instance, we consider the momentum transported with the current j_+; the momentum taken across is the one existing at $z = -\kappa\, l$. The total momentum current density going with (19.08) thus equals

$$\tfrac{1}{4}\, n\, \bar{v}\, (m\, c_0)_{z=-\kappa l} \tag{19.13}$$

This may be approximated as

$$\tfrac{1}{4}\, n\, \bar{v}\, m \left(c_0 - \kappa l\, \frac{\partial c_0}{\partial z} \right)$$

A similar contribution having opposite sign will arise from the current j_-. The total current density of momentum thus comes out to be

$$\tfrac{1}{4} n \bar{v} m \left\{ \left(c_0 - \kappa l \frac{\partial c_0}{\partial z} \right) - \left(c_0 + \kappa l \frac{\partial c_0}{\partial z} \right) \right\} = - \tfrac{1}{2} \kappa n \bar{v} m \, l \frac{\partial c_0}{\partial z}$$

The current density of momentum equals the shearing stress across the plane. The definition (19.01) therefore enters into play, and we get

$$\eta = \tfrac{1}{2} \kappa n \bar{v} m \, l \tag{19.14}$$

Substituting for the mean free path from (19.12), we obtain

$$\eta = \frac{1}{\sqrt{8}} \frac{\kappa m \bar{v}}{\sigma} \tag{19.15}$$

Substituting further for $\bar{v}$ according to Problem 3, Chapter 4,

$$\bar{v} = \left(\frac{8 \, k \, T}{\pi \, m} \right)^{\frac{1}{2}} \tag{19.16}$$

we get, on the assumption that $\kappa = 1$,

$$\eta = \left(\frac{m \, k \, T}{\pi} \right)^{\frac{1}{2}} \frac{1}{\sigma} \tag{19.17}$$

The remarkable thing about (19.14) or (19.17) is that it shows for η no dependence on pressure, only dependence on temperature. This prediction, which runs counter to intuitive feeling, was one of the first surprising results Maxwell predicted from his theory. It is in complete agreement with experiment. Modern, more accurate theories show that the κ in (19.17) is very close to 1.

The calculation just given needs only minor alterations to yield the conductivity of heat. The quantity which is being transported across the plane shown in Fig. 19.2 is now energy instead of momentum. We therefore replace the quantity $m \, c_0$ in (19.13) by ε and proceed as previously. When it comes to taking derivatives, we write

$$\frac{\partial \varepsilon}{\partial z} = c_v \frac{\partial T}{\partial z}$$

where c_v is the heat capacity of a single molecule at constant volume. Interpretation is now in terms of the definition (19.02) and yields

$$\lambda = \tfrac{1}{2} \kappa' n \bar{v} c_v \, l \tag{19.18}$$

where κ' is another pure number, similar to but not equal to κ. Comparison with (19.14) shows that this can be simply written as a proportion

$$\frac{\lambda}{\eta} = \frac{\kappa'}{\kappa}\frac{c_v}{m} \tag{19.19}$$

The quantity c_v/m is the heat capacity per unit mass, that is, the specific heat in the usual experimental units. A more careful computation shows that κ'/κ is not as close to 1 as κ was. The reason is that there is a strong correlation between the energy of a molecule and the contribution it makes to the current j_+ worked out in (19.08). The correlation is positive, the molecules having the most energy also being the most rapid in crossing the z plane. We must therefore expect that κ', and also κ'/κ, are numbers substantially larger than 1. The formulas just given show that the conductivity of heat in gases shares with the viscosity the property of being dependent on temperature only, and not on pressure.

As our third transport phenomenon we take up diffusion. Diffusion demands in practice two species 1 and 2 which diffuse against each other. Otherwise we have a net difference in total density which equilibrates under a driving force according to the laws of mechanics. Once this observation is made we can proceed very much as before, basing ourselves on (19.08) for either species. Let us discuss the situation from the point of view of species 1. The quantity which is to be transported this time, in analogy to the model case (19.13), is the particle density itself. We therefore have to work out

$$i_1 = \tfrac{1}{4}\bar{v}_1(n_1)_{z=-\kappa_1 l_1} - \tfrac{1}{4}\bar{v}_1(n_1)_{z=+\kappa_1 l_1} \tag{19.20}$$

Developing as after (19.13) and using the definition (19.03), we find without difficulty

$$D_{12} = \tfrac{1}{2}\kappa_1 \bar{v}_1 l_1 \tag{19.21a}$$

It follows from the nature of the phenomenon that an opposite current exists of particles of species 2, and that the currents and the density gradients of the two species are approximately equal in magnitude and direction, and opposite in sense. We get from this that

$$D_{12} = D_{21} = \tfrac{1}{2}\kappa_2 \bar{v}_2 l_2 \tag{19.21b}$$

For both relations (19.21) to hold simultaneously, the constants κ_1 and κ_2 must clearly depend on the properties of both gases and on their proportion in the mixture. Substitution of (19.12) into (19.21) yields for the diffusion coefficient an expression varying inversely as the pressure at fixed temperature, and thus sets it apart from the first two coefficients.

Our fourth transport phenomenon, thermal diffusion, is not amenable to simple treatment. A clear indication of this is that the sign of the effect is not immediately obvious. We can therefore not give a simplified theory of the manner in which it operates. In fact we shall not treat it at all here. It is, however, the very fact of its ambiguity which makes it a valuable research tool for investigating forces between molecules.

19-2. Determination of transport coefficients from the Boltzmann equation

It is interesting that the results of the previous section could be obtained with the help of the Maxwellian velocity distribution function. For if that distribution is exactly valid at any position in space the mean velocity of the molecules in any direction is zero; the same is true for the various components of current which we discussed in that section. From this point of view the essential feature of the previous computations is the disturbance of the distribution due to external conditions. This disturbance was introduced in a very intuitive way by the device of a mean free path. It has been found since that one gets more reliable answers by calculating the modification of the Maxwellian distribution first, and then computing the transport phenomenon as an average over the disturbed distribution. This is the method we shall be following in the future.

In order to render this chapter independent of the preceding one, a short explanatory introduction of the Boltzmann transport equation will be given here. For all deeper questions associated with that equation, consultation of Chapter 18 is, however, essential.

The Boltzmann transport equation has as its object the Boltzmann distribution function which is a density $f(\mathbf{v}, \mathbf{r})$ in six-dimensional space indicating how many molecules have velocities within $dv_x\, dv_y\, dv_z$ of $\mathbf{v}$ and positions within $dr_x\, dr_y\, dr_z$ of $\mathbf{r}$. We then estimate the rate of change of this function through binary collisions and set

$$\frac{df}{dt} = \left(\frac{\partial f}{\partial t}\right)_{\text{coll}} \tag{19.22}$$

Even without collisions, the function f changes: the velocity $\mathbf{v}$ makes the representative points change their position and an acceleration $\mathbf{a}$ makes them change their velocity. In hydrodynamics, this is taken care of by the introduction of the *hydrodynamic total derivative*, which indicates changes taking place at a point moving with the fluid. The left-hand side of (19.22) is such a derivative; it has the form

$$\frac{df}{dt} = \frac{\partial f}{\partial t} + \mathbf{v} \cdot \frac{\partial f}{\partial \mathbf{r}} + \mathbf{a} \cdot \frac{\partial f}{\partial \mathbf{v}} \tag{19.23}$$

The tendency of the collision term is to restore thermal equilibrium locally. Even if only binary collisions are assumed, its exact form is very complicated. Since the term tends to make the distribution relax toward thermal equilibrium the relaxation approximation is suggestive:

$$\left(\frac{\partial f}{\partial t}\right)_{coll} = -\frac{f-f_0}{\tau} \tag{19.24}$$

Here f_0 is an equilibrium distribution locally, having the same mean density, mean velocity, and mean energy as f; τ is a number, the *relaxation time*, which is sometimes made a function of v. The discussion of the previous chapter indicates that with some judicious handling the approximation (19.24) is quite acceptable. With the substitutions (19.23) and (19.24), (19.22) becomes a simplified Boltzmann equation which reads

$$\frac{\partial f}{\partial t} + \mathbf{v} \cdot \frac{\partial f}{\partial \mathbf{r}} + \mathbf{a} \cdot \frac{\partial f}{\partial \mathbf{v}} = -\frac{f-f_0}{\tau} \tag{19.25}$$

Let us now start discussing viscosity with the help of (19.25). As our zeroth approximation we take a gas having Maxwellian velocity distribution, with a drift term in the y direction. This drift velocity varies linearly with z, in response to a shearing motion of the bounding surfaces. In other words, we assume the function f modified so as to take the form

$$f(\mathbf{v}, \mathbf{r}) = n\left(\frac{m\beta}{2\pi}\right)^{3/2} \exp[-\tfrac{1}{2}m\beta(v_x^2+(v_y-\gamma z)^2+v_z^2)] \cdot \{1 + g(v_x,v_y-\gamma z,v_z)\} \tag{19.26}$$

Since the terms on the left of (19.25) vanish at equilibrium they can be treated as small. The modification $g(\mathbf{v})$ of the Maxwellian distribution can therefore be neglected on that side. This means that the unknown function g appears only in the relaxation term on the right, and can be solved for directly. We find without difficulty

$$g(\mathbf{v}) = -\gamma \tau m \beta (v_y - \gamma z) v_z \tag{19.27}$$

Equation (19.27) shows, incidentally, that the treatment given is exact for Maxwellian molecules. For the $g(\mathbf{v})$ shown is the eigenfunction $g_{2,2}^0$ of the linearized collision operator, as is verified by consultation of (18.54) and (18.56). The relaxation rate for this eigenfunction is given by (18.71). It equals

$$R_2^0 = n \iiint \{1 - \sin^2 \tfrac{1}{2}\chi\, P_2(\sin \tfrac{1}{2}\chi) - \cos^2 \tfrac{1}{2}\chi\, P_2(\cos \tfrac{1}{2}\chi)\}\, w\, d\sigma$$

which works out as

$$\frac{1}{\tau} = R_2^0 = \tfrac{3}{2}n \iiint \{1 - \sin^4 \tfrac{1}{2}\chi - \cos^4 \tfrac{1}{2}\chi\}\, w\, d\sigma \tag{19.28}$$

Table 18-1 shows that R_2^0 is not the slowest relaxation rate, but that R_0^2 and R_1^1 are smaller by a factor 2/3. Using simply the slowest rate would therefore bring about a substantial error. However, a refinement using the slowest rate going with the correct spherical harmonic removes the error here and probably does so in most applications. For the spherical harmonics index l is a good "quantum number" also for other laws of force, and insertion of the slowest τ for $l = 2$ would presumably make (19.27) correct in all cases.

The calculation of viscosity is completed by using (19.26) and (19.27) to compute the transport of y momentum across the z plane, as in the elementary calculation of the preceding section. We can do this now in a clean way, multiplying the distribution function $f(\mathbf{v}, \mathbf{r})$ with $m\,(v_y - \gamma z)\,v_z$ and integrating over $\mathbf{v}$. As expected, the undisturbed Maxwellian part of the distribution makes no contribution to the momentum current density, which is also a shearing stress π_{yz}. The stress arises therefore because of the presence of g and equals

$$\pi_{yz} = n \left(\frac{m\beta}{2\pi}\right)^{3/2} \gamma\,\tau\,m^2\,\beta\,\iiint (v_y - \gamma z)^2\,v_z^2\,\exp[-\tfrac{1}{2}m\{v_x^2 + (v_y - \gamma z)^2 + v_z^2\}]d\mathbf{v}$$

which works out to be

$$\pi_{yz} = \frac{n\,\gamma\,\tau}{\beta} \tag{19.29}$$

By construction γ is the velocity gradient in the system. The definition (19.01) therefore enters into play and yields for the viscosity

$$\eta = \frac{n\,\tau}{\beta} \tag{19.30}$$

or, since $n/\beta = p$,

$$\eta = p\,\tau \tag{19.31}$$

This formula was made use of in Table 18-1 where relaxation rates of Maxwellian molecules were quoted in units of p/η.

Formulas (19.30) and (19.31) have the disadvantage of obscuring the fact that η is independent of pressure at fixed temperature. For τ varies inversely as the density for most laws of force, as explained in connection with (18.80). It would be nice to make the substitution for a realistic case such as the hard sphere model. Unfortunately, the slowest relaxation time for $l = 2$ is not now in the literature for hard spheres. The quantity should certainly be bracketed by the estimates (18.79) and (18.80). The first formula gives the slowest relaxation rate altogether and should give a time which is too large; the second formula gives the mean free time between collisions, which should be too small. Indeed, the two estimates yield

$$\eta_{\mathrm{I}} = \frac{15\,\pi}{64\,\sqrt{2}}\frac{m\,\bar{v}}{\sigma} = 0.520\,\frac{m\,\bar{v}}{\sigma} \tag{19.32a}$$

$$\eta_{\mathrm{II}} = \frac{\pi}{8\,\sqrt{2}}\frac{m\,\bar{v}}{\sigma} = 0.278\,\frac{m\,\bar{v}}{\sigma} \tag{19.32b}$$

while, actually the estimate (19.15) is about right with a κ equal to 1. It reads

$$\eta_{\mathrm{III}} = \frac{1}{\sqrt{8}}\frac{m\,\bar{v}}{\sigma} = 0.3535\,\frac{m\,\bar{v}}{\sigma} \tag{19.32c}$$

There can be little doubt that if the correct relaxation time for $l = 2$ were available it would reproduce this last result fairly closely.

As our second example we take up heat conduction in a gas. The formalism for heat conduction is essentially the same as for viscosity. We must replace in (19.26) the idea of a variable body velocity by the idea of a variable temperature; this variable temperature will be written as

$$\beta = \beta_0 + \gamma\,z$$

The density n must be varied concurrently so as to keep the pressure constant. This means that

$$n = \frac{n_0}{\beta_0}(\beta_0 + \gamma\,z)$$

In the place of the modification (19.26) of (19.06) we now have a distribution function of the form

$$f = \frac{n_0}{\beta_0}\left(\frac{m}{2\,\pi}\right)^{3/2}(\beta_0 + \gamma z)^{5/2}\exp[-\tfrac{1}{2}\,m\,(\beta_0 + \gamma z)\,v^2]\,\{1 + g(\mathbf{v})\} \tag{19.33}$$

Insertion of this ansatz into the Boltzmann equation (19.25) yields in the usual first-order approximation

$$g = \gamma\,\tau\left\{\tfrac{1}{2}\,m\,v^2 - \frac{5}{2}\frac{1}{\beta_0 + \gamma z}\right\}v_z \tag{19.34}$$

This result shows again, like the corresponding result (19.27), that the treatment given is exact for Maxwellian molecules, for the g shown is just equal to the eigenfunction $g_{1,0}^1$ of the linearized collision operator. It has the slowest relaxation rate R_1^1, which is two-thirds of the rate shown in (19.28). In the case of other force laws the result is also of importance. Since the term is just g_1^1, it is orthogonal to g_1^0, which is an eigenfunction with eigenvalue zero for any type of collision, that is, it does not damp out. Equation (18.75) shows that the same sort of orthogonality relation also holds for eigenfunctions generally. Hence (19.34), when developed in eigenfunctions, never contains any g_1^0. It is interesting to trace this orthogonality through the calculation. If we had held n constant instead of p, then we would have had in (19.33) a factor $(\beta_0 + \gamma z)^{3/2}$ instead of $(\beta_0 + \gamma z)^{5/2}$; this would produce a coefficient 3/2 instead of 5/2 in g. It would then not be orthogonal to g_1^0 but would contain some of it as an undamped component. This undamped component represents a motion down the pressure gradient. This undesirable effect is eliminated in the calculation by seeing to it that the pressure is held constant, not the density.

We shall finish the calculation only for monatomic molecules having no internal structure. The energy then equals $\frac{1}{2} m v^2$, and the energy current density h equals

$$h = \iiint (\tfrac{1}{2} m v^2) \, v_z f \, d\mathbf{v} \tag{19.35}$$

This yields

$$h = n \gamma \tau \left(\frac{m \beta}{2 \pi}\right)^{3/2} \iiint e^{-\frac{1}{2} m \beta v^2} \left\{\tfrac{1}{3} m v^2 - \frac{5}{2 \beta}\right\} (\tfrac{1}{2} m v^2) \, v_z^2 \, d\mathbf{v}$$

which works out to be

$$h = \frac{5}{2} \frac{n \gamma \tau}{m \beta^3} \tag{19.36}$$

The quantity γ equals $- k \beta^2 (dT/dz)$. The definition (19.02) can therefore be applied, yielding for the conductivity of heat λ,

$$\lambda = \frac{5}{2} \frac{n \tau k^2 T}{m} \tag{19.37}$$

It is tempting to divide this equation by (19.30) and to write

$$\lambda = \frac{5}{2} \frac{k}{m} \eta$$

This equation is not right because the τ entering here is $1/R_1^1$, which is longer than $1/R_2^0$. For Maxwellian molecules the ratio is 3/2, so that we have

$$\lambda = \frac{15}{4} \frac{k}{m} \eta \tag{19.38}$$

$\frac{3}{2} k$ is the heat capacity c_v of a molecule. Equation (19.38) can therefore be written as

$$\lambda = \frac{5}{2} \left(\frac{c_v}{m}\right) \eta \qquad (19.39)$$

Comparing this with (19.19), we see that the approximate calculation has missed a substantial factor here. In the case of hard spheres we need the slowest nonzero relaxation rate for $l = 1$, which is also not published. We can hopefully assume, however, that there is approximate degeneracy between R_1^1 and R_0^2 as in the case of Maxwellian molecules. We may then substitute expression (18.80) for $1/\tau$ into (19.37). This yields

$$\lambda = \frac{75}{64} \frac{1}{\sigma} \left(\frac{\pi k^3 T}{m}\right)^{\frac{1}{2}} \qquad (19.40)$$

The result is in exact agreement with the first approximation of the rigorous theory.* This approximation subsequently is raised by 2% in later approximations. Division of this result by (19.17) confirms equation (19.39) closely for hard spheres as well as for Maxwellian molecules. The factor in (19.39) comes out in our calculation as

$$\tfrac{2}{3}\tfrac{5}{2} \pi = 2.46$$

The rigorous theory yields exactly $5/2$ for this factor in first approximation. Deviations in later approximations are in the upward direction and amount to less than 1%.

The derivation of the diffusion coefficient for two gases from the Boltzmann equation will not be carried out in this book because it necessitates the simultaneous consideration of two distribution functions and three types of encounters. For the same reason we shall not treat here thermal diffusion.

19-3. Discussion of empirical viscosity data

In order to gain some notion about the value of kinetic theory we shall pick out the viscosity of gases for comparison between theory and experiment. We are starting out here with an excellent piece of agreement, namely, the independence of viscosity of pressure, predicted and verified

* See S. Chapman and T. G. Cowling. *The Mathematical Theory of Non-uniform Gases*, Cambridge University Press, 1952, p. 169.

by Maxwell. To test the theory further we shall examine the dependence of viscosity upon temperature.

TABLE 19–1. Viscosity of helium as function of temperature, empirical molecular diameter, and predicted values, using a power law with $n = 13.6$

T ($^\circ$K)	$\eta \times 10^7$ (gm/cm sec)	$10^8 d$ (cm)	$\eta \times 10^7$ (calc.)
15.1	294.6	2.667	288.7
95.6	817.6	2.370	821.3
170.6	1392	2.226	1389
290.8	1967	2.140	1965
457	2681	2.052	2632
665	3388	2.005	3360
1088	4703	1.966	4640

For a first rough estimate we may think of molecules as hard spheres. Our formula for viscosity is then (19.17) with $\kappa = 1$, and $\sigma = \pi d^2$. The formula predicts a viscosity varying as $T^{1/2}$. This is very roughly true; however, all gases deviate from this law in the same direction, namely, towards a more rapid variation of viscosity with temperature. The reason is of course that all real cross sections decrease slightly as the speed increases. To make this notion quantitative we use (19.17) backward from the viscosity to compute a molecular diameter. This is done for helium gas in Table 19–1. One observes that the computed diameter tends to drift downward as temperature increases. The drift is not severe enough to make the concept useless or incorrect.

As the next refinement, we can try to take care of the "softness" of molecules by replacing the hard sphere idea by a repulsive potential varying inversely as some power of the intermolecular distance. The model has the advantage of providing some flexibility while yet giving η versus T in closed form. This is seen from an analysis of deflection angles, supplemented by dimensional analysis. Mechanics tells us that the entire pattern of trajectories can be scaled with the energy, leaving the relative probabilities for scattering in a given direction untouched. We therefore need only consider the variation of the cross section with speed or temperature. This is determined by dimension alone. For an energy expression such as

$$E = \tfrac{1}{2} m v^2 + \frac{\alpha}{r^n} \tag{19.41}$$

contains only the two parameters α/m and $E/m = \frac{1}{2} v_0^2$, out of which the cross section has to be made up. The only combination possible is

$$\sigma \propto \left(\frac{\alpha}{m\, v_0^2}\right)^{\frac{2}{n}} = \left(\frac{\alpha}{k\, T}\right)^{\frac{2}{n}}$$

We may insert this into (19.17) to get the temperature dependence of η. We find

$$\eta \propto T^{\frac{1}{2}+\frac{2}{n}} \tag{19.42}$$

For some gases this approach works very well. In Table 19-1 calculated values of η are added in the last column with a temperature exponent of 0.647, which means $n = 13.6$. Agreement is quite satisfactory.

For many other gases agreement with a power law is not as good. Indications are that what is missing in the theory is the attractive force which molecules experience at large distance. A semiempirical formula involving essentially the assumptions for the van der Waals equation is the viscosity formula of Sutherland. It reads

$$\eta \propto \frac{T^{\frac{1}{2}}}{1 + \dfrac{S}{T}} \tag{19.43}$$

where S is an empirical constant, the Sutherland constant. In Table 19–2 the theory is compared with experimental data for nitrogen. The constant S was taken as 102.7. Agreement is better than it could be with any simple power law.

TABLE 19–2. Viscosity versus temperature for nitrogen, and comparison with the formula of Sutherland

T ($^\circ$K)	$\eta \times 10^7$ (obs.)	$\eta \times 10^7$ (calc.)
196.9	1275	1269
235.3	1465	1469
289.3	1728	1728
324.8	1880	1884
373.4	2084	2086
473.2	2461	2461

More elaborate formulas have been constructed occasionally for the analysis of viscosity data. However, the formulas given pretty well exhaust what the experiment can provide. The verification of kinetic theory from viscosity data must be considered as altogether satisfactory.

A. RECOMMENDED PROBLEMS

1. In a molecular beam experiment the source is a tube containing oxygen at a pressure of 0.3 mm of mercury and a temperature of 300° K. The tube has a slit 20 mm × 0.01 mm. The receiver slit is 1 m away and of the same dimension.

(a) How many oxygen molecules leave the source slit per second?

(b) How many arrive at the receiver slit?

(c) If the receiver slit is part of a low-pressure chamber, with the intervening path even more highly evacuated, what is the steady state pressure in the receiving chamber?

2. Show that the mean relative speed $\langle |\mathbf{u} - \mathbf{v}| \rangle$ of two molecules in a Maxwellian distribution equals $\sqrt{2}$ times the mean speed.

B. GENERAL PROBLEMS

3. What fraction of the gas molecules in a Maxwellian distribution has a kinetic energy smaller than the mean kinetic energy?

4. Show from general principles that for any velocity distribution the inequality

$$\langle v \rangle \left\langle \frac{1}{v} \right\rangle > 1$$

must hold. Verify this result for the Maxwellian and the fully degenerate Fermi gas.

5. Would you expect the most probable kinetic energy of a set of gas molecules to be smaller or larger than their mean? Carry out the calculation for a Maxwellian gas and a completely degenerate Fermi gas. Comment.

6. In a Pirani gauge for measuring low pressures a thin wire is held at an elevated temperature by passing a current through it. It may be assumed that the gas molecules hit the wire with their natural temperature, and leave it with the temperature of the wire. How much power per unit length is needed to maintain a filament of diameter $\frac{1}{10}$ mm at a temperature of 100° above the ambient temperature of 300° K in helium gas at a pressure of 10^{-2} mm of mercury?

7. At an elevated temperature T the vapor pressure of a metal of atomic

weight $\mathcal{M}$ can be measured by enclosing a wire of radius r of the material in a loosely fitting glass sleeve. One then heats the wire to the temperature T, cools the sleeve, and observes m', the mass of the metal deposited on the sleeve per unit length and unit time. Derive an expression for the vapor pressure in terms of these measurable quantities.

8. A molecule of methane (molecular weight 16) can be considered a sphere of about 5 times the volume of an argon atom (atomic weight 40). What should be the ratio of the viscosities of methane and argon at equal temperature?

9. It was observed in the text that the shifts from the Maxwellian distribution associated with simple transport processes stand in a close relation to the harmonic oscillator eigenfunctions. Verify this statement for the shift (19.34) arising in heat conduction by showing that the function

$$\Psi = e^{-\frac{1}{2}r^2} \left(r^2 - \tfrac{5}{2} \right) r \cos \vartheta$$

satisfies the equation

$$\nabla^2\Psi + r^2\,\Psi = (2n + 3)\,\Psi$$

20

Kinetics of charge carriers in solids and liquids

The investigation of charged particles having gas-like behavior is a special and very fruitful branch of kinetic theory. The remark made at the beginning of Chapter 14 applies here: even though electric currents are marginal among the natural phenomena surrounding us, we attach to them special importance for technical reasons.

From the point of view of insight into the structure of matter, a study of electric conduction in the condensed phases is of greater value than a study of electric conduction in gases. We already know fundamentally the structure of a gas, as was shown in several preceding chapters. In the condensed phases, on the other hand, electrons play a fundamental role; in the elucidation of this role, electric conduction has played a decisive part. Even if this historical angle is ignored, it is important for the reader to realize that gaseous kinetics has its place in the study of solids and liquids. We shall therefore treat this aspect of electric conduction first, and discuss electric conduction in gases in the following chapter.

20-1. Kinetic theory of Ohmic conduction

A particular advantage of the kinetics of charged particles is that it puts at our disposal a transport phenomenon which is easily handled in theory and experiment; the phenomenon is *Ohmic conduction*. If an electric field is applied to a sample containing electric carriers a current will be produced in it. In most circumstances the current density **i** is related to the field **E** by the relation

$$\mathbf{i} = \sigma\,\mathbf{E} \tag{20.01}$$

Here σ is the *conductivity* of the sample. The number of current-carrying devices not obeying (20.01) is, however, quite substantial. In some cases there are trivial reasons, such as lack of temperature control. Many cases are more serious. A very common reason why (20.01) can fail is that the current as a process produces additional carriers, or that the number of carriers is controlled by an injection device. In such cases Ohmic conduction as a linear transport phenomenon can be salvaged by the decomposition of the current density into two parts, as follows:

$$\mathbf{i} = n\,e\,\langle\mathbf{v}\rangle \tag{20.02}$$

Here n is the number density of carriers, which may depend on the field and the current, and $\langle\mathbf{v}\rangle$ is the mean velocity or *drift velocity*. Very often, it obeys a linear law of the form

$$\langle\mathbf{v}\rangle = \mu\,\mathbf{E} \tag{20.03}$$

Under these circumstances, μ is called the *mobility* of the carriers. From the equations given, the two coefficients mobility and conductivity are related through

$$\sigma = n\,e\,\mu \tag{20.04}$$

Nonlinearity in σ thus is reduced to nonlinearity in the number density n of the carriers. This can go so far in some cases that some carriers or all carriers are absent from certain sections of an electric device. The depletion layer of an n–p junction studied in Chapter 14 is an example of this kind. Other examples are the space between grid and anode in a vacuum tube or the cathode fall in a glow discharge. There is a limited number of situations in which kinetic theory applies, and yet even the form (20.03) of Ohm's law fails. Such cases need a more fundamental investigation than the one given above. We shall discuss some of these cases in Section 21-2.

For the discussion of Ohm's law in general, the form (19.25) of the Boltzmann equation is very appropriate. The full Boltzmann equation assumes a very detailed mechanism for the scattering of charged particles. Sometimes, this picture is valid. In many cases it is not, particularly for the scattering of electrons in solids. The form (19.25) does not commit us as deeply with regard to the scattering mechanism as (18.16). One modification is, however, useful, namely, to make τ a function of v. When originally introduced, it was justified as being the particular eigenvalue of an integral operator, which has the most decisive influence on the process under consideration. This is a natural procedure in a medium in which both the scattered object and the target are in motion. Even there, it probably produces poor convergence in cases where τ is strongly dependent on speed. For Maxwellian molecules, there is no such dependence, and the idea works well. For hard spheres, τ varies inversely as the speed and ignoring this fact may well make computation unnecessarily awkward. In solids, in particular, many scattering centers are stationary and the relative speed occurring in the Boltzmann equation is just the speed of the carriers. There is then no good reason not to bring out an eventual dependence on speed immediately. The mathematical complication of this step is small, and the original theoretical foundation can always be regained by making τ a constant. The assumption $\tau \propto 1/v$ will, however, assume almost equal practical importance as it subsumes the usual mean free path formalism.

In an idealized arrangement for d–c Ohmic conduction the sample may be assumed in uniform condition throughout, and stationary in time. For a start, we shall simply treat the carriers as free particles. Equation (19.25) then reads

$$\mathbf{a} \cdot \frac{\partial f}{\partial \mathbf{v}} + \frac{1}{\tau(v)} (f - f_0) = 0 \tag{20.05}$$

with

$$\mathbf{a} = \frac{e}{m} \mathbf{E} \tag{20.06}$$

Ohmic conduction is a linear transport process which is clearly associated with a treatment of $\mathbf{a}$ as a first-order perturbation. We may therefore replace f by f_0 in the first term of (20.05). Solution for f is then elementary and yields

$$f = f_0 - \tau(v) \mathbf{a} \cdot \frac{\partial f_0}{\partial \mathbf{v}} \tag{20.07}$$

The equation is worth some comments. If we substitute for f_0 the Maxwellian distribution, then the deviation from the Maxwellian form is just the eigenfunction $g_1^0(\mathbf{v})$ of the collision operator. By (18.53) this function has a relaxation rate which is zero for all interactions. Physically, this corresponds to the fact that a field transmits momentum to the charge carriers, and that they cannot get rid of this momentum by colliding with each other. The $\tau(v)$ thus must refer to collisions with objects which can absorb momentum: the framework of a solid body, or in the case of liquids and gases, neutral molecules, or ions of opposite charge. The entire question of relaxation will have to be examined afresh for such situations.

Following the standard procedure developed in Chapter 19, we compute from (20.07) the electric current density **i** by multiplying f with the elementary charge and the component of velocity parallel to **a**. We then integrate over **v**. As usual the undisturbed distribution f_0 makes no contribution to the result, and we get

$$e\, n \, \langle v_z \rangle = i_z = -\, e\, a \iiint \tau(v)\, v_z \frac{\partial f_0}{\partial v_z}\, d\mathbf{v}$$

By (20.01) and (20.06) this yields for the conductivity

$$\sigma = -\frac{e^2}{m} \iiint \tau(v)\, v_z \frac{\partial f_0}{\partial v_z}\, d\mathbf{v} \tag{20.08}$$

Further handling of this expression is influenced by the fact that, in solids at least, the carriers may obey Fermi or Boltzmann statistics, depending on the degree of degeneracy; this degree of degeneracy is often not well known in a particular case. It so happens that an integration by parts on v_z will lead to a result independent of statistics. This result is

$$\sigma = n \frac{e^2}{m} \left\langle \frac{\partial}{\partial v_z} [v_z\, \tau(v)] \right\rangle \tag{20.09}$$

and

$$\mu = \frac{e}{m} \left\langle \frac{\partial}{\partial v_z} [v_z\, \tau(v)] \right\rangle \tag{20.10}$$

Here the pointed brackets indicate an average over the thermal equilibrium distribution. The formulas are valid independently of statistics. Two special cases of the formulas are $\tau = $ constant and $\tau = l/v$. The first case corresponds to a potential which is too "soft," the second one somewhat too hard, as compared to actual cases observed in nature. Real cases are therefore bracketed by these two limiting assumptions. They yield

$$\mu = \frac{e\, \tau}{m} \tag{20.11a}$$

$$\sigma = n \frac{e^2\, \tau}{m} \tag{20.11b}$$

and

$$\mu = \frac{2}{3} \frac{e\, l}{m} \left\langle \frac{1}{v} \right\rangle \tag{20.12a}$$

and

$$\sigma = \frac{2}{3} n \frac{e^2\, l}{m} \left\langle \frac{1}{v} \right\rangle \tag{20.12b}$$

In view of the importance of Ohm's law in physics, it is worth while elaborating on these formulas in several directions. The first elaboration is concerned with electrons in solids and is specifically designed for solid state applications. Conduction electrons in solids occur in allowed energy bands. The states in those bands have a certain similarity to free particle states: they are labeled by their wave vector **k**. The quantity $\hbar$ **k** acts very much like a momentum, and its conjugate variable, the lattice vector, like a coordinate. In particular a field **E** tends to modify **k** according to the relation

$$\hbar \frac{d\mathbf{k}}{dt} = e\,\mathbf{E} \qquad (20.13)^*$$

The main distinction between the band picture and the case of free particles is the relation connecting the energy with the wave vector

$$\mathscr{E} = W(\mathbf{k}) \qquad (20.14)$$

It resembles a free particle parabola only near an energy minimum within a band. In addition, a maximum in a band leads to a parabolic relationship for holes or defect electrons. A direct consequence of (20.14) is a modified relation between velocity and momentum, which takes the form (14.13):

$$\mathbf{v} = \frac{1}{\hbar} \frac{\partial W}{\partial \mathbf{k}} \qquad (20.15)$$

A very simple way of taking into account the band character of electronic energy states in solids is to modify a relation such as (20.12) by replacing the mass by an effective mass $m^\star$. The substitution is appropriate for semiconductors with low carrier density; the carriers, depending on their sign, occupy only a small region of the band near its energy minimum or maximum; even then, as we shall see below, the correct procedure is the use of a reciprocal mass tensor. In addition to these cases we have the alkali metals with their half-filled s valence bands. The bottom half of these bands differs remarkably little from a spherically symmetric paraboloid.

It is not worth while dwelling on the effective mass approximation very much because the kinetic formalism adapts with great ease to a more complicated band shape. Overlapping or mutually degenerate bands pose perhaps a more serious problem. If we can consider one band at a time then we may define a distribution function $f(\mathbf{k},\mathbf{r})$ in much the same way as was done for free particles. If we think specifically of the case of Ohm's

---

* For a detailed analysis of this point see G. H. Wannier, *Elements of Solid State Theory*, Cambridge Univesity Press, 1959, pp. 190–201; also *Revs. Modern Phys.* **34**, 645 (1962).

law we need only be concerned with a function f of $\mathbf{k}$ alone and the modified form of (20.05) which it obeys. Because of (20.13) this modified form reads

$$\frac{e\,\mathbf{E}}{\hbar} \cdot \frac{\partial f}{\partial \mathbf{k}} + \frac{1}{\tau(\mathbf{k})}\,(f - f_0) = 0 \tag{20.16}$$

Solving in the usual first-order approximation, we get from this

$$f(\mathbf{k}) = f_0(\mathbf{k}) - \frac{e\,\mathbf{E}\,\tau(\mathbf{k})}{\hbar} \cdot \frac{\partial f_0(\mathbf{k})}{\partial \mathbf{k}} \tag{20.17}$$

To find the mean drift in the field we have to multiply (20.17) with (20.15) and integrate over $\mathbf{k}$. Because of the anisotropy of the medium the drift is not necessarily in the direction of $\mathbf{E}$, and the relation (20.01) takes on a tensorial character. To take care of this we use tensorial index notation with the summation convention for indices appearing twice. We get then for the electric current density

$$i_s = \frac{e}{\hbar} \int\int\int f(\mathbf{k}) \frac{\partial W}{\partial k_s}\, d\mathbf{k} \tag{20.18}$$

Not too surprisingly, f_0 again makes no contribution to this integral. The reason, valid even for complicated band shapes, is time reversal; it expresses itself in the relation (14.12):

$$W(\mathbf{k}) = W(-\mathbf{k}) \tag{20.19}$$

The two states linked by this equation have opposite velocities and make contributions to the current which cancel mutually. The result arises therefore from the correction to f_0. It equals, from (20.17) and (20.18),

$$i_s = -\frac{e^2}{\hbar^2}\, E_t \int\int\int \frac{\partial W}{\partial k_s} \frac{\partial f_0}{\partial k_t}\, \tau(\mathbf{k})\, d\mathbf{k}$$

The conductivity tensor equals therefore

$$\sigma_{st} = -\frac{e^2}{\hbar^2} \int\int\int \frac{\partial W}{\partial k_s} \frac{\partial f_0}{\partial k_t}\, \tau(\mathbf{k})\, d\mathbf{k} \tag{20.20}$$

Equation (20.20) shows that the conductivity tensor is a symmetric tensor. For f_0, whether of the Fermi or Boltzmann type, depends on $\mathbf{k}$ only through the energy $W(\mathbf{k})$ as an intermediary; substitution of this feature into (20.20) makes the integral symmetric in s and t.

Further treatment of (20.20) is inspired by the same consideration as before, namely, that it is possible to write down an expression for the conductivity which is independent of statistics. The result is

$$\sigma_{st} = \frac{e^2}{\hbar^2} \int\int\int \frac{\partial}{\partial k_t} \left[\frac{\partial W}{\partial k_s}\, \tau(\mathbf{k}) \right] f_0\, d\mathbf{k} \tag{20.21}$$

In other words, we have an expectation value on the equilibrium distribution. We get from this for the mobility tensor μ_{st}

$$\mu_{st} = \frac{e}{\hbar^2} \left\langle \frac{\partial}{\partial k_t} \left[\frac{\partial W}{\partial k_s} \tau(\mathbf{k}) \right] \right\rangle \tag{20.22}$$

Another direction in which one can branch out on Ohm's law is to consider mobility in alternating fields, the production of such fields being so extremely easy. The relevant change is then an addition of a term $\partial f/\partial t$ to (20.05), yielding

$$\frac{\partial f}{\partial t} + \mathbf{a} \cdot \frac{\partial f}{\partial \mathbf{v}} + \frac{1}{\tau(v)} (f - f_0) = 0 \tag{20.23}$$

The first-order approximation now works as follows. Although $\partial f/\partial t$ is not small, f_0 makes no contribution to it. In the second term, replacement of f by f_0 is again justified; the time dependence comes here from $\mathbf{a}$, which acts as the driving force. We may then set

$$\mathbf{a} = \mathbf{a}_0 \, e^{i\omega t}; \; (f - f_0) = g_0 \, e^{i\omega t} \tag{20.24}$$

and find

$$\left(\frac{1}{\tau} + i\omega \right) g_0 = \mathbf{a}_0 \cdot \frac{\partial f}{\partial \mathbf{v}} \tag{20.25}$$

The result differs from the previous equation (20.07) only by the substitution

$$\frac{1}{\tau} \to \frac{1}{\tau} + i\omega \tag{20.26}$$

We therefore get a complex conductivity equal to

$$\sigma = \frac{e^2 n}{m} \left\langle \frac{\partial}{\partial v_z} \left(\frac{v_z}{\frac{1}{\tau(v)} + i\omega} \right) \right\rangle \tag{20.27}$$

or in the solid state case (20.21)

$$\sigma_{st} = \frac{e^2 n}{\hbar^2} \left\langle \frac{\partial}{\partial k_t} \left[\frac{\frac{\partial W}{\partial k_s}}{\frac{1}{\tau(\mathbf{k})} + i\omega} \right] \right\rangle \tag{20.28}$$

These last two formulas become very simple when the frequency is made high enough so that $1/\tau$ can be neglected in comparison to the frequency.

The average is then the second derivative of W with respect to $\mathbf{k}$, which is the best definition of the reciprocal mass tensor. We may then write

$$\sigma_{st} = -i \frac{e^2 n}{\omega} \left\langle \left(\frac{1}{m}\right)_{st} \right\rangle \tag{20.29}$$

Equation (20.29) is of course a formula of mechanics, all kinetic features having disappeared. It simply describes the free oscillations of electrons in an alternating field; for such oscillations, the velocity is out of phase with the field. This type of motion is not easy to observe in practice because there are other inertial effects for electrons in metals, notably self-induction, which are apt to mask the inertia due to mass. What is, however, significant in these considerations is that kinetic effects tend to disappear as the frequency is raised, and the resultant motion can be described more and more by straight mechanics.

20-2. Nature of the charge carriers in matter; Nernst relation

In the preceding section we have given a modern account of Ohm's law as it is understood today. We now step back a bit to 1880, when the nature of the carriers of electric charge began to be suspected. The first glimpse came from Faraday's laws of electrolysis for ions in aqueous solution. These laws made it highly likely that the Faraday divided by Avogadro's number was the charge carried by an individual monovalent ion. When looking for a proof of this supposition, Nernst hit upon the idea that the size of the charge carried by an individual charge carrier would reveal itself if diffusion were studied together with mobility: the smaller the particles the higher their diffusivity. This relationship can be very well demonstrated with the Boltzmann equation (19.25). We have so far reduced it to the form (20.05) and studied the motion induced by a field. A concentration gradient can induce a similar motion by Fick's law. If we compare

$$\mathbf{a} \cdot \frac{\partial f_0}{\partial \mathbf{v}} + \frac{1}{\tau}(f - f_0) = 0 \tag{20.30}$$

and

$$\mathbf{v} \cdot \frac{\partial f_0}{\partial \mathbf{r}} + \frac{1}{\tau}(f - f_0) = 0 \tag{20.31}$$

then we find the two disturbances producing entirely equivalent effects on the velocity distribution, if the particles obey Boltzmann statistics.

For the ions in aqueous solution studied by Nernst this was the case. Substituting

$$f_0 = n \left(\frac{2\pi}{m\beta}\right)^{3/2} \exp[-\tfrac{1}{2}\beta m v^2]$$

into (20.30) and

$$f_0 = n(x) \left(\frac{2\pi}{m\beta}\right)^{3/2} \exp[-\tfrac{1}{2}\beta m v^2]$$

into (20.31), the perturbing terms become respectively

$$- e E \beta v_x f_0(\mathbf{v})$$

and

$$\frac{d \ln n}{dx} v_x f_0(\mathbf{v})$$

The structure of the perturbed distribution is therefore the same for the two phenomena, the equivalence being

$$e E \beta \approx - \frac{d \ln n}{dx} \tag{20.32}$$

On the other hand, we have the empirical equivalence for the production of current

$$n \mu E \approx - D \frac{dn}{dx} \tag{20.33}$$

Dividing the two relations by each other, we get

$$\mu = \frac{e}{kT} D \tag{20.34}$$

The formula was used by Nernst to determine the magnitude e of the charge carried by an individual ion in a solution.[†] Equation (20.34) is commonly known as the Einstein relation, because Einstein rediscovered it in connection with his studies of Brownian motion.

20-3. Nature of the electric carriers in metals; law of Wiedemann and Franz

The discovery of the atomic nature of the electric current in electrolytes was a historic first, preceding a similar analysis for metals by several

† W. Nernst. Z. Phys. Chem. **9**, 613 (1884).

decades. Shortly after the discovery of the electron, Drude surmised that electrons were the carriers of current in metals. He made the natural assumption that they formed a gas and arrived at a formula equivalent to (20.12b). He naturally assumed further that the electron gas had a Maxwellian distribution, in which case one finds without difficulty

$$\left\langle \frac{1}{v} \right\rangle = \left(\frac{2}{\pi} m \beta \right)^{1/2}$$

If one applies these ideas to copper, assumed monovalent, one obtains the following results. Copper has a resistivity of 1.72 microohm-cm, which is a conductivity σ of

$$\sigma = 3.31 \times 10^{17} \text{ statmho/cm}$$

The estimate from (20.12b) becomes then

$$3.31 \times 10^{17} = \frac{\sqrt{8}}{3\sqrt{\pi}} \frac{0.93 \times 10^{23} \times (4.8)^2 \times 10^{-20} \times l}{(0.91 \times 10^{-27} \times 1.38 \times 10^{-16} \times 0.3 \times 10^3)^{1/2}}$$

which yields

$$l = 1.78 \times 10^{-7} \text{ cm}$$

From the point of view of the times one might perhaps have expected a slightly smaller mean free path, but the value found was not considered unreasonable.

Naturally, Drude looked around for a confirmation of his hypothesis. This is best found by forming the ratio of two transport coefficients, somewhat on the model of (20.34). Diffusion experiments being out of the question, he took up the empirical law discovered some time earlier by Wiedemann and Franz, according to which the electrical conductivity σ and the conductivity of heat λ of metals are related by the formula

$$\frac{\lambda}{\sigma T} = \text{constant} \qquad (20.35)$$

The law is an expression of the empirical fact that good electric conductors are also good thermal conductors. Drude guessed correctly that this meant that, in metals, heat is also conducted by electrons, and proceeded to compute this quantity for his gas of free electrons. It is a curious accident that this law is confirmed for a gas of electrons in either statistical limit: for Boltzmann or degenerate Fermi statistics. This accident has prevented a true understanding of electronic conditions in metals for some time. We shall now give this double derivation.

For either type of statistics the Boltzmann equation for the problem is (19.25). It reads here

$$\mathbf{a} \cdot \frac{\partial f}{\partial \mathbf{v}} + \mathbf{v} \cdot \frac{\partial f}{\partial \mathbf{r}} + \frac{v}{l}(f-f_0) = 0 \qquad (20.36)$$

The presence of an electric field is necessary to prevent an electric current. The dependence on $\mathbf{r}$ is through a variable temperature, which shall be assumed to vary along the x direction. The equation is to be solved in the usual first-order manner; that is, the solution of (20.36) is taken in the form

$$f = f_0 - \frac{l}{v}\left\{ a\frac{\partial f_0}{\partial v_x} + v_x \frac{\partial f_0}{\partial x} \right\} \qquad (20.37)$$

We start out by presenting the older derivation which takes f_0 to be a Boltzmann distribution

$$f_0 = n\left(\frac{2\pi}{m\beta}\right)^{3/2} \exp[-\tfrac{1}{2}m\beta v^2]$$

There is a slight complication in this calculation, which arises from the side condition that $\mathbf{a}$ shall be such that no total current flows:

$$\iiint v_x f\, d\mathbf{v} = 0 \qquad (20.38)$$

Under these conditions the heat flow h is to be evaluated:

$$h = \iiint \tfrac{1}{2}m v^2\, v_x f\, d\mathbf{v} \qquad (20.39)$$

Substitution of (20.37) into (20.38) yields

$$a = \frac{7}{2}\frac{k}{m}\frac{dT}{dx} \qquad (20.40)$$

Substitution of (20.40) into (20.39) yields in turn after some elementary transformations

$$h = -\frac{4}{3}\frac{n\, l\, k^2\, T}{m}\left\langle \frac{1}{v} \right\rangle \frac{dT}{dx} \qquad (20.41)$$

or with the definition (19.02) for the conductivity of heat λ

$$\lambda = \frac{4}{3}\frac{n\, l\, k^2\, T}{m}\left\langle \frac{1}{v} \right\rangle$$

The Wiedemann–Franz ratio (20.35) now follows from comparison of the above with (20.12b):

$$\frac{\lambda}{\sigma T} = 2\frac{k^2}{e^2} \qquad (20.42)$$

We now pass to the modern derivation which takes the electrons to be a degenerate Fermi gas. Now f_0 equals

$$f_0 = C\, F[\beta\, (\tfrac{1}{2}mv^2 - \mu)] \tag{20.43}$$

Here F is the Fermi function (14.31)

$$F(x) = \frac{1}{e^x + 1}$$

Expression (20.37) now takes the form

$$f = C\, F - l\left\{a\, m\, \beta + \frac{d\beta}{dx}\left(\frac{1}{2}m\, v^2 - \mu\right)\right\}\frac{v_x}{v}\, C\, F'$$

Integrals containing this expression can be carried out by integration by parts on F'; the result is then expressible as an equilibrium expectation value. In this way, the constraint (20.38) becomes

$$a = \left\{\frac{\langle v\rangle}{\left\langle\dfrac{1}{v}\right\rangle} - \frac{\mu}{m}\right\}\frac{1}{T}\frac{dT}{dx} \tag{20.44}$$

and from this the heat flow (20.39)

$$h = -\, n\frac{1}{6}\, l\, m\, \frac{1}{T}\frac{dT}{dx}\left\{3\langle v^3\rangle - 4\frac{\langle v\rangle^2}{\left\langle\dfrac{1}{v}\right\rangle}\right\} \tag{20.45}$$

The coefficient of $-\,(dT/dx)$ in this expression is the conductivity of heat λ. We go on immediately, forming the Wiedemann–Franz ratio (20.35) by dividing through with (20.12b). The result is

$$\frac{\lambda}{\sigma\, T} = \frac{1}{4}\frac{m^2}{e^2\, T^2}\frac{\left\{3\langle v^3\rangle\left\langle\dfrac{1}{v}\right\rangle - 4\langle v\rangle^2\right\}}{\left\langle\dfrac{1}{v}\right\rangle^2} \tag{20.46}$$

There remains the task of evaluation of these averages for an almost completely degenerate Fermi distribution. Denoting the argument of the Fermi function in (20.43) by x, we can express the speed v in terms of it and write

$$v = \left\{\frac{2}{m}\,(\mu + x\, k\, T)\right\}^{\frac{1}{2}} \tag{20.47}$$

It is advantageous to make all calculations in the neighbourhood of the Fermi energy; this means reintroducing F' and reversing the integrations by parts on F. We get thus

$$\langle v^s \rangle = \frac{4 \pi C}{n} \int_0^\infty v^{s+2} \, F[\beta(\tfrac{1}{2}mv^2 - \mu)] \, dv$$

$$= \frac{4 \pi C}{n} \frac{\beta m}{s + 3} \int_0^\infty v^{s+4} \, F'[\beta(\tfrac{1}{2}mv^2 - \mu)] \, dv$$

$$\approx \frac{4 \pi C}{n} \frac{1}{s + 3} \int_{-\infty}^{+\infty} \frac{\left\{ \frac{2}{m}(\mu + x k T) \right\}^{(s+3)/2}}{4 \cosh^2 \tfrac{1}{2} x} \, dx$$

Here the integration limits were extended in the last step, because the denominator enforces rapid convergence for points away from the Fermi surface. For the same reason expansion in powers of $k T$ is indicated in the numerator. In such an expansion the linear term gives no result by symmetry, so expansion must be to quadratic terms. For the averages entering into (20.46) we find

$$\left\langle \frac{1}{v} \right\rangle = \frac{4 \pi C}{n} \frac{\mu}{m} I_1$$

$$\langle v \rangle = \frac{4 \pi C}{n} \left(\frac{\mu}{m} \right)^2 \left\{ I_1 + \frac{k^2 T^2}{\mu^2} I_2 \right\}$$

$$\langle v^3 \rangle = \frac{16 \pi C}{3 n} \left(\frac{\mu}{m} \right)^3 \left\{ I_1 + 3 \frac{k^2 T^2}{\mu^2} I_2 \right\}$$

Here I_1 is the integral

$$I_1 = \int_{-\infty}^{+\infty} \frac{dx}{4 \cosh^2 \tfrac{1}{2} x} = \tfrac{1}{2} \tanh \tfrac{1}{2} x \Big|_{-\infty}^{+\infty} = 1$$

and I_2 equals

$$I_2 = \int_{-\infty}^{+\infty} \frac{x^2 \, dx}{4 \cosh^2 \tfrac{1}{2} x} = 2 \int_0^\infty x \, (1 - \tanh \tfrac{1}{2} x) \, dx$$

$$= 4 \int_0^\infty \frac{x \, d x}{e^x + 1} = 4 \int_0^\infty x \, dx \, \{ e^{-x} - e^{-2x} + e^{-3x} - \cdots \}$$

$$= \frac{\pi^2}{3}$$

Substitution of these results yields for the combination of velocity averages in (20.46) the expression $4 \pi^2 k^2 T^2/3 m^2$, and thus for the Wiedemann–Franz ratio

$$\frac{\lambda}{\sigma T} = \frac{\pi^2}{3} \frac{k^2}{e^2} \tag{20.48}$$

The close agreement of (20.48) and (20.42) is indeed an astonishing result. One can argue that (20.42) must have the dimensional form shown because there are no other constants which could enter the velocity averages. For the result in Fermi statistics this argument cannot be made

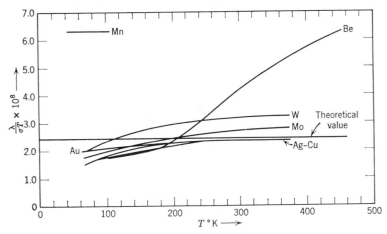

Fig. 20.1. The Wiedemann–Franz ratio plotted as function of temperature for several metals. The theoretical value is indicated in heavy outline as a horizontal line (after Seitz).

because a second energy, the Fermi energy μ, could have appeared in it. Such a modification does occur, for instance, for the electronic heat capacity as exhibited in (14.38). The absence of the Fermi energy in (20.48) is thus an accident. The accident arises as follows. Even though electrons transport a great deal less energy from a hot to a cold region than classical theory suspects, they move a lot faster doing it. The resultant changes in the conductivity of heat happen to cancel out exactly.

Before discussing the implication of these results we shall show a comparison of experimental data with (20.48) and (20.42). The agreement of either formula with experiment must be classed as very good. Equation (20.48) fits a little better, but deviations from either formula do occur for some metals. The results can therefore be considered a confirmation of the theory.

But now comes the point. What theory do they confirm? Naturally Drude felt that they confirmed his theory of the electrons as a Maxwell gas of free particles. It was pointed out in Chapter 14 that Drude's theory fails because the heat capacity which such a gas would have is not present. From the point of view of theoretical development, the near coincidence of (20.42) and (20.48) must be considered an unfortunate circumstance. The result seemed to confirm a theory which was then brought to a standstill by the heat capacity problem. Sommerfeld's theory of the Fermi gas retains the first results and corrects the second in a very satisfactory way.

20-4. Separation of carrier density and carrier velocity; Hall effect

The previous discussion still leaves open one essential point about electrons as carriers of current in metals and semiconductors, namely,

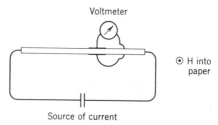

Fig. 20.2. Standard arrangement for the observation of the Hall effect.

their number. The conductivity contains only the product of their number and speed, and the effects discussed up to now only determine their charge. This leaves the question just how many particles move to produce the observed current. In metals one can give arguments (which are not very strong) that the number of electrons per atom is of the order unity. In semiconductors this is not the case, and counting the number of carriers becomes a serious business. Even in metals a split-up of density and speed proceeding from the experiment is preferable to a priori theoretical assumptions.

There are today several methods to measure the drift velocity of carriers directly. Under certain circumstances, time-of-travel measurements can be made to work. But among these methods, measurement of the *Hall effect* has achieved primacy as the most widely used.

The arrangement for the Hall effect is shown in Fig. 20.2. A current is sent through a rectangular slab of material. The slab is placed into a uniform magnetic field. The field is perpendicular to the current and to one pair of faces of the slab. A voltage then develops across the remaining two faces, at right angles to the current and the magnetic field. The voltage is called the Hall voltage, and the associated electric field the Hall field. It is defined as

$$E_\perp = R\,i\,H \qquad (20.49)$$

Here $E_\perp$ is the component of the electric field perpendicular to the current, and R is the *Hall constant* of the material.

It is not intended, in this book, to even enumerate the large number of effects and corresponding coefficients which arise when voltages, currents, magnetic fields, temperature gradients, and heat fluxes are combined in all

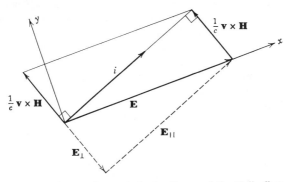

Fig. 20.3. Axes and vectors in the theory of the Hall effect. **H** is assumed perpendicular to the paper. The Hall field, designated as $E_\perp$, is such as to cancel exactly the magnetic part of Lorentz force. The resultant force is thus parallel to **i**.

possible combinations of angular arrangements. A specific exception is made here for the Hall effect because of the instrumental role it plays in untangling the two components, density and drift velocity, which make up the normal Ohmic current.

In view of the use which is made of the Hall effect it is desirable to derive this effect under as general conditions as possible. What is compared is the Hall voltage and an Ohmic current. The Ohmic current, by (20.01), is controlled by a symmetric tensor of rank 2, while the Hall field, by (20.49), is controlled by an antisymmetric tensor of rank 3. Both these tensors have the property of reducing to a single constant for cubic crystals. It is therefore possible to retain the simple isotropic definition of the Hall effect in any cubic crystal, and characterize it by a single number. We may in particular choose to analyze the effect in a Cartesian coordinate system with equivalent axes. We shall choose the magnetic field along the z axis and the electric field along the x axis. The current then lies in the x–y plane, and the Hall field is defined as the component of **E** at right angles to **i**; this arrangement is shown in Fig. 20.3.

The Boltzmann equation to be used is (20.16), amended to include the entire Lorentz force, with the velocity given by (20.15).‡ The equation then takes the form

$$\frac{e}{\hbar}\left(\mathbf{E} + \frac{1}{\hbar c}\frac{\partial W}{\partial \mathbf{k}} \times \mathbf{H}\right) \cdot \frac{\partial f}{\partial \mathbf{k}} + \frac{1}{\tau}(f - f_0) = 0 \tag{20.50}$$

In a simple first-order perturbation approach, as used previously, the magnetic term drops out. For f_0 depends on $\mathbf{k}$ only through the intermediary of the energy band function $W(\mathbf{k})$, as defined in (20.14); the magnetic term then contains a cross product of a vector with itself and vanishes. We must therefore treat the magnetic term in second-order perturbation. This means that we write f in the form

$$f = f_0 + g_1 + g_2 \tag{20.51}$$

where g_1 is proportional to E, and g_2 proportional to E and H. For g_1 we find the expression (20.17), namely,

$$g_1 = -\tau \frac{e}{\hbar} E \frac{\partial f_0}{\partial k_x} \tag{20.52}$$

and for g_2

$$g_2 = -\frac{e\tau}{c\hbar^2} H \left\{ \frac{\partial g_1}{\partial k_x}\frac{\partial W}{\partial k_y} - \frac{\partial g_1}{\partial k_y}\frac{\partial W}{\partial k_x} \right\}$$

or

$$g_2 = \frac{e^2 \tau^2 E H}{c\hbar^3} \left\{ \frac{\partial^2 f_0}{\partial k_x^2}\frac{\partial W}{\partial k_y} - \frac{\partial^2 f_0}{\partial k_x \partial k_y}\frac{\partial W}{\partial k_x} \right\} \tag{20.53}$$

The formulas contain a change of conductivity with the magnetic field. This is the so-called *magnetoresistance*, which we shall ignore. We therefore calculate the current in the x direction only for zero magnetic field, reproducing the result (20.21)

$$i_x = n \frac{e^2 \tau}{\hbar^2} E_x \left\langle \frac{\partial^2 W}{\partial k_x^2} \right\rangle \tag{20.54}$$

where the pointed bracket represents an equilibrium average. For the current in the y direction we employ g_2

$$i_y = \frac{e}{\hbar} \iiint g_2(\mathbf{k}) \frac{\partial W}{\partial \mathbf{k}_y} d\mathbf{k}$$

Inserting (20.53) into this and integrating by parts once, we get

$$i_y = -\frac{e^3 \tau^2 E H}{c\hbar^2} \iiint \frac{\partial f_0}{\partial k_x} \left\{ \frac{\partial W}{\partial k_y}\frac{\partial^2 W}{\partial k_y \partial k_x} - \frac{\partial W}{\partial k_x}\frac{\partial^2 W}{\partial k_x^2} \right\} d\mathbf{k}$$

‡ For a justification of this feature see G. H. Wannier, *Elements of Solid State Theory*, Cambridge University Press, 1959, pp. 194–196 and 200–201.

For further simplification it is useful to remember that f_0, in either statistics, depends on **k** only through the intermediary of W, and therefore

$$\frac{\partial f_0}{\partial k_x}\frac{\partial W}{\partial k_y} = \frac{\partial f_0}{\partial k_y}\frac{\partial W}{\partial k_x}$$

If we insert this above and integrate by parts once more so as to have f_0 in the integrand, we obtain finally

$$i_y = n\,\frac{e^3\,\tau^2\,E\,H}{c\,\hbar^4}\left\langle\left(\frac{\partial^2 W}{\partial k_x\,\partial k_y}\right)^2 - \frac{\partial^2 W}{\partial k_x^2}\frac{\partial^2 W}{\partial k_y^2}\right\rangle \tag{20.55}$$

The angle between the electric field and the current, the so-called *Hall angle*, is given by i_y/i_x. The hall field $E_\perp$ is then

$$E_\perp = -\frac{i_y}{i_x}\,E$$

and the Hall constant, from the definition (20.49), becomes

$$R = -\frac{i_y\,E}{i_x^2\,H} \tag{20.56}$$

Insertion of the averages (20.54) and (20.55) into this yields finally

$$R = \frac{1}{n\,e\,c}\,\frac{\left\langle\dfrac{\partial^2 W}{\partial k_x^2}\dfrac{\partial^2 W}{\partial k_y^2} - \left(\dfrac{\partial^2 W}{\partial k_x\,\partial k_y}\right)^2\right\rangle}{\left\langle\dfrac{\partial^2 W}{\partial k_x^2}\right\rangle^2} \tag{20.57}$$

The classical case for (20.57) arises if the effective mass approximation is valid. We have then

$$W = \frac{\hbar^2}{2\,m^\star}\,(k_x^2 + k_y^2 + k_z^2) \tag{20.58}$$

The two averages are then equal, and we have simply

$$R = \frac{1}{n\,e\,c} \tag{20.59}$$

In order to test possible discrepancies for actual bands we shall take up the case of n-type silicon and germanium. These semiconductors have cubic structures and thus a well-defined Hall coefficient. The band minima in which the conduction electrons are located do not have cubic symmetry, however, but have only axial symmetry. Cubic symmetry is restored over-all by the location of other such minima at cubically equivalent positions. For an individual minimum the energy has the form

$$W = \tfrac{1}{2} A k_x^2 + \tfrac{1}{2} B k_y^2 + \tfrac{1}{2} C k_z^2 + D k_y k_z + E k_z k_x + F k_x k_y$$

The numerator of (20.57) becomes then

$$(A B - F^2)$$

Because of over-all cubic symmetry this must average out to

$$\tfrac{1}{3}[(A B - F^2) + (B C - D^2) + (C A - E^2)]$$

The same reasoning yields for the average in the denominator

$$\tfrac{1}{3} (A + B + C)$$

The two averages are invariants for the ellipsoid whose interpretation follows if we transform to principal axes. Then W takes the form

$$W = \frac{\hbar^2}{2} \left(\frac{k_1^2}{m_1} + \frac{k_2^2}{m_2} + \frac{k_3^2}{m_3} \right)$$

Equation (20.57) then becomes

$$R = \frac{1}{n e c} \frac{\dfrac{1}{3} \left(\dfrac{1}{m_2 m_3} + \dfrac{1}{m_3 m_1} + \dfrac{1}{m_1 m_2} \right)}{\dfrac{1}{9} \left(\dfrac{1}{m_1} + \dfrac{1}{m_2} + \dfrac{1}{m_3} \right)^2} \tag{20.60}$$

The ellipsoids in silicon and germanium are prolate ellipsoids of revolution which are extremely thin. They have therefore two small masses which are equal and one large mass which is negligible in this problem. Under these circumstances the ratio of averages in (20.60) reduces to 3/4, and we get

$$R = \frac{3}{4} \frac{1}{n e c} \tag{20.61}$$

Relation (20.59) is therefore salvaged in a qualitative way. It is clear, however, that if the band shape becomes very complicated, the calculation of the number of carriers by the Hall effect becomes unreliable.

A very important feature of the Hall constant is that it contains the charge e in the first power. Measurement of the constant thus determines the sign of the carriers. In particular, for semiconductors it means that one can tell whether the carriers are electrons or holes. This aspect is easily checked qualitatively. When a current flows in a slab and a magnetic field is applied, the carriers are deflected toward the same side of the slab regardless of the sign of their charge, because e and $\mathbf{v}$ are reversed simultaneously in the Lorentz force. The tendency toward an accumulation of a surplus charge is thereupon cancelled by an electrostatic field arising from the space charge. The direction of this field is different for positive and negative charges.

The real shortcoming of the Hall effect as a method to count electric carriers lies exactly in this dependence of the Hall voltage on the sign of the carriers. Many semiconductors have both positive and negative carriers present, and their Hall voltages tend to cancel. Even in specimens having only carriers of one sign, results become difficult to interpret if the carriers form two or more distinct groups of specific characteristics.

A. RECOMMENDED PROBLEMS

1. The mobility of an electron in a metal or semiconductor varies inversely as the number of scattering events it experiences per second. This number, in turn, is proportional to the mean square amplitude of oscillation of the atoms and the mean speed of the electron. What temperature dependence do you infer from this for each of the following?

(*a*) The electron mobility in a metal.
(*b*) The electron mobility in a semiconductor.
(*c*) The resistivity of a metal.
(*d*) The resistivity of a semiconductor.

B. GENERAL PROBLEMS

2. The Hall constant of aluminum is -0.3×10^{-10} volt m^2/amp weber. Its atomic volume is 10 cm^3/mole. What value results for the sign and number of charge carriers per atom?

3. Derive the structure of the high-frequency conductivity tensor (20.29) by straight mechanics.

4. The text deals with only one species of carrier while, in semiconductors, electrons and holes participate in conduction. As a preliminary to the description below it should be mentioned that the passage from extrinsic to intrinsic conduction can be identified by a sharp break in a conductivity versus temperature plot. The mobility of both types of carriers is then determined in *p*-type material by two Hall measurements and two conductivity measurements as follows:

(*a*) In the extrinsic range, Hall and conductivity data yield the mobility of the holes.

(b) In the break the temperature for which the Hall effect is zero is determined and the associated conductivity σ_0 measured. For the same temperature a value σ_e can be found which is extrapolated from the region (a).

Calculate the mobility of the electrons, assuming simple effective masses and relaxation times for both carriers.

5. On page 446 the integrals I_n, defined as

$$I_n = \int_{-\infty}^{+\infty} \frac{x^n \, dx}{4 \cosh^2 \frac{1}{2}x}$$

were evaluated by individual methods. A more efficient procedure is to evaluate the integral

$$J(k) = \int_{-\infty}^{+\infty} \frac{e^{ikx} \, dx}{4 \cosh^2 \frac{1}{2}x}$$

since powers series expansion of $J(k)$ yields

$$J(k) = \sum_{\nu=0} \frac{(i\,k)^\nu}{\nu!} I_\nu$$

Evaluate the integral $J(k)$ by contour integration.

★21

Kinetics of charge carriers in gases

For the modern physicist, electric conduction in gases lacks the fundamental interest of electric conduction in solids: the carriers are just like other gaseous particles, except that they happen to have a charge. This lack of fundamental interest is reinforced by the practical difficulties of making a good theory which applies to carriers in a gas discharge and which can be realized in practice. Ohm's law is rarely verified for conducting gases, because a gas in equilibrium has no electric carriers. The number of carriers in a discharge is thus not a fixed quantity, and the problem of their creation and annihilation during the discharge assumes primary importance. The motion of the carriers is also complicated by the long-range nature of the Coulomb force. This feature gives rise to *plasma oscillations;* they are better handled by a combination of electromagnetism and mechanics than by the methods of kinetic theory.

On the other hand, it is true, although not generally known, that if suitable care is taken in the arrangement of the experiment, gaseous discharges can offer the most beautiful and perfect application of the Boltzmann transport equation. Examination of the preceding chapter shows that the kinetics of charge carriers in metals and semiconductors is not a complete theory. The methods of kinetic theory proper are usually not sufficient for an understanding of the scattering phenomena. Kinetic theory can therefore make predictions only about the ratios of transport coefficients. In the case of gaseous ions, the collision process itself can be analyzed by kinetic methods, and thus a complete theory can be offered.

21-1. Kinetics of the polarization force

The Boltzmann equation for ions in a weakly ionized gas is very attractive mathematically because it is naturally linear, for most encounters take place with neutral molecules. It is usually also possible to assume the neutral molecules to have an undisturbed Maxwellian distribution. Denote by M (V) the velocity distribution of these molecules. We shall assume both $M(V)$ and $f(v)$ normalized to unity on v alone as we are not going to deal with spatial distributions. We also shall only consider stationary situations, so that (18.16) takes the form

$$\mathbf{a} \cdot \frac{\partial f}{\partial \mathbf{v}} = - n \int \int \int \int \int \{f(\mathbf{v}) \, M(\mathbf{V}) - f(\mathbf{v}') \, M(\mathbf{V}')\} \, d\,\mathbf{V} \; \sigma' \, w \, d\Omega_{\mathbf{w}'} \quad (21.01)$$

Here n is the number density of the neutral gas molecules.

Another very agreeable feature of the theory of ions or electrons in gases is that they satisfy the postulate Maxwell made originally for his gas molecules, at least if the speed is not too high: namely, they attract the gas molecules with a force that varies inversely as the fifth power of the distance. Kinetic calculations are very easy in this case, and we shall have occasion to become acquainted with Maxwell's original method of computing averages without knowledge of the distribution function. The field also gives us the occasion to calculate results completely to the end, rather than having to break off with some unknown "mean free path."

The force which predominates between ions and molecules at moderate speeds is the so-called polarization force, a force acting between the ionic charge and the dipole moment induced by the charge on the molecule. Its potential equals

$$\Phi = - \frac{1}{2} \frac{e^2 \, P}{r^4} \quad (21.02)$$

where P is the polarizability of the molecule and e the charge of the ion. We shall use the classical theory of this interaction, partly because angular momenta are usually high, and partly because quantum discrepancies from the classical pattern seem not very pronounced.* We can then follow

* Erich Vogt and G. H. Wannier. *Phys. Rev.* **95**, 1190 (1954).

classical methods† for the computation of the angle of deflection χ due to a potential of the form (21.02). We find

$$\chi = \pi - 2 \int_0^{u_1} \frac{du}{\left\{ \dfrac{1}{b^2} - u^2 + \dfrac{e^2 \, P \, (M+m)}{M \, m \, b^2 \, w^2} u^4 \right\}^{1/2}} \tag{21.03}$$

Here the variable of integration u is the reciprocal radius and b is the impact parameter. The quantity u_1 is the lower of the two positive roots of the polynomial in the denominator; if the polynomial has no real root, the integration goes from 0 to ∞. The question whether the denominator has a real root or not is tied up with the nature of the orbit. If b is sufficiently large, a root exists, and the orbit looks like a hyperbola; for small b, no root exists and the two particles are "sucked" towards each other in a spiraling orbit. The two regimes are separated by a limiting orbit in which the particles spiral asymptotically into a circular orbit. This limiting orbit is found by setting the discriminant of the square root in (21.03) equal to 0. We find

$$b_{\lim}^4 = \frac{4 \, e^2 \, P \, (M+m)}{M \, m \, w^2} \tag{21.04}$$

The cross section σ going with spiraling collisions is of course related to this quantity $b_{\lim}$ through

$$\sigma = \pi \, b_{\lim}^2 \tag{21.05}$$

and the associated mean free time τ_s between spiraling collisions equals

$$\tau_s = \frac{1}{n \, w \, \pi \, b_{\lim}^2} \tag{21.06}$$

which yields

$$\tau_s = \frac{1}{2 \, \pi \, e \, n} \left\{ \frac{M \, m}{(M+m) \, P} \right\}^{1/2} \tag{21.07}$$

It is to be observed that this time is independent of the particle speed as predicted for Maxwellian molecules.

Whenever we have to compute an angular average of a function $\varphi(\chi)$ in kinetic theory we are supposed to evaluate a reciprocal time of the form

$$n \, w \int \int \varphi(\chi) \, d\sigma = 2 \, \pi \, n \, w \int_0^\infty \varphi(\chi) \, b \, db$$

We render b dimensionless by setting

$$\beta = \frac{b}{b_{\lim}} \tag{21.08}$$

† P. Langevin. *Ann. Chim. et Phys.* **5**, 245 (1905).

$b_{\lim}$ being given by (21.04). The average of $\varphi(\chi)$ becomes then

$$\pi\, n\, w\, b_{\lim}^{2} \int_{0}^{\infty} \varphi(\chi)\, d(\beta^{2})$$

The outside factor is just $1/\tau_s$ as defined by (21.06) so that finally

$$n \iint \phi(\chi)\, \sigma'\, w\, d\Omega = \left\langle \frac{\varphi(\chi)}{\tau} \right\rangle = \frac{1}{\tau_s} \int_{0}^{\infty} \varphi(\chi)\, d(\beta^{2}) \qquad (21.09)$$

The dimensionless form of (21.03) which goes with (21.09) is

$$\chi = \pi - 2 \int_{0}^{v_1} \frac{dv}{\left\{1 - v^2 + \dfrac{v^4}{4\,\beta^4}\right\}^{1/2}} \qquad (21.10)$$

Equations (21.09) and (21.10) provide a straightforward numerical averaging procedure for any function $\varphi(\chi)$. As usual in such cases, $\varphi(\chi)$ has to provide convergence for large impact parameters; that is, it must vanish for $\chi = 0$. The two most common expressions which arise are‡

$$\left\langle \frac{1 - \cos\chi}{\tau} \right\rangle = \frac{1}{\tau_s}\, 1.1052 \qquad (21.11)$$

and

$$\left\langle \frac{\sin^2\chi}{\tau} \right\rangle = \frac{1}{\tau_s}\, 0.772 \qquad (21.12)$$

Forearmed with this unusually complete knowledge of the relevant collision integrals, we can proceed, getting results from (21.01) in the way Maxwell did. To find the drift velocity in a field we multiply (21.01) with v_z, the z direction being parallel to $\mathbf{a}$; thereupon, we integrate over $\mathbf{v}$. The equation reads then, with (18.13),

$$\iiint a\, v_z\, \frac{\partial f}{\partial v_z}\, d\mathbf{v} = -\, n \iiint f(\mathbf{v})\, M(\mathbf{V})\, v_z\, d\mathbf{v}\, d\mathbf{V} \iint \sigma'\, w\, d\Omega'$$
$$+\, n \iiint f(\mathbf{v}')\, M(\mathbf{V}')\, d\mathbf{v}'\, d\mathbf{V}' \iint v_z\, \sigma'\, w\, d\Omega \qquad (21.13)$$

The problem is the presence of v_z in the second integral. Starting out with the momentum conservation law

$$m\, \mathbf{v} + M\, \mathbf{V} = m\, \mathbf{v}' + M\, \mathbf{V}' \qquad (21.14)$$

and the definitions (18.06) and (18.07) for the relative velocities $\mathbf{w}$ and $\mathbf{w}'$, we can get the law in the form

$$(M+m)\, \mathbf{v} = (M+m)\, \mathbf{v}' + M\, (\mathbf{w}-\mathbf{w}') \qquad (21.15)$$

‡ See G. H. Wannier. *Bell System Tech. J.* **32**, 170 (1953).

with **w** and **w**′ of equal magnitude. We next take components along **a** of this equation, and apply the spherical law of cosines to $\cos \vartheta$, as shown in the upper triangle in Fig. 21.1. The result is

$$(M+m)v_z = (M+m)v_z' + Mw\{\cos \vartheta' \cos \chi + \sin \vartheta' \sin \chi \cos \varepsilon\} - Mw \cos \vartheta'$$

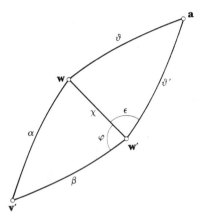

Fig. 21.1. Spherical triangles occurring in the computation of the mean drift and the mean energy of gaseous ions. Vectors show up through their piercing points, polar angles as sides, azimuths as angles.

Here $w \cos \vartheta'$ equals w_z', $w \sin \vartheta'$ equals $|\mathbf{w}' \times \mathbf{z}_1|$, where $\mathbf{z}_1$ is a unit vector along z. The law of conservation of momentum becomes thus

$$v_z = \frac{m+M \cos \chi}{m+M} v_z' + \frac{M}{m+M} (1-\cos \chi) V_z' + M |\mathbf{w}' \times \mathbf{z}_1| \sin \chi \cos \varepsilon$$

$$(21.16)$$

When this expression for v_z is substituted into the last term in (21.13), integration over $d\varepsilon$, being the azimuthal part of the integration over $d\Omega$, annuls the last term of (21.16). Also, as the function $M(\mathbf{V}')$ is Maxwellian and hence spherically symmetric, the second term averages to zero. So only the first term above remains. In the integral (21.13) we save the collision integral to the last, for it is independent of speed, by (21.09). As for the left side of (21.13), we integrate it by parts with respect to v_z. It becomes then the normalization integral for $f(\mathbf{v})$. The result is therefore

$$- a = - n\langle v_z \rangle \int \int \left\{ 1 - \frac{m+M \cos \chi}{m+M} \right\} \sigma' w \, d\Omega$$

The definition (21.09) reduces this to

$$\langle \mathbf{v} \rangle = \frac{\dfrac{m+M}{M}\,\mathbf{a}}{\left\langle \dfrac{1 - \cos \chi}{\tau} \right\rangle} \tag{21.17}$$

The results (20.06), (21.11), and (21.07) bring this into the explicit form

$$\langle \mathbf{v} \rangle = \frac{0.9048}{2\pi} \left(\frac{1}{M} + \frac{1}{m} \right)^{\frac{1}{2}} \frac{1}{n\,P^{\frac{1}{2}}}\,\mathbf{E} \tag{21.18}$$

The mobilities at standard density of the noble gas molecular ions in the parent gas are shown in Table 21–1. They were measured by a direct time-of-flight technique which allows the experimenter to distinguish between different ions. The calculated results are taken from (21.18) and are in complete agreement with the data.§

TABLE 21–1. Mobilities at standard density of the noble gas molecular ions in the parent gas

Gas	μobs. (cm²/volt sec)	μcalc. (cm²/volt sec)
He	18	18.2
Ne	6.5	6.21
Ar	1.9	2.09
Kr	1.2	1.18
Xe	0.7	0.74

Note. Observations by Hornbeck and Varney compared with formula (21.18).

Having computed and verified the mobilities of ions in gases in a favorable instance, we do well to get some further information concerning ions, which is of qualitative interest. This concerns the mean energy of traveling ions. For this purpose we multiply (21.01) with v^2. We then obtain expression (21.13) again, this time with the factor v^2 replacing v_z in all terms. Thereby v^2 becomes the "wrong" multiplier in the second integral instead of v_z. We get an expression for it by squaring (21.15) and picking angles from the lower triangle in Fig. 21.1. This gives

$$(M+m)^2 v^2 = (M+m)^2 v'^2 + 2 M^2 w^2 (1 - \cos \chi)$$
$$+ 2 M (M+m) v' w (\cos \alpha - \cos \beta)$$

§ Atomic ions in the parent gas do not obey the present formula because of their large cross section for charge exchange with neutral atoms.

The angle to be eliminated by the spherical law of cosines is $\cos \alpha$. We also use

$$v' \, w \cos \beta = \mathbf{v}' \cdot (\mathbf{v}' - \mathbf{V}') = v'^2 - \mathbf{v}' \cdot \mathbf{V}'$$

and

$$v' \, w \sin \beta = |\mathbf{v}' \times (\mathbf{v}' - \mathbf{V}')| = |\mathbf{v}' \times \mathbf{V}'|$$

We get then

$$(M+m)^2 v^2 = (M^2 + m^2 + 2Mm \cos \chi) \, v'^2 + 2M^2 (1-\cos \chi) \, V'^2$$
$$+ 2M(m-M)(1-\cos \chi) \, \mathbf{v}' \cdot \mathbf{V}' + 2 \, M(M+m) \, |\mathbf{v}' \times \mathbf{V}'| \sin \chi \cos \varphi \quad (21.19)$$

This time the *two* first terms on the right survive the integrations. The last term dies because of $\cos \varphi$, φ being a collision azimuth, and the third term because $M(\mathbf{V}')$ is spherically symmetric. Integration by parts on the left-hand side of the modified equation (21.13) yields this time twice the drift velocity. We find thus

$$- 2 \, a \, \langle v_z \rangle = n \int \int \left\{ - \frac{2 \, M \, m}{(M+m)^2} \, (1 - \cos \chi) \, \langle v^2 \rangle \right.$$
$$\left. + \frac{2 \, M^2}{(M+m)^2} \, (1 - \cos \chi) \langle V^2 \rangle \right\} \sigma' \, w \, d\Omega$$

Our unknown is $\langle v^2 \rangle$. If we isolate it and make use of (21.09) and (21.17) the result can be given the attractive form

$$\langle \tfrac{1}{2} \, m \, v^2 \rangle = \langle \tfrac{1}{2} \, M \, V^2 \rangle + \tfrac{1}{2} \, (M+m) \, \langle v_z \rangle^2 \quad (21.20a)$$

or

$$\langle \tfrac{1}{2} \, m \, v^2 \rangle = \tfrac{3}{2} \, k \, T + \tfrac{1}{2} \, m \, \langle v_z \rangle^2 + \frac{M}{m} \, \tfrac{1}{2} \, m \, \langle v_z \rangle^2 \quad (21.20b)$$

All terms in (21.20) are self explanatory except one. The first term on the right of (21.20b) is the equipartition value arising from the gas temperature, and the second is the energy visible in the drift motion. The third term thus represents an additional energy beyond these two. It is particularly large for electrons in a gas where M/m equals several thousand; it represents the capacity of these light particles to store energy gained from the field in the form of random motion. This is sometimes expressed by saying that the electrons have a higher "temperature" than the surrounding gas. For ions whose mass is of the same order as that of the gas molecules the term is also present, but of moderate proportions.

It is important to realize that this is the first kinetic theory calculation which we did not solve in perturbation theory, but exactly. Consequently, the results do not depend on the assumption that the field produces only a small perturbation of thermal equilibrium. The linear relationship between the drift velocity and the field thus happens to be a particular property of the polarization force which is valid at all fields. In the energy relation (21.20) the field does in fact enter not linearly but quadratically. Because of the "storage" aspect of this equation it is relatively easy to get out of the linear range for electrons, but it is also possible to do so for ions if the circumstances are favorable. These circumstances are of some interest because they teach us something about the intrinsic limitations of Ohm's law.

21-2. "High field" velocity distribution of ions and electrons in gases

The preceding calculation is remarkable in that it permits kinetic analysis of ionic mobilities in gases, without the need of examining their velocity distribution. The possibility of a departure of this distribution from Maxwellian form is of sufficient interest to be taken up in its own right. Such a departure will arise if the energy gained by the ions from the field predominates so completely over thermal energy that the latter can be neglected. Let us denote this "high field" distribution function by $h(\mathbf{v})$. It obeys the limiting form of equation (21.01) which is reached when the Maxwellian functions $M(\mathbf{V})$ are replaced by δ functions. The equation becomes then

$$\mathbf{a} \cdot \frac{\partial h(\mathbf{v})}{\partial \mathbf{v}} + n \, \sigma \, v \, h(\mathbf{v}) = n \int \int \int \int \int h(\mathbf{v}') \, \delta(\mathbf{V}') \, \sigma' \, v' \, d\mathbf{V} \, d\Omega_{\mathbf{v}'} \quad (21.21)$$

Some information about high field behavior of ions can be gained from (21.21) by dimensional considerations. We have seen earlier that simple power laws of interaction provide an angular pattern independent of speed, so that the dependence on speed is concentrated in the cross section. Suppose we take an interaction

$$\Phi = \pm \frac{\alpha}{r^s} \quad (21.22)$$

then a cross section can be made out of α/m and v in only one way, namely, by setting

$$\sigma = \left(\frac{\alpha}{m \, v^2}\right)^{2/s} \quad (21.23)$$

The law includes the hard sphere case for $s = \infty$. Now the high field Boltzmann equation (20.80) contains only one dimensional parameter for all velocities, namely,

$$v \propto \left(\frac{a}{n \, \sigma}\right)^{1/2} \quad (21.24)$$

Insertion of σ from (21.23) into this yields

$$v \propto \left(\frac{a}{n}\right)^{\frac{s}{2s-4}} \tag{21.25}$$

The combination a/n becomes E/p in experimental parlance. The law checks the previously found linear law for $s = 4$. The hard sphere model

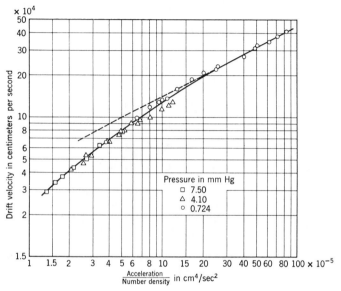

Fig. 21.2. Drift velocity in an electric field for Ne^+ ions in neon gas. Comparison of observed results with an asymptotic straight line of slope 1/2.

is, however, much closer to the truth. Its dimensional result is written out in (21.25). If the analysis applies, the formula gives the drift velocity at high fields, apart from a constant. This quantity should increase as the square root of the field for high fields. This is confirmed for the noble gas atomic ions in the parent gas. Figure 21.2 shows on a log–log plot results for Ne^+ ions in neon gas. The asymptotic straight line of slope 1/2 is shown in dashed outline. What makes the graph particularly impressive is that it combines data at different pressures, testing in effect the dependence on the combination a/n, as well as the square root law. At low fields the curve bends over to a 45° slope, as is proper when the field acts as a perturbation only. The large cross section independent of velocity, which is found here for collisions of atomic ions in the parent gas, seems at first somewhat at variance with the results for molecular ions collected in Table 21-1. The difference is brought about by the charge exchange reaction

$$Ne^+ + Ne = Ne + Ne^+$$

which depends only on a very slight wave function overlap to go. This is a fixed cross section process which is kinetically equivalent to a head-on collision. Assuming that such is the case, we can in fact finish evaluating

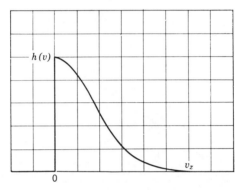

Fig. 21.3. High field velocity distribution function for noble gas ions in the parent gas. In this graph charge exchange without momentum transfer is assumed as the scattering mechanism.

equation (21.21) to get the charge exchange cross section. The condition permitting solution of the equation is

$$\mathbf{v}' = \mathbf{V}$$
$$\mathbf{v} = \mathbf{V}'$$
$$\iint \sigma' \, d\Omega_v = \sigma$$

Then (21.21) becomes

$$a \frac{\partial h(\mathbf{v})}{\partial v_z} + n \sigma v \, h(\mathbf{v}) = n \sigma \, \delta(\mathbf{v}) \iiint h(\mathbf{v}') \, v' \, d\mathbf{v}' \qquad (21.26)$$

This solves to

$$h(\mathbf{v}) = C\delta(v_x)\delta(v_y) \, s(v_z) \exp\left[-\frac{n \sigma v_z^2}{2 a}\right] \qquad (21.27)$$

where $s(v_z)$ is a step function which is zero for negative v_z and 1 for positive v_z. The curve is shown in Fig. 21.3. Evaluation of $\langle v_z \rangle$ from this distribution function is elementary. We get

$$\langle v_z \rangle = \left(\frac{2 a}{\pi n \sigma}\right)^{\frac{1}{2}} \qquad (21.28)$$

Interest lies in this case in computing the charge exchange cross section from the observed drift velocity, that is, to reverse (21.28) and solve for σ

$$\sigma = \frac{2a}{\pi n \langle v_z \rangle^2} \tag{21.29}$$

On Fig. 21.2 we read off a $\langle v_z \rangle$ of 3.5×10^5 cm/sec against an a/n of 6×10^{-4} cm^4/sec^2. This yields

$$\sigma = 31 \times 10^{-16} \text{ cm}^2$$

The value is considerably in excess of atom–atom cross sections determined from viscosity. For neon, this cross section is 21×10^{-16} cm^2. Table 21–2 lists in the first column the ion–atom charge exchange cross section for the noble gases, and in the third the viscosity cross section of the gaseous

TABLE 21–2. Ion–atom exchange cross sections and atom–atom cross sections in comparison

Gas	σ (Å^2) formula (21.29)	σ (Å^2) formula (21.30)	Viscosity Cross Section (Å^2)
He	26	54	15
Ne	31	65	21
Ar	65	134	42
Kr	77	157	49
Xe	93	192	67

atoms in comparison. The approximate proportionality of the two columns indicates that wave function overlap is probably responsible for either type of phenomenon.

The preceding analysis, culminating in (21.29), and the first column of Table 21-2 should be accepted with some reserve. It is based on the hypothesis that all collisions are "head on." Such an analysis is in violation of the argument following (18.18), according to which all angles of scattering are possible in a scattering event. The practical difference between the two lines of argument is that a collision is probably coupled with the charge exchange process so that the new carrier gains part of the momentum in the field direction from the old one. An exact analysis of this process is scarcely feasible, but we can go to the other extreme and analyze the data in terms of a "hard sphere" cross section. The dimensional features being the same, this viewpoint yields only different numerical constants in the old formulas. The one relevant to the experimental data is (21.28), which reads now‖,¶

$$v_z = 1.1467 \left(\frac{a}{n\,\sigma}\right)^{\frac{1}{2}} \tag{21.30}$$

Determination of the cross section from this equation yields the second column in Table 21-2. It is seen that the deduced cross section is considerably larger. The actual cross section is probably somewhere between these numbers. In cases for which charge exchange plays no role, (21.30) is obviously the correct expression to use at high fields.

21-3. Velocity distribution functions for electrons; formulas of Davydov and Druyvesteyn

The type of velocity distribution going with the data of Table 21-2 is very far removed from a Maxwellian distribution. Figure 13 of the reference indicated by ‖ gives a crude tentative outline of it. It has the shape of an onion, with the tip of the onion pointing in the field direction. It has almost certainly no simple analytical structure. The one case which has analytical possibilities is that of electrons in a gas. Electrons store large amounts of energy gained in the field in the form of random motion, as was pointed out earlier in connection with (21.20). This is a consequence of their small mass in comparison to that of the molecules with which they collide. As a result of this, the velocity distribution remains very nearly spherically symmetric, even in the high field region. This insight may be combined with the insight that, since electrons change their speed only very little in an encounter, it should be possible to bring the integral operator in (21.01), which represents encounters, into differential form. We shall now carry out this transformation, which was first performed by B. Davydov.**

Our starting equation is (21.01), into which the assumption

$$f(\mathbf{v}) = f_0(v) + f_1(v)\,\frac{v_z}{v} \tag{21.31}$$

is immediately introduced. The derivative term becomes then

$$\frac{\partial f}{\partial v_z} = \frac{v_z}{v}\frac{df_0}{dv} + \frac{v_z^2}{v^2}\frac{df_1}{dv} + \frac{1}{v}\,f_1 - \frac{v_z^2}{v^3}\,f_1$$

‖ G. H. Wannier. *Bell System Tech. J.* **32**, 170 (1953), Equation (92).

¶ J. A. Hornbeck and G. H. Wannier. *Phys. Rev.* **82**, 458 (1951).

** B. Davydov. *Phys. Z. Sov.* **8**, 59 (1939). See also S. Chapman and T. G. Cowling, *The Mathematical Theory of Non-uniform Gases*, Cambridge University Press, 1952, pp. 349–350.

Contributions arising from $P_2(v_z/v)$ are negligible by assumption, and we can set

$$\frac{v_z^2}{v^2} \approx \frac{1}{3}$$

As was explained in Sections 18-5 and 18-6, the right-hand side of (21.01) is diagonal with respect to an expansion in spherical harmonics. Spherically symmetric terms and terms in v_z/v can therefore be separated; this yields the two equations

$$a\left(\frac{1}{3}\frac{df_1}{dv} + \frac{2}{3}\frac{1}{v}f_1\right) = -n \iiiint \{f_0(v) M(V) - f_0(v') M(V')\} \, dV \, \sigma' \, w \, d\Omega_{w'}$$

(21.32)

$$a\frac{df_0}{dv}\cos\vartheta = -n \iiiint \{f_1(v)M(V)\cos\vartheta - f_1(v')M(V')\cos\vartheta'\}dV \, \sigma'w \, d\Omega_{w'}$$

(21.33)

The transformation of the right-hand sides into differential form is based on the notion that v and v' differ very little from each other. The reduction is somewhat easier for (21.33). The temperature is assumed not to be extremely low so that molecular recoil velocities are small compared to thermal velocities. This means that

$$M V \gg m v$$

(21.34a)

while, because of the energy storage phenomenon,

$$M V^2 \ll m v^2$$

(21.34b)

Under these circumstances, we have

$$\mathbf{V} \approx \mathbf{V'}$$

(21.34c)

$$v \approx v' \approx w$$

(21.34d)

If relations (21.34) are valid the only parameter in (21.33) which has to be expressed in terms of other quantities is $\cos \vartheta'$. The spherical triangle shown in Fig. 21.4 yields for it

$$\cos \vartheta' = \cos \vartheta \cos \chi + \sin \vartheta \sin \chi \cos \varepsilon$$

The second term will die upon integration over the scattering azimuth. Therefore (21.33) becomes

$$a\frac{df_0(v)}{dv} = \frac{v}{l(v)} f_1(v)$$

(21.35)

where the abbreviation

$$\frac{1}{l(v)} = n \int \int \sigma' \left(1 - \cos \chi\right) d\Omega \qquad (21.36)$$

has been introduced.

The same degree of approximation yields a null result for the right-hand side of (21.32). It is therefore essential to estimate carefully the small velocity

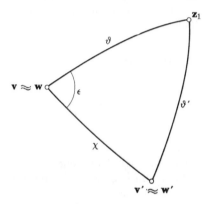

Fig. 21.4. Spherical triangle used in the reduction of (21.33).

shifts which occur during a collision. As a first step we multiply the equation with dv and integrate over all values for which $v < c$. We get

$$\tfrac{4}{3}\pi a c' f_1(c) = -n \iiint\!\!\iiint\int_{v<c} \{f_0(v)M(V) - f_0(v')M(V')\} \, dv d\mathbf{V} \, \sigma' w \, d\Omega_{\mathbf{w}'} \quad (21.37)$$

The eight integrations occurring in (21.37) are the symmetric combination (18.13). The first and the second half of the integral are therefore the same integral, except for the fact that the limits are not quite the same. For the first half the limit is

$$v \leqq c$$

For the second half the limit will temporarily be given the form

$$v' \leqq c + \Delta c$$

with the equality sign holding simultaneously in both relations. Δc will be treated as a small quantity.

The formal identity of the two halves of the collision integral in (21.37) permits writing it as a single integral, and giving to the equation the form

$$\tfrac{1}{3} a c^2 f_1(c) = n \iiint\!\!\int\int M(V) \left[\int_c^{c+\Delta c} f_0(v) \, v^2 \, dv \right] d\mathbf{V} \, \sigma' \, w \, d\Omega_{\mathbf{w}'}$$

The range of integration on v is assumed small. We set therefore

$$f_0(v) = f_0(c) + (v-c) \frac{df_0}{dc}$$

and find to the lowest order in Δc

$$\tfrac{1}{3} a c^2 f_1(c) = \tfrac{1}{3} n f_0(c) \int\int\int\int\int 3 c^2 \Delta c \, M(V) \, dV \, \sigma' \, w \, d\Omega_{w'} \tag{21.38}$$

$$+ n \frac{df_0}{dc} \int\int\int\int\int \tfrac{1}{2} c^2 \Delta c^2 \, M(V) \, dV \, \sigma' \, w \, d\Omega_{w'}$$

To proceed further we must investigate the small quantity Δc. We write down the conservation laws

$$m c^2 + M V^2 = m c'^2 + M V'^2 \tag{21.39}$$

$$m \mathbf{c} + M \mathbf{V} = m \mathbf{c}' + M \mathbf{V}' \tag{21.40}$$

and assume the unprimed variables to be the initial ones. Then we eliminate $\mathbf{V}'$, writing

$$m c^2 + M V^2 = m c'^2 + \frac{1}{M} (m \mathbf{c} + M \mathbf{V} - m \mathbf{c}')^2$$

Under these circumstances, we have $c = c' + \Delta c$. The above equation reads therefore

$$m \{(c' + \Delta c)^2 - c'^2\} = \frac{1}{M} (M\mathbf{V} + m\mathbf{c} - m\mathbf{c}')^2 - M V^2$$

which yields in first order

$$2 m c' \Delta c = 2 m \mathbf{V} \cdot (\mathbf{c}'-\mathbf{c}) + \frac{m^2}{M} (\mathbf{c}-\mathbf{c}')^2$$

or

$$\Delta c = (\mathbf{V} \cdot \mathbf{k}_1) \sqrt{2 (1-\cos \chi)} + \frac{m}{M} c (1-\cos \chi) \tag{21.41}$$

Here $\mathbf{k}_1$ is a unit vector along the bisector of $\mathbf{c}$ and $\mathbf{c}'$. It follows from (21.34a) that the first term on the right of (21.41) is the larger one. It is the one to be kept unless it happens to give a zero result upon integration.

We now substitute (21.41) into (21.38). In the first integral, the linear term in $\mathbf{V}$ integrates to zero, so that we get, replacing w also by c, and using (21.36),

$$\tfrac{1}{3} a c^2 f_1(c) = \frac{m c^4}{M \, l(c)} f_0(c) + \frac{k T c^3}{M \, l(c)} \frac{df_0}{dc}$$

or

$$f_0 + \frac{1}{\beta m c} \frac{df_0}{dc} - \frac{1}{3} \frac{a M \, l(c)}{m c^2} f_1 = 0 \qquad (21.42)$$

Before applying (21.42) it is good to observe that no error has crept in through the neglect of higher terms in Δc. Indeed, for $a = 0$, the equation has the Maxwellian solution

$$f_0 = \text{constant} \cdot \exp[- \tfrac{1}{2} \beta m c^2]$$

as required by equilibrium theory.

The solution for the velocity distribution in the presence of a field results from combination of (21.35) and (21.42). Elimination of f_1 yields

$$\frac{df_0}{dc} \left\{ \frac{k T}{m c} + \frac{1}{3} \frac{a^2 M \, l^2(c)}{m c^3} \right\} + f_0 = 0$$

which solves to

$$f_0(v) = C \exp\left[-m \int_0^v \frac{c \, dc}{k T + \frac{1}{3} \frac{M a^2 l^2(c)}{c^2}} \right] \qquad (21.43)$$

Expression (21.43) is the so-called *Davydov distribution function*, a closed-form expression for the distribution of electronic speeds in gases. For small fields, the function is simply the Maxwellian distribution; for increasing a, however, it shifts gradually into a distribution controlled by the field. If we make the mean free path approximation

$$l(c) = \text{constant} \qquad (21.44)$$

the high field limiting distribution of (21.43) has the form

$$h_0(v) = C \exp\left[-\frac{3}{4} \frac{m}{M} \frac{c^4}{a^2 l^2} \right] \qquad (21.45)$$

This is the so-called *Druyvesteyn distribution*. The formula has a sharper drop-off at high speeds than the Maxwellian distribution. It fits experimental electron velocity distributions a little better than the Maxwellian formula, but it provides by no means perfect agreement. A gas discharge is a very complex condition in which many processes take place. It is therefore difficult to provide "clean" experimental conditions to test a formula like (21.45). It is good, however, to have a look at other possible velocity distributions, which may arise in steady state phenomena well away from equilibrium.

There are no problems for this chapter.

22

Fluctuations and Brownian motion

Fluctuations have been discussed incidentally several times in earlier parts of this book. Whenever the mean value of a quantity is computed statistically, the mean fluctuation of the quantity is also a result of the same statistical reasoning.

A widely used measure of fluctuation is obtained in the following way. Suppose we deal with a quantity x whose statistical average is $\bar{x}$. Then we have of course by definition

$$\langle x - \bar{x} \rangle = 0 \tag{22.01}$$

However, if we take the square of this difference first, and then the average of that square, we find ourselves dealing with an average over nonnegative numbers. This average can only vanish if all quantities x are equal. In general, it will be a positive number:

$$\langle \delta x^2 \rangle = \langle (x - \bar{x})^2 \rangle = \langle x^2 \rangle - \langle x \rangle^2 \tag{22.02}$$

This *mean square deviation* gives an indication of the statistical spread of the data. The *root mean square deviation*, which is defined as

$$\sqrt{\langle \delta x^2 \rangle} = \sqrt{\langle x^2 \rangle - \langle x \rangle^2} \tag{22.03}$$

has the same dimension as x itself and is conveniently compared with the original average. The ratio of these two quantities is a pure number; it is generally used as an indication of the scatter of the data.

Fluctuation plays also an important role of principle in statistical physics. It has been pointed out repeatedly that the type of averages formed are often not in strict accordance with mechanics. It is then necessary to invoke the idea of an ensemble of systems in order to take the kind of average which seems to be indicated. There is then a danger that the average so formed is meaningless. If the fluctuation is computed in the same manner and is found small, the procedure receives thereby some additional support. No proof of the correctness does of course result from this, but at least there is a proof of internal consistency.

Chapters 2 and 4 of this book contain fluctuations computed for particularly simple situations. Thereafter, it was found convenient to ignore the fluctuation aspect of statistics, and to treat the averages as if they were the quantities themselves. This sacrifice of rigor was emphasized in the text by leaving off any symbol denoting average formation from the left sides of formulas such as (5.02) and (5.32). The sacrifice is worth while not only because an enormous amount of applications can be carried out within this outlook, but also because it permits the perfect fusion of statistics and thermodynamics; this fusion was carried through in Chapter 7. It is the business of this chapter to come back on this sacrifice and to investigate directly the nature of fluctuations.

22-1. Equilibrium theory of fluctuations

Many discussions of fluctuations leave an impression that fluctuations are a phenomenon which is outside equilibrium statistical mechanics. This is only partly true. Fluctuations tend to vary rapidly in time, and this aspect of their nature is usually not incorporated into equilibrium statistics. However, fluctuations also have a static aspect. Whenever this static aspect is sufficient for the theoretical study of a particular problem, then equilibrium statistical mechanics is also sufficient. A perfect example of this is formula (4.39), which relates the energy fluctuation of a sample to its heat capacity. The formula reads

$$\langle \delta U^2 \rangle = C_0\, k\, T^2 \tag{22.04}$$

Here C_0 is the anergetic heat capacity of the sample. The connection is a natural one, for a given temperature fluctuation will produce a large energy fluctuation if the specific heat is high. By the same argument, density fluctuations should be large in a material having high compressibility. This is indeed the case, as we shall show now.

In general, density fluctuations are more easily computed in the grand ensemble formalism than in the canonical one; for in the canonical one the number of particles of each member is assumed fixed. In the grand ensemble this is not the case, and fluctuations in the number of particles are thus not suppressed artificially. The relevant formulas for our purpose are (8.57) and (8.58). They read

$$P(N) = \frac{1}{\mathscr{F}} e^{\beta\mu N} F(N) \tag{22.05}$$

$$\mathscr{F} = \sum_{N=0}^{\infty} e^{\beta\mu N} F(N) \tag{22.06}$$

Here $F(N)$ is the canonical partition function for a system containing N particles. This yields for the fluctuation in the number of particles

$$\langle N^2 \rangle - \langle N \rangle^2 = \frac{1}{\beta^2} \left\{ \frac{1}{\mathscr{F}} \frac{\partial^2 \mathscr{F}}{\partial \mu^2} - \frac{1}{\mathscr{F}^2} \left(\frac{\partial \mathscr{F}}{\partial \mu} \right)^2 \right\}$$

With (8.08)

$$\langle N \rangle = \frac{1}{\beta} \frac{1}{\mathscr{F}} \frac{\partial \mathscr{F}}{\partial \mu} \tag{22.07}$$

this becomes

$$\langle \delta N^2 \rangle = \frac{1}{\beta} \left(\frac{\partial \langle N \rangle}{\partial \mu} \right)_{\beta, x_i} \tag{22.08}$$

In accordance with our previous thermodynamic habits the pointed bracket on the right will be omitted below.

Further reduction of (22.08) is restricted to perfect fluids, for which the distinction between extensive and intensive variables, explained in Section 7-2, is valid, and the only form of reversible work is $- p\, dV$. We can then also postulate the existence of an equation of state of the form (1.03). Relation (22.08) then takes the form

$$\langle \delta N^2 \rangle = k\, T \left(\frac{\partial N}{\partial \mu} \right)_{T, V} \tag{22.09}$$

We now apply the chain rule twice, introducing the number density n and the pressure p as intermediate variables. Equation (22.09) then becomes

$$\langle \delta N^2 \rangle = k\, T \left(\frac{\partial N}{\partial n} \right)_{T, V} \left(\frac{\partial n}{\partial p} \right)_T \left(\frac{\partial p}{\partial \mu} \right)_T \tag{22.10}$$

Here the index V was left off on the last two derivatives because they are derivatives connecting intensive variables, whose relationship is not affected by the value of the extensive variable V. The first of the three derivatives in (22.10) is taken care of by the identity

$$N = n V$$

which yields

$$\left(\frac{\partial N}{\partial n} \right)_{T,V} = V$$

The same identity permits us to write the second of the three derivatives in (22.10) as

$$\left(\frac{\partial n}{\partial p} \right)_T = - \frac{N}{V^2} \left(\frac{\partial V}{\partial p} \right)_{T,N}$$

Finally, the third of the derivatives in (22.10) follows from (7.14) and (7.27) as

$$\left(\frac{\partial p}{\partial \mu} \right)_T = \frac{N}{V}$$

Combination of the three results yields

$$\langle \delta N^2 \rangle = - k T \left(\frac{N}{V} \right)^2 \left(\frac{\partial V}{\partial p} \right)_T \qquad (22.11\text{a})$$

or in terms of the isothermal bulk modulus B_T defined in (1.13)

$$\langle \delta N^2 \rangle = \frac{N^2 k T}{V B_T} \qquad (22.11\text{b})$$

As expected, the mean square density fluctuation of a fluid varies inversely as its bulk modulus. Beside this point, there are several other features worthy of note in (22.11). First, the square fluctuation of the number of particles varies with the first power of the size of the sample taken unless B_T vanishes. Second, the form (22.11a) shows particularly that the derivative $(\partial V/\partial p)_T$ cannot be positive for any fluid in equilibrium, because the left-hand side is intrinsically positive. Third, for perfect gases we have, by (1.41), that $B_T = p$ and hence

$$\langle \delta N^2 \rangle = N \qquad (22.12)$$

This result was established earlier by elementary methods in (2.20).

22-2. Brownian motion

The phenomenon of Brownian motion was discovered in 1827 by Robert Brown on a suspension of pollen in water. What he noticed was that the pollen particles were in a state of very agitated motion at all times. At first it was thought that the phenomenon had something to do with pollen being alive, but it was soon found that many other suspensions of fine particles behave in the same fashion. We know now that Brownian motion is simply kinetic theory becoming visible; it needs only the right circumstances—low mass, weak binding, and small frictional forces—to make its appearance.

Perrin* first proved experimentally that the above explanation of Brownian motion is quantitatively correct. He realized that the phenomenon should permit the determination of Avogadro's number by visual observation. The observation of random speeds being very difficult experimentally, he hit upon the idea of checking the barometric formula (4.31) instead. It is incidental to Brownian motion that a suspension does not settle at the bottom of a vessel, but shows a distribution in height obeying that relation. He took particles of volume $V = 9.8 \times 10^{-15}$ cm³ and density 1.35 gm/cm³, dispersed them in water, and got the following series of observations:

x (cm)	Mean Number	Ratio of Successive Numbers
0.0000	200	1.18
0.0025	170	1.16
0.0050	146	1.26
0.0075	116	1.16
0.0100	100	

The effective mass of the suspended particles is $(1.35 - 1) \times 9.8 \times 10^{-15}$, the -1 arising from the buoyancy of water. We therefore find for Boltzmann's constant

$$k = \frac{9.8 \times 10^{-15} \times 0.35 \times 980 \times 0.0100}{\ln 2 \times 300} = 1.6 \times 10^{-16} \text{ erg/deg}$$

The most interesting aspect of Brownian motion is of course the "motion" part of it. This motion is similar to the motion of gas molecules but shows some characteristic differences. The differences arise because an object executing Brownian motion is as a rule sufficiently large to be interacting with the surrounding medium at all times.

Langevin analyzed the nature of Brownian motion in the following way. A particle moving in a medium, say water, is subject to a fluctuating force arising from its surroundings. This fluctuating force averages out for a

* J. Perrin. *Compt. Rend.* **146,** 967 (1908).

reasonably prolonged time interval to a viscous drag, but differs from this at any instant. We thus can write an equation of motion of the form

$$\dot{\mathbf{v}} = -\gamma \mathbf{v} + \mathbf{A}(t) \qquad (22.13)$$

Here γ is a frictional constant and $\mathbf{A}(t)$ is a function which averages to zero in time. In detail $\mathbf{A}(t)$ may depend on $\mathbf{v}$, but it will be assumed independent of position; in other words, the medium is supposed uniform.

We now form the scalar product of the position vector $\mathbf{r}$ with (22.13). For the products we use the identities

$$\mathbf{r} \cdot \mathbf{v} = \frac{1}{2}\frac{d}{dt} r^2$$

$$\mathbf{r} \cdot \dot{\mathbf{v}} = \frac{1}{2}\frac{d^2}{dt^2} r^2 - v^2$$

Then (22.13) becomes

$$\frac{1}{2}\frac{d^2}{dt^2} r^2 + \frac{1}{2}\gamma \frac{d}{dt} r^2 = v^2 + \mathbf{r} \cdot \mathbf{A}(t) \qquad (22.14)$$

We now form the average of the terms over a substantial span of time. The last term then gives zero because $\mathbf{r}$ and $\mathbf{A}(t)$ are assumed uncorrelated by assumption, and $\mathbf{A}(t)$ averages out to zero also by assumption. For v^2 we can write the equipartition value $3\,k\,T/M$. If we denote by $\langle r^2 \rangle$ the value of r^2 so averaged over time, we get (22.14) in the form

$$\frac{d^2}{dt^2} \langle r^2 \rangle + \gamma \frac{d}{dt} \langle r^2 \rangle = \frac{6\,k\,T}{M} \qquad (22.15)$$

Equation (22.15) is a sort of "coarse-grained" equation of motion in the sense that passage from (22.14) to (22.15) already implies an integration over time. The interpretation is that v^2 and $\mathbf{A}(t)$ reach their time average within a period for which r^2 does not change appreciably.

Equation (22.15) is easily integrated. The solution is

$$\langle r^2 \rangle = \frac{6\,k\,T}{\gamma\,M} t + C_1 + C_2\, e^{-\gamma t}$$

Assume the particle initially at $\mathbf{r} = 0$. Then the first derivative $2\mathbf{r} \cdot \mathbf{v}$ is also zero initially. The constants C_1 and C_2 must therefore be determined so that the function and its first derivative vanish at $t = 0$. The result is

$$\langle r^2 \rangle = \frac{6\,k\,T}{\gamma\,M} \left\{ t - \frac{1}{\gamma} + \frac{1}{\gamma}\, e^{-\gamma t} \right\} \qquad (22.16)$$

For times small compared to the "braking time" $1/\gamma$ the solution obtained is simply

$$\langle r^2 \rangle = \langle v^2 \rangle \, t^2$$

but thereafter $\langle r^2 \rangle$ falls below that value. For large times it increases only as the first power of the time, yielding finally

$$\langle r^2 \rangle \sim \frac{6 \, k \, T}{\gamma \, M} \, t \tag{22.17}$$

Relation (22.17) is intimately connected with the diffusion equation. According to Fick's law (19.03), the diffusion coefficient D is defined by the empirical relation between particle current $\mathbf{j}$ and particle density n:

$$\mathbf{j} = - \, D \frac{\partial n}{\partial \mathbf{r}}$$

Combining this with the conservation equation

$$\operatorname{div} \mathbf{j} + \frac{\partial n}{\partial t} = 0$$

we get

$$\frac{\partial n}{\partial t} = D \, \nabla^2 \, n \tag{22.18}$$

Among the various integrals of this equation there is one which has the form

$$n = \frac{1}{(4 \, \pi \, D \, t)^{3/2}} \exp\left[- \frac{r^2}{4Dt} \right] \tag{22.19}$$

The solution integrates to 1 at all times and equals a δ function at the origin at the time $t = 0$. Thereafter the solution spreads out in space. The mean square value of the displacement $\langle r^2 \rangle$ is obtained as

$$\langle r^2 \rangle = \int \int \int n \, r^2 \, d\mathbf{x}$$

and comes out to be

$$\langle r^2 \rangle = 6 \, D \, t \tag{22.20}$$

Equations (22.20) and (22.17) have the same form even though they were obtained by different means. Clearly they deal with the same phenomenon, and therefore the coefficients must be the same. $1/\gamma$ is connected with the mobility of the particle, as is seen from (22.13). If $\mathbf{A}(t)$ were not fluctuating, but were a fixed external force, the terminal velocity of the particle would be A/γ. For a charged particle in a field $\mathbf{A} = e\,\mathbf{E}/M$, and thus the mobility is $e/M\,\gamma$. Substitution of this into (22.17) and (22.20) yields again (20.34)

$$ D = \frac{k\,T}{e}\,\mu \qquad (22.21a) $$

This is the connection in which Einstein derived the relation. Actually, the charge concept is extraneous to the relationship found. If we define mobility as the terminal velocity per unit force, then (22.13) yields a definition of mobility as $1/M\,\gamma$, regardless of the nature of the force. Thereupon relation (22.21a) becomes

$$ D = k\,T \times \text{Mobility} \qquad (22.21b) $$

as a relation valid for any sort of particle which is free to move but is subject to a frictional drag.

22-3. Spectral decomposition of Brownian motion; Wiener–Khinchin theorem

The preceding section has brought the time element into the theory of fluctuations, but only for a special circumstance. Very little was said about the rapidly oscillating motion of the Brownian particles. Instead, equation (22.15) was constructed which governs the behavior of these particles over "reasonably extended" time intervals. The motion was then shown to be equivalent to diffusion. These time intervals are obviously such as to smooth out the rapid random motion of the particles which is their most characteristic feature.

The use of high-frequency circuits in physics has focused attention on this part of Brownian motion. The random oscillation of the current in a circuit is picked up as "noise" by high-frequency devices, while the drift of the carriers as a whole is controlled by applied voltages.† It is therefore of interest to consider a modified form of Brownian motion in which the variable is held to a fixed average value and the random fluctuations about that value are the object of study.

Figure 22.1 shows oscilloscope recordings of the current versus time for two different circuits. Even if the mean square displacements were normalized to be the same, the patterns would not look alike. The second

† This statement is not universally true. Motion of carriers across the central layer of the "sandwich" in a transistor is controlled by diffusion, and external intervention primarily influences their number.

pattern is more jagged, whereby we mean that its high-frequency components are more prominent. At the same time there is a greater predictability of the first pattern because it is represented by a smoother curve. These two aspects are in fact linked. We shall now discuss their mathematical connection. The establishment of this connection must be preceded by a more precise formulation of the two properties.

Let the quantity whose fluctuation is under study be called x. A record of the type shown in Fig. 22.1 is then represented by a function $x(t)$.

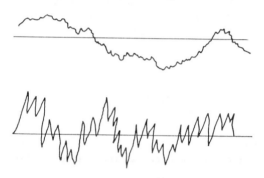

Fig. 22.1. Variation in time of a variable which fluctuates about an equilibrium. Two patterns showing different frequency composition.

Taking up first the "jaggedness" aspect of these curves, it seems reasonable to use for it the formalism of Fourier decomposition. The method is particularly appropriate if $x(t)$ is an electric current, for Fourier analysis is the customary way of analysis for an electrical response anyway. For convenience we first subtract out the average of x, if any, and then write for the remainder

$$x(t) = \int_{-\infty}^{+\infty} y(f)\, e^{2\pi i f t}\, df \tag{22.22}$$

and take note of two properties of Fourier integrals.‡ One is the inversion property

$$y(f) = \int_{-\infty}^{+\infty} x(t)\, e^{-2\pi i f t}\, dt \tag{22.23}$$

and the other the theorem of Parseval concerning the norm

$$\int_{-\infty}^{+\infty} |x(t)|^2\, dt = \int_{-\infty}^{+\infty} |y(f)|^2\, df \tag{22.24}$$

‡ See P. M. Morse and H. Feshbach. *Methods of Theoretical Physics*. New York: McGraw-Hill, 1953, pp. 453–471.

Some caution is required in applying the last three formulas to the type of fluctuating function to be analyzed. For $x(t)$ is such as not to be square integrable even after its average is subtracted out. Equation (22.24) is thus certainly invalid, and even the two other relations are questionable. We circumvent this difficulty by taking a "window" out of the total x versus t record. We introduce the Dirichlet function, which is defined as

$$D(u) = 1, \quad \text{if } -1 < u < +1 \tag{22.25a}$$

$$D(u) = 0, \quad \text{if } |u| > 1 \tag{22.25b}$$

Instead of analyzing $x(t)$, we now analyze the function

$$x(t) \, D\left(\frac{2t}{T}\right)$$

which satisfies all requirements needed in the theory of Fourier integrals. We now can write safely

$$x(t) \, D\left(\frac{2t}{T}\right) = \int_{-\infty}^{+\infty} Y(f, T) \, e^{2\pi i f t} \, df \tag{22.26}$$

$$Y(f, T) = \int_{-\frac{1}{2}T}^{+\frac{1}{2}T} x(t) \, e^{-2\pi i f t} \, dt \tag{22.27}$$

$$\int_{-\frac{1}{2}T}^{+\frac{1}{2}T} |x(t)|^2 \, dt = \int_{-\infty}^{+\infty} |Y(f, T)|^2 \, df \tag{22.28}$$

We observe that $|Y(f, T)|$ is even in f. For as $x(t)$ is real, we have

$$x(t) \, D\left(\frac{2t}{T}\right) = x^\star(t) \, D\left(\frac{2t}{T}\right)$$

$$= \int_{-\infty}^{+\infty} Y^\star(f, T) \, e^{-2\pi i f t} \, df$$

$$= \int_{-\infty}^{+\infty} Y^\star(-f, T) \, e^{2\pi i f t} \, df$$

However, the Fourier decomposition is unique. Therefore the last expression is identical with (22.26), and we have

$$Y^\star(-f, T) = Y(f, T) \tag{22.29a}$$

$$|Y(-f, T)|^2 = |Y(f, T)|^2 \tag{22.29b}$$

Equation (22.28) can therefore be given the form

$$\int_{-\frac{1}{2}T}^{+\frac{1}{2}T} |x(t)|^2 \, dt = 2 \int_0^{\infty} |Y(f, T)|^2 \, df \tag{22.30}$$

It would appear that the need to introduce an arbitrary time interval T into the analysis robs the procedure of some of its usefulness. We can restore this usefulness by introducing the physical notion that any other time window would reveal the same properties of $x(t)$. Hence it is reasonable to postulate that the result of analyzing a record covering a time interval of length T tends to become independent of T if only T is chosen long enough. The postulate made is strong enough to imply that, for the left side of (22.30), the time interval could be shifted without altering the result. Furthermore, if the integral is first divided by T, and then T is made large, the limiting expression should be independent of T. In other words we assume that we have

$$\lim_{T=\infty} \frac{1}{T} \int_{t'}^{t'+T} [x(t)]^2 \, dt = \langle x^2 \rangle = \text{independent of } t' \text{ and } T \quad (22.31)$$

Actually our postulate implies more than that. It is not just the mean square fluctuation which should come out independent of the position and the extension of the time window used for its observation, but the fluctuation as viewed through any frequency filter. Therefore the process of dividing by T and letting T go to infinity can be carried out not only on the right-hand side of (22.30) (which is implicit in (22.31)), but on any spectral part of it. In other words, the process can be carried out under the integral sign as follows

$$\lim_{T=\infty} \frac{1}{T} |Y(f, T)|^2 = \tfrac{1}{2} G(f) = \text{independent of } T \quad (22.32)$$

$G(f)$ is the *spectral distribution function* of the noise associated with $x(t)$. It is related to the mean square fluctuation through

$$\langle x^2 \rangle = \int_0^\infty G(f) \, df \quad (22.33)$$

Furthermore, if the derivation is gone through again with the assumption that a spectral filter transparent from f_1 to f_2 is inserted, we get

$$\langle x^2 \rangle_{f_1 \leftrightarrow f_2} = \int_{f_1}^{f_2} G(f) \, df \quad (22.34)$$

More complicated formulas can of course be written down for more elaborate receiving devices. Anyhow, the spectral distribution function $G(f)$ is clearly the proper measure of the "jaggedness" of the fluctuating function $x(t)$.

We now have to subject the "predictability" of the fluctuating function $x(t)$ to a similar analysis. If the value of the function $x(t+\tau)$ can be predicted with a good expectation of success when the value $x(t)$ is known, then the product $x(t) \cdot x(t+\tau)$ does not become zero when averaged over t. In the simplest case the product is likely to be positive, and perhaps close to $\langle x^2 \rangle$. We capture these notions in defining the *autocorrelation* of $x(t)$ as

$$C(\tau) = \lim_{T = \infty} \frac{1}{T} \int_{t'}^{t'+T} x(t)\, x(t+\tau)\, dt \qquad (22.35)$$

That the autocorrelation is independent of the quantities t' and T used in its definition is part of the idea we have of a random record of fluctuations.

Having defined the spectral distribution function $G(f)$ of the randomly varying quantity $x(t)$ and its autocorrelation $C(\tau)$, we can enunciate the Wiener–Khinchin theorem as follows:

The autocorrelation of a quantity which varies randomly about the mean value zero is the Fourier cosine transform of its spectral distribution function.

The proof requires a few manipulations and one difficult passage to the limit. We start out by writing the definition (22.35) so that (22.26) can be applied to the factor $x(t)$ within it:

$$C(\tau) = \lim_{T = \infty} \frac{1}{T} \int_{-\frac{1}{2}T}^{+\frac{1}{2}T} x(t+\tau)\, x(t)\, dt$$

$$= \lim_{T = \infty} \frac{1}{T} \int_{-\frac{1}{2}T}^{+\frac{1}{2}T} x(t+\tau)\, dt \int_{-\infty}^{+\infty} Y(f, T)\, e^{2\pi i f t}\, df$$

Exchanging integrations, and bringing the limit inside, we transform it into

$$C(\tau) = \int_{-\infty}^{+\infty} e^{-2\pi i f \tau}\, df \left\{ \lim_{T = \infty} \frac{Y(f, T)}{T} \int_{-\frac{1}{2}T}^{+\frac{1}{2}T} x(t+\tau)\, e^{2\pi i f(t+\tau)}\, dt \right\} \qquad (22.36)$$

The inner integral is essentially taken care of by the Fourier inversion formula (22.27), except for the difficulty that the "record" which is being Fourier analyzed has been shifted by an amount τ. The discrepancy is, however, small. Indeed, the difference between the inner integral and the definition (22.27), written with f replaced by $-f$, equals

$$\int_{-\frac{1}{2}T+\tau}^{\frac{1}{2}T+\tau} x(t)\, e^{2\pi i f t}\, dt - \int_{-\frac{1}{2}T}^{+\frac{1}{2}T} x(t)\, e^{2\pi i f t}\, dt = -\int_{-\frac{1}{2}T}^{-\frac{1}{2}T+\tau} + \int_{\frac{1}{2}T}^{\frac{1}{2}T+\tau}$$

If the function $x(t)$ is bounded such that

$$|x(t)| < A$$

then this difference can be majorized in the following way:

$$\left| -\int_{-\frac{1}{2}T}^{-\frac{1}{2}T+\tau} x(t)\, e^{2\pi i f t}\, dt + \int_{\frac{1}{2}T}^{\frac{1}{2}T+\tau} x(t)\, e^{2\pi i f t}\, dt \right| <$$

$$\int_{-\frac{1}{2}T}^{-\frac{1}{2}T+\tau} A\, dt + \int_{\frac{1}{2}T}^{\frac{1}{2}T+\tau} A\, dt = 2\, A\, \tau$$

This result is finite and proportional to τ, which is fixed. In taking the limit of the curly bracket in (22.36), we have to divide the result by T and let T tend to infinity. In the course of this operation the difference between the integral and the definition (22.27) will tend to zero. The integral can therefore be replaced by (22.27). Thus (22.36) becomes

$$C(\tau) = \int_{-\infty}^{+\infty} e^{-2\pi i f \tau} \, df \left\{ \lim_{T=\infty} \frac{1}{T} \, Y(f, T) \, Y(-f, T) \right\}$$

In view of the symmetry relation (22.29), the curly bracket now equals exactly the definition (22.32) of $\frac{1}{2} G(f)$. We get therefore

$$C(\tau) = \int_{-\infty}^{+\infty} e^{-2\pi i f \tau} \, \tfrac{1}{2} \, G(f) \, df \qquad (22.37a)$$

or because of (22.29)

$$C(\tau) = \int_{0}^{\infty} \cos 2\pi f \tau \; G(f) \, df \qquad (22.37b)$$

which is the statement made in the Wiener–Khinchin theorem.

A. RECOMMENDED PROBLEMS

1. In the text the Langevin equation (22.13) was exploited to study the dependence of the mean square displacement on time. Instead, integrate the equation so as to get the dependence of the velocity on time.

2. Use the result of Problem 1 to get an expression for the damping constant γ in terms of the autocorrelation of the force m **A**.

3. Most solid state junction devices are noisy, with a spectrum going as $1/f$. A naive picture of the origin of this noise is the following. The current in a junction fluctuates because at uncorrelated times t_ν a section of the junction breaks down or blocks up. If such a unit is placed in series with a source of d.c. voltage, a unit results whose voltage varies by random steps s_ν at random times t_ν. To prevent drift we place the source in series with a noise-free resistance R and a capacitance C, and measure the voltage across R.

(a) Find the mean square voltage across the resistor.
(b) Find the spectral distribution function of that voltage.
(c) Is the model a success?

B. GENERAL PROBLEMS

4. In an experiment by Kappler [*Ann. Phys.* **11**, 233 (1931)], Boltzmann's constant was determined from visual observation of a galvanometer mirror. At 287.1°K the mean square angle of deflection was 4.178 × 10⁻⁶. The harmonic restoring torque for the mirror was 1.8856 × 10⁻⁸ φ gm cm² sec⁻². Find the resultant value of Boltzmann's constant.

5. The law of atmospheres is hard to observe on suspended particles because of the small sedimentation velocity, which is roughly given by Stokes' formula.

$$v = \frac{F}{6 \pi a \eta}$$

Compute the velocity for particles of aluminum of 1 micron diameter suspended in water.

6. Show that for fermions the fluctuation $\langle \delta n^2 \rangle$ of the occupation number of a state equals

$$\langle \delta n^2 \rangle = \langle n \rangle \langle 1 - n \rangle$$

7. Repeat the study of the random walk introduced in Problem 3, Chapter 2. Adopt a modified procedure and do the following:

(a) Derive an equation giving the probability after $s + 1$ steps, assuming known the solution after s steps.

(b) Verify that the old solution satisfies the equation under (a).

(c) Derive the differential equation which results if all differences are approximated by differentials (Fokker–Planck equation). What equation have you reproduced for $\pi = \frac{1}{2}$?

8. What must the relation between the spectral densities of the force per mass **A** and the velocity **v** be if the Langevin formula (22.13) is to hold?

9. If we add to the assumptions of the text the further assumption that the random force in **A** is not correlated with **v**, a simple expression for the autocorrelation of **v** can be obtained from (22.13). Write down that expression.

10. Combine the results of the two preceding problems to get an expression for the spectral distribution function of the force m **A**.

*23

Connection between transport coefficients and equilibrium statistics

Ever since the discovery of the Nernst relation it has been vaguely realized that the major linear transport coefficients are essentially a property of the equilibrium system. They are listed as such in numerical tables. However, kinetic theory did not establish such a connection for many decades. Concepts foreign to equilibrium statistical mechanics, such as the mean free path, seemed to dominate kinetics. And, of course, the mean free paths are never quite the same in different processes so that each process finally seemed to require its own kinetic theory. The real breakthrough came in the Nyquist theorem,* which will be explained in the next section. To understand it we must first realize that the work of the preceding chapter associates with the equilibrium state quantities which vary in time: Brownian motion and other types of fluctuations. The modern relationships associating transport coefficients with equilibrium properties connect the former actually to these time-dependent features of the equilibrium state; these are not always easy to compute. Thus we arrive at the paradox that even though transport coefficients can now be expressed in terms of equilibrium properties, their computation has not been facilitated thereby as much as might have been expected.

* H. Nyquist. *Phys. Rev.* **32,** 110 (1928).

23-1. Nyquist relation

Nyquist first showed that an impedance, taken as a circuit element, produces a noise computable from its impedance; this noise arises from thermal agitation. His proof was based on circuit theory. We shall give a slightly modified circuit theoretical proof first, and then branch out into its microscopic implications in the next section.

We start out by trying to estimate the noise in a pure resistance R in a frequency range in the neighborhood of f. For this purpose, we place it in series with an L-C tuned circuit, as shown in diagram (a) of Fig. 23.1.

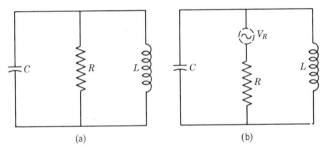

(a)   (b)

Fig. 23.1. Actual (a) and symbolic (b) circuit for the computation of the noise in a pure resistance R. The resistance is placed in series with a resonant L–C circuit.

We represent the effect of the resistor on the remainder of the circuit by a noise generator of voltage V_R placed in series with the resistance R. This arrangement is shown in diagram (b) of Fig. 23.1. We shall assume that the noise generator has a very flat frequency spectrum when compared to the sharply tuned resonant circuit.

Let us at first assume that the potential output V_R of the equivalent noise generator can be given the form

$$V_R = V_0 \, e^{2\pi i f t} \tag{23.01}$$

It follows then from circuit theory that the potential V_r across the resonant circuit equals

$$V_r = \frac{V_R}{\left\{1 + R^2 \left(2\pi f C - \dfrac{1}{2\pi f L}\right)^2\right\}^{\frac{1}{2}}} \tag{23.02}$$

Now, in reality, we do not want to think of the noise generator as generating a particular frequency. It shall on the contrary be a source of random noise, whose voltage output has the properties of the variable x in Section

22-3. What can be defined for it is a spectral distribution function as was done in (22.32).

The statistical element of the calculation consists in saying that the L-C circuit is a harmonic oscillator. The frequency of the oscillator is usually such that its quantum aspect may be ignored; the electric energy stored in the mean in its capacitance is then $\frac{1}{2} k T$. In other words we have

$$C \langle V_r^2 \rangle = k T \tag{23.03}$$

With the help of (23.02) we can write this in the form

$$C \left\langle \frac{V_R^2}{1 + R^2 \left(2\pi f C - \dfrac{1}{2\pi f L} \right)^2} \right\rangle \approx k T \tag{23.04}$$

Equation (23.04) is not entirely logical. V_R is not a pure source of frequency f, but is composed according to equation (22.34), with x identified with V_R. The logical statement associated with (23.04) is therefore

$$C \int_0^\infty \frac{G(f)\, df}{1 + R^2 \left(2\pi f C - \dfrac{1}{2\pi f L} \right)^2} = k T \tag{23.05}$$

Evaluation of the integral is clearly quite difficult in the general case. However, the assumption made earlier about the flatness of $G(f)$, as compared to the sharpness of the resonance denominator, comes to our aid. This should permit us to take $G(f)$ outside the integral sign and replace f by the resonant frequency f_0. This frequency equals

$$f_0 = \frac{1}{2\pi \sqrt{LC}} \tag{23.06}$$

Relation (23.05) thereupon takes the form

$$C\, G(f_0) \int_0^\infty \frac{df}{1 + R^2 \dfrac{C}{L} \left(\dfrac{f}{f_0} - \dfrac{f_0}{f} \right)^2} = k T \tag{23.07}$$

The value of the definite integral entering here is known:

$$\int_0^\infty \frac{dz}{1 + a^2 \left(z - \dfrac{1}{z} \right)^2} = \frac{\pi}{2a}$$

The identity (23.07) reduces therefore to

$$G(f_0) = 4 R k T \tag{23.08}$$

We can make use of the definition (22.34) to render the formula more concrete. It takes then the form

$$\langle V_R^2 \rangle = 4\,R\,k\,T\,\Delta f \qquad (23.09)$$

where Δf is the band width under which the noise voltage is observed.

Equation (23.09) is the form which the Nyquist relation takes for a pure resistance. It was first discovered empirically by J. B. Johnson[†] and then

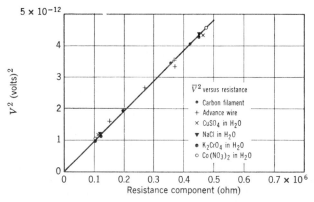

Fig. 23.2. Experimental plot of noise mean square voltage versus resistance in various resistors at fixed temperature and band width (after J. B. Johnson).

derived by Nyquist. Figure 23.2 shows a plot of noise voltages versus resistance of various resistors at fixed temperature, as observed through a band pass filter having a fixed band width of 1000 cycles per second. The plot is a good straight line and verifies the dependence of $\langle V_R^2 \rangle$ on R only, as predicted by (23.09). The analysis is carried further in Fig. 23.3, which shows a plot of the apparent power $\langle V_R^2 \rangle / R$ versus temperature. The plot exhibits strict linearity. Having checked these two essential features and being given the band width of 1000 cycles, we can directly solve for Boltzmann's constant k from the graph, using (23.09). It comes out to be

$$k = \frac{10^{-17}}{1000 \times 4 \times 300} = 0.8 \times 10^{-23} \text{ watt/deg}$$

This is a reasonable, though not extremely accurate, determination of k. With a more careful analysis involving the true admittance of the band pass amplifier a more precise determination of k can be found. Values of k within 30 % of the accepted value result. Thermal noise of resistors is often referred to as *Johnson noise* in honor of its discoverer.

† J. B. Johnson. *Phys. Rev.* **32**, 97 (1928).

Extension of the Nyquist formula to any passive linear circuit element is a simple matter, which can be accomplished by invoking the second law of thermodynamics. Equations (23.08) and (23.09) give us the noise

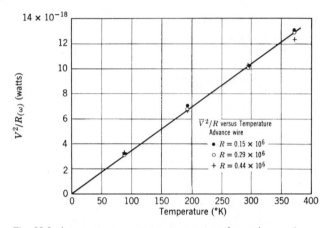

Fig. 23.3. Apparent power versus temperature for various resistors, as observed through a band pass filter of width of 1000 cycles at room temperature. The graph permits a direct determination of Boltzmann's constant (after J. B. Johnson).

associated with a pure resistance in the form of an equivalent mean square voltage in a circuit of which the resistor is a part. In order to extend the argument to an arbitrary impedance Z we place the impedance in series

Fig. 23.4. Circuit arrangement to determine the noise associated with an arbitrary impedance Z.

with the resistance R and a wave filter, which lets through only a frequency band Δf and blocks the remainder; this arrangement is shown in Fig. 23.4. We assume that the current fluctuation in Z is also equivalent to a voltage source V_Z for portions outside the impedance Z, just as was the case for the resistance R.

The argument which determines V_Z is the Second Law in the form of the postulate of Clausius (not discussed in Chapter 6): since the circuit of Fig. 23.4 is assumed in equilibrium, no net heat flow can take place from one of the circuit elements to the other. The power dissipated in the impedance Z by the voltage V_R must therefore be equal to the power dissipated in R by the voltage V_Z. Let us start with the power dissipated in Z, which arises from the voltage fluctuation originating in R; we call it $(P_Z)_R$. The voltage $(V_Z)_R$ across Z arising from V_R equals

$$(V_Z)_R = \frac{|Z|}{|R + Z|}\, V_R$$

The power $(P_Z)_R$ dissipated in Z equals its conductance Y times $(V_Z)_R^2$. Thus, when we take the average, a mean square voltage enters which we can take from (23.09). We thus get the result

$$\langle (P_Z)_R \rangle = Y \frac{|Z|^2}{|R + Z|^2}\, 4\, R\, k\, T\, \Delta f \tag{23.10}$$

In the reverse calculation the noise potential V_Z arising in the impedance Z is our unknown. The part $(V_R)_Z$ lying across R is

$$(V_R)_Z = \frac{R}{|Z + R|}\, V_Z$$

The mean power dissipation $(P_R)_Z$ in R, which arises from the noise in Z, thus equals

$$\langle (P_R)_Z \rangle = \frac{R}{|Z + R|^2}\, \langle V_Z^2 \rangle \tag{23.11}$$

Equating (23.10) and (23.11) on the basis of the Second Law, and remembering that

$$Y = \frac{\mathscr{R}(Z)}{|Z|^2}$$

we find for $\langle V_Z^2 \rangle$

$$\langle V_Z^2 \rangle = 4\, \mathscr{R}(Z)\, k\, T\, \Delta f \tag{23.12}$$

which is the proper generalization of (23.09).

To conclude this section on the Nyquist theorem, it is well to throw out the concept of "equivalent voltage," which is clearly meaningless physically. Noise arises from fluctuating currents, and voltages deal with path integrals of the electric field. Johnson noise clearly does not arise from such fields; whatever fields are present are likely to be opposed to the flow of the current. It is only to the *outside* of a resistor that noise may be

conveniently represented as an electromotive force. For physical interpretation, equation (23.12) is thus best transformed into a current fluctuation. If a circuit consists of an impedance Z and a frequency filter of
band width Δf, the current within the band width will be

$$|I| = \frac{|V|}{|Z|}$$

If this relation is squared and averaged, and (23.12) is substituted, the end
result is

$$\langle I^2 \rangle = 4\, Y\, k\, T\, \Delta f \qquad (23.13)$$

23-2. Kubo's equilibrium expression for electrical conductivity

Early studies of (23.09) or (23.13) were mainly concerned with prediction
of noise with the help of known material properties. From a fundamental
viewpoint, the situation is reversed. Noise is a phenomenon associated
with equilibrium, and formulas (23.09) or (23.13) can therefore be
regarded as expressions for the transport coefficients R or Y in terms of
properties of the equilibrium state. It took some time for this reversed
attitude to find its way into theoretical physics. The Nyquist relation then
suggests a widespread possibility that the transport coefficients of a substance might be expressible in terms of its equilibrium properties. The first
such calculation for a transport coefficient is due to M. S. Green.‡ We
shall restrict ourselves here to the study of Ohmic conduction because it
permits a simple form of derivation; this simple form is due to Kubo.§
The essential point about his derivation or any of the others is that it is an
atomic theory, which is from the start a many-body derivation.

Suppose we deal with a set of particles whose coordinates are q_1, q_2,
. . . q_N and whose momenta are p_1 . . . p_N. Let the hamiltonian $\mathscr{H}$ of these
particles be

$$\mathscr{H} = \sum_\nu \frac{1}{2\, m_\nu} p_\nu^2 + V(q_1, \ldots q_N) \qquad (23.14)$$

In addition to this zero-order hamiltonian let there be an "applied"
electrical potential $\Phi(q, t)$ which acts on the charges of the particles
through

$$V'(q_1 \ldots q_N) = \sum_\nu e_\nu\, \Phi(q_\nu, t) \qquad (23.15)$$

‡ M. S. Green. *J. Chem. Phys.* **20**, 1281 (1952).
§ R. J. Kubo. *J. Phys. Soc. Japan* **12**, 570 (1957).

The time dependence of the perturbing potential Φ may contain a frequency, but is meant in addition to imply a "switching on" process. At $t = -\infty$ only (23.14) is supposed to be acting, and the system can be described in terms of the equilibrium density matrix as follows:

$$\rho_0 = \rho(-\infty) = \exp[\alpha - \beta\mathcal{H}] \tag{23.16a}$$

with

$$\exp[-\alpha] = \mathrm{Tr}\,(\exp[-\beta\mathcal{H}]) \tag{23.16b}$$

Thereupon the perturbing term (23.15) begins to act, and the density matrix undergoes change in time. The equation of motion for this change is (3.24)

$$i\,\hbar\,\frac{d\rho}{dt} = (\mathcal{H}\rho - \rho\,\mathcal{H}) + (V'\rho - \rho\,V') \tag{23.17}$$

As before in the computation of transport coefficients the equation is only solved to first order in the perturbing potential V'. One sets therefore

$$\rho = \rho_0 + \rho' \tag{23.18}$$

and actually solves the equation

$$i\,\hbar\,\frac{d\rho'}{dt} = (\mathcal{H}\,\rho' - \rho'\,\mathcal{H}) + (V'\,\rho_0 - \rho_0 V') \tag{23.19}$$

It may be verified by direct differentiation that (23.19) has the following solution

$$\rho' = \frac{1}{i\,\hbar}\int_{-\infty}^{t} dt'\,\exp\left[\frac{i\,\mathcal{H}(t'-t)}{\hbar}\right]\{V'(t')\rho_0 - \rho_0 V'(t')\}\exp\left[-\frac{i\,\mathcal{H}(t'-t)}{\hbar}\right] \tag{23.20}$$

Further progress in the analysis of ρ' is made with the help of *Kubo's identity*, which reads

$$V'\,e^{-\beta\mathcal{H}} - e^{-\beta\mathcal{H}}\,V' = -\,e^{-\beta\mathcal{H}}\int_{0}^{\beta} d\lambda\,e^{\lambda\mathcal{H}}\,(V'\,\mathcal{H} - \mathcal{H}\,V')\,e^{-\lambda\mathcal{H}} \tag{23.21}$$

The identity may be checked by starting with the preliminary observation that both sides vanish for $\beta = 0$. Thereupon if the equation is multiplied on the left with $e^{\beta\mathcal{H}}$ and then differentiated with respect to β, the derivatives come out to be equal; this makes the two expressions in (23.21) equal. Now expression (23.20), because of (23.16), contains the left side of (23.21) in it. The identity reduces this to the commutator of V' and $\mathcal{H}$

which is more easily evaluated. The evaluation of the commutator of expressions (23.14) and (23.15) goes as follows:

$$V' \mathscr{H} - \mathscr{H} V' = [V', \mathscr{H}]$$

$$= \left[\sum_\nu e_\nu \Phi(\mathbf{q}_\nu, t), \sum_\mu \frac{1}{2 m_\mu} p_\mu^2 + V(\mathbf{q}) \right]$$

$$= \sum_\nu \frac{e_\nu}{2 m_\nu} [\Phi(\mathbf{q}_\nu, t), p_\nu^2]$$

$$= \sum_\nu \frac{e_\nu}{2 m_\nu} \{ [\Phi(\mathbf{q}_\nu), \mathbf{p}_\nu] \cdot \mathbf{p}_\nu + \mathbf{p}_\nu \cdot [\Phi(\mathbf{q}_\nu), \mathbf{p}_\nu] \}$$

$$= \frac{i \hbar}{2} \sum_\nu \frac{e_\nu}{m_\nu} \left\{ \frac{\partial \Phi}{\partial \mathbf{q}_\nu} \cdot \mathbf{p}_\nu + \mathbf{p}_\nu \cdot \frac{\partial \Phi}{\partial \mathbf{q}_\nu} \right\}$$

We rewrite this as an integral in the following way:

$$V' \mathscr{H} - \mathscr{H} V' = \frac{i\hbar}{2} \iiint d\mathbf{x} \sum_\nu \left\{ \frac{e_\nu \mathbf{p}_\nu}{m_\nu} \delta(\mathbf{q}_\nu - \mathbf{x}) + \delta(\mathbf{q}_\nu - \mathbf{x}) \frac{e_\nu \mathbf{p}_\nu}{m_\nu} \right\} \cdot \frac{\partial \Phi(\mathbf{x}, t')}{\partial \mathbf{x}}$$

The sum in the integral is just twice the current density operator at the position $\mathbf{x}$, while the derivative at the end is the negative of the electric field. We find therefore

$$V' \mathscr{H} - \mathscr{H} V' = - i \hbar \iiint d\mathbf{x} \, \mathbf{i}(\mathbf{x}) \cdot \mathbf{E}(\mathbf{x}, t') \qquad (23.22)$$

Insertion of (23.22) and (23.21) into (23.20) puts the density matrix at the time t into the form

$$\rho(t) = \rho_0 \left\{ 1 + \int_{-\infty}^t dt' \int_0^\beta d\lambda \exp\left[\frac{i \mathscr{H} (t'-t)}{\hbar} + \lambda \mathscr{H} \right] \right.$$

$$\qquad (23.23)$$

$$\left. \iiint d\mathbf{x}' \, \mathbf{i}(\mathbf{x}') \cdot \mathbf{E}(\mathbf{x}', t') \exp\left[- \frac{i \mathscr{H} (t'-t)}{\hbar} - \lambda \mathscr{H} \right] \right\}$$

The first thing to remark about (23.23) is that $\mathbf{E}(\mathbf{x}', t')$ is no longer an operator, but an ordinary space-time function describing the disturbing field. The only operator quantity beside $\mathscr{H}$ which is present is the current density operator $\mathbf{i}(\mathbf{x}')$. It is multiplied in front and behind with an exponential in $\mathscr{H}$. If the exponential were purely imaginary, the two exponentials would simply yield $\mathbf{i}(\mathbf{x}', t')$, as observed with a density matrix at the time t. The real part of the exponent arises from the Kubo transform. As it is in competition with a term in $1/h$, the classical limit is easily taken at this point by simply setting $\lambda = 0$. Expression (23.23) then reduces to

$$\rho(t) = \rho_0 \left\{ 1 + \beta \int_{-\infty}^{t} dt' \int \int \int d\mathbf{x}' \, \mathbf{i}(\mathbf{x}', t') \cdot \mathbf{E}(\mathbf{x}', t') \right\} \quad (23.24)$$

Having obtained the density matrix as a function of time to first order in the perturbing field $\mathbf{E}(\mathbf{x}', t')$, we can now ask the question to which we desire an answer: namely, what is the current density at position $\mathbf{x}$ and time t? For this purpose we multiply (23.24) with the current density operator and take the trace

$$\langle \mathbf{i}(\mathbf{x}, t) \rangle = \mathrm{Tr} \, \{\mathbf{i}(\mathbf{x}) \, \rho(t)\} \quad (23.25)$$

As usual the term ρ_0 gives a null result and the correction term yields the entire answer. This answer can be expressed as a trace over the undisturbed density matrix ρ_0

$$\mathrm{Tr} \, \{\mathbf{i} \, \rho(t)\} = \mathrm{Tr} \left\{ \beta \rho_0 \int_{-\infty}^{t} dt' \int \int \int d\mathbf{x}' \, [\mathbf{i}(\mathbf{x}', t') \cdot \mathbf{E}(\mathbf{x}', t')] \, \mathbf{i}(\mathbf{x}, t) \right\} \quad (23.26)$$

Such a trace over the undisturbed density matrix is an equilibrium expectation value.

As remarked earlier, $\mathbf{E}$ does not enter into the quantum-mechanical trace and can therefore be taken outside. The resultant relation between current density and field is linear, but nonlocal and nonsimultaneous. There may very well be materials in which (23.26) has to be used in full generality. We adapt ourselves to the current state of experimental knowledge when we say that the nonlocal aspect of conductivity does not seem to play a role. It must mean that we usually deal with situations where the correlation between currents is so short range that $\mathbf{E}$ does not vary significantly within the correlation distance. A similar assumption about the time would be quite unwise, as it would leave high frequency out of the picture. However, the correlation expression may be assumed sensitive to time differences only. We therefore take $\mathbf{E}$ out of the volume integral, but not the time integral, and replace t' by $t - \tau$. To get a numerical answer for the conductivity, we also take at this point the specimen to be of uniform constitution. Only the difference $\mathbf{x}' - \mathbf{x}$ has then any significance; we denote it by $\boldsymbol{\xi}$. Equation (23.26) thus becomes

$$i^\nu(t) = \beta \int_0^\infty d\tau \, E^\mu(t-\tau) \left\langle \int \int \int i^\mu(\boldsymbol{\xi}, -\tau) \, i^\nu(\mathbf{0}, 0) \, d\boldsymbol{\xi} \right\rangle \quad (23.27)$$

This equation achieves the general purpose of relating the current density and the field in a way which involves only equilibrium properties of the system. The relation is linear, but does not reduce to a simple proportion. Even after the nonlocal character in space has been removed artificially, it remains nonlocal in time: the current at time t depends on the electric field which existed at all previous times. This type of relationship is conveniently analyzed in terms of frequency. To bring this out we set

$$E^\mu = E^\mu(\omega)\, e^{i\omega t}; \quad i^\nu = i^\nu(\omega)\, e^{i\omega t} \tag{23.28}$$

and find

$$i^\nu(\omega) = E^\mu(\omega)\, \sigma^{\nu\mu}(\omega) \tag{23.29}$$

with

$$\sigma^{\nu\mu}(\omega) = \beta \int_0^\infty e^{-i\omega\tau}\, d\tau \left\langle \int\int\int i^\mu(\boldsymbol{\xi}, -\tau)\, i^\nu\,(\mathbf{0}, 0)\, d\boldsymbol{\xi} \right\rangle \tag{23.30}$$

23-3. Reduction of the Kubo relation to those of Nernst and Nyquist

We clearly have in the Kubo relation (23.30) a result of such great generality that it is to be expected that earlier equilibrium expressions for electric transport coefficients must be special cases of this result. Such is indeed the situation. Both the Nernst relation (22.21) and the Nyquist relation (23.13) arise from the above result as special cases.

We shall start out with the Nernst relation (22.21). It appears at first sight as a relation between two transport coefficients. This defect is easily remedied, as diffusion is a process occurring at equilibrium, when uncorrelated particles move under no forces. So an equilibrium expression for the diffusion coefficient must exist. We get it in the following way.

We start out by writing the following identity, valid for a particle located at the origin at the time $t = 0$:

$$\mathbf{r}(t) = \int_0^t \mathbf{v}(t')\, dt'$$

We conclude from it that

$$\{\mathbf{r}(t)\}^2 = \int_0^t \int_0^t \mathbf{v}(t') \cdot \mathbf{v}(t'')\, dt'\, dt'' \tag{23.31}$$

We now assume that we deal with a large number of identical particles. Their motion shall be uncorrelated, both as between particles and as for the same particle after a certain rather small time. These assumptions are essential in order to have diffusion take place. Under these conditions we average (23.31) over a large number of such equivalent particles. The equation then takes the form

$$\langle r^2(t) \rangle = \int_0^t \int_0^t \langle \mathbf{v}(t') \cdot \mathbf{v}(r'') \rangle \, dt' \, dt'' \qquad (23.32)$$

The area of integration is the square shown in Fig. 23.5. By assumption, the two velocities appearing in the integrand are only correlated when the

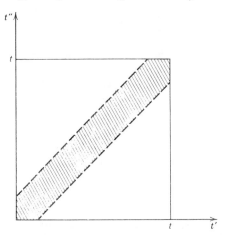

Fig. 23.5. Domain of integration in estimating the integral (23.32), which yields the mean square displacement in diffusive motion. Most of the contribution comes from the cross-hatched area.

times are nearly equal. This means that most of the contribution to the integral arises from the cross-hatched region in Fig. 23.5. We bring this out by setting

$$t'' = t' + \tau$$

and extending the limits on τ to infinity. This brings (23.32) into the form

$$\langle r^2(t) \rangle = \int_0^t dt' \int_{-\infty}^{+\infty} d\tau \, \langle \mathbf{v}(t') \cdot \mathbf{v}(t'+\tau) \rangle$$

The absolute time t' has clearly no effect on the autocorrelation of the velocity. It can therefore be dropped from the argument in the integrand. Thereupon the integration over t' can be performed and the following result obtained:

$$\langle r^2(t) \rangle = t \int_{-\infty}^{+\infty} \langle \mathbf{v}(0) \cdot \mathbf{v}(\tau) \rangle \, d\tau \qquad (23.33)$$

Comparison of this result with (22.20) shows then that

$$D = \tfrac{1}{6} \int_{-\infty}^{+\infty} d\tau \, \langle \mathbf{v}(0) \cdot \mathbf{v}(\tau) \rangle \tag{23.34a}$$

or

$$D = \tfrac{1}{3} \int_{0}^{\infty} d\tau \, \langle \mathbf{v}(0) \cdot \mathbf{v}(\tau) \rangle \tag{23.34b}$$

Having thus reduced the diffusion coefficient to an equilibrium average, we can proceed to get the Nernst relation from the Kubo relation (23.30). Whenever the motion of electric carriers is internally uncorrelated, as is, for instance, the case for ions in aqueous solution, then the only part of the equilibrium current which is correlated to the current at an earlier time is the current arising from the same carrier. So if $i^{\nu}(\mathbf{0}, 0)$ in (23.30) is decomposed into the contributions made by individual carriers, the only correlated part in $i^{\mu}(\boldsymbol{\xi}, -\tau)$ is the one referring to the same carrier. Integration over $\boldsymbol{\xi}$ then yields $e\,\mathbf{v}(-\tau)$ for the carrier chosen. If one also sets $\nu = \mu$ and sums over μ, one gets then (23.30) in the form

$$3\,\sigma = n\,e^2\,\beta \int_{0}^{\infty} d\tau \, \langle \mathbf{v}(-\tau) \cdot \mathbf{v}(0) \rangle$$

Substitution of (20.04) into the left of this, and (23.34) into the right, yields then

$$\mu = \frac{e}{kT}\,D \tag{23.35}$$

which is the Nernst relation (22.21).

We shall now proceed to derive the Nyquist relation from the Kubo formula, assuming the object under consideration to be a pure resistance. We shall not assume, however, that the field, sample, or current is uniform.

To make the nonuniformity perfectly clear, we take a resistor with an extremely nonuniform current pattern such as the salt-water tank in Fig. 23.6. Current enters through the electrode at A and leaves at B. The dotted lines represent field lines, and the solid lines surfaces of constant potential. The tank is characterized by its frequency-dependent resistance $R(f)$, and we are trying to relate this quantity to equilibrium properties of the tank.

The starting equation is (23.26). We introduce curvilinear coordinates $q, q^{\star}, q^{\star\star}$ with q running along the field lines and $q^{\star}, q^{\star\star}$ along an equipotential surface. dA shall denote an element of area on such a surface. We now take (23.26) and integrate it over such an equipotential surface of fixed q. We also break up the volume integration into one over A and q. The equation becomes then

$$I(q, t) = \beta \int_{-\infty}^{t} dt' \left\langle \iint d\mathbf{A} \cdot \mathbf{i}(\mathbf{x}, t) \iint d\mathbf{A}' \cdot \int \mathbf{i}(\mathbf{x}', t') \, dq' \, E(\mathbf{x}', t') \right\rangle \quad (23.36)$$

Here the vector sign was transferred from E to $d\mathbf{A}'$. The last integral can now be carried out in the following way. Although $\mathbf{i}$ depends on q', $q'^{\star}$,

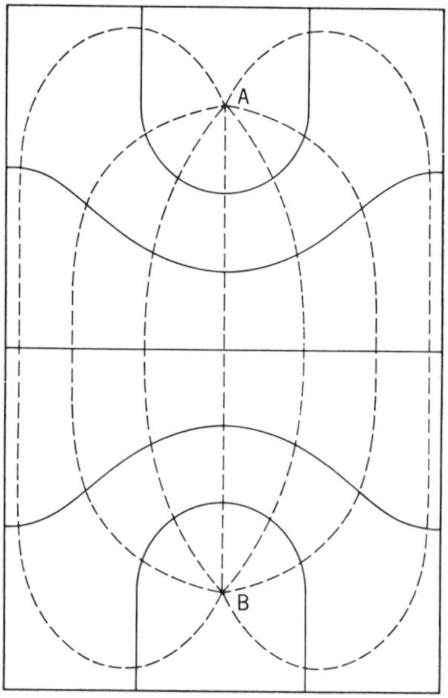

Fig. 23.6. Salt-water tank as an example of a lumped, non-uniform resistance.

$q'^{\star\star}$, E can be integrated over q' from one equipotential surface to a neighboring one; it yields thereby their potential difference, which is independent of the point chosen on the surface. Thereafter we can integrate $\mathbf{i}$ over $d\mathbf{A}'$; if we deal with a "pure resistance" in the sense of circuit theory, this integral yields the current I, which depends on no variable except the time. The same argument applies to the other surface integral. Proceeding this way from one equipotential surface to the other, covering all space, we get (23.36) in the form

$$I(t) = \beta \int_{-\infty}^{t} dt' \, V(t') \left\langle I(t) \, I(t') \right\rangle \quad (23.37)$$

Like (23.27), this relation between lumped circuit quantities is of the aftereffect type, with the equilibrium autocorrelation of the current acting as the kernel of the integral relation. Ohm's law is obtained from this at fixed frequency, with t' replaced by $t - \tau$:

$$I(t) = I(f)\, e^{2\pi i f t}; \quad V(t) = V(f)\, e^{2\pi i f t} \tag{23.38}$$

whereupon we get

$$I(f) = \frac{V(f)}{R(f)} \tag{23.39}$$

with

$$\frac{1}{R(f)} = \beta \int_0^\infty d\tau\, e^{-2\pi i f \tau} \langle I(0)\, I(\tau) \rangle$$

or because of the symmetry of the autocorrelation with respect to the sign of τ

$$\frac{1}{R(f)} = \tfrac{1}{2}\,\beta \int_{-\infty}^{+\infty} d\tau\, e^{-2\pi i f \tau} \langle I(0)\, I(\tau) \rangle \tag{23.40}$$

To get the Nyquist relation in its usual form we apply the Wiener–Khinchin theorem to the integral on the right. By (22.37), the Fourier transform of the autocorrelation of the current is one half the spectral distribution function $G(f)$ of this current, and hence

$$\frac{1}{R(f)} = \tfrac{1}{4}\,\beta\, G(f) \tag{23.41}$$

For the concrete interpretation of the spectral distribution function, we finally turn to (22.34). Viewed through a narrow band pass filter, the mean square current equals

$$\langle I^2 \rangle = \frac{4\,k\,T}{R(f)}\, \Delta f \tag{23.42}$$

which is equivalent to (23.13).

23-4. Onsager relations

The Kubo–Nyquist type of formulas make a statement about the rate of change produced in a quantity by the conjugate driving force. It is possible to do this for any pair of conjugate variables, although the detailed proof is often quite tedious. The general model of such an equation is (23.27) or (23.37). In generalized form it reads

$$\dot{x}(t) = \beta \int_0^\infty d\tau\, X(t-\tau) \langle \dot{x}(0)\, \dot{x}(\tau) \rangle \tag{23.43}$$

Here X is the force conjugate to x in the sense of the First Law, and the average in pointed brackets is an equilibrium autocorrelation which contains no reference to the flow process. Even heat flow problems can be handled on the model of (23.43). The terms $\dot{x}$ are then heat currents, and the conjugate force X is the difference of the logarithms of the absolute temperature. The proofs are not always straightforward, and an occasional modification may well be required. Our problem of a nonlocal current response in (23.26) was of this type; it was resolved by an appeal to experiment. One must not forget that the present theory is meant to predict all possible responses of all possible systems. So a certain generality in the answer is to be expected.

There is one possible generalization which we have avoided in the preceding section. We might have proceeded with the calculation as far as (23.24), but then, when asking the question to which we want an answer, we might have asked, not for the expectation value of the velocity conjugate to the driving force, but for the expectation value of some other velocity. The calculation could have followed the same steps as before, but the final answer, instead of taking the general aspect (23.43), would have read

$$\dot{y}(t) = \beta \int_0^\infty d\tau \; X(t-\tau) \left\langle \dot{x}(0) \, \dot{y}(\tau) \right\rangle \qquad (23.44)$$

In other words we would have ended with a cross-coupling coefficient which would have involved the cross correlation of two velocities at equilibrium. We had in fact a case of this type in (20.20) for the off-diagonal components of the conductivity tensor. However, the principle is obviously of much wider application.

What is interesting about (23.44) is that the same coefficient appears if x and y are interchanged:

$$\dot{x}(t) = \beta \int_0^\infty d\tau \; Y(t-\tau) \left\langle \dot{x}(0) \, \dot{y}(\tau) \right\rangle \qquad (23.45)$$

The essential argument is microscopic reversibility, which allows us to write

$$\left\langle \dot{y}(0) \, \dot{x}(\tau) \right\rangle = \left\langle \dot{y}(-\tau) \, \dot{x}(0) \right\rangle = \left\langle \dot{y}(\tau) \, \dot{x}(0) \right\rangle \qquad (23.46)$$

The equivalence (23.46) is the essential content of the Onsager relations. They apply to linear processes, that is, processes in which generalized forces produce generalized velocities which are proportional to the forces. The Onsager relations state that the matrix of the response coefficients is symmetric. In our schematic outline this symmetry expresses itself in the identity of the pointed brackets occurring in (23.44) and (23.45).

It is not pretended that the above considerations constitute a proof of the Onsager relations. A proof would require the simultaneous consideration of several flow processes and would never have full generality. We turn therefore to a quasi-thermodynamic method similar to the one used by Onsager originally.‖

We assume that we deal with a system close to thermal equilibrium which is adiabatically isolated. Its entropy is then close to a maximum. We take the Second Law in the form (5.37)

$$\frac{1}{k} dS = \beta \, dU - \beta \sum_s X_s \, dx_s \tag{23.47}$$

According to our convention X_s is the external force necessary to hold the system at a given condition. Therefore $- X_s$ is the internally produced force responding to any departure from it. To simplify (23.47), we define

$$\frac{1}{k} dS = d\Sigma \tag{23.48a}$$

$$- \beta X_s = F_s, \quad s \neq 0 \tag{23.48b}$$

$$\beta = F_0 \tag{23.48c}$$

$$U = x_0 \tag{23.48d}$$

The Second Law then takes the form

$$d\Sigma = \sum_j F_j \, dx_j \tag{23.49}$$

We can, without loss of generality, assume that the equilibrium state is reached for the values $x_j = 0$; similarly $\Sigma = 0$ may be taken as the value of the dimensionless entropy under those conditions. Then, unless there is something singular about one of the coordinates, the maximum condition on Σ means that it can be expanded as a quadratic form in the coordinates. Therefore we can write, employing the summation convention,

$$\Sigma = - \tfrac{1}{2} s_{ij} x_i x_j \tag{23.50}$$

The quadratic form is negative definite, which is expressed in the minus sign; for convenience we also take $s_{ij} = s_{ji}$. Combination of (23.49) with (23.50) yields then

$$F_i = - s_{ij} x_j \tag{23.51}$$

It is to be noticed that in the system (23.51) s_{ij} and s_{ji} no longer play the same role. Their equality thus has now physical content. The content is

‖ L. Onsager. *Phys. Rev.* **37**, 405 (1931); **38**, 2265 (1931).

that, unless the equality holds, work can be extracted reversibly from our system, and Kelvin's form of the Second Law is violated.

Onsager's relations apply only to linear systems. By this we mean that the presence of "forces" F_i is assumed to produce velocities $\dot{x}_j$ which are related linearly to the forces, thus:

$$\dot{x}_i = l_{ij} F_j \tag{23.52}$$

In terms of the matrix L the Onsager relations state that it is symmetric The proof follows from statistical mechanics. The probability of a given set of coordinates being realized is proportional to

$$P(x_0, x_1, \ldots x_T) \, dx_0 \, dx_1 \ldots dx_T \propto e^{\Sigma} \, dx_0 \, dx_1 \ldots dx_T \tag{23.53}$$

We use the probability (23.53) to calculate the average

$$\langle x_i \, F_j \rangle = \frac{\iiiint \ldots \iint e^{\Sigma} \, x_i \, \dfrac{\partial \Sigma}{\partial x_j} \, dx_0 \, dx_1 \ldots dx_T}{\iiint \ldots \iint e^{\Sigma} \, dx_0 \, dx_1 \ldots dx_T}$$

Because of the form (23.50) assumed for Σ the integrands are such as to converge rapidly for large x_i. We may therefore take the integrals from $-\infty$ to $+\infty$ on all variables with negligible error; for the same reason we may integrate the numerator by parts on x_j. The result is

$$\langle x_i \, F_j \rangle = -\, \delta_{ij} \tag{23.54}$$

Equation (23.54) is essentially a lemma. We can now pass to the key point of the argument. Although the phenomenological equations (23.52) deal with irreversible processes, the basic mechanics underlying them obeys time reversal. This means that time correlations must be the same whether measured forward or backward. In particular we have

$$\langle x_i(\tau) \, x_j(0) \rangle = \langle x_i(-\tau) \, x_j(0) \rangle$$

or with a shift in the absolute time scale

$$\langle x_i(\tau) \, x_j(0) \rangle = \langle x_i(0) \, x_j(\tau) \rangle$$

Subtract on both sides of this equation $\langle x_i(0) \, x_j(0) \rangle$, divide by τ, and let τ tend to zero. The result is

$$\langle \dot{x}_i(0) \, x_j(0) \rangle = \langle x_i(0) \, \dot{x}_j(0) \rangle \tag{23.55}$$

If we substitute into this the development (23.52) of the velocities, we find

$$\langle l_{i\nu} \, F_\nu(0) \, x_j(0) \rangle = \langle x_i(0) \, l_{j\mu} \, F_\mu(0) \rangle$$

Substitution of (23.54) into this brings it into the final form

$$l_{ij} = l_{ji} \tag{23.56}$$

which was to be proved.

The most widely used case to which the Onsager relations apply is that of tensors governing linear dissipative processes such as the conductivity tensor. Previously we had to employ special structural arguments to show that such tensors are symmetric. Such an argument was used, for instance, in connection with (20.20). We see now that the argument was in fact not necessary, but that the symmetry is a consequence of the Onsager relations.

Onsager's relations find their most spectacular application in the cross coupling of processes of different nature. The most famous case of this

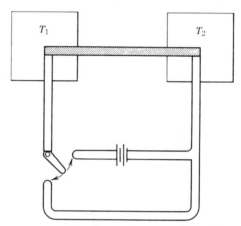

Fig. 23.7. Schematic diagram of an arrangement showing the Seebeck and Peltier effects.

kind is the cross coupling of electrical and thermal currents under a combination of a voltage and a temperature difference. Figure 23.7 shows a schematic picture of the arrangement in which the cross coupling can be observed. A circuit is built consisting of two different types of metallic wires in series. The circuit may or may not have a built-in source of electromotive force. In addition, the junctions between the two metals are placed in temperature baths which may or may not be kept at different temperatures. The flow is best thought of as taking place between the two thermal reservoirs at the two junctions. The term $\beta\, dU$ in (23.47) should then be amended to read

$$\beta_1\, dU_1 + \beta_2\, dU_2$$

which, because of the side condition,

$$U_1 + U_2 = \text{constant}$$

becomes

$$(\beta_1 - \beta_2)\, dU$$

We therefore have to modify (23.48) slightly and set

$$x_0 = U$$
$$F_0 = \beta_2 - \beta_1$$
$$x_1 = Q \text{ (total charge passed at a}$$
$$\text{reference point)}$$
$$F_1 = \beta_1 V_1 - \beta_2 V_2 \text{ (applied voltage)}$$

The linear equation system then becomes

$$\dot{U}_{1\to2} = l_{00}(\beta_2 - \beta_1) + l_{01}(\beta_1 V_1 - \beta_2 V_2) \tag{23.57a}$$

$$\dot{Q}_{1\to2} = l_{10}(\beta_2 - \beta_1) + l_{11}(\beta_1 V_1 - \beta_2 V_2) \tag{23.57b}$$

Here l_{00} is proportional to the heat conductivities of various parts of the circuit, while $1/l_{11}\beta$ is the circuit resistance. Neither one is of special interest here. We are interested in the two cross effects known respectively as Seebeck and Peltier effects.

In the Seebeck effect the two ends of a thermocouple are held at different temperature. An open circuit voltage then arises in the circuit which is proportional to the temperature difference

$$\Delta V = \Psi \, \Delta T \tag{23.58}$$

In the Peltier effect the two temperature baths are kept at equal temperature, and a battery is inserted. A fixed amount of heat per unit charge is now transported from one bath to the other, that is, we have

$$\dot{U} = \Pi \, \dot{Q} \tag{23.59}$$

Ψ is the Seebeck coefficient of the arrangement, Π the Peltier coefficient.

In terms of our fundamental equation pair (23.57) the Seebeck arrangement corresponds to $\dot{Q} = 0$. We have then

$$V_1 - V_2 = -\frac{l_{10}}{l_{11}}\frac{\beta_2 - \beta_1}{\beta} = \frac{l_{10}}{l_{11}}\frac{T_2 - T_1}{T}$$

This means that the Seebeck constant in (23.58) equals

$$\Psi = -\frac{l_{10}}{l_{11}T} \tag{23.60}$$

The Peltier arrangement, on the other hand, arises from (23.57) if $\beta_1 = \beta_2$. We get then

$$\frac{\dot{U}}{\dot{Q}} = \frac{l_{01}}{l_{11}} \tag{23.61}$$

which yields for the Peltier coefficient

$$\Pi = \frac{l_{01}}{l_{11}} \tag{23.62}$$

The Onsager relation now enters into play, telling us that $l_{10} = l_{01}$. From this it follows that

$$\Pi = -T\Psi' \tag{23.63}$$

The relation (23.63) was known for many decades before Onsager's derivation. It is known as *Kelvin's relation*. Kelvin derived it thermodynamically, assuming the two effects to be reversible, even though the diagonal effects in the system (23.57) are not. Opinions are divided as to whether Kelvin's derivation is fundamentally correct or not.

PROBLEM

1. Derive the Nernst–Einstein relation by computing the average defining D in (23.34b) for the case of the Langevin equation (22.13).

Supplementary Literature

General texts may be listed twice, once in the general list, once in a speciality in which they supplement particularly well the work presented here.

Thermodynamics

Herbert B. Callen, *Thermodynamics*, Wiley, New York, 1961
Enrico Fermi, *Thermodynamics*, Prentice-Hall, Englewood Cliffs, N.J., 1937 (Dover, New York, 1956)
E. A. Guggenheim, *Thermodynamics*, North Holland, Amsterdam, 1949
P. T. Landsberg, *Thermodynamics*, Interscience, New York, 1961
M. Planck, *Vorlesungen über Thermodynamik*, De Gruyter, Berlin-Leipzig, 1939
J. C. Slater, *Introduction to Chemical Physics*, McGraw-Hill, New York, 1939
A. Sommerfeld, *Vorlesungen über theoretische Physik*, Dieterich, Wiesbaden, 1952, Vol. V
Mark W. Zemansky, *Heat and Thermodynamics*, McGraw-Hill, New York, 1957

Statistical Mechanics

Richard Becker, *Theorie der Wärme*, Springer, Berlin, 1955
Norman Davidson, *Statistical Mechanics*, McGraw-Hill, New York, 1962
H. Eyring, D. Henderson, B. J. Stover, and E. Eyring, *Statistical Mechanics and Dynamics*, Wiley, New York, 1964

R. H. Fowler and E. A. Guggenheim, *Statistical Thermodynamics*, Cambridge University Press, Cambridge, 1952

Terrell L. Hill, *Statistical Mechanics*, McGraw-Hill, 1956 (Dover, New KıYork, 1987)

Kerson Huang, *Statistical Mechanics*, Wiley, New York, 1963

C. Kittel, *Elementary Statistical Physics*, Wiley, New York, 1958

R. Kubo, H. Ichimura, T. Usui, and N. Hashizume, *Statistical Mechanics*, Wiley, New York, 1965

L. D. Landau and E. M. Lifshitz, *Statistical Physics*, Pergamon, London, 1958

J. E. Mayer and M. G. Mayer, *Statistical Mechanics*, Wiley, New York, 1940

F. Reif, *Fundamentals of Statistical and Thermal Physics*, McGraw-Hill, 1965

D. Ter Haar, *Elements of Statistical Mechanics*, Holt, Rinehart & Winston, New York, 1961

Fundamentals of Statistics

Brandeis Summer Institute Lectures, *Statistical Physics*, Benjamin, New York, 1962

M. Delbrück and G. Molière, "Statistiche Quantenmechanik and Thermodynamik," *Abhandl. preuss. Akad. Wiss.*, De Gruyter, Leipzig, 1936

R. H. Fowler, *Statistical Mechanics*, Cambridge University Press, Cambridge, 1936

J. W. Gibbs, "Elementary Principles of Statistical Mechanics," *Collected Works of J. Willard Gibbs*, Longmans Green, New York, 1931

A. Khinchin, *Mathematical Foundations of Statistical Mechanics*, Dover, New York, 1949

A. Sommerfeld and L. Waldmann, "Die Boltzmannsche Statistik und ihre Modification durch die Quantentheorie," *Jahrbuch der chemischen Physik*, III/2, 1939

R. C. Tolman, *Principles of Statistical Mechanics*, Oxford Press, Oxford, 1938 (Dover, New York, 1980)

Properties of Gases

S. Chapman and T. G. Cowling, *The Mathematical Theory of Non-uniform Gases*, Cambridge University Press, Cambridge, 1952

R. H. Fowler and E. A. Guggenheim, *Statistical Thermodynamics*, Cambridge University Press, Cambridge, 1950, Chapter III

J. O. Hirschfelder, C. F. Curtiss, and R. B. Bird, *Molecular Theory of Gases and Liquids*, Wiley, New York, 1954

Earle H. Kennard, *Kinetic Theory of Gases*, McGraw-Hill, New York, 1938

J. E. Mayer and M. G. Mayer, *Statistical Mechanics*, Wiley, New York, 1940, Chapters 5–8, also Appendix IX

G. E. Uhlenbeck and G. W. Ford, *The Theory of Linear Graphs with Applications to the Theory of the Virial Development of the Properties of Gases*, Vol. I of *Studies in Statistical Mechanics*, North Holland, Amsterdam, 1962, Part B

G. E. Uhlenbeck and G. W. Ford, *Lectures in Statistical Mechanics*, American Mathematical Society, Providence, 1963

Theory of Solids

M. Born and K. Huang, *Dynamical Theory of Crystal Lattices*, Oxford Press, Oxford, 1954

Handbuch der Physik (1933 edition), Vol. XXIV, 2, Springer, Berlin, 1933. Chapters 3 (Sommerfeld and Bethe) and 4 (Born and M. Göppert Mayer)

A. K. Jonscher, *Principles of Semiconductor Device Operation*, G. Bell & Sons, London, 1960

Charles Kittel, *Introduction to Solid State Physics*, Wiley, New York, 1956

G. F. Newell and E. W. Montroll, "On the Theory of the Ising Model of Ferromagnetism," *Revs. Modern Phys.*, **25**, 353, 1953

Frederick Seitz, *The Modern Theory of Solids*, McGraw-Hill, New York, 1940 (Dover, New York, 1987)

J. H. Van Vleck, *Theory of the Electric and Magnetic Susceptibilities*, Oxford Press, Oxford, 1932

Gregory H. Wannier, *Elements of Solid State Theory*, Cambridge University Press, Cambridge, 1960

J. M. Ziman, *Principles of the Theory of Solids*, Cambridge University Press, Cambridge, 1964

Theory of Solutions

Enrico Fermi, *Thermodynamics*, Prentice-Hall, Englewood Cliffs, N.J., 1937, Chapter VII (Dover, New York, 1956)

E. A. Guggenheim, *Thermodynamics*, North-Holland, Amsterdam, 1949, Chapter IV–X

H. S. Harned and B. B. Owen, *The Physical Chemistry of Electrolytic Solutions*, Reinhold, New York, 1950

Irreversible Processes and Noise

H. B. Callen, *Thermodynamics*, Wiley, New York, 1960, Part III
S. Chandrasekhar, "Stochastic Problems in Physics and Astronomy," *Revs. Modern Phys.*, **15**, 1, 1943
Sidney Chapman and T. G. Cowling, *The Mathematical Theory of Non-uniform Gases*, Cambridge University Press, Cambridge, 1952
Aldert van der Ziel, *Noise*, Prentice-Hall, Englewood Cliffs, N.J., 1956
Ludwig Waldmann, "Transporterscheinungen in Gasen von mittlerem Druck," *Handbuch der Physik*, Springer, Berlin, 1958, Vol. XII, fourth article
Satori Watanabe, "Symmetry of Physical Laws," *Revs. Modern Phys.*, **27**, 26, 1955

Answers to Problems

Chapter 1

1. $x = 0.236$.   2. $4.58 \times 10^{-4}\ \text{deg}^{-1}$.   3. $\alpha = \dfrac{1}{T}$.   4. 333 j; 405 j.

5. 0.01 j; 0.75 j.   6. /.   7. (a) $W = \dfrac{1}{2}\dfrac{l}{Yq}(f_2^2 - f_1^2)$;

(b) $W(l) = Y\Omega\left\{\dfrac{l}{l_0} - 1 - \ln\dfrac{l}{l_0}\right\}$; (c) /.   8. 66 tons.   9. /.

10. $B(T) = b - \dfrac{a}{RT}$.   11. $B_T = p\left(1 + \dfrac{pb}{RT}\right)$; $\alpha = \dfrac{1}{T\left(1 + \dfrac{pb}{RT}\right)}$;

at given pressure and temperature the volume is larger by a fixed amount. Both the thermal expansion coefficient and the compressibility are smaller, making the gas stiffer.   12. $\Delta U = 582\ \text{j}$; $W = 117\ \text{j}$; $Q = 699\ \text{j}$.

13. $\dfrac{dV}{dt} = \dfrac{W}{5\,p_0\{1 - (\frac{1}{2})^{2/5}\}} = 2.19\ l/\text{sec}$; $T_1 = 2^{2/5}\,T_0 = 366°\ \text{K}$.

14. 0; $1.2 \times 10^{-3}\ \text{deg/sec}$; $1.2 \times 10^{-3}\ \text{deg/sec}$.   15. $C = \dfrac{N\,\mu^2}{3\,k}$.

16. /.   17. (a) $\dfrac{T^{4.03}\,e^{0.001\,T}}{p} = \text{constant}$; (b) $T^4/p = \text{constant}$.

18. $p_c = \dfrac{a}{27\,b^2}$, $V_c = 3b$, $T_c = \dfrac{8a}{27\,Rb}$, $\dfrac{p_c V_c}{R\,T_c} = 0.375$;

$\dfrac{p_c V_c}{R\,T_c} = 2\,e^{-2} = 0.270$.   19. /.   20. (a) yes; (b) $p\left(V + \dfrac{a}{T^2}\right) = R\,T$.

21. $\dfrac{1}{\gamma - 1} = \dfrac{f_1}{\gamma_1 - 1} + \dfrac{f_2}{\gamma_2 - 1}$.   22. /.   23. /.

Chapter 2

1. (a) 0.125, 0.500, 0.375; (b) 0.096, 0.497, 0.407.

2. (a) $\dfrac{N!}{(N-M)!M!}$; (b) 2^{N-1}.   3. (a) /; (b) $(2\pi - 1)\,s\,d$;

(c) /.   4. /.   5. /.   6. $2.8 \times 10^{-3}\,\text{cm}^{-3}$.   7. 2.69×10^{19};
0.52×10^{10}.   8. $0.37 \times 10^{-15}\,\text{cm}^3$.   9. $3:2$; 0.00467, 0.00309.

10.

AKK	0.125	0.110	0.096
AK	0.250	0.260	0.249
KK	0.125	0.130	0.124
A	0.125	0.130	0.136
K	0.250	0.260	0.271
–	0.125	0.110	0.124

11. $8/11 = 0.727$.   12. /.   13. /.   14. /.   15. /.   16. /.
17. /.   18. $p\,V = \frac{1}{3}\,U$.   19. /.

Chapter 3

None.

Chapter 4

1. 0.742, 0.208, 0.050.   2. /.

3. $\bar{v} = \sqrt{\dfrac{8}{\pi}\dfrac{kT}{m}} = \sqrt{\dfrac{8}{3\pi}}\,\sqrt{\langle v^2 \rangle} = 0.921\,\sqrt{\langle v^2 \rangle}$.   4. /.

5. $\mathcal{M} = 1.3\,\text{gm/mole}$.   6. $v_{\max} = \sqrt{\dfrac{2kT}{m}} = \sqrt{\tfrac{2}{3}}\,\sqrt{\langle v^2 \rangle} = 0.817\,\sqrt{\langle v^2 \rangle}$.

7. /.   8. /.   9. /.   10. $T = \frac{3}{4}\,\Theta/\ln 2 = 1.08\,\Theta$.   11. /.
12. /.   13. /.   14. $10^{-318}\,\text{gm/cm}^3$.

15. $\dfrac{dT}{dx} = -\dfrac{2}{7}\dfrac{\mathcal{M}\,g}{R} = -10\,\text{deg/km}$.

Chapter 5

1. /.   2. $\mathcal{U} = \mathcal{U}_0(T) - \dfrac{a}{\mathcal{V}}$; $\mathcal{S} = \displaystyle\int \dfrac{d\mathcal{U}_0(T)}{T} + R\ln(\mathcal{V} - b) + \text{constant}$;

$T(\mathcal{V} - b)^{\frac{R}{\mathcal{C}_v}} = \text{constant}$.   3. (a) $\Delta S = (N_1 + N_2)\,k\ln\dfrac{V_1\,c_1 + V_2\,c_2}{V_1^{c_1}\,V_2^{c_1}}$;

(b) An amount of heat $Q = T\,\Delta S$ must be furnished, and an equal
amount of work is gained.   4. 18.5 j/gm; 24 j/gm.   5. All p_i's
become equal and then equal also $1/M(E_0)$; from this the claimed result
follows.   6. In the Ehrenfest principle, only the energy of motion in
that direction is at first involved; the equation $p\,V = N\,k\,T$ is confirmed

as to the force on that moving wall if the energy distribution is Boltzmannian; after the motion the energy shift is redistributed over the other degrees of freedom.   7. $\dfrac{3}{2}\dfrac{\Delta E}{E_{\text{trans}}} = -\dfrac{\Delta V}{V}$; but $E_{\text{trans}} = \dfrac{3}{5}E$;

hence $\dfrac{5}{2}\dfrac{\Delta E}{E} = \dfrac{5}{2}\dfrac{\Delta T}{T} = -\dfrac{\Delta V}{V}$.   8. As the ratio on the left becomes 1 by assumption, the ratio on the right must do likewise.

9. $C_0 = Nk \ln 2 = Nk \times 0.693$.   10. $/$.

Chapter 6

1. 0.5 atm and 20 l, 1.25 atm and 8 l.   2. 14.9 times.   3. $/$.

4. $x\, y\, f\!\left(\dfrac{x}{y}\right)$; $(y-x-1)f((y-x-1)\,e^{-x})$; $y^3 f\!\left(y\, e^{-\frac{1}{2}\frac{x^2}{y^2}}\right)$; the functions f are arbitrary.   5. $/$.   6. 600° K.   7. 4.08° K.

8. $\eta = 1 - \dfrac{T_2}{T_1} = 1 - \left(\dfrac{V_1}{V_2}\right)^\gamma$; $\dfrac{T_2}{T_1} > \dfrac{T_2'}{T_1}$.

Chapter 7

1. $-0.75°$ C.   2. $/$.   3. $p/T = \varphi\,(\mathcal{V})$.

4. $\dfrac{dT}{dh} = -\dfrac{T_{\text{steam}}^2}{T_{\text{air}}}\dfrac{g\,\mathcal{M}_{\text{air}}}{\mathcal{H}_{\text{steam}}} = -3.3$ deg/km.   5. $/$.   6. $/$.   7. $/$.

8. When mechanical and thermal properties are decoupled the two relations predict the same relation between p and V under isentropic and isothermal conditions.   9. $/$.   10. 330.0° C.   11. $/$.

12. $\dfrac{c_p}{c_v} = 1 + \dfrac{\alpha^2\, T\, u^2}{c_p}$; 4.126, 0.1207 j/deg gm; 8.94R, 2.99R. Mercury obeys the law of Dulong and Petit quite accurately, even in the liquid state; water obeys it approximately, if you treat its three atoms as separate units.

13. $p = p_0\, e^{-\frac{\mathcal{H}}{RT}}$; 0.65 atm as against an experimental value of 0.69 atm.

14. $/$.   15. $/$.   16. $U/A = 130.8$ ergs/cm².

Chapter 8

1. $\varepsilon = 1 + \dfrac{4\,\pi\, p\, \mu^2}{3\,(k\,T)^2} = 1.00550$.   2. $/$.   3. The equation of state is the same, the heat capacity is $\frac{1}{2}(C_p + C_v)$.   4. $T = 5.9 \times 10^{-3\,°}$ K.

5. $\dfrac{p\,V}{N\,k\,T} = f\!\left(\dfrac{N\,a^3}{V}\right)$.   6. $V = V_0\, e^{-x/L}$; $L = \left(\dfrac{k\,T}{4\,\pi\, e^2\, n_0}\right)^{\frac{1}{2}}$.

Chapter 9

1. Boltzmann: $2.09°$ K, -0.17×10^{-16} erg; Fermi: $1.05°$ K, 0 erg; Bose: $3.08°$ K, -2.90×10^{-16} erg.   2. /.   3. /.   4. /.   5. /.   6. /.

7. The formulas of Problems 4, 5, and 6 were obtained by the grand ensemble technique, and thus apply exactly only to a subsystem of a large system or to a system which can be divided into a large number of subsystems. As soon as the total number of particles becomes a significant constraint affecting the population of an individual level, (9.34) cannot possibly hold. This invalidates the derivation of the three formulas.

Chapter 10

1. If the energy density of black body radiation is Ψ the amount traveling in unit time into unit solid angle is $\dfrac{c}{4\pi}\Psi$. A unit surface of a black body does not have to emit this entire amount, but only the amount incident upon it: nothing at all into one hemisphere, and an amount reduced by $\cos\vartheta$ at an angle ϑ. Hence

$$E_b(\nu) = \frac{c}{4\pi}\,\Psi\, 2\pi \int_0^{\pi/2} \cos\vartheta\,\sin\vartheta\,d\vartheta = \tfrac{1}{4}\,c\,\Psi$$

2. /.   3. 1630 watts/m².   4. $T_e = \tfrac{1}{2}\sqrt{\gamma}\,T_s = 360°$ K.
5. $p\,V^{4/3} = $ constant.   6. /.

Chapter 11

1. (a) $\omega_1\sqrt{1-\lambda^2}$, ω_2; (b) $\omega^2 = \tfrac{1}{2}(\omega_2^2+\omega_1^2) \pm \{\tfrac{1}{4}(\omega_2^2-\omega_1^2)^2 + \lambda^2\,\omega_1^2\,\omega_2^2\}^{1/2}$.
2. /.   3. $\Delta r/r = 0.00078\,J(J+1)$.
4. $f_{rot} = 6\sum\limits_{J\ even}(2J+1)\exp[-J(J+1)\,y] + 3\sum\limits_{J\ odd}(2J+1)\exp[-J(J+1)\,y]$.
5. $f \propto p^{-\frac{1}{3}}$.   6. /.

7. $K = K^\star = 0.0260\exp\left[-\dfrac{543}{T}\right]\dfrac{\sinh^2\dfrac{1594}{T}}{\sinh\dfrac{3034}{T}\sinh\dfrac{148.1}{T}}$; $K = K^\star = 0.020$.

Chapter 12

1. $T = \dfrac{2a}{Rb}\left(1 - \dfrac{b}{V}\right)^2$.   2. /.

3. $-\beta\mu = [-\beta\mu(T,V)]_{\text{perfect}} - \dfrac{2B(T)}{V}$.

4. We deal with a Joule–Thomson expansion; in such an expansion, the

specific entropy of the material remaining behind is constant. 95 atm.
5. /.   6. /.

Chapter 13

1. $E \propto T^{5/2}$.   2. 2.97 eV; 2.97 eV; 2.96 eV.   3. /.   4. /.   5. /.

6. $p = -\dfrac{dU(0)}{dV} - \dfrac{U}{\Theta}\dfrac{d\Theta}{dV}; \quad -\dfrac{1}{\Theta}\dfrac{d\Theta}{dV} = -\dfrac{\Gamma}{V}.$

Chapter 14

1. /.   2. The problem is separable, and bands can be labeled by a double index (j, j'). If j and j' are of equal parity, extrema are at $(0, 0)$ and (π, π); if of opposite parity, at $(0, \pi)$ and $(\pi, 0)$. If j and j' are different, the bands (j, j') and (j', j) always overlap. For $j = j'$ the masses are isotropic.   3. 0.50 ohm^{-1} cm^{-1}.   4. 0.29 ohm^{-1} cm^{-1}.   5. The curve $W = 0$ form a regular network of straight lines, inclined at 45° to the axes.   6. The factor $m^{3/2}$ is to be replaced by $\{m_1 \, m_2 \, m_3\}^{1/2}$.   7. The key formula reads $B \, A_1 \, n_2 \, (N_1 - n_1) = B \, A_2 \, n_1 \, (N_2 - n_2)$; the factors $N_i - n_i$, counting the unoccupied states, arise from the Pauli principle.

Chapter 15

1. $\chi = \dfrac{N \, \mu^2}{V \, k \, T} \dfrac{2}{e^{\beta \triangle} + 3}.$ 2. /.   3. 0.027°.

4. $\dfrac{C}{N k} = 1 - x^2 \operatorname{cosech}^2 x$; at low temperature this expression has C approach Nk, while in (15.19) it approaches zero. The Third Law is thus violated. The origin of this is that (15.05) neglects quantization and dormancy. $k \, T$ is the equipartition value for the potential energy of a two-dimensional harmonic oscillator.   5. The quantity Φ written out in (15.41) is smaller for the other branch.   6. Follows directly from (5.55).   7. Expression (15.40) is reproduced.   8. /.   9. $\dfrac{dH}{dT}$ must be negative; $H \dfrac{dH}{dT}$ must vanish for $H = 0$.   10. $E = -\frac{1}{2} N z J$

$- g \, N \, s \, \mu d \, H + 1.34 \, \frac{3}{16} \, N \, \dfrac{\omega}{\rho^3} \left(\dfrac{6 \, s}{\pi \, J \, z}\right)^{3/2} (k \, T)^{5/2}$; when compared with standard thermal results of the type $N \, k \, T$, this is small because of a factor $\left(\dfrac{k \, T}{J}\right)^{3/2}$.

Chapter 17

1. A temperature difference of 0.0074° arises from the pressure, by Clausius–Clapeyron. The remainder, 0.0024°, yields 0.0013 mole/liter of

air dissolved.   2. /.   3. $\dfrac{d \ln (N_1/N_0)}{dT} = \dfrac{h_1 - h^\star}{k\,T^2}$; h_1 is defined in

Problem 2, $h^\star$ is the enthalpy per molecule of the crystalline material.
4. (a) 1.16×10^{-2} gm; (b) 5.2×10^{-4} gm; (c) $e^{-85} = 10^{-37}$ gm.

5. $\dfrac{d}{dT} \ln \left(\dfrac{N_1}{N_0\,p} \right) = -\dfrac{h^\star - h}{k\,T^2}$; $h^\star$ now is the enthalpy of the gaseous phase.

6. (a) 0.87 gm per 100 gm; (b) orthorhombic sulfur will be precipitated

out.   7. $\dfrac{d \ln (N_1/N_0)}{dp} = -\beta\,(v_1 - v^\star)$.

8. $\dfrac{1}{5} \displaystyle\sum_{m=-2}^{+2} \exp[2.05\,\beta\,\mu_B\,m\,H] = 1.000189$.

9. $e^{R/L} = 2 \left(1 + \dfrac{R}{L} \right)$; $R = 1.72 \times 10^{-6}$ cm; 257 ions.

Chapter 18

1. $\dfrac{S}{k} = -V\,H - N \ln \left(\dfrac{h^3}{N\,m^3} \right)$.   2. /.   3. (a) /; (b) If $\phi = \dfrac{\alpha}{r^s}$, then
$\sigma \propto v^{-4/s}$.   4. /.   5. /.

Chapter 19

1. (a) 2.14×10^{17} sec^{-1}; (b) 1.36×10^{10} sec^{-1}; (c) 1.9×10^{-8} mm.

2. /.   3. 0.608.   4. Maxwell $\langle v \rangle \left\langle \dfrac{1}{v} \right\rangle = \dfrac{4}{\pi}$, Fermi $\langle v \rangle \left\langle \dfrac{1}{v} \right\rangle = \dfrac{9}{8}$.

5. Maxwell: $\dfrac{\text{Most probable}}{\text{mean}} = \dfrac{2}{3}$, Fermi: $\dfrac{\text{Most probable}}{\text{mean}} = \dfrac{5}{3}$. The density
of states tends to produce the Fermi result, but the Maxwellian tail of very
fast particles reverses this.

6. $P = (T_2 - T_1)\,p\,r \left(\dfrac{8\,\pi\,k}{m\,T_1} \right)^{1/2} = 0.88$ mw/cm.   7. $p = \dfrac{m'}{r} \left(\dfrac{R\,T}{2\,\pi\,M} \right)^{1/2}$.

8. 0.22.   9. /.

Chapter 20

1. (a) as T^{-1}; (b) as $T^{-3/2}$; (c) as T; (d) not predictable from the data given.
2. 3.4 negative electrons.   3. The Hamiltonian pair of equations of

motion is (20.13) and (20.15).   4. $\mu_n = \mu_p \dfrac{\sigma_0}{\sigma_0 - \sigma_e}$.

5. $J(k) = \pi\,k\,\operatorname{cosech} \pi\,k$.

Chapter 22

1. $v(t) - v(0)e^{-\gamma t} = e^{-\gamma t} \int_0^t e^{\gamma t'} \mathbf{A}(t') \, dt.$ 2. Square result 1, use the short-time correlation property of $\mathbf{A}$, and let t go to infinity:

$\gamma = \dfrac{1}{6} m \beta \displaystyle\int_{-\infty}^{+\infty} \langle \mathbf{A}(0) \cdot \mathbf{A}(\tau) \rangle \, d\tau.$ 3. (a) $\langle V^2 \rangle = \dfrac{1}{2} R C \dfrac{1}{T} \displaystyle\sum_{|t_\nu| < \frac{1}{2}T} s_\nu^2$

$= \frac{1}{2} R C n \langle s_\nu^2 \rangle$; here n is the mean frequency of the steps in time.

(b) $G(f) = \dfrac{2 \langle s_\nu^2 \rangle n}{\dfrac{1}{R^2 C^2} + 4 \pi^2 f^2}$; it is most easily obtained from the auto-

correlation via the Wiener–Khinchin theorem. (c) Not quite; the model gives a noise spectrum varying essentially as $1/f^2$.

4. 1.372×10^{-16} erg/deg.

5. 0.92×10^{-4} cm/sec.   6. /.   7. (a) $f(m, s+1) = \pi f(m-1, s)$

$(1-\pi) f(m+1, s)$;   (b) /.   (c) $\dfrac{\partial f}{\partial s} = (1-2\pi) \dfrac{\partial f}{\partial m} + \dfrac{1}{2} \dfrac{\partial^2 f}{\partial m^2}$; the diffusion

equation.   8. $G_v(f) = \dfrac{G_A(f)}{\gamma^2 + 4 \pi^2 f^2}.$ 9. $\langle v(0) \cdot (\tau) \rangle = \dfrac{k T \gamma}{m} e^{-\gamma \tau}.$

10. $G_A(f) = \dfrac{12 k T \gamma}{m}.$

Chapter 23

None.

Index

A CATALOG OF SELECTED
DOVER BOOKS
IN SCIENCE AND MATHEMATICS

A CATALOG OF SELECTED

DOVER BOOKS

IN SCIENCE AND MATHEMATICS

QUALITATIVE THEORY OF DIFFERENTIAL EQUATIONS, V.V. Nemytskii and V.V. Stepanov. Classic graduate-level text by two prominent Soviet mathematicians covers classical differential equations as well as topological dynamics and ergodic theory. Bibliographies. 523pp. 5⅜ × 8½. 65954-2 Pa. $14.95

MATRICES AND LINEAR ALGEBRA, Hans Schneider and George Phillip Barker. Basic textbook covers theory of matrices and its applications to systems of linear equations and related topics such as determinants, eigenvalues and differential equations. Numerous exercises. 432pp. 5⅜ × 8½. 66014-1 Pa. $10.95

QUANTUM THEORY, David Bohm. This advanced undergraduate-level text presents the quantum theory in terms of qualitative and imaginative concepts, followed by specific applications worked out in mathematical detail. Preface. Index. 655pp. 5⅜ × 8½. 65969-0 Pa. $14.95

ATOMIC PHYSICS (8th edition), Max Born. Nobel laureate's lucid treatment of kinetic theory of gases, elementary particles, nuclear atom, wave-corpuscles, atomic structure and spectral lines, much more. Over 40 appendices, bibliography. 495pp. 5⅜ × 8½. 65984-4 Pa. $12.95

ELECTRONIC STRUCTURE AND THE PROPERTIES OF SOLIDS: The Physics of the Chemical Bond, Walter A. Harrison. Innovative text offers basic understanding of the electronic structure of covalent and ionic solids, simple metals, transition metals and their compounds. Problems. 1980 edition. 582pp. 6⅛ × 9¼. 66021-4 Pa. $16.95

BOUNDARY VALUE PROBLEMS OF HEAT CONDUCTION, M. Necati Özisik. Systematic, comprehensive treatment of modern mathematical methods of solving problems in heat conduction and diffusion. Numerous examples and problems. Selected references. Appendices. 505pp. 5⅜ × 8½. 65990-9 Pa. $12.95

A SHORT HISTORY OF CHEMISTRY (3rd edition), J.R. Partington. Classic exposition explores origins of chemistry, alchemy, early medical chemistry, nature of atmosphere, theory of valency, laws and structure of atomic theory, much more. 428pp. 5⅜ × 8½. (Available in U.S. only) 65977-1 Pa. $11.95

A HISTORY OF ASTRONOMY, A. Pannekoek. Well-balanced, carefully reasoned study covers such topics as Ptolemaic theory, work of Copernicus, Kepler, Newton, Eddington's work on stars, much more. Illustrated. References. 521pp. 5⅜ × 8½. 65994-1 Pa. $12.95

PRINCIPLES OF METEOROLOGICAL ANALYSIS, Walter J. Saucier. Highly respected, abundantly illustrated classic reviews atmospheric variables, hydrostatics, static stability, various analyses (scalar, cross-section, isobaric, isentropic, more). For intermediate meteorology students. 454pp. 6⅛ × 9¼. 65979-8 Pa. $14.95

CATALOG OF DOVER BOOKS

RELATIVITY, THERMODYNAMICS AND COSMOLOGY, Richard C. Tolman. Landmark study extends thermodynamics to special, general relativity; also applications of relativistic mechanics, thermodynamics to cosmological models. 501pp. 5⅜ × 8½. 65383-8 Pa. $13.95

APPLIED ANALYSIS, Cornelius Lanczos. Classic work on analysis and design of finite processes for approximating solution of analytical problems. Algebraic equations, matrices, harmonic analysis, quadrature methods, much more. 559pp. 5⅜ × 8½. 65656-X Pa. $13.95

INTRODUCTION TO ANALYSIS, Maxwell Rosenlicht. Unusually clear, accessible coverage of set theory, real number system, metric spaces, continuous functions, Riemann integration, multiple integrals, more. Wide range of problems. Undergraduate level. Bibliography. 254pp. 5⅜ × 8½. 65038-3 Pa. $8.95

INTRODUCTION TO QUANTUM MECHANICS With Applications to Chemistry, Linus Pauling & E. Bright Wilson, Jr. Classic undergraduate text by Nobel Prize winner applies quantum mechanics to chemical and physical problems. Numerous tables and figures enhance the text. Chapter bibliographies. Appendices. Index. 468pp. 5⅜ × 8½. 64871-0 Pa. $12.95

ASYMPTOTIC EXPANSIONS OF INTEGRALS, Norman Bleistein & Richard A. Handelsman. Best introduction to important field with applications in a variety of scientific disciplines. New preface. Problems. Diagrams. Tables. Bibliography. Index. 448pp. 5⅜ × 8½. 65082-0 Pa. $12.95

MATHEMATICS APPLIED TO CONTINUUM MECHANICS, Lee A. Segel. Analyzes models of fluid flow and solid deformation. For upper-level math, science and engineering students. 608pp. 5⅜ × 8½. 65369-2 Pa. $14.95

ELEMENTS OF REAL ANALYSIS, David A. Sprecher. Classic text covers fundamental concepts, real number system, point sets, functions of a real variable, Fourier series, much more. Over 500 exercises. 352pp. 5⅜ × 8½. 65385-4 Pa. $11.95

PHYSICAL PRINCIPLES OF THE QUANTUM THEORY, Werner Heisenberg. Nóbel Laureate discusses quantum theory, uncertainty, wave mechanics, work of Dirac, Schroedinger, Compton, Wilson, Einstein, etc. 184pp. 5⅜ × 8½. 60113-7 Pa. $6.95

INTRODUCTORY REAL ANALYSIS, A.N. Kolmogorov, S.V. Fomin. Translated by Richard A. Silverman. Self-contained, evenly paced introduction to real and functional analysis. Some 350 problems. 403pp. 5⅜ × 8½. 61226-0 Pa. $10.95

PROBLEMS AND SOLUTIONS IN QUANTUM CHEMISTRY AND PHYSICS, Charles S. Johnson, Jr. and Lee G. Pedersen. Unusually varied problems, detailed solutions in coverage of quantum mechanics, wave mechanics, angular momentum, molecular spectroscopy, scattering theory, more. 280 problems plus 139 supplementary exercises. 430pp. 6½ × 9¼. 65236-X Pa. $13.95

ASYMPTOTIC METHODS IN ANALYSIS, N.G. de Bruijn. An inexpensive, comprehensive guide to asymptotic methods—the pioneering work that teaches by explaining worked examples in detail. Index. 224pp. 5⅜ × 8½.   64221-6 Pa. $7.95

OPTICAL RESONANCE AND TWO-LEVEL ATOMS, L. Allen and J.H. Eberly. Clear, comprehensive introduction to basic principles behind all quantum optical resonance phenomena. 53 illustrations. Preface. Index. 256pp. 5⅜ × 8½.
65533-4 Pa. $8.95

COMPLEX VARIABLES, Francis J. Flanigan. Unusual approach, delaying complex algebra till harmonic functions have been analyzed from real variable viewpoint. Includes problems with answers. 364pp. 5⅜ × 8½.   . 61388-7 Pa. $9.95

ATOMIC SPECTRA AND ATOMIC STRUCTURE, Gerhard Herzberg. One of best introductions; especially for specialist in other fields. Treatment is physical rather than mathematical. 80 illustrations. 257pp. 5⅜ × 8½.   60115-3 Pa. $6.95

APPLIED COMPLEX VARIABLES, John W. Dettman. Step-by-step coverage of fundamentals of analytic function theory—plus lucid exposition of five important applications: Potential Theory; Ordinary Differential Equations; Fourier Transforms; Laplace Transforms; Asymptotic Expansions. 66 figures. Exercises at chapter ends. 512pp. 5⅜ × 8½.   64670-X Pa. $12.95

ULTRASONIC ABSORPTION: An Introduction to the Theory of Sound Absorption and Dispersion in Gases, Liquids and Solids, A.B. Bhatia. Standard reference in the field provides a clear, systematically organized introductory review of fundamental concepts for advanced graduate students, research workers. Numerous diagrams. Bibliography. 440pp. 5⅜ × 8½.   64917-2 Pa. $11.95

UNBOUNDED LINEAR OPERATORS: Theory and Applications, Seymour Goldberg. Classic presents systematic treatment of the theory of unbounded linear operators in normed linear spaces with applications to differential equations. Bibliography. 199pp. 5⅜ × 8½.   64830-3 Pa. $7.95

LIGHT SCATTERING BY SMALL PARTICLES, H.C. van de Hulst. Comprehensive treatment including full range of useful approximation methods for researchers in chemistry, meteorology and astronomy. 44 illustrations. 470pp. 5⅜ × 8½.   64228-3 Pa. $11.95

CONFORMAL MAPPING ON RIEMANN SURFACES, Harvey Cohn. Lucid, insightful book presents ideal coverage of subject. 334 exercises make book perfect for self-study. 55 figures. 352pp. 5⅜ × 8¼.   64025-6 Pa. $11.95

OPTICKS, Sir Isaac Newton. Newton's own experiments with spectroscopy, colors, lenses, reflection, refraction, etc., in language the layman can follow. Foreword by Albert Einstein. 532pp. 5⅜ × 8½.   60205-2 Pa. $11.95

GENERALIZED INTEGRAL TRANSFORMATIONS, A.H. Zemanian. Graduate-level study of recent generalizations of the Laplace, Mellin, Hankel, K. Weierstrass, convolution and other simple transformations. Bibliography. 320pp. 5⅜ × 8½.   65375-7 Pa. $8.95

THE ELECTROMAGNETIC FIELD, Albert Shadowitz. Comprehensive undergraduate text covers basics of electric and magnetic fields, builds up to electromagnetic theory. Also related topics, including relativity. Over 900 problems. 768pp. 5⅜ × 8¼. 65660-8 Pa. $18.95

FOURIER SERIES, Georgi P. Tolstov. Translated by Richard A. Silverman. A valuable addition to the literature on the subject, moving clearly from subject to subject and theorem to theorem. 107 problems, answers. 336pp. 5⅜ × 8½. 63317-9 Pa. $9.95

THEORY OF ELECTROMAGNETIC WAVE PROPAGATION, Charles Herach Papas. Graduate-level study discusses the Maxwell field equations, radiation from wire antennas, the Doppler effect and more. xiii + 244pp. 5⅜ × 8½. 65678-0 Pa. $6.95

DISTRIBUTION THEORY AND TRANSFORM ANALYSIS: An Introduction to Generalized Functions, with Applications, A.H. Zemanian. Provides basics of distribution theory, describes generalized Fourier and Laplace transformations. Numerous problems. 384pp. 5⅜ × 8½. 65479-6 Pa. $11.95

THE PHYSICS OF WAVES, William C. Elmore and Mark A. Heald. Unique overview of classical wave theory. Acoustics, optics, electromagnetic radiation, more. Ideal as classroom text or for self-study. Problems. 477pp. 5⅜ × 8½. 64926-1 Pa. $12.95

CALCULUS OF VARIATIONS WITH APPLICATIONS, George M. Ewing. Applications-oriented introduction to variational theory develops insight and promotes understanding of specialized books, research papers. Suitable for advanced undergraduate/graduate students as primary, supplementary text. 352pp. 5⅜ × 8½. 64856-7 Pa. $9.95

A TREATISE ON ELECTRICITY AND MAGNETISM, James Clerk Maxwell. Important foundation work of modern physics. Brings to final form Maxwell's theory of electromagnetism and rigorously derives his general equations of field theory. 1,084pp. 5⅜ × 8½. 60636-8, 60637-6 Pa., Two-vol. set $23.90

AN INTRODUCTION TO THE CALCULUS OF VARIATIONS, Charles Fox. Graduate-level text covers variations of an integral, isoperimetrical problems, least action, special relativity, approximations, more. References. 279pp. 5⅜ × 8½. 65499-0 Pa. $8.95

HYDRODYNAMIC AND HYDROMAGNETIC STABILITY, S. Chandrasekhar. Lucid examination of the Rayleigh-Benard problem; clear coverage of the theory of instabilities causing convection. 704pp. 5⅜ × 8¼. 64071-X Pa. $14.95

CALCULUS OF VARIATIONS, Robert Weinstock. Basic introduction covering isoperimetric problems, theory of elasticity, quantum mechanics, electrostatics, etc. Exercises throughout. 326pp. 5⅜ × 8½. 63069-2 Pa. $8.95

DYNAMICS OF FLUIDS IN POROUS MEDIA, Jacob Bear. For advanced students of ground water hydrology, soil mechanics and physics, drainage and irrigation engineering and more. 335 illustrations. Exercises, with answers. 784pp. 6⅛ × 9¼. 65675-6 Pa. $19.95

NUMERICAL METHODS FOR SCIENTISTS AND ENGINEERS, Richard Hamming. Classic text stresses frequency approach in coverage of algorithms, polynomial approximation, Fourier approximation, exponential approximation, other topics. Revised and enlarged 2nd edition. 721pp. 5⅜ × 8½.
65241-6 Pa. $15.95

THEORETICAL SOLID STATE PHYSICS, Vol. I: Perfect Lattices in Equilibrium; Vol. II: Non-Equilibrium and Disorder, William Jones and Norman H. March. Monumental reference work covers fundamental theory of equilibrium properties of perfect crystalline solids, non-equilibrium properties, defects and disordered systems. Appendices. Problems. Preface. Diagrams. Index. Bibliography. Total of 1,301pp. 5⅜ × 8½. Two volumes. Vol. I 65015-4 Pa. $16.95
Vol. II 65016-2 Pa. $14.95

OPTIMIZATION THEORY WITH APPLICATIONS, Donald A. Pierre. Broad-spectrum approach to important topic. Classical theory of minima and maxima, calculus of variations, simplex technique and linear programming, more. Many problems, examples. 640pp. 5⅜ × 8½. 65205-X Pa. $14.95

THE CONTINUUM: A Critical Examination of the Foundation of Analysis, Hermann Weyl. Classic of 20th-century foundational research deals with the conceptual problem posed by the continuum. 156pp. 5⅜ × 8½. 67982-9 Pa. $6.95

ESSAYS ON THE THEORY OF NUMBERS, Richard Dedekind. Two classic essays by great German mathematician: on the theory of irrational numbers; and on transfinite numbers and properties of natural numbers. 115pp. 5⅜ × 8½.
21010-3 Pa. $5.95

THE FUNCTIONS OF MATHEMATICAL PHYSICS, Harry Hochstadt. Comprehensive treatment of orthogonal polynomials, hypergeometric functions, Hill's equation, much more. Bibliography. Index. 322pp. 5⅜ × 8½. 65214-9 Pa. $9.95

NUMBER THEORY AND ITS HISTORY, Oystein Ore. Unusually clear, accessible introduction covers counting, properties of numbers, prime numbers, much more. Bibliography. 380pp. 5⅜ × 8½. 65620-9 Pa. $9.95

THE VARIATIONAL PRINCIPLES OF MECHANICS, Cornelius Lanczos. Graduate level coverage of calculus of variations, equations of motion, relativistic mechanics, more. First inexpensive paperbound edition of classic treatise. Index. Bibliography. 418pp. 5⅜ × 8½. 65067-7 Pa. $12.95

MATHEMATICAL TABLES AND FORMULAS, Robert D. Carmichael and Edwin R. Smith. Logarithms, sines, tangents, trig functions, powers, roots, reciprocals, exponential and hyperbolic functions, formulas and theorems. 269pp. 5⅜ × 8½. 60111-0 Pa. $6.95

THEORETICAL PHYSICS, Georg Joos, with Ira M. Freeman. Classic overview covers essential math, mechanics, electromagnetic theory, thermodynamics, quantum mechanics, nuclear physics, other topics. First paperback edition. xxiii + 885pp. 5⅜ × 8½. 65227-0 Pa. $21.95

HANDBOOK OF MATHEMATICAL FUNCTIONS WITH FORMULAS, GRAPHS, AND MATHEMATICAL TABLES, edited by Milton Abramowitz and Irene A. Stegun. Vast compendium: 29 sets of tables, some to as high as 20 places. 1,046pp. 8 × 10½. 61272-4 Pa. $24.95

MATHEMATICAL METHODS IN PHYSICS AND ENGINEERING, John W. Dettman. Algebraically based approach to vectors, mapping, diffraction, other topics in applied math. Also generalized functions, analytic function theory, more. Exercises. 448pp. 5⅜ × 8¼. 65649-7 Pa. $10.95

A SURVEY OF NUMERICAL MATHEMATICS, David M. Young and Robert Todd Gregory. Broad self-contained coverage of computer-oriented numerical algorithms for solving various types of mathematical problems in linear algebra, ordinary and partial, differential equations, much more. Exercises. Total of 1,248pp. 5⅜ × 8½. Two volumes. Vol. I 65691-8 Pa. $14.95
Vol. II 65692-6 Pa. $14.95

TENSOR ANALYSIS FOR PHYSICISTS, J.A. Schouten. Concise exposition of the mathematical basis of tensor analysis, integrated with well-chosen physical examples of the theory. Exercises. Index. Bibliography. 289pp. 5⅜ × 8½. 65582-2 Pa. $8.95

INTRODUCTION TO NUMERICAL ANALYSIS (2nd Edition), F.B. Hildebrand. Classic, fundamental treatment covers computation, approximation, interpolation, numerical differentiation and integration, other topics. 150 new problems. 669pp. 5⅜ × 8½. 65363-3 Pa. $15.95

INVESTIGATIONS ON THE THEORY OF THE BROWNIAN MOVEMENT, Albert Einstein. Five papers (1905–8) investigating dynamics of Brownian motion and evolving elementary theory. Notes by R. Fürth. 122pp. 5⅜ × 8½. 60304-0 Pa. $4.95

CATASTROPHE THEORY FOR SCIENTISTS AND ENGINEERS, Robert Gilmore. Advanced-level treatment describes mathematics of theory grounded in the work of Poincaré, R. Thom, other mathematicians. Also important applications to problems in mathematics, physics, chemistry and engineering. 1981 edition. References. 28 tables. 397 black-and-white illustrations. xvii + 666pp. 6⅛ × 9¼. 67539-4 Pa. $17.95

AN INTRODUCTION TO STATISTICAL THERMODYNAMICS, Terrell L. Hill. Excellent basic text offers wide-ranging coverage of quantum statistical mechanics, systems of interacting molecules, quantum statistics, more. 523pp. 5⅜ × 8½. 65242-4 Pa. $12.95

STATISTICAL PHYSICS, Gregory H. Wannier. Classic text combines thermodynamics, statistical mechanics and kinetic theory in one unified presentation of thermal physics. Problems with solutions. Bibliography. 532pp. 5⅜ × 8½. 65401-X Pa. $12.95

ORDINARY DIFFERENTIAL EQUATIONS, Morris Tenenbaum and Harry Pollard. Exhaustive survey of ordinary differential equations for undergraduates in mathematics, engineering, science. Thorough analysis of theorems. Diagrams. Bibliography. Index. 818pp. 5⅜ × 8½. 64940-7 Pa. $18.95

STATISTICAL MECHANICS: Principles and Applications, Terrell L. Hill. Standard text covers fundamentals of statistical mechanics, applications to fluctuation theory, imperfect gases, distribution functions, more. 448pp. 5⅜ × 8½. 65390-0 Pa. $11.95

ORDINARY DIFFERENTIAL EQUATIONS AND STABILITY THEORY: An Introduction, David A. Sánchez. Brief, modern treatment. Linear equation, stability theory for autonomous and nonautonomous systems, etc. 164pp. 5⅜ × 8¼. 63828-6 Pa. $6.95

THIRTY YEARS THAT SHOOK PHYSICS: The Story of Quantum Theory, George Gamow. Lucid, accessible introduction to influential theory of energy and matter. Careful explanations of Dirac's anti-particles, Bohr's model of the atom, much more. 12 plates. Numerous drawings. 240pp. 5⅜ × 8½. 24895-X Pa. $6.95

THEORY OF MATRICES, Sam Perlis. Outstanding text covering rank, non-singularity and inverses in connection with the development of canonical matrices under the relation of equivalence, and without the intervention of determinants. Includes exercises. 237pp. 5⅜ × 8½. 66810-X Pa. $8.95

GREAT EXPERIMENTS IN PHYSICS: Firsthand Accounts from Galileo to Einstein, edited by Morris H. Shamos. 25 crucial discoveries: Newton's laws of motion, Chadwick's study of the neutron, Hertz on electromagnetic waves, more. Original accounts clearly annotated. 370pp. 5⅜ × 8½. 25346-5 Pa. $10.95

INTRODUCTION TO PARTIAL DIFFERENTIAL EQUATIONS WITH AP-PLICATIONS, E.C. Zachmanoglou and Dale W. Thoe. Essentials of partial differential equations applied to common problems in engineering and the physical sciences. Problems and answers. 416pp. 5⅜ × 8½. 65251-3 Pa. $11.95

BURNHAM'S CELESTIAL HANDBOOK, Robert Burnham, Jr. Thorough guide to the stars beyond our solar system. Exhaustive treatment. Alphabetical by constellation: Andromeda to Cetus in Vol. 1; Chamaeleon to Orion in Vol. 2; and Pavo to Vulpecula in Vol. 3. Hundreds of illustrations. Index in Vol. 3. 2,000pp. 6⅛ × 9¼. 23567-X, 23568-8, 23673-0 Pa., Three-vol. set $44.85

CHEMICAL MAGIC, Leonard A. Ford. Second Edition, Revised by E. Winston Grundmeier. Over 100 unusual stunts demonstrating cold fire, dust explosions, much more. Text explains scientific principles and stresses safety precautions. 128pp. 5⅜ × 8½. 67628-5 Pa. $5.95

AMATEUR ASTRONOMER'S HANDBOOK, J.B. Sidgwick. Timeless, comprehensive coverage of telescopes, mirrors, lenses, mountings, telescope drives, micrometers, spectroscopes, more. 189 illustrations. 576pp. 5⅜ × 8¼. (Available in U.S. only) 24034-7 Pa. $11.95

SPECIAL FUNCTIONS, N.N. Lebedev. Translated by Richard Silverman. Famous Russian work treating more important special functions, with applications to specific problems of physics and engineering. 38 figures. 308pp. 5⅜ × 8½.
60624-4 Pa. $9.95

OBSERVATIONAL ASTRONOMY FOR AMATEURS, J.B. Sidgwick. Mine of useful data for observation of sun, moon, planets, asteroids, aurorae, meteors, comets, variables, binaries, etc. 39 illustrations. 384pp. 5⅜ × 8¼. (Available in U.S. only)
24033-9 Pa. $8.95

INTEGRAL EQUATIONS, F.G. Tricomi. Authoritative, well-written treatment of extremely useful mathematical tool with wide applications. Volterra Equations, Fredholm Equations, much more. Advanced undergraduate to graduate level. Exercises. Bibliography. 238pp. 5⅜ × 8½.
64828-1 Pa. $8.95

POPULAR LECTURES ON MATHEMATICAL LOGIC, Hao Wang. Noted logician's lucid treatment of historical developments, set theory, model theory, recursion theory and constructivism, proof theory, more. 3 appendixes. Bibliography. 1981 edition. ix + 283pp. 5⅜ × 8½.
67632-3 Pa. $8.95

MODERN NONLINEAR EQUATIONS, Thomas L. Saaty. Emphasizes practical solution of problems; covers seven types of equations. ". . . a welcome contribution to the existing literature. . . ."—*Math Reviews.* 490pp. 5⅜ × 8½. 64232-1 Pa. $11.95

FUNDAMENTALS OF ASTRODYNAMICS, Roger Bate et al. Modern approach developed by U.S. Air Force Academy. Designed as a first course. Problems, exercises. Numerous illustrations. 455pp. 5⅜ × 8½.
60061-0 Pa. $9.95

INTRODUCTION TO LINEAR ALGEBRA AND DIFFERENTIAL EQUATIONS, John W. Dettman. Excellent text covers complex numbers, determinants, orthonormal bases, Laplace transforms, much more. Exercises with solutions. Undergraduate level. 416pp. 5⅜ × 8½.
65191-6 Pa. $10.95

INCOMPRESSIBLE AERODYNAMICS, edited by Bryan Thwaites. Covers theoretical and experimental treatment of the uniform flow of air and viscous fluids past two-dimensional aerofoils and three-dimensional wings; many other topics. 654pp. 5⅜ × 8½.
65465-6 Pa. $16.95

INTRODUCTION TO DIFFERENCE EQUATIONS, Samuel Goldberg. Exceptionally clear exposition of important discipline with applications to sociology, psychology, economics. Many illustrative examples; over 250 problems. 260pp. 5⅜ × 8½.
65084-7 Pa. $8.95

LAMINAR BOUNDARY LAYERS, edited by L. Rosenhead. Engineering classic covers steady boundary layers in two- and three-dimensional flow, unsteady boundary layers, stability, observational techniques, much more. 708pp. 5⅜ × 8½.
65646-2 Pa. $18.95

LECTURES ON CLASSICAL DIFFERENTIAL GEOMETRY, Second Edition, Dirk J. Struik. Excellent brief introduction covers curves, theory of surfaces, fundamental equations, geometry on a surface, conformal mapping, other topics. Problems. 240pp. 5⅜ × 8½.
65609-8 Pa. $8.95

CATALOG OF DOVER BOOKS

ROTARY-WING AERODYNAMICS, W.Z. Stepniewski. Clear, concise text covers aerodynamic phenomena of the rotor and offers guidelines for helicopter performance evaluation. Originally prepared for NASA. 537 figures. 640pp. 6⅛ × 9¼.
64647-5 Pa. $15.95

DIFFERENTIAL GEOMETRY, Heinrich W. Guggenheimer. Local differential geometry as an application of advanced calculus and linear algebra. Curvature, transformation groups, surfaces, more. Exercises. 62 figures. 378pp. 5⅜ × 8½.
63433-7 Pa. $9.95

INTRODUCTION TO SPACE DYNAMICS, William Tyrrell Thomson. Comprehensive, classic introduction to space-flight engineering for advanced undergraduate and graduate students. Includes vector algebra, kinematics, transformation of coordinates. Bibliography. Index. 352pp. 5⅜ × 8½. 65113-4 Pa. $9.95

A SURVEY OF MINIMAL SURFACES, Robert Osserman. Up-to-date, in-depth discussion of the field for advanced students. Corrected and enlarged edition covers new developments. Includes numerous problems. 192pp. 5⅜ × 8½.
64998-9 Pa. $8.95

ANALYTICAL MECHANICS OF GEARS, Earle Buckingham. Indispensable reference for modern gear manufacture covers conjugate gear-tooth action, gear-tooth profiles of various gears, many other topics. 263 figures. 102 tables. 546pp. 5⅜ × 8½. 65712-4 Pa. $14.95

SET THEORY AND LOGIC, Robert R. Stoll. Lucid introduction to unified theory of mathematical concepts. Set theory and logic seen as tools for conceptual understanding of real number system. 496pp. 5⅜ × 8¼. 63829-4 Pa. $12.95

A HISTORY OF MECHANICS, René Dugas. Monumental study of mechanical principles from antiquity to quantum mechanics. Contributions of ancient Greeks, Galileo, Leonardo, Kepler, Lagrange, many others. 671pp. 5⅜ × 8½.
65632-2 Pa. $14.95

FAMOUS PROBLEMS OF GEOMETRY AND HOW TO SOLVE THEM, Benjamin Bold. Squaring the circle, trisecting the angle, duplicating the cube: learn their history, why they are impossible to solve, then solve them yourself. 128pp. 5⅜ × 8½. 24297-8 Pa. $4.95

MECHANICAL VIBRATIONS, J.P. Den Hartog. Classic textbook offers lucid explanations and illustrative models, applying theories of vibrations to a variety of practical industrial engineering problems. Numerous figures. 233 problems, solutions. Appendix. Index. Preface. 436pp. 5⅜ × 8½. 64785-4 Pa. $11.95

CURVATURE AND HOMOLOGY, Samuel I. Goldberg. Thorough treatment of specialized branch of differential geometry. Covers Riemannian manifolds, topology of differentiable manifolds, compact Lie groups, other topics. Exercises. 315pp. 5⅜ × 8½. 64314-X Pa. $9.95

HISTORY OF STRENGTH OF MATERIALS, Stephen P. Timoshenko. Excellent historical survey of the strength of materials with many references to the theories of elasticity and structure. 245 figures. 452pp. 5⅜ × 8½. 61187-6 Pa. $12.95

GEOMETRY OF COMPLEX NUMBERS, Hans Schwerdtfeger. Illuminating, widely praised book on analytic geometry of circles, the Moebius transformation, and two-dimensional non-Euclidean geometries. 200pp. 5⅜ × 8¼.
63830-8 Pa. $8.95

MECHANICS, J.P. Den Hartog. A classic introductory text or refresher. Hundreds of applications and design problems illuminate fundamentals of trusses, loaded beams and cables, etc. 334 answered problems. 462pp. 5⅜ × 8½. 60754-2 Pa. $10.95

TOPOLOGY, John G. Hocking and Gail S. Young. Superb one-year course in classical topology. Topological spaces and functions, point-set topology, much more. Examples and problems. Bibliography. Index. 384pp. 5⅜ × 8¼.
65676-4 Pa. $10.95

STRENGTH OF MATERIALS, J.P. Den Hartog. Full, clear treatment of basic material (tension, torsion, bending, etc.) plus advanced material on engineering methods, applications. 350 answered problems. 323pp. 5⅜ × 8½. 60755-0 Pa. $9.95

ELEMENTARY CONCEPTS OF TOPOLOGY, Paul Alexandroff. Elegant, intuitive approach to topology from set-theoretic topology to Betti groups; how concepts of topology are useful in math and physics. 25 figures. 57pp. 5⅜ × 8½.
60747-X Pa. $3.95

ADVANCED STRENGTH OF MATERIALS, J.P. Den Hartog. Superbly written advanced text covers torsion, rotating disks, membrane stresses in shells, much more. Many problems and answers. 388pp. 5⅜ × 8½. 65407-9 Pa. $10.95

COMPUTABILITY AND UNSOLVABILITY, Martin Davis. Classic graduate-level introduction to theory of computability, usually referred to as theory of recurrent functions. New preface and appendix. 288pp. 5⅜ × 8½. 61471-9 Pa. $8.95

GENERAL CHEMISTRY, Linus Pauling. Revised 3rd edition of classic first-year text by Nobel laureate. Atomic and molecular structure, quantum mechanics, statistical mechanics, thermodynamics correlated with descriptive chemistry. Problems. 992pp. 5⅜ × 8½. 65622-5 Pa. $19.95

AN INTRODUCTION TO MATRICES, SETS AND GROUPS FOR SCIENCE STUDENTS, G. Stephenson. Concise, readable text introduces sets, groups, and most importantly, matrices to undergraduate students of physics, chemistry, and engineering. Problems. 164pp. 5⅜ × 8½. 65077-4 Pa. $7.95

THE HISTORICAL BACKGROUND OF CHEMISTRY, Henry M. Leicester. Evolution of ideas, not individual biography. Concentrates on formulation of a coherent set of chemical laws. 260pp. 5⅜ × 8½. 61053-5 Pa. $7.95

THE PHILOSOPHY OF MATHEMATICS: An Introductory Essay, Stephan Körner. Surveys the views of Plato, Aristotle, Leibniz & Kant concerning propositions and theories of applied and pure mathematics. Introduction. Two appendices. Index. 198pp. 5⅜ × 8½. 25048-2 Pa. $8.95

THE DEVELOPMENT OF MODERN CHEMISTRY, Aaron J. Ihde. Authoritative history of chemistry from ancient Greek theory to 20th-century innovation. Covers major chemists and their discoveries. 209 illustrations. 14 tables. Bibliographies. Indices. Appendices. 851pp. 5⅜ × 8½. 64235-6 Pa. $18.95

DE RE METALLICA, Georgius Agricola. The famous Hoover translation of greatest treatise on technological chemistry, engineering, geology, mining of early modern times (1556). All 289 original woodcuts. 638pp. 6¾ × 11.

60006-8 Pa. $18.95

SOME THEORY OF SAMPLING, William Edwards Deming. Analysis of the problems, theory and design of sampling techniques for social scientists, industrial managers and others who find statistics increasingly important in their work. 61 tables. 90 figures. xvii + 602pp. 5⅜ × 8½.

64684-X Pa. $15.95

THE VARIOUS AND INGENIOUS MACHINES OF AGOSTINO RAMELLI: A Classic Sixteenth-Century Illustrated Treatise on Technology, Agostino Ramelli. One of the most widely known and copied works on machinery in the 16th century. 194 detailed plates of water pumps, grain mills, cranes, more. 608pp. 9 × 12.

28180-9 Pa. $24.95

LINEAR PROGRAMMING AND ECONOMIC ANALYSIS, Robert Dorfman, Paul A. Samuelson and Robert M. Solow. First comprehensive treatment of linear programming in standard economic analysis. Game theory, modern welfare economics, Leontief input-output, more. 525pp. 5⅜ × 8½.

65491-5 Pa. $14.95

ELEMENTARY DECISION THEORY, Herman Chernoff and Lincoln E. Moses. Clear introduction to statistics and statistical theory covers data processing, probability and random variables, testing hypotheses, much more. Exercises. 364pp. 5⅜ × 8½.

65218-1 Pa. $10.95

THE COMPLEAT STRATEGYST: Being a Primer on the Theory of Games of Strategy, J.D. Williams. Highly entertaining classic describes, with many illustrated examples, how to select best strategies in conflict situations. Prefaces. Appendices. 268pp. 5⅜ × 8½.

25101-2 Pa. $7.95

CONSTRUCTIONS AND COMBINATORIAL PROBLEMS IN DESIGN OF EXPERIMENTS, Damaraju Raghavarao. In-depth reference work examines orthogonal Latin squares, incomplete block designs, tactical configuration, partial geometry, much more. Abundant explanations, examples. 416pp. 5⅜ × 8¼.

65685-3 Pa. $10.95

THE ABSOLUTE DIFFERENTIAL CALCULUS (CALCULUS OF TENSORS), Tullio Levi-Civita. Great 20th-century mathematician's classic work on material necessary for mathematical grasp of theory of relativity. 452pp. 5⅜ × 8½.

63401-9 Pa. $11.95

VECTOR AND TENSOR ANALYSIS WITH APPLICATIONS, A.I. Borisenko and I.E. Tarapov. Concise introduction. Worked-out problems, solutions, exercises. 257pp. 5⅜ × 8¼.

63833-2 Pa. $8.95

THE FOUR-COLOR PROBLEM: Assaults and Conquest, Thomas L. Saaty and Paul G. Kainen. Engrossing, comprehensive account of the century-old combinatorial topological problem, its history and solution. Bibliographies. Index. 110 figures. 228pp. 5⅜ × 8½. 65092-8 Pa. $6.95

CATALYSIS IN CHEMISTRY AND ENZYMOLOGY, William P. Jencks. Exceptionally clear coverage of mechanisms for catalysis, forces in aqueous solution, carbonyl- and acyl-group reactions, practical kinetics, more. 864pp. 5⅜ × 8½. 65460-5 Pa. $19.95

PROBABILITY: An Introduction, Samuel Goldberg. Excellent basic text covers set theory, probability theory for finite sample spaces, binomial theorem, much more. 360 problems. Bibliographies. 322pp. 5⅜ × 8½. 65252-1 Pa. $9.95

LIGHTNING, Martin A. Uman. Revised, updated edition of classic work on the physics of lightning. Phenomena, terminology, measurement, photography, spectroscopy, thunder, more. Reviews recent research. Bibliography. Indices. 320pp. 5⅜ × 8¼. 64575-4 Pa. $8.95

PROBABILITY THEORY: A Concise Course, Y.A. Rozanov. Highly readable, self-contained introduction covers combination of events, dependent events, Bernoulli trials, etc. Translation by Richard Silverman. 148pp. 5⅜ × 8¼. 63544-9 Pa. $6.95

AN INTRODUCTION TO HAMILTONIAN OPTICS, H. A. Buchdahl. Detailed account of the Hamiltonian treatment of aberration theory in geometrical optics. Many classes of optical systems defined in terms of the symmetries they possess. Problems with detailed solutions. 1970 edition. xv + 360pp. 5⅜ × 8½. 67597-1 Pa. $10.95

STATISTICS MANUAL, Edwin L. Crow, et al. Comprehensive, practical collection of classical and modern methods prepared by U.S. Naval Ordnance Test Station. Stress on use. Basics of statistics assumed. 288pp. 5⅜ × 8½. 60599-X Pa. $7.95

DICTIONARY/OUTLINE OF BASIC STATISTICS, John E. Freund and Frank J. Williams. A clear concise dictionary of over 1,000 statistical terms and an outline of statistical formulas covering probability, nonparametric tests, much more. 208pp. 5⅜ × 8½. 66796-0 Pa. $7.95

STATISTICAL METHOD FROM THE VIEWPOINT OF QUALITY CONTROL, Walter A. Shewhart. Important text explains regulation of variables, uses of statistical control to achieve quality control in industry, agriculture, other areas. 192pp. 5⅜ × 8½. 65232-7 Pa. $7.95

THE INTERPRETATION OF GEOLOGICAL PHASE DIAGRAMS, Ernest G. Ehlers. Clear, concise text emphasizes diagrams of systems under fluid or containing pressure; also coverage of complex binary systems, hydrothermal melting, more. 288pp. 6½ × 9¼. 65389-7 Pa. $10.95

STATISTICAL ADJUSTMENT OF DATA, W. Edwards Deming. Introduction to basic concepts of statistics, curve fitting, least squares solution, conditions without parameter, conditions containing parameters. 26 exercises worked out. 271pp. 5⅜ × 8½. 64685-8 Pa. $9.95

CATALOG OF DOVER BOOKS

TENSOR CALCULUS, J.L. Synge and A. Schild. Widely used introductory text covers spaces and tensors, basic operations in Riemannian space, non-Riemannian spaces, etc. 324pp. 5⅜ × 8¼. 63612-7 Pa. $9.95

A CONCISE HISTORY OF MATHEMATICS, Dirk J. Struik. The best brief history of mathematics. Stresses origins and covers every major figure from ancient Near East to 19th century. 41 illustrations. 195pp. 5⅜ × 8½. 60255-9 Pa. $7.95

A SHORT ACCOUNT OF THE HISTORY OF MATHEMATICS, W.W. Rouse Ball. One of clearest, most authoritative surveys from the Egyptians and Phoenicians through 19th-century figures such as Grassman, Galois, Riemann. Fourth edition. 522pp. 5⅜ × 8½. 20630-0 Pa. $11.95

HISTORY OF MATHEMATICS, David E. Smith. Nontechnical survey from ancient Greece and Orient to late 19th century; evolution of arithmetic, geometry, trigonometry, calculating devices, algebra, the calculus. 362 illustrations. 1,355pp. 5⅜ × 8½. 20429-4, 20430-8 Pa., Two-vol. set $26.90

THE GEOMETRY OF RENÉ DESCARTES, René Descartes. The great work founded analytical geometry. Original French text, Descartes' own diagrams, together with definitive Smith-Latham translation. 244pp. 5⅜ × 8½. 60068-8 Pa. $7.95

THE ORIGINS OF THE INFINITESIMAL CALCULUS, Margaret E. Baron. Only fully detailed and documented account of crucial discipline: origins; development by Galileo, Kepler, Cavalieri; contributions of Newton, Leibniz, more. 304pp. 5⅜ × 8½. (Available in U.S. and Canada only) 65371-4 Pa. $9.95

THE HISTORY OF THE CALCULUS AND ITS CONCEPTUAL DEVELOPMENT, Carl B. Boyer. Origins in antiquity, medieval contributions, work of Newton, Leibniz, rigorous formulation. Treatment is verbal. 346pp. 5⅜ × 8½. 60509-4 Pa. $9.95

THE THIRTEEN BOOKS OF EUCLID'S ELEMENTS, translated with introduction and commentary by Sir Thomas L. Heath. Definitive edition. Textual and linguistic notes, mathematical analysis. 2,500 years of critical commentary. Not abridged. 1,414pp. 5⅜ × 8½. 60088-2, 60089-0, 60090-4 Pa., Three-vol. set $31.85

GAMES AND DECISIONS: Introduction and Critical Survey, R. Duncan Luce and Howard Raiffa. Superb nontechnical introduction to game theory, primarily applied to social sciences. Utility theory, zero-sum games, n-person games, decision-making, much more. Bibliography. 509pp. 5⅜ × 8½. 65943-7 Pa. $12.95

THE HISTORICAL ROOTS OF ELEMENTARY MATHEMATICS, Lucas N.H. Bunt, Phillip S. Jones, and Jack D. Bedient. Fundamental underpinnings of modern arithmetic, algebra, geometry and number systems derived from ancient civilizations. 320pp. 5⅜ × 8½. 25563-8 Pa. $8.95

CALCULUS REFRESHER FOR TECHNICAL PEOPLE, A. Albert Klaf. Covers important aspects of integral and differential calculus via 756 questions. 566 problems, most answered. 431pp. 5⅜ × 8½. 20370-0 Pa. $8.95

CHALLENGING MATHEMATICAL PROBLEMS WITH ELEMENTARY SOLUTIONS, A.M. Yaglom and I.M. Yaglom. Over 170 challenging problems on probability theory, combinatorial analysis, points and lines, topology, convex polygons, many other topics. Solutions. Total of 445pp. 5⅜ × 8½. Two-vol. set.

Vol. I 65536-9 Pa. $7.95
Vol. II 65537-7 Pa. $7.95

FIFTY CHALLENGING PROBLEMS IN PROBABILITY WITH SOLUTIONS, Frederick Mosteller. Remarkable puzzlers, graded in difficulty, illustrate elementary and advanced aspects of probability. Detailed solutions. 88pp. 5⅜ × 8½.

65355-2 Pa. $4.95

EXPERIMENTS IN TOPOLOGY, Stephen Barr. Classic, lively explanation of one of the byways of mathematics. Klein bottles, Moebius strips, projective planes, map coloring, problem of the Koenigsberg bridges, much more, described with clarity and wit. 43 figures. 210pp. 5⅜ × 8½.

25933-1 Pa. $6.95

RELATIVITY IN ILLUSTRATIONS, Jacob T. Schwartz. Clear nontechnical treatment makes relativity more accessible than ever before. Over 60 drawings illustrate concepts more clearly than text alone. Only high school geometry needed. Bibliography. 128pp. 6⅛ × 9¼.

25965-X Pa. $7.95

AN INTRODUCTION TO ORDINARY DIFFERENTIAL EQUATIONS, Earl A. Coddington. A thorough and systematic first course in elementary differential equations for undergraduates in mathematics and science, with many exercises and problems (with answers). Index. 304pp. 5⅜ × 8½.

65942-9 Pa. $8.95

FOURIER SERIES AND ORTHOGONAL FUNCTIONS, Harry F. Davis. An incisive text combining theory and practical example to introduce Fourier series, orthogonal functions and applications of the Fourier method to boundary-value problems. 570 exercises. Answers and notes. 416pp. 5⅜ × 8½.

65973-9 Pa. $11.95

AN INTRODUCTION TO ALGEBRAIC STRUCTURES, Joseph Landin. Superb self-contained text covers "abstract algebra": sets and numbers, theory of groups, theory of rings, much more. Numerous well-chosen examples, exercises. 247pp. 5⅜ × 8½.

65940-2 Pa. $8.95

Prices subject to change without notice.
Available at your book dealer or write for free Mathematics and Science Catalog to Dept. GI, Dover Publications, Inc., 31 East 2nd St., Mineola, N.Y. 11501. Dover publishes more than 175 books each year on science, elementary and advanced mathematics, biology, music, art, literature, history, social sciences and other areas.